Domesticating Empire

Domesticating Empire

Egyptian Landscapes in Pompeian Gardens

CAITLÍN EILÍS BARRETT

OXFORD
UNIVERSITY PRESS

OXFORD
UNIVERSITY PRESS

Oxford University Press is a department of the University of Oxford. It furthers the University's objective of excellence in research, scholarship, and education by publishing worldwide. Oxford is a registered trade mark of Oxford University Press in the UK and certain other countries.

Published in the United States of America by Oxford University Press
198 Madison Avenue, New York, NY 10016, United States of America.

CIP data is on file at the Library of Congress
ISBN 978–0–19–064135–1

For Dave Zarzycki

Contents

List of Figures

Acknowledgments

DURING THE PAST seven years, many people have lent invaluable support to this project. I began the research for this manuscript as a postdoctoral fellow at the University of Pennsylvania, and I am grateful to colleagues there, especially Brian Rose, Tom Tartaron, and Ann Kuttner, for providing feedback on my early drafts and paper presentations. Going even further back in time, my interest in Greco-Roman receptions of Egyptian material culture dates back to my days as a graduate student at Yale, where I wrote a Ph.D. thesis (subsequently adapted into my first book, Barrett 2011) on the use of Egyptian imagery in terracotta figurines from Hellenistic Delos. I thus owe many thanks to the scholars with whom I worked on that earlier project, especially the members of my dissertation committee: John Coleman Darnell, Richard Burger, Tom Tartaron, and Milette Gaifman. My current work on the "cultural biography" of Nilotic imagery would not have been possible without the training in Egyptian and Hellenistic art, archaeology, and religion that I obtained through my graduate studies and first book project. The varied disciplinary backgrounds and perspectives of my graduate mentors also provided me with permanent encouragement to cross traditional disciplinary lines in my own research, drawing on perspectives, approaches, and datasets from Classics, Egyptology, archaeology, art history, and anthropology.

More recently, I am thankful to multiple colleagues for their support, feedback, and productive discussions about the role of "Egypt" in Roman visual and material culture. Verity Platt and Éric Rebillard were both kind enough to read near-complete drafts of the manuscript for this book, and their helpful and perceptive comments were important in sharpening my thinking about Roman adaptations of Egyptian visual culture. Annetta Alexandridis suggested some helpful parallels for the "Bacchic" herms from the Casa di Acceptus e Euhodia; Kathryn Gleason offered thoughtful responses to my work on the role of gardens in Roman thought and social practice; and Ross Thomas offered useful feedback on the coroplastic material discussed in chapter 2. Additionally, I am grateful to the two colleagues who have served as Classics Department chair during my time at Cornell, Sturt Manning and Charles Brittain, for their mentorship, advice, and unflagging support of my research.

I am also very grateful to Miguel John Versluys, who kindly read and commented on a draft of the manuscript. Versluys' pioneering work on Roman Aegyptiaca has reshaped the field and charted new directions for research on cross-cultural encounters in antiquity. He also offered feedback on an earlier publication from which the content of chapter 1 is partly drawn (Barrett 2017a). I thank him heartily for his generosity with his time and for his careful and thoughtful reading of both documents. His comments and questions on the manuscripts were very helpful in strengthening the final texts.

Over the years, I have presented a range of conference papers and invited lectures on different aspects of this project. For comments, questions, and suggestions that helped shape the final text, I thank the audiences who attended my talks at Cornell University, New York University, Tufts University, Yale University, the University of Pennsylvania, Bryn Mawr College, the Franklin Institute, the Memorial Art Gallery of Rochester, the Edmonton Society of the Archaeological Institute of America, the 112th and 114th Annual Meetings of the Archaeological Institute of America (San Antonio, Texas, 2011, and Seattle, 2013), and the 62nd Annual Meeting of the American Research Center in Egypt (Chicago, 2011).

At Oxford University Press, I thank my editor, Sarah Pirovitz, for her support of this project, her helpful feedback, and her keen eye for detail. I also thank the two anonymous readers of the manuscript for their careful reading and useful comments; their advice has greatly improved the final text.

Two chapters of this book are adapted from my previously published articles. Substantial portions of chapter 1 draw from an article published in *Oxford Handbooks Online in Classical Studies* (Barrett 2017a), while chapter 3, part of chapter 2 ("Parallel Landscapes in the Garden"), and the final sentence of chapter 7 are adapted from an article published in the *American Journal of Archaeology* (Barrett 2017b). I thank the editors at OUP (Jenna Vaccaro and Ben Leonard) and at AJA (Jane B. Carter, Sheila Dillon, and Naomi J. Norman) for permission to republish this material, as well as for their guidance during the writing and production of the original articles.

As always, responsibility for any errors and all opinions in the present volume remains with me.

This project would not have been possible without on-site research at Pompeii and in-person examination of objects in the collections of the Museo Archeologico Nazionale di Napoli (MANN). For permission to conduct on-site research at Pompeii, I thank Teresa Elena Cinquantaquattro (former Soprintendente, Soprintendenza Speciale per i Beni Archeologici di Napoli e Pompei) and Massimo Osanna (Direttore Generale, Parco Archeologico di Pompei). For permission to examine objects in the MANN and to publish images of artifacts, I thank Paolo Giulierini, Valeria Sampaolo, and Giuseppe Proietti. I am also very grateful to Fortunato Stefanizzi and Giorgio Albano for their kind assistance in my research at the MANN.

Numerous other institutions and individuals were generous enough to provide me with permission to publish copyrighted images or reproduce previously published material. For image permissions, I thank the following museums (in alphabetical order): Ägyptisches Museum und Papyrussammlung, Staatliche Museen zu Berlin; British Museum; Brooklyn Museum; Martin von Wagner Museum, Würzburg; Museo Archeologico Nazionale di Napoli; Museum für Kunst und Gewerbe, Hamburg; Ny Carlsberg Glyptotek, Copenhagen; Rijksmuseum van Oudheden, Leiden; Szépmüveszeti Múzeum, Budapest. I also thank the following publishers for permission to republish images: Egypt Exploration Society; "L'Erma" di Bretschneider; the Oriental Institute of the University of Chicago; and Società Tiburtina di Storia e d'Arte. The Jashemski Archives at the University of Maryland granted permission for me to republish a photograph originally taken by Stanley Jashemski. Finally, many individual scholars kindly granted permission for me to republish images they had created. John Clarke allowed me to reproduce his plans of the Casa del Medico and Casa del Fauno; Gemma Jansen permitted me to use her plan of the Casa del Medico hydraulic system; and Katherine von Stackelberg allowed me to use her plan of the Casa di Octavius Quartio. I also received permission to reproduce photographs from Dieter Arnold, Véronique Dasen, Barry Kemp, Arielle Kozloff Brodkey, Karol Myśliwiec, Klaus Parlasca, Christina Riggs, Mary Ann Sullivan, Kaja Tally-Schumacher, Eleni Vassilika, and Helen Whitehouse. The illustrations of this book provide ample testimony to the generosity of all these scholars and institutions, to whom I am most sincerely grateful.

Financial support for this project came from Cornell University, the University of Pennsylvania, and the Andrew W. Mellon Foundation. At Cornell, I received grants from the Classics Department Gift Fund, the Provost's Special Research Fund, the Cornell Institute for Archaeology and Material Studies (CIAMS), and the Society for the Humanities (Humanities Collaborative Research Grant).

Finally, I want to thank my family for their love and support throughout the process of research and writing. My husband, Dave Zarzycki, has now spent the better part of a decade traveling with me to Pompeii and Naples, helping photograph archaeological sites and artifacts, advising me on photography techniques, shouldering an extra share of work at home whenever my research schedule became particularly demanding, encouraging me to catch up on sleep when absolutely necessary, exchanging ideas and sharing in my excitement about my work, and always offering his constant love and kindness. This book is dedicated to him.

Abbreviations

IN THIS VOLUME, abbreviations for Classical authors, works, and standard reference sources (e.g., *LSJ, TLG*, etc.) follow those found in the *Oxford Classical Dictionary*, 4th edition. In addition to these standard abbreviations, I also use *RICIS* for *Recueil des inscriptions concernant les cultes isiaques* (Bricault 2005) and MANN for Museo Archeologico Nazionale di Napoli. In the References section, I use the abbreviations employed by the *American Journal of Archaeology*, with two additions: *PPM* for the series *Pompei: Pitture e mosaici*, ed. I. Baldassarre; and *SAK* for the journal *Studien zur Altägyptischen Kultur*.

References to houses in Pompeii use either the commonly used modern name of the house (e.g., Casa dell'Efebo, Casa del Medico, et cetera), the house address, or both. As is standard at Pompeii, house addresses are numbered according to Fiorelli's system of *regio*, insula, and doorway: e.g., I 7.11, VIII 5.24, et cetera.

I have used standard published translations of Latin, Greek, and Egyptian texts where appropriate, with modifications from those translations marked in the footnotes where relevant. When no published translation is cited, the translations are my own.

Domesticating Empire

1

Introduction: Egypt in the Garden

Prologue: Visiting the Casa dell'Efebo

In the late 70s CE, dinner guests arriving at the large Pompeian house known today as the "Casa dell'Efebo" would have found themselves navigating a landscape at once both foreign and familiar.[1] Within the garden stood a pleasant outdoor dining area: a masonry triclinium shaded from the afternoon heat by a vine-covered canopy (figure 1.1). Splashing water from a nearby fountain filled a small channel to the triclinium benches (figure 1.2), creating the impression that the diners were reclining on a riverbank. Outdoor dining areas of this type were a familiar sight in Roman Italy, but certain aspects of this particular installation simultaneously suggest more distant landscapes. As visitors came closer to the triclinium, they would have noticed a painted frieze on its benches depicting scenes of festival, cult, and banqueting along the banks of the Nile (figures 1.3–1.8). The characters populating this painted landscape would have struck Pompeian viewers as unusual in a number of ways. Depicted as dwarfs or pygmies, their very bodies mark them as fundamentally different from their presumed viewers. Their behaviors, too, seem intended to be read as "exotic," "foreign" customs: they perform worship before Egyptian animal gods, do battle with crocodiles, and even engage in public sex on the riverbank.

Yet at the same time, other aspects of the garden installation challenge this sense of alterity, suggesting that this landscape is neither so inaccessible nor so far away. As they, too, recline and feast on the banks of a model "river," the dinner guests effectively join the painted pygmies in their festivities. Domestic water canals were sometimes called *nilus*, "Nile."[2] Flowing between the painted scenes of Nilotic riverbanks, this canal itself becomes a model Nile, causing the distant

---

1. This case study will be discussed in depth in chapter 3, this volume.

2. Cic. *Q.Fr.* 3.9.7, *Leg.* 2.2; Dessales 2013: 59–62; Zarmakoupi 2014: 157–163.

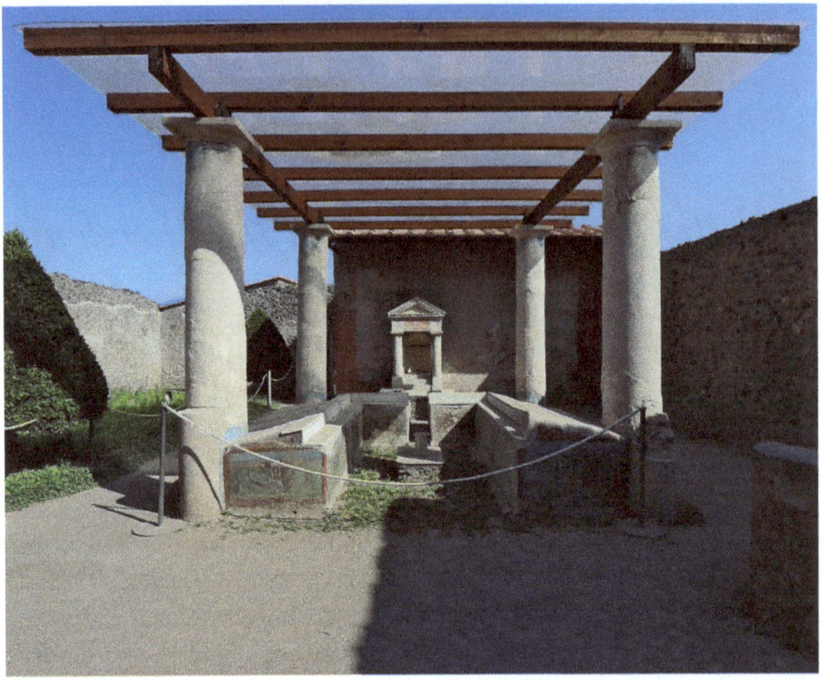

FIGURE 1.1 Triclinium and water installation in the garden of the Casa dell'Efebo (I 7.10–12), Pompeii. Root casts at the column bases suggest that vines once grew around the columns.
Photo: C.E. Barrett.

Egyptian landscape to become present and immediate. The fresco even creates visual analogies between the painted pygmies and the dinner guests themselves, as several vignettes show pygmies banqueting and carousing in circumstances very much like their viewers': outdoors, beneath a canopy of vines, alongside running water. Feasting on the banks of a domestic, Italian Nile, the dinner guests appear to become the very figures they are viewing.

Such an assemblage sets up a series of unresolved tensions between foreign and familiar, "Self" and "Other," reality and representation. Rather than providing a single authoritative solution, the garden invites viewers to engage with the questions it raises. Are visitors to identify with the pygmies' carnivalesque exuberance, or to treat them as negative *exempla* modeling behaviors to avoid? The diners' own subsequent behavior becomes their response to—and performative commentary on—the imagery. However they choose to act, those choices constitute participation in, and creative collaboration with, the domestic visual culture surrounding them.

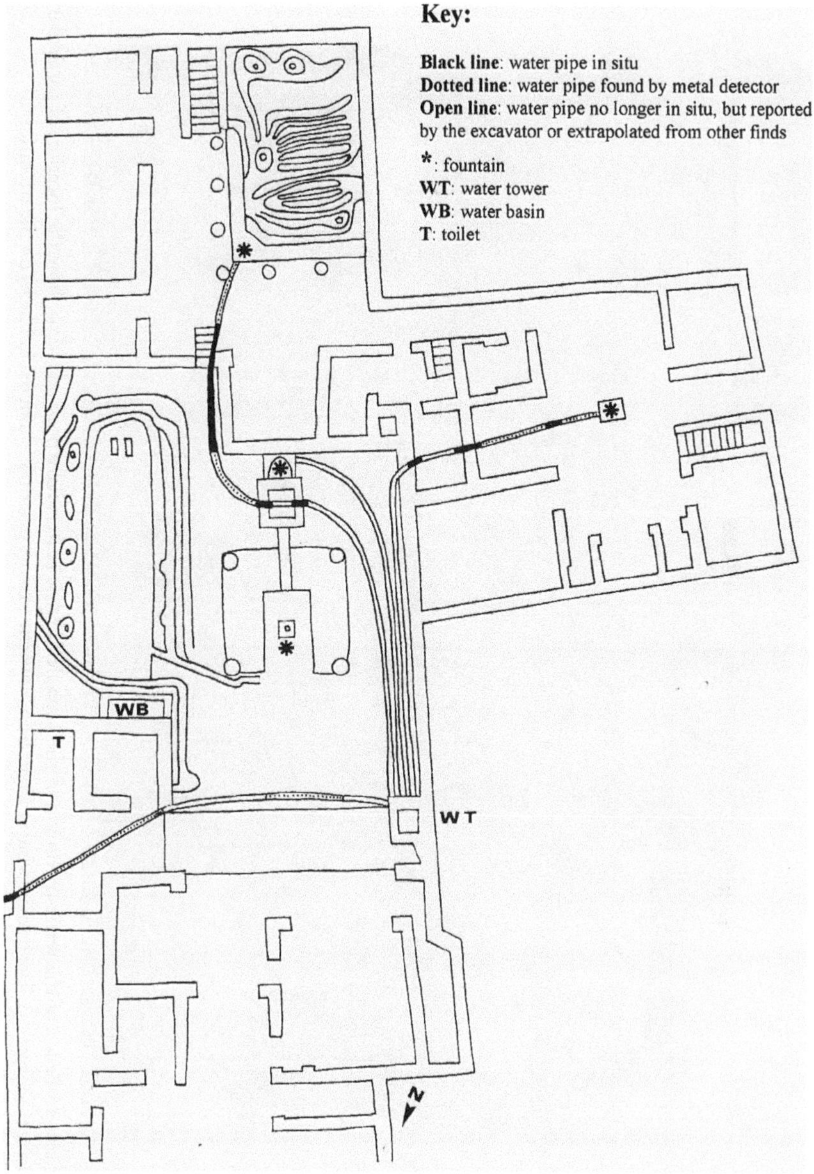

FIGURE 1.2 Plan of the Casa dell'Efebo garden and adjacent rooms, illustrating Gemma Jansen's reconstruction of the water system. Hydraulic features (e.g., water pipes, channels, outlets) are not shown in the drawings that accompany figures 1.3–1.8. The rooms to the south and west of the garden-triclinium, which do not appear in Maiuri's plan of the Casa dell'Efebo (figure 3.1), originally belonged to House I 7.19, which was being renovated for addition to the Casa dell'Efebo.

Source: Drawing by G. Jansen, reproduced with permission; modified from Jansen 2005: 283, fig. 12.

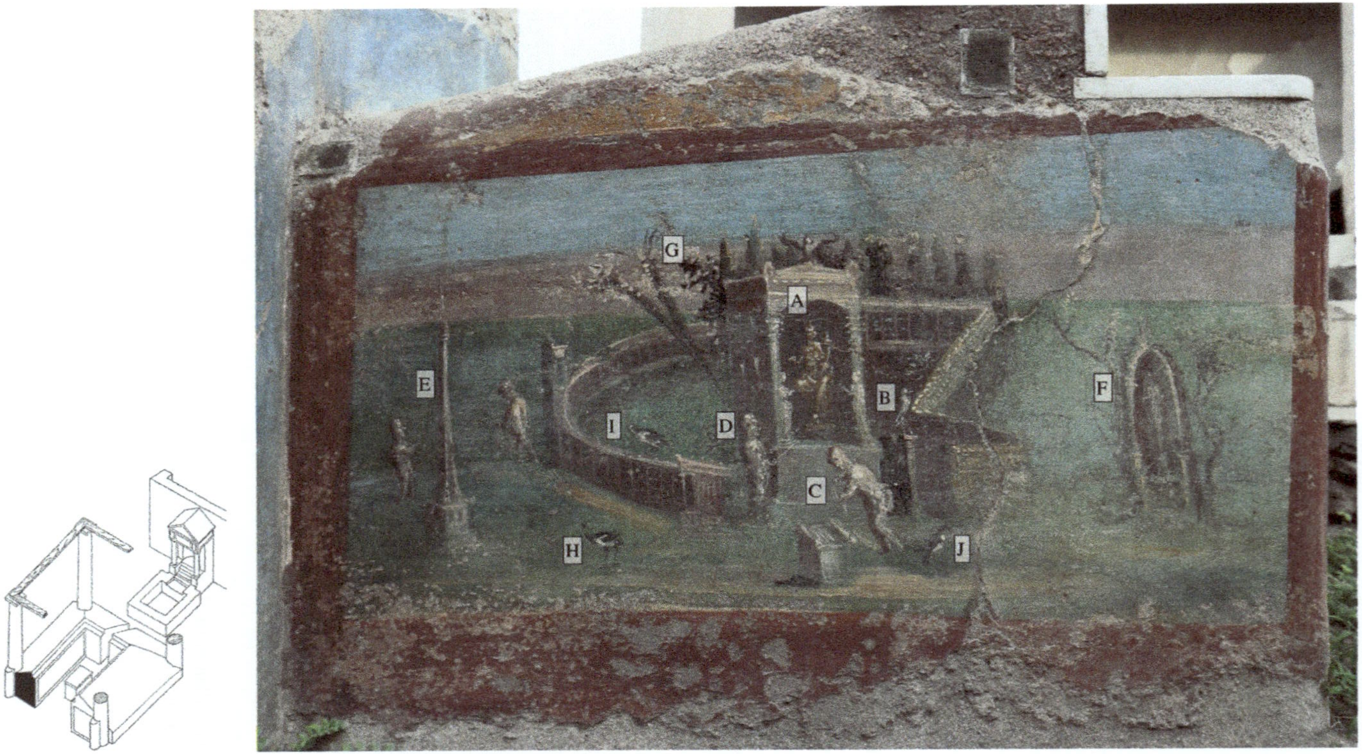

FIGURE 1.3 North face of eastern bench of the water-triclinium in the Casa dell'Efebo. Letters have been added to show the location of particular features discussed in the main text (chapter 3). The line drawing at lower left shows the image's position (rendered in black) on the triclinium.

Sources: Photo by C.E. Barrett; drawing modified from Soprano 1950: fig. 31.

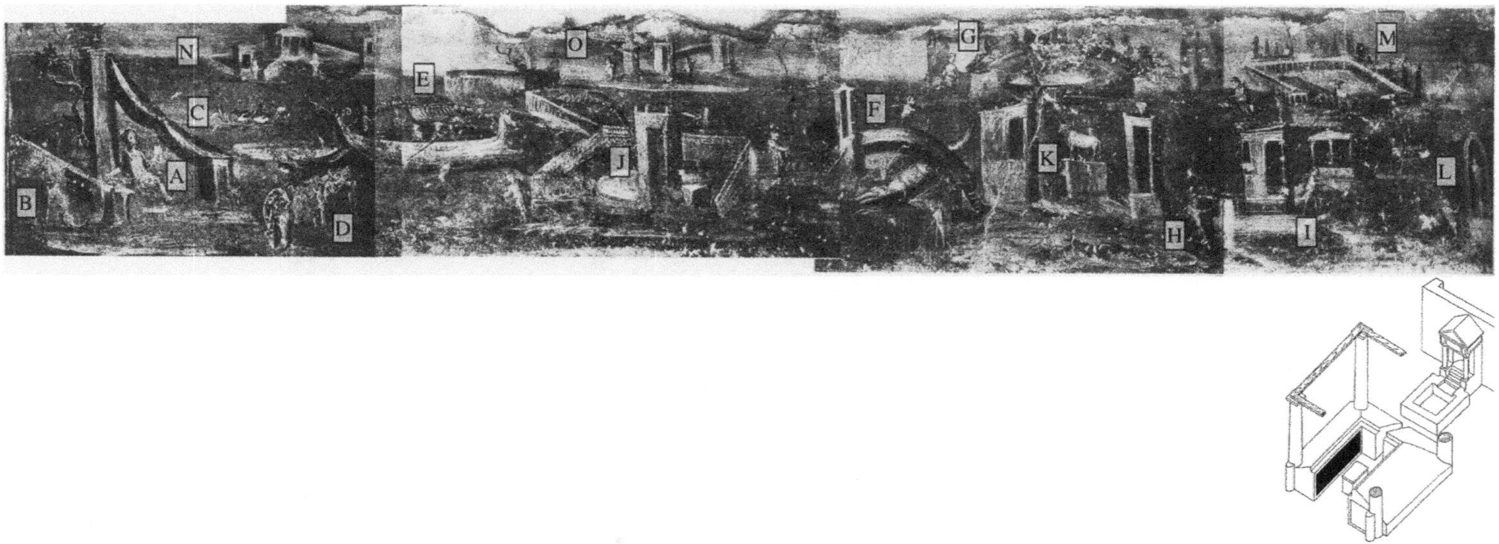

FIGURE 1.4 Interior face of eastern bench of the water-triclinium in the Casa dell'Efebo. The archival photographs, originally published by Maiuri (1927: pl. 8, figs. 25, 26), preserve more detail than is visible today. Letters have been added to show the location of particular features discussed in the main text (chapter 3). The line drawing at lower left shows the image's position (rendered in black) on the triclinium.

Source: Composite photograph adapted from A. de Vos 1990: figs. 169b–c, 173 a–b. Line drawing modified from Soprano 1950: fig. 31.

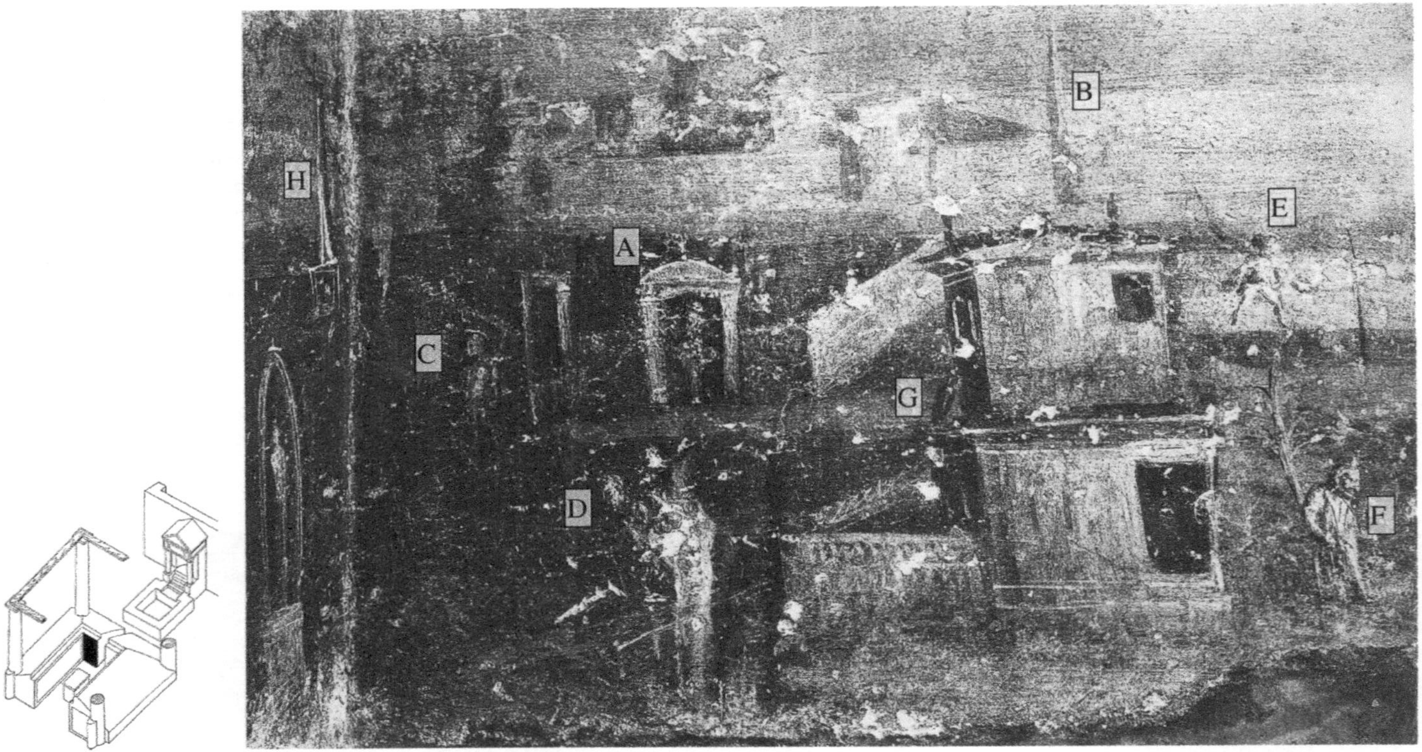

FIGURE 1.5 Interior face of eastern side of southern bench of the water-triclinium in the Casa dell'Efebo. The archival photograph, originally published by Maiuri (1927: fig. 27), preserves more detail than is visible today. Note the edge of the eastern bench (figure 1.4) visible at the left edge of the photograph. Letters have been added to show the location of particular features discussed in the main text (chapter 3). The line drawing at lower left shows the image's position (rendered in black) on the triclinium.

Sources: Photograph after A. de Vos 1990: fig. 176. Line drawing modified from Soprano 1950: fig. 31.

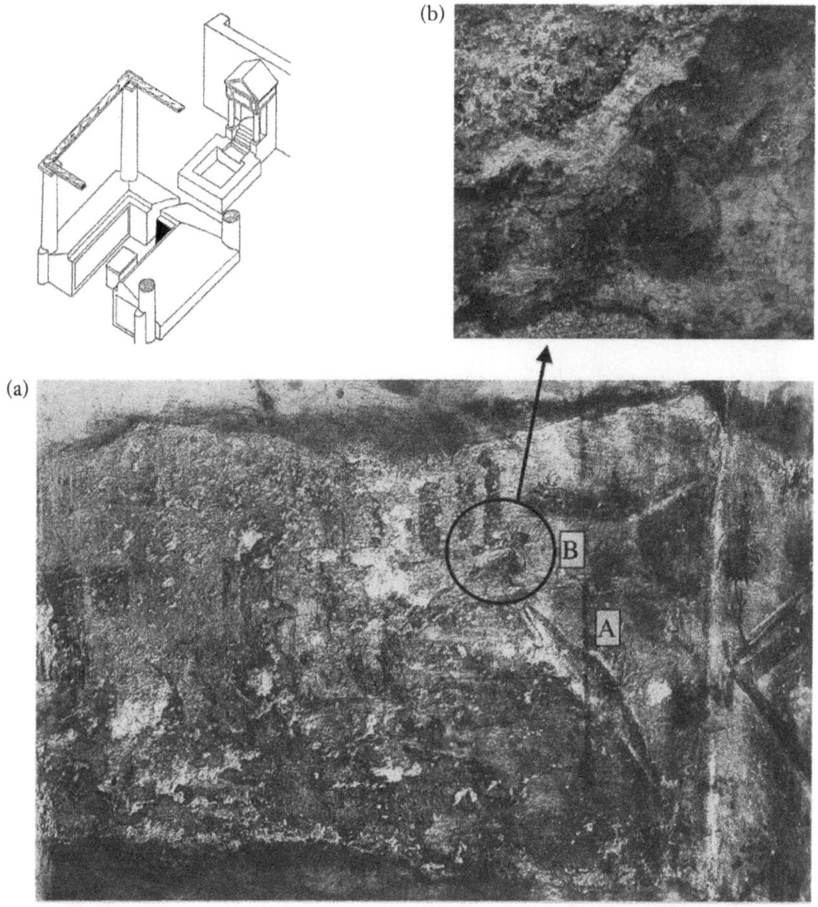

FIGURE 1.6 Interior face of western side of southern bench of the water-triclinium in the Casa dell'Efebo. (a) Entire bench face. The archival photograph, taken during Maiuri's excavations in the 1920s, preserves more detail than is visible today. Note the edge of the painting on the western bench (figure 1.7) visible at the right edge of (a). (b) State in 2012 of the figure circled in (a). Letters have been added to show the location of particular features discussed in the main text (chapter 3). The line drawing at upper left shows the position of the bench face (rendered in black) on the triclinium.

Sources: (a) After A. de Vos 1990: fig. 177. (b) Photo by C.E. Barrett. Line drawing in upper left modified from Soprano 1950: fig. 31.

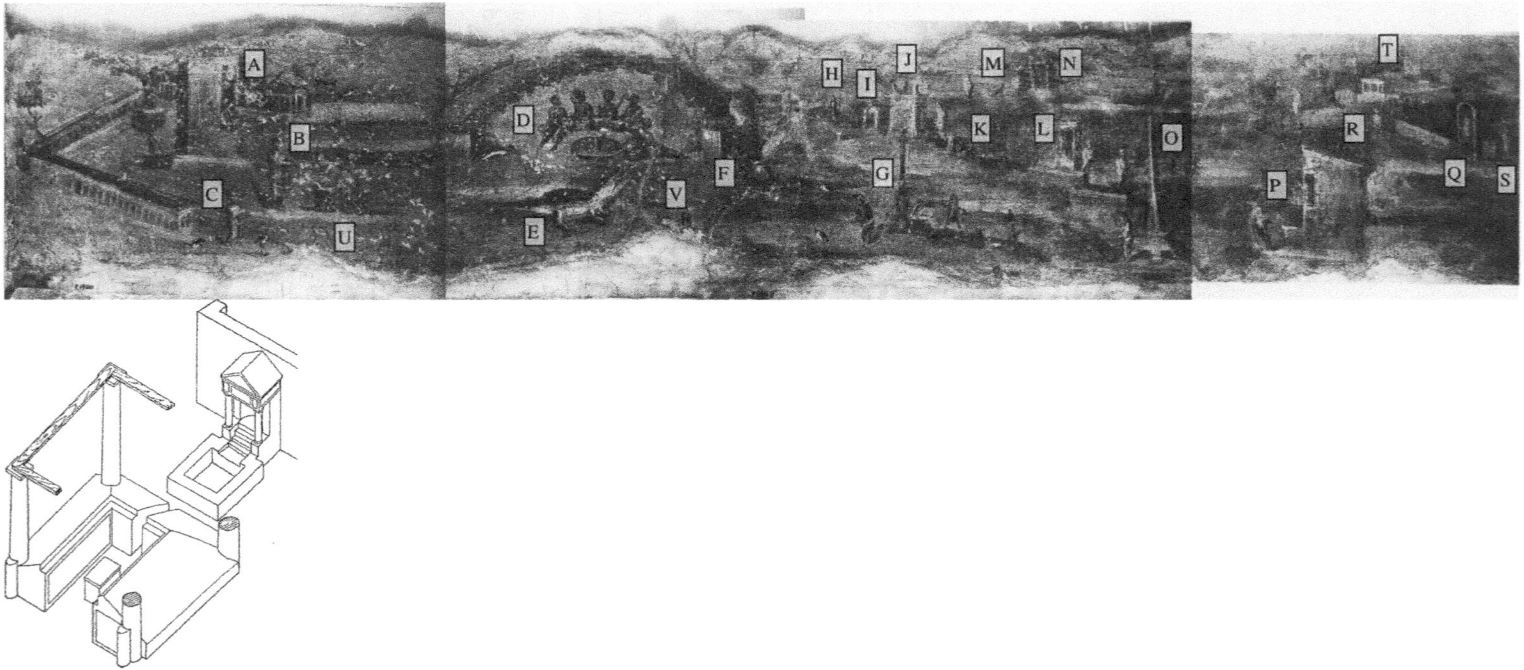

FIGURE 1.7 Interior face of western bench of the water-triclinium in the Casa dell'Efebo. The archival photographs, originally published by Maiuri (1927: fig. 28, pls. 7, 8), preserve more detail than is visible today. Letters have been added to show the location of particular features discussed in the main text (chapter 3). This image appears on the only interior bench face not visible in the line drawing at lower left.

Sources: Composite photograph adapted from A. de Vos 1990: figs. 179a–b, 182a–b. Line drawing modified from Soprano 1950: fig. 31.

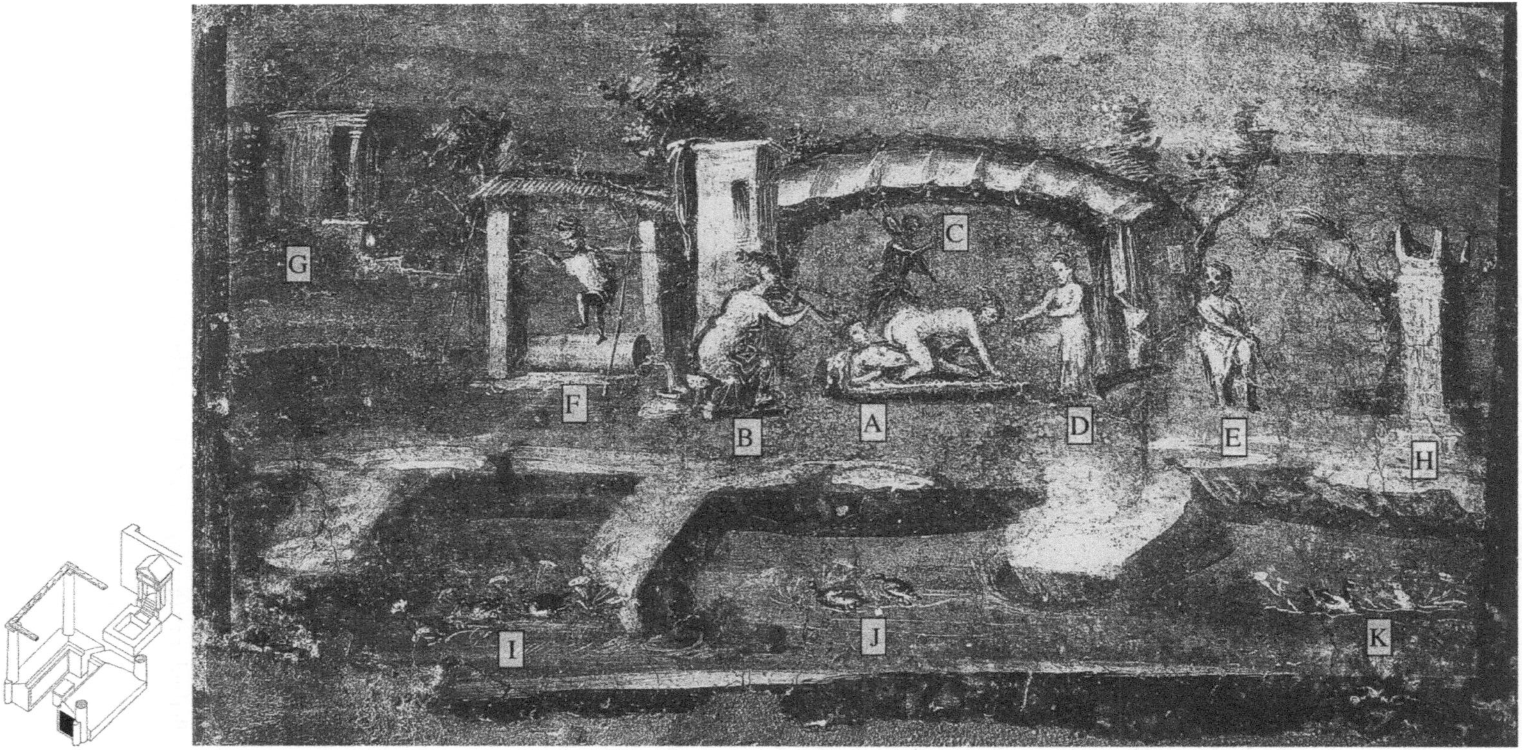

FIGURE 1.8 Northern face of western bench of the water-triclinium in the Casa dell'Efebo. The archival photograph, taken during Maiuri's excavations in the 1920s, preserves more detail than is visible today. Letters have been added to show the location of particular features discussed in the main text (chapter 3). The line drawing at lower left shows the image's position on the triclinium.

Sources: Photograph after A. de Vos 1990: fig. 186. Line drawing modified from Soprano 1950: fig. 31.

Conceptualizing Egypt in Roman Visual and Material Culture

The Nilotic dining installation at the Casa dell'Efebo is one of many evocations of Egypt in Roman domestic settings.[3] Imagery and objects associated with the new province of Aegyptus were hugely popular in the early Roman Empire. Scholars often describe such objects and images as "Aegyptiaca," a wide-ranging term covering many different types of material culture: imported objects from Egypt; objects produced outside of Egypt that visually allude to Egyptian styles or prototypes; and depictions of places, people, or deities typically associated with Egypt. The term "Aegyptiaca" is thus both maximally flexible and maximally vague, conflating a wide range of variables that include place of production, style of manufacture, and iconographic content. Such broad labels have both advantages and disadvantages, and research of the past decade has subjected analytical categories such as "Egyptian," "Egyptianizing," and "Aegyptiaca" to extensive critical reevaluation. Scholars have traditionally reserved the term "Egyptian" for Egyptian-made imports in Italy, using a different term—"Egyptianizing"—for Italian-made objects that allude to Egypt or employ Egyptian styles. However, Molly Swetnam-Burland argues that Italian consumers may not always have attached different valuations to imported versus locally produced objects. Seeking to avoid stigmatizing Italian-made "Egyptianizing" material culture according to "modern notions of authenticity," Swetnam-Burland interprets all Roman material and visual references to Egypt, regardless of place of production, as "legitimate *Roman* cultural products."[4] Accordingly, she advocates for the use of one term, "Aegyptiaca," for all Egyptian-looking artifacts or images from Italy.[5] More recently, Eva Mol has also criticized scholarly constructions of "Aegyptiaca." Mol argues that not all objects perceived by modern scholars as "Egyptian"-looking would have been so perceived in antiquity, and ancient consumers may have valued such artifacts for reasons unconnected to their Egyptian associations.

3. Significant portions of the rest of this chapter have been adapted, with modifications, from a previous article by the author (Barrett 2017a) and are here republished by permission of Oxford University Press, USA. See especially the sections here entitled "Conceptualizing Egypt in Roman Visual and Material Culture," "'Aegyptiaca' in the Household: Contextualizing Interactive Ensembles," "Material Culture, Identity, and 'Authenticity,'" "Rethinking Things," and "Rethinking Contacts."

4. Swetnam-Burland 2007 (quoting 114, 116), followed by Pearson 2015: 2–3.

5. Swetnam-Burland 2007: 119 (further nuanced in Swetnam-Burland 2015: 12); cf. Versluys 2015b: 147, 150; van Aerde 2015: 32–33, 61, 292. For earlier discussions of Egyptian or Egyptian-style material culture in Italy as "Aegyptiaca," see Swetnam-Burland 2002; Versluys 2002.

Accordingly, she seeks to destabilize constructions of "Aegyptiaca" as a "homogeneous category of material culture" in Italy.[6]

Both critiques highlight the under-theorization of traditional scholarship on Egypto-Roman visual culture, as well as the need to reconsider the analytical categories that such scholarship employs. Among other things, these critiques emphasize the need to distinguish between emic and etic perspectives on Egypt in Roman visual culture.[7] However, there is nothing inherently problematic about etic categories as long as we explicitly recognize them as such. For example, even if distinctions between imported and locally produced goods were not relevant to ancient consumers, they may still be relevant to modern researchers who seek to investigate craft production or reconstruct trade networks. In other words, distinctions between local products and imports may still be useful from etic research perspectives, even if they were not always important within emic frameworks of value.

This book thus retains the term "Egyptianizing" for objects made in Italy whose style, form, or iconography makes reference to parallels in Egyptian material culture. However, this usage explicitly acknowledges that "Egyptianizing" is an etic term without precise emic equivalent, and that ancient consumers may not necessarily have valued or recognized such distinctions.[8] Furthermore, when we do examine considerations of production and distribution, we need to acknowledge that not all objects fall neatly into binary categories of "Egyptian" versus "Italian" products. Some objects were produced by Italian workshops using imported Egyptian raw materials, while other imports were produced by Egyptian workshops but then reworked or modified after reaching Italy.[9] Additionally, production *outside Egypt* need not always imply *local* production; an Egyptian-style object at, for example, Pompeii might just as well be imported from Rome as from Alexandria.

This book also retains "Aegyptiaca" as a conceptual category, albeit one requiring certain qualifications. Arguments against the existence of an emic category corresponding to "Aegyptiaca" should not be overstated. The Latin

---

6. Mol 2013 (quoting 118), 2015, 2017; cf. Müskens 2017: 13–28.

7. Anthropologists differentiate "emic" concepts, which are internal to a culture, from "etic" concepts, which are employed by modern scholars but may not have existed within the cultures they study.

8. Greek (though not Latin) does possess a verb *aigyptiazein*, loosely translatable as "to Egyptianize." However, this verb typically describes human actions (e.g., speaking Egyptian or engaging in supposedly Egyptian-like behavior), rather than the qualities of consumer goods (*LSJ Online*, s.v. Αἰγυπτιάζω; *TLG*).

9. For various examples, see Swetnam-Burland 2015: 18–64.

adjectives *Aegyptiacus* and *Aegyptius* can both describe objects, as well as people (as can the Greek *Aigyptios*), and Roman literary and visual sources clearly associate the idea of "Egypt" with a variety of tropes, concepts, and images.[10] Roman culture thus offered individuals a range of ways to conceptualize "Egyptianness," and individuals might associate such concepts with a range of material culture. However, modern constructions of "Egyptianness" may not necessarily map onto ancient ideas; images or objects that evoke "Egypt" for modern scholars may not always have done so for ancient Romans, and vice versa.[11] Additionally, even when people perceived an object as alluding to "Egypt," such allusions need not have wholly determined that object's valuation. People might also have sought out such objects for other reasons, such as their functional properties, aesthetic qualities, or the value of their raw materials.

In this book, the term "Aegyptiaca" describes material culture found in Italy that makes emically recognizable visual reference to Egypt, Egyptians, or cultural practices associated with Egypt.[12] However, I explicitly recognize that this covering term conceals numerous different possible functions, values, and cultural referents; that ancient notions of what constituted a "visual reference to Egypt" might differ from modern expectations; and that the hypothesis of an intended visual reference to Egypt can only be a starting point, rather than an end goal, of analysis. Rather than assuming that certain images or artifacts always evoked the same ideas about Egypt (or, indeed, recalled Egypt at all) in any setting, we should proceed contextually, examining the interactions of objects with larger domestic assemblages.

"Aegyptiaca" in the Household: Contextualizing Interactive Ensembles

Material culture alluding to Egypt, or drawing in some way on Egyptian precedents, appeared in a wide range of Roman settings from public to private and conveyed an equally wide range of possible meanings and values to different

---

10. For example, the Latin word *aegyptiaca* can describe books or histories (*OLD*, s.v. "Aegyptiaca"; *TLL*, s.v. "Aegyptiaca"). Textual attestations of *aegyptiacus* and *aegyptius*: Swetnam-Burland 2007: 119. Cf. Greek *Aigyptios*, which can similarly describe both objects and people (e.g., a medication called *aigyptia*: Gal. *Comp. Med. per Gen.* 13.643 K). ***Roman tropes concerning "Egypt":*** see, e.g., Smelik and Hemelrijk 1984; Fowden 1993; Hornung 2001; Maehler 2003; Tracy 2014: 1–12; Leemreize 2016.

11. Swetnam-Burland 2007: 116; Mol 2015.

12. On differences between this definition of "Aegyptiaca" and that of Malaise (2005: 201–204), see note 98, this chapter.

viewers. Within early imperial domestic contexts, Egyptian imagery might appear in almost any medium: wall paintings or mosaics, statuary, terracotta lamps, and even objects of personal adornment, such as amulets, jewelry, and seals.[13] Such objects might be present in practically any part of a house—and indeed, portable artifacts such as jewelry would have traveled anywhere their wearers went.

Since the later twentieth century, scholars have frequently interpreted Italian Aegyptiaca as evidence for "Roman" conceptions of "Egypt" and have assigned such objects to three major interpretive categories:[14] they are said to be (1) religious objects associated with Isis cult, (2) politically motivated celebrations of Rome's conquest of Egypt, or (3) fashion statements testifying to Roman "Egyptomania" or a taste for the "exotic." However, many artifacts do not fit neatly into these conceptual boxes, and meanings may often have been much more flexible and shifting. Rather than simply assigning labels such as "politics," "religion," or "fashion" to Roman images of Egypt, we need to seek more nuanced approaches that recognize material culture's "meanings" to be multiple, fluid, and contextual.

A contextual analysis of domestic Aegyptiaca tends to problematize rather than resolve tensions between "Self" and "Other," "religious" and "exotica," "foreign" and "familiar." This book argues that instead of making a single definitive statement about Egypt or Egyptian culture, interactive assemblages like that of the Casa dell'Efebo garden enable viewers to grapple for themselves with these contradictions, explore the tensions between different possible identities, and act out the images' complex games of identification and rejection. Within the context of a garden whose material culture evokes the breadth of the Roman world, such interactive assemblages enabled individuals to perform cosmopolitanism within the household and to participate in the ongoing creation and negotiation of a Roman-controlled, but pan-Mediterranean, cultural *koine*.[15]

It may seem hard to claim that Roman Aegyptiaca are understudied, given the incredible proliferation of bibliography on this subject within the past twenty years.[16] However, much work remains to be done on the *contextualization* of

13. Examples of this range of material culture: Arslan 1997; De Caro 2006b; Mol 2015.

14. My critique here is in sympathy with that of Pearson 2015: 5–12.

15. For more discussion of visual *koine* as a counterpart to linguistic *koine*, see chapter 5, this volume, "From Hellenistic *Koine* to Roman *Imperium*."

16. E.g. (focusing here on published works, not including unpublished PhD dissertations), Takács 1995; Arslan 1997; Bricault 2000a; 2000b; 2004; 2005; 2008; 2013; Versluys 2002; 2015a; 2017a; Ashton 2004; Malaise 2005; De Caro 2006b; Bricault et al. 2007; Kleibl 2009; Bricault and Veymiers 2011; Davies 2011; Bricault and Versluys 2012; 2014b; Swetnam-Burland 2012; 2015; Manolaraki 2013; Rose 2013; Poole 2016; Barrett 2017a; 2017b; Merrills 2017; Spier et al. 2018. *Aegyptiaca in Campania*: see notes 55 and 58, this chapter.

Egyptian or Egyptian-style material culture from domestic settings. While previous work has frequently addressed the roles of Aegyptiaca within imperial, public, and temple contexts,[17] household Aegyptiaca still remain comparatively understudied. A number of synthetic studies or catalogs of Aegyptiaca have included material from domestic contexts, but detailed contextualization has usually not been such works' primary purpose.[18] There is thus a great need for more studies placing domestic Aegyptiaca into explicit dialogue with the entire repertoire of architectural features, paintings, mosaics, statues, stuccoes, and household artifacts that constituted the rooms from which they came.[19]

This need for contextualization is particularly important in the light of extensive recent work on contextual appropriateness (*decor, decorum*) in Roman art.[20] Both textual sources and archaeological finds suggest that people judged certain types of decoration (sculpted, painted, or otherwise) to be more or less appropriate to certain settings.[21] Similar notions of "appropriateness" appear to inform Roman adaptations of styles from earlier periods and cultures.[22] Hölscher's

17. E.g., Lembke 1994; Petersen 2006: 17–56; Kleibl 2009; selected articles in Bricault and Veymiers 2011; Bülow Clausen 2012; 2015; Swetnam-Burland 2015; Versluys 2017a; Versluys et al. 2018. Müskens 2014 uses contextual analysis to demonstrate that a group of Aegyptiaca need *not* necessarily have come from a shrine in the *Domus Flavia*.

18. E.g., Arslan 1997; De Caro 2006b. Tran Tam Tinh 1964 draws on material from households as well as sanctuaries to study Isis worship at Pompeii, but his book has been rightly criticized for taking nearly all Egyptian imagery as evidence for "Isis cult" (Petersen 2006: 42, 246n80). De Vos's 1980 catalogue of "Egyptianizing" wall paintings includes much domestic material but provides little contextual analysis. Versluys's 2002 study of Nilotic scenes similarly includes many images from houses, but the empire-wide, nearly millennium-long scope of his corpus inevitably restricts the opportunities for detailed contextual analysis.

19. Some welcome recent examples: Swetnam-Burland 2002: 70–76; von Stackelberg 2009: 101–106; Petersen 2012; Mol 2013, 2015; Parslow 2013. John Clarke's work offers many stimulating contextual interpretations of Egyptian motifs in wall paintings and mosaics, although the primary focus remains on the architecture, wall paintings, and mosaics of the rooms, as opposed to statuary or small finds (e.g., Clarke 1991: 170–207; 2003: 191–195; 2007a; 2007b: 87–107; 2007c). This emphasis on wall paintings and architecture also characterizes studies of Egyptian motifs in Augustus's house on the Palatine (e.g., Söldner 1999).

20. *Decor* and *decorum:* Pollitt 1974: 68–70, 341–347; Perry 2005; cf. Carey 2002: 56; Fullerton 2003: 111; Tronchin 2012a: 270; Platt and Squire 2017a: 45–46. While using the word "art," I acknowledge the existence of extensive debate about the appropriateness of applying this term to ancient material culture (Platt and Squire 2010).

21. Cic. *Att.* 2.2, 4.2, 10.5; Cic. *Fam.* 7.23.2; Plin. *HN* 35.2.7–19; Vitr. 7.4.4–7.5.7; Marvin 1989; Tybout 1989: 70–92; Ling 1991: 21, 50, 135–138, 148–149; Leach 1997: 70; Dickmann 1999: 240–252; Carey 2002: 56–58; Leach 2004, esp. 9–10, 17, 53–54; Barbet 2009: 57–77; Squire 2012: 616–617.

22. *"Style" in art history:* Elsner 2003. *Archaeological approaches to "style":* Dietler and Herbich 1998; Barrett 2009: 223. *"Egyptian style" in Roman art:* Swetnam-Burland 2015: 53–63.

now-classic study characterizes Roman art as a "semantic system" in which allusions to various Greek precedents—"modes of representation, figural types and formulae for detail which may be traced back to different epochs in Greek art"[23]—conveyed different associations and meanings, and were thus suited to different subjects and settings. Some have critiqued the inflexibility of Hölscher's schema, and one might also question his reliance on linguistic metaphors for material culture.[24] However, evidence does suggest that Roman audiences generally expected art to seem appropriate to its context, and that constructions of "appropriateness" might involve formal and stylistic features, iconographic content, or all of the above.[25]

Accordingly, if we are to understand how Roman material culture created meaning for its viewers (or derived meaning from them), we need to understand it in context. When we encounter wall paintings, mosaics, or statues in museums or as photographs in publications, we may be tempted to treat them as self-contained works to be viewed in isolation.[26] However, in their original settings, these images and objects were part of much larger, three-dimensional installations: designed to be walked around, dined upon, or even (in the case of the painted benches in the Casa dell'Efebo) sat upon. While much art-historical research has productively interrogated ancient modes of viewing and visuality,[27] gazing was only one of many ways in which people interacted with domestic space. Recent work on the "archaeology of the senses" further emphasizes the importance of moving beyond exclusively visual approaches to material culture.[28] In order to understand the functioning of any individual aspect of domestic visual culture—a wall painting, mosaic, statue, figurine, shrine, or anything else—we need to examine its interactions with larger multimedia environments.[29] Such

23. Hölscher 2004: 18.

24. *Hölscher's system as useful, but overly formulaic:* Varner 2006: 280; van Aerde 2015: 46–53; van Eck et al. 2015: 10–12; Versluys 2015b: 154. *On his linguistic metaphors:* van Aerde 2015: 46–53; Elsner 2004: xxv–xxvii.

25. Juxtapositions of painted motifs may often have emphasized "formal analogies" rather than "narrative coherence" (Mayer 2012: 298; cf. Hodske 2007: 132, 138; Lorenz 2008: 262).

26. Cogent critiques of such temptations: Clarke 1991: xxiii; 2007b: 119–120; Bergmann 2002a, esp. 15–18.

27. E.g., Nelson 2000; Fredrick 2002; Platt 2002; Zanker 2004; Elsner 1995, 2007.

28. E.g., Hamilakis 2002, 2013.

29. Important predecessors for this approach include Clarke's 1991 examination of the roles of paintings within "decorative ensembles" and Bergmann's emphasis on the interactions of painted images with built environments, landscapes, and the motion of viewers through space (see esp. 2002a, 2002b).

environments include not only the permanent built structures of a room but also the object assemblages[30] contained within those structures, as well as the living beings whose movement enlivened the space.

The recontextualization of Egyptian imagery further enables us to reframe questions about what such images "represented." As the much-discussed "material turn" in archaeology has emphasized (see "Rethinking Things" later in this chapter), material culture is not just a passive reflection of human behavior; it also takes an active role in shaping that behavior. Beyond asking how representations of Egypt *reflect* ancient attitudes toward cultural identity, we should thus also consider how these objects and images might have helped *shape* behavior and attitudes. In other words, we need to ask not only what material culture *represents* but also what it *does*. This question is particularly relevant for domestic material culture, with which the inhabitants of a house would have interacted frequently and repeatedly; its use and viewing would have been thoroughly integrated into daily routines. Accordingly, we need to investigate what meanings and relationships that domestic setting constructs for its users. What relationships does a particular context create between the viewers of an image, the imagined "Egypt" that image evokes, and the various other types of imagined landscape that the image's setting may suggest? How might interaction with these particular representations, in this particular setting, have helped shape those viewers' own behaviors, attitudes, and performances?

This book thus offers an explicitly contextually oriented analysis of domestic material culture. Focusing on the relationships between artifacts and larger multimedia ensembles, I treat "meaning" not as an inherent property of objects but, rather, as an emergent phenomenon resulting from the shifting interactions of people and things in temporally and spatially specific settings. This emphasis on human–thing entanglements[31] as *shifting* is important, as the functions, contents, and inhabitants of a given space would have varied not only from house to house but also even over the course of a single day. For example, at different times or from different vantage points, the same garden might have served as an outdoor dining/entertainment area for invited guests, a work space for domestic slaves, a

30. My use of "assemblage" follows that of Smith, who defines assemblages as "arrays of material forms that, to use Nicholas Thomas's (1991) evocative term, 'entangle.' Assemblages make our practices social by mediating links among humans and material forms that are at once spatial—generating relationships of position and movement—and historical—demanding continual investment and attention. . . . Assemblages can embrace the organic and inorganic, the durable and ephemeral, and may envelop human bodies as well in forms such as mortal remains, chattel slaves, and transplanted organs" (Smith 2015: 32).

31. On this use of "entanglement," see Hodder 2012.

play area for children, and an appealing vista glimpsed through a door by clients in the atrium (or even by passersby in the street outside).

Thus, when I employ the term "context" to describe the larger assemblages within which domestic artifacts were entangled, these contexts should not be construed as static or immobile, but instead as relational, fluid, interactive, and constantly changing. Within art history and literary studies, the concept of "context" has sometimes attracted critique for its supposed implications of inertness or inflexibility.[32] In archaeological usage, however, the "context" of an artifact is explicitly conceptualized as ever-changing, even post-depositionally. As Lucas puts it, for archaeologists, "no context is static."[33] In contrast to the immobile constructions of "context" that Culler, Bal, and others have critiqued, the archaeological conceptualization of "context" is theoretically much closer to recent art-historical discourse on "framing," which similarly emphasizes the dynamic interactions between images and "frames" and characterizes houses, cities, and even empires as a series of nested framing devices.[34]

In its archaeological sense, the context of an artifact, ecofact, or feature possesses three components: *matrix* (the material, typically sediment, around an object), *provenience* (the three-dimensional location of the object within the matrix), and *association* (the spatial relationship of the object to other artifacts, features, and ecofacts). Artifacts may further come from primary contexts or secondary contexts. In situ deposits, or de facto assemblages, are those left behind when people abandoned an activity area.[35] More common are refuse deposits, which Schiffer divides into primary refuse (deposited at the location of use) and secondary refuse (deposited elsewhere).[36] Even when an object is in situ within an activity area, however, that context reflects only one moment of the use-life of that object; people may have used it in many other places or ways earlier. Furthermore, even after burial, archaeological contexts continue to change. Numerous cultural and natural transformations ("C-transforms" and "N-transforms"[37]), from human disturbance to erosion and decay, ensure that "context" remains eternally in flux.

---

32. Culler 1988, esp. viii–ix; Leja 1993: 11–12; Bal 2002: 133–173, esp. 133–138; Dillon 2006: 190n21; Platt and Squire 2017a.

33. Lucas 2012: 13.

34. On "framing" in Classical art, see Platt and Squire 2017b.

35. Schiffer 1987: 89–97; Ault and Nevett 1999.

36. Schiffer 1972: 161; 1987: 58.

37. Schiffer 1987: 22.

Domestic Niles: Egyptian Landscapes in Roman Gardens

The need to position Aegyptiaca within larger domestic assemblages is particularly pressing for the substantial corpus often described as "Nilotica": wall paintings and mosaics depicting the Nile River within a rural Egyptian landscape.[38] First attested in the second or first centuries BCE, such landscapes became popular throughout much of the Roman world in the early empire.[39] Although they appear in various settings, they are most common in household contexts, especially domestic gardens. Indeed, these paintings and mosaics represent one of the most common forms of Egyptian imagery in the gardens of Pompeian houses. As with many other Roman representations of Egyptian themes, their interpretation has inspired great controversy; for different researchers, these landscapes have represented everything from expressions of Isiac devotion to depictions of Roman xenophobia and anti-Egyptian prejudice.

Much previous research has focused on the iconography of such scenes (for which see further chapter 2, this volume). In antiquity, however, Nilotic frescoes and mosaics were not just self-contained images but also often formed part of larger three-dimensional installations designed for viewer participation. While a few studies have laid important initial groundwork for the contextualization of domestic Nilotica, such work has been limited to general observations about room types.[40] In particular, these studies have shown that Nilotic imagery is especially common in rooms associated with *otium* (cultured relaxation) and more limited access, such as gardens and triclinia.[41] This identification of general architectural room types has made a major contribution to scholarship on Nilotic scenes. However, Roman rooms were multifunctional, and artifact distribution often indicates uses more varied than architectural form suggests.[42] If we wish to

38. Definitions and descriptions of the genre: Meyboom 1995; Versluys and Meyboom 2000; Versluys 2002; Meyboom and Versluys 2007; Barrett 2018d. On Malaise's (2005: 201–214) proposed distinctions between "Aegyptiaca," "Pharaonica," and "Nilotica," see note 98, this chapter.

39. Versluys 2002: 241–248. On the earliest Nilotic mosaics and the still-controversial dating of the "Palestrina mosaic," see note 23, chapter 2, this volume.

40. Versluys and Meyboom 2000; Versluys 2002: 248–261; Mol 2015: 256–293.

41. *Prevalence of Nilotic scenes in gardens (including peristyles):* Versluys and Meyboom 2000: 123–125; Versluys 2002: 256–258; Clarke 2007b: 93–95; Mol 2015: 287–290; Pearson 2015: 60; Merrills 2017: 139–140. *Associations of these contexts with* otium: Versluys 2002: 257–258; Myers 2005. *Access to gardens:* von Stackelberg 2009: 67–72. Cf. Clarke (2003: 192–193; 2007b: 94–96) on the contexts of Pompeian pygmy scenes (which are often, but not always, Nilotic).

42. Leach 1997; 2004: 18–54; Allison 2004a; Nevett 2010: 89–118; Dickmann 2015: 217.

understand how paintings and mosaics functioned as part of larger assemblages, we need to go beyond short descriptions of findspots to analyze interactions between these images and the other material culture with which they shared space.

As subsequent chapters will demonstrate, people appear to have viewed domestic gardens as particularly appropriate settings for many forms of Egyptian imagery. From paintings or mosaics of Nilotic landscapes, to sculpture representing Egyptian animals or dwarf-gods, to shrines dedicated to Egyptian deities, numerous visual allusions to Egypt appear in garden settings.[43] Furthermore, representations of Egyptian landscapes are often larger and more prominent in garden settings than in other parts of the *domus*. In other areas of domestic space, Egyptian imagery typically serves as a subsidiary element of larger painted or statuary compositions (e.g., as small ornaments surrounding larger painted panels), but it rarely forms the focal point of an entire room.[44] Within gardens, however, representations of Egyptian themes may cover an entire dining installation (as in the Casa dell'Efebo), dominate a whole wall, serve as the focal point of a composition, or form a major component of a statuary assemblage.[45]

Accordingly, this book investigates the ways that representations of Egyptian landscapes would have functioned within and around domestic garden spaces to influence the experiences of those spaces' occupants. Rather than taking a more traditional art-historical approach focused on compiling all examples of a single iconographic genre (e.g., representations of a particular deity or scene type—a task already admirably completed for "Nilotica" by Versluys[46]), or a single artistic medium (e.g., paintings, mosaics, statuary, etc.), I thus focus on representations of Egyptian landscapes in multiple media within a particular *type of household space*.

Some clarification is necessary for the term "garden." The Latin language has a variety of terms for gardens and green space.[47] I follow here von Stackelberg's minimal definition of a Roman garden as a cultivated, enclosed, and irrigated space

43. For previous observations of the association of Egyptian imagery with garden contexts or paintings of gardens, see Grimal 1969: 183–184; Versluys and Meyboom 2000; Swetnam-Burland 2002: 73–75; 2007: 131–132; Swetam-Burland 2015: 28; Versluys 2002: 253–258; Clarke 2003: 192; 2007b: 94–95; Hales 2003: 153–162; von Stackelberg 2009: 122; Mol 2013: 127.

44. See chapter 5, this volume, "Introduction."

45. See chapter 5, this volume, "The Garden as Margin."

46. Recent examples of such approaches to Aegyptiaca: Versluys 2002 ("Nilotic scenes"); Beaurin 2008 (representations of Isis-Fortuna).

47. On Latin terms for garden spaces, see La Rocca 1986; Allison 2004a: 169–170; von Stackelberg 2009: 10–21. Note Purcell's cautions (1996: 151n29) against modern applications of certain ancient terms, such as "viridiarium," to spaces that might not have fallen under such headings in antiquity (and cf. Allison 2004a: 169).

attached to a man-made structure and "associated with activities which add value to the space above its productive potential."[48] This definition thus encompasses a wide variety of planted domestic spaces, ranging from tiny vegetable plots to more architecturally distinctive garden forms such as peristyles (four-sided colonnaded courts, adapted originally from Greek architecture and modified into planted spaces in Roman houses) and partially colonnaded pseudo-peristyles.[49]

In these settings, it may have been less necessary to mark Egyptian motifs as "marginal" vis-à-vis a particular wall or painted panel; the garden context itself coded these images as belonging to the margins of the *domus*. Much recent research has explored Roman gardens as artificial "landscapes" within the home—liminal spaces simultaneously domestic and yet also evocative of vistas beyond the household.[50] Gardens thus create a productive environment for exploring—and reshaping—the lines between domestic and exotic, familiar and fantastic. Certain forms of imagery, including Egyptian themes, Bacchic themes, and representations of wild animals in rocky outdoor environments, are particularly common in gardens.[51] Hales has characterized all these images as evocations of the "fantastic" or "transgressive";[52] but what does it really mean to label Egyptian imagery as "transgression" or "fantasy," and are these terms always appropriate? As images of the edges of empire, Nilotic landscapes may well be appropriate to the edges of domestic space: the borders of painted walls, or the borders of the *domus* itself. Yet even as the association of this imagery with liminal space might appear to reinforce the idea of Egypt as far away, strange, or "Other," such domestic border zones simultaneously provide a way to incorporate the "Other" within the household. The appearance of Nilotic imagery within a Pompeian domestic garden does not just "Egyptianize" the garden; it also "Italianizes" and domesticates the Nile.

48. Von Stackelberg 2009: 62.

49. *On peristyles:* Grimal 1969: 204–244; Dickmann 1997, 1999; von Stackelberg 2009: 21–23; Zarmakoupi 2010: 621–622; 2014: 104; Bergmann 2012: 229–232; Trentin 2014. *Planting of peristyle gardens:* Jashemski 1981. Moormann emphasizes the difference between peristyles proper, and gardens with porticoes on fewer than four sides (1996: 393; 2002: 432n15). Partially-colonnaded gardens may be termed a "loose *porticus* structure" (Zarmakoupi 2010: 622), "Rumpfperistyl" (Dickmann 1999: 141–144, 203, 207, 307), "pseudo-peristyle" (Farrar 1998: 28–29), "truncated peristyle" (Trentin 2014), or "partial peristyle." The modern use of "peristyle" differs somewhat from Vitruvius's use; his *peristylum* is not just any colonnaded court but also follows certain guidelines for placement and proportion (Vitr. 6.3.7; Allison 2007: 270).

50. Hales 2003: 153–162; Pagán 2006: 20–21, 36; von Stackelberg 2009: 53, 122; Marzano 2014: 198; Merrills 2017: 139–140; cf. Grimal 1969: 440–442.

51. Hales 2003: 153–162.

52. Hales 2003: 155.

The resulting visual entanglement of imagined Egyptian landscapes and real Italian households speaks to a number of ongoing debates within Roman archaeology and art history concerning changing conceptions of identity, "Self," and "Other" in the early empire; the nature and functions of Roman artistic emulation of earlier visual cultures; and the integration of individual households into broader, empire-wide cultural and economic networks. More broadly, interactive installations like the Casa dell'Efebo garden also speak directly to one of the central concerns of archaeology as a discipline: the ways in which material culture not only reflects but also actively shapes constructions of cultural and social identity.

Flavian Pompeii as Case Study

Rather than attempting to assemble a comprehensive corpus of garden Nilotica from all over the Roman world or throughout Roman history, this book offers an in-depth examination of case studies from one particular site and period, which—perhaps unsurprisingly—is Pompeii in 79 CE. The selection of Pompeii relies on several considerations. First is, of course, the remarkable preservation of the site, which facilitates the type of contextual analysis necessary for this study. As is now well known, the "Pompeii premise" does not really hold true even for Pompeii itself; both ancient and early modern looting altered the assemblages in the houses, and the earliest excavations at the site were not documented according to modern methods.[53] Nonetheless, the Vesuvian cities still offer an unparalleled level of fine-grained detail about the provenience and associations of artifacts, making it possible to reconstruct domestic assemblages in a way not often feasible outside of Campania.[54]

Another advantage of focusing on Pompeii is the wealth of existing data on Egypto-Campanian, including specifically Egypto-Pompeian, economic and cultural connections. Such data enable us to situate domestic artifacts within broader webs of economic and cultural connectivity. Aegyptiaca at Pompeii circulated within, interacted with, and drew meaning from networks of economic, political, and cultural exchange that spanned the local (Pompeii itself), the regional (Campania), and the trans-Mediterranean. Campania was an important region for Italian trade with Egypt, with many Egyptian imports—most crucially grain, but also manufactured goods such as textiles—passing through the port

---

53. See discussions in Allison 2004a: 21–26, 30–36; Berry 2007: 292–294; Dwyer 2012: 306–307; Dickmann 2015.

54. Berry 1997, 2007; Allison 2004a, 2004b, 2006; Dickmann 2015.

of Puteoli.[55] Pompeii was also a port city in its own right, described by Strabo as a regional entrepôt controlling commerce along the Sarno River.[56] Along with trade came opportunities to engage with new ideas, objects, and practices. The first evidence for cults of Egyptian deities in Italy comes from Campania, where Pompeii had an Isis temple and Puteoli a Sarapis temple, and shrines in a number of Pompeian houses contained images of Egyptian gods.[57] Also popular in domestic contexts were numerous imports, imitations, and adaptations of Egyptian material culture, including statues and statuettes of Egyptian deities, faience figurines, lamps, metal vessels and cult implements, amulets, and wall paintings and mosaics that incorporate Egyptian motifs, imitate Pharaonic artistic styles, or depict Egyptian landscapes.[58]

A focus on Pompeii in 79 CE enjoins certain cautions concerning the generalizability of the data. As Dyson puts it, "the Pompeian house should be read as a Pompeian house and not as a Roman house."[59] Pompeians' attitudes toward Egypt—or, for that matter, toward Romanness[60] or Hellenicity—might well be quite different from those of their contemporaries in Rome or elsewhere. Pompeii

---

55. Sen. *Ep.* 77.1–3; Suet. *Aug.* 96, *Ner.* 20; Plin. *HN* 19.1.3; D'Arms 1974; Casson 1980: 26–29; De Caro 1983; Meyboom 1995: 343–344n28, 346–347n34; Erdkamp 2005: 252; Jones 2006: 26–33, 92–102; Zevi 2006; Manolaraki 2013: 186n8. On the commercial importance of Puteoli, cf. Strabo 5.4.6. Besides grain, imports from Egypt likely included textiles, papyrus, stone, and various luxury goods (Dalby 2000: 174–177; De Caro 2006a: 15–16; Pearson 2015: 43–46; Soto Marín 2018).

56. Strabo 5.4.8; Laurence 2007: 51–69. References to an Alexandrian trader and Egyptian imports (specifically, textiles) also appear in the archives of the Pompeian banker L. Caecilius Iucundus (*CIL* IV 3340.100; Andreau 1974: 125, 218, 289, 293, 306).

57. An inscription indicates the Puteoli temple was functioning in 105 BCE (*CIL* X 1781), and the first phase of the Pompeii Iseum may date to about 100–80 BCE (Kleibl 2009: 283). Blanc et al. (2000; followed by Saragoza 2017: 374) prefer to date the Pompeian temple to the Augustan period; Gasparini (2008: 67) is unconvinced by the redating, and Swetnam-Burland (2015: 106–107) leaves the question open. *Egyptian cults in Campania*: Tran tam Tinh 1964, 1971, 1972; Poole 2004; Steuernagel 2004: 40, 212–227; De Caro 2006b; Bassani 2008: 84–86. *The Isis temple at Pompeii*: De Caro 1992; Hoffmann 1993; Blanc et al. 2000; Swetnam-Burland 2002, esp. 94–163; Petersen 2006: 17–56; Sampaolo 2006; Moormann 2007; Kleibl 2009: 277–286; Gasparini 2013; Nielsen 2014: 69–70; Saragoza 2017.

58. On Egyptian imports and locally produced objects incorporating Egyptian imagery or stylistic features at Pompeii, see Tran tam Tinh 1964: 154–175, 181–185; de Vos 1983; Arslan 1997: 425–444; Swetnam-Burland 2002: 54–93; Bragantini 2006, 2012; Mol 2015. On paintings, see de Vos 1980 ("Egyptomania" in paintings and mosaics throughout Campania); Versluys 2002: 94–160 ("Nilotic scenes" from Pompeii).

59. Dyson 1997: 154.

60. On using "Romanness" rather than *Romanitas* (a term Revell critiques as generating a false "aura of authenticity" and sense of emic significance, given that it was in fact "little used in the Roman period"), see Revell 2008: xi.

was itself a complex palimpsest of cultures, languages, and identities. Before Sulla's conquest in 89 BCE led to its becoming a Roman colony, Pompeii was an Oscan-speaking Samnite city located in a part of Italy with longstanding, deep-seated cultural connections to the Greek world.[61] Pompeii's history of economic and cultural ties to Egypt also differed from Rome's. Campanian merchants were already active in the Hellenistic period at the cosmopolitan port of Delos: a site frequented by merchants from Ptolemaic Egypt and a major center for the Greco-Roman worship of Egyptian gods.[62] Perhaps in connection with these economic relationships, Campanians appear to have erected temples to Egyptian gods at an earlier period, and with less attendant controversy, than in Rome itself.[63] In late Republican and early imperial Rome, the worship of Isis and Sarapis periodically inspired resistance and even scandal, and Isis and Sarapis only entered into the Roman *sacra publica* in the mid-to-later first century CE (perhaps not until Vespasian's reign).[64] However, such controversies largely concerned issues specific to the physical site of Rome—e.g., whether or not Isis temples could stand within the *pomerium* of that city[65]—and there is no indication that they made much of any impression in Campania.

So Pompeii is by no means a stand-in for the Roman Empire as a whole—but this very cultural and geographic specificity also contributes to the importance of

61. See the useful recent discussions in Wallace-Hadrill 2008: 128–137; 2011.

62. *For Italians, Egyptians, and other ethnic groups on Delos*, see Martzavou 2010; Barrett 2011: 120ns365–371. Many Italians on Delos came from Campania or elsewhere in southern Italy (Tran tam Tinh 1964: 25–26; Malaise 1972: 271, 303; D'Arms 1974: 108–109; Poole 2004: 222, 224; Martzavou 2010: 183), though others had central Italian origins (Coarelli et al. 1982; Wallace-Hadrill 2008: 84, 113; see overview of debate in Roselaar 2012: 97-98). Hellenistic trade connections between Egypt and Italy may have followed both an "eastern" route via Delos and a "western" route via the African coast and Sicily (Meyboom 1995: 346–347n34). Based on an apparent absence of "Nilotic scenes" on Delos, Meyboom suggested such motifs reached Pompeii via the "western" route (1995: 352n 44; followed by Malaise 2005: 214). However, more recently studied finds from Delos indicate the presence of Nilotica there as well (Versluys 2002: 223–224, no. 121). *For Italian merchants at other Hellenistic sites associated with Isis cult*, see Martzavou 2010. *On Egyptian gods on Delos*, see most recently Barrett 2011; Moyer 2011: 142–207; Mazurek 2016.

63. Petersen 2006: 41–42.

64. Roman receptions of Isis and Sarapis: Takács 1995. Scheid contends that neither the Iseum Campense nor Isis's inclusion in the *sacra publica* can be securely demonstrated before Vespasian's reign (Scheid 2004; supported by Van Haeperen 2006: 44; Belayche and Rebillard 2008: 148; Bülow Clausen 2015: 140, 190–191; resisted by Gasparini 2009: 349). On the Iseum Campense, see Versluys et al. 2018.

65. On Octavian's prohibition on Egyptian cults within the *pomerium* (Cass. Dio 53.2.4), see Orlin's recent reassessment (2008).

Pompeii as a case study. Aegyptiaca in Italy are often viewed through the binary lens of "Roman" versus "Egyptian." Scholars discuss these objects as evidence for "Roman" and "Egyptian" interactions, debate the degree to which they convey "Roman" versus "Egyptian" meanings, and analyze their implications for "Roman" constructions of "Egyptian" Others. However, Pompeii provides a useful corrective to this binary perspective. Constructions of identity in Pompeii would have been much more complex than such dichotomies imply—as, indeed, would constructions of identity (cultural, ethnic, political, and social) throughout the empire. Campanian local identities did not necessarily disappear after the Social War, at which point the population of Pompeii included, among others, old Samnite families, Roman colonists, and foreign transplants of various origins and social statuses.[66] The cultural and ethnic origins of specific house owners in 79 CE are seldom transparent.[67] Wall paintings and domestic furnishings may reflect a range of attitudes both positive and negative[68] toward Roman official ideology, but such attitudes may not correlate neatly with families' origins or histories. Evidence for conflict between colonist families and old Pompeian families is limited.[69] Rather than disappearing, Pompeian elite families maintained high social status and roles in local administration, while also adopting "Roman" fashions.[70] Roman and Pompeian identities appear to have been complementary rather

66. Wallace-Hadrill 2011. On Italic identities, see Wallace-Hadrill 2008.

67. Dickmann 1999: 252–255. Certain terrace houses are sometimes associated with Roman colonists (Zanker 1998: 74), but this identification remains uncertain (Hales 2003: 100; Tybout 2007: 417).

68. *Roman state ideology in Pompeian houses:* see Giacobello 2008: 78 and Petersen 2012: 328 on the shrine to the Capitoline triad in the Casa degli Amorini Dorati. Domestic representations of Aeneas may recall the divine origins of the Julii (Clarke 2007b: 147–151). *Possible evidence for resistance to "official" ideologies*: A painting of Aeneas and Romulus as dog-headed apes is sometimes taken as mocking Augustan dynastic mythology and expressing Pompeian resentment of Roman colonists (Clarke 2007b: 151–157). However, do anti-Augustan sentiments necessarily require the patron to have a local background—and, for that matter, is opposition to Augustus necessarily the same thing as opposition to Rome? Since the Italian allies were likely fighting for Roman citizenship rather than independence (Wallace-Hadrill 2008: 81), an "Oscan" or "Samnite" identity may not have required anti-Roman sentiments. Even before the Social War, Pompeian elites' self-presentation was compatible with that of Roman elites (Wallace-Hadrill 2008: 136–137).

69. Hales suggests that Romans would have interpreted the 59 CE riot in Pompeii's amphitheater (Tac. *Ann.* 14.17) as a sign of Pompeians' inability to perform Romanness successfully (2003: 265n82). However, while Tacitus notes the Pompeians' punishment, he does not say anything about this making them less Roman.

70. *Status and administrative roles:* Mouritsen 1988: 87–89, 112–122 (contra Castrén 1975: 52–55); cf. Dyson 1997: 154. *Adoption of Roman fashions:* Dickmann 1999: 255. *On "waves of fashion":* Wallace-Hadrill 2008: 356–440.

than oppositional, and the members of Pompeii's diverse population would have navigated a sea of potential cultural, ethnic, political, legal, and linguistic identities ranging from local to pan-Mediterranean.

Additionally, the chronological specificity of Pompeii's destruction layer enables us to reconsider some other longstanding assumptions about "Roman" views of "Egypt." There is often a strong temptation to interpret Italian Aegyptiaca through reference to specific events known from the historical record, and Augustus's conquest of Egypt in 30 BCE has exerted a particularly strong gravitational force on scholarly research.[71] For example, many of the painted Egyptian landscapes discussed in chapter 2 have been interpreted through the lens of Actium as celebrations of conquest and derision of the conquered.[72] The canonical status of what was once called "Golden Age" Latin literature has also encouraged the privileging of literary evidence from Augustus's principate as a source for Roman attitudes toward Egypt.[73] However, Augustan literary references to Egypt are deeply embedded in a particular social, political, and chronological context; these are largely texts from an elite sphere closely associated with the circle of the *princeps*, responding among other things to the political necessity of presenting the conflict between Octavian and Antony as a foreign war rather than a civil one.[74] As Swetnam-Burland has recently pointed out, "watershed events" such as Actium "shaped public interest [in Egypt] but do not wholly explain it."[75] Material culture is more than an illustration of the historical record, and popular culture is more than an echo of elite activities.[76]

Certainly, putting material and textual evidence into dialogue is essential for understanding the cultural, historical, and social context of Pompeian

71. As also critiqued by Davies 2011: 354, 359; Pearson 2015: 8–9; Swetnam-Burland 2015: 67, 84–85; Mol 2017: 172–173. Cf. the similar gravitational force of Augustan politics on scholarly understandings of Third Style wall paintings (Pearson 2011). On interpretation of Aegyptiaca through reference to historical events, cf. Bricault and Versluys 2014a: 6; Swetnam-Burland 2015: 1.

72. E.g., Versluys 2002, esp. p. 438 (critiqued by Tybout 2003, esp. 509–515); Merrills 2017: 106–149.

73. Cf. the observations of Manolaraki 2013: 4, 29.

74. *Augustan poets on Egypt:* Maehler 2003: 205–211; Leemreize 2016: 79–111, 129–135, 163–201. *Octavian's political motives:* Davies 2011: 357.

75. Swetnam-Burland 2015: 1.

76. Cf. Squire's 2009 critique of logocentrism. On popular and elite culture, see Tronchin's recent critique (2011a; cf. 2012a: 275–278) of Zanker's (1979, 1998) explanation of Pompeian houses as emulations of wealthy villas.

Aegyptiaca—and indeed, the study of domestic Aegyptiaca has much to reveal about the intersections of visual and literary culture within the household (see especially chapter 3). However, the date of Pompeii's destruction should encourage us to move away from interpretive frameworks focused on the moment of Egypt's conquest and, instead, to understand Pompeian Aegyptiaca within the social, political, and cultural context of the Flavian period.

In Flavian Italy, anti-Egyptian sentiment had much less political currency than at the start of Augustus's principate.[77] Egypt was part of the empire and no longer a political threat, and Egyptian gods were thoroughly integrated into the Roman pantheon. The Flavian emperors associated their own dynastic fortunes with Egypt, patronized Egyptian cults, and incorporated Egyptian deities into official iconography.[78] Additionally, the Flavian period was in many ways a time of unprecedented connectivity, mobility, and center–periphery integration for the Roman world.[79] Contemporary conceptions of empire reflected this changing sociopolitical environment. Much early imperial material culture attempts to present the populations of conquered provinces as, in Brian Rose's words, "contributors to peace rather than its opponents."[80] In literature, too, some scholars have identified a vein of would-be "positive imperialism" in Flavian sources, emphasizing not only the glory of conquest but also the shared well-being of Rome and its provinces.[81] Within this context, some contemporary literary references present Egypt and the Nile as productive contributors to the Roman world, destabilizing older Augustan literary tropes that treat Egypt primarily as exotic "Other."[82] Recent research on

---

77. Tybout 2003: 510–512.

78. *Flavian attitudes toward Egypt*: Takács 1995: 94–104; Gasparini 2009; 2014: 297–298; Pfeiffer 2010; Manolaraki 2013: 121–132; Capriotti Vittozzi 2014; Versluys 2017a. Pfeiffer (2010) cautions against overstating Egypt's importance to Flavian imperial strategies, although Versluys (2013b: 256n67) characterizes Pfeiffer's approach as "misconceived."

79. Ando 2003: 326–327; Manolaraki 2013: 126.

80. Rose 2013: 544; cf. Kuttner 1995: 86–93. Less benevolent representations of conquered peoples certainly also persisted: e.g., Ferris 2000: 48–60; Cody 2003; Rodgers 2003.

81. Fear 2011 (quoting 34); Manolaraki 2013: 131n43. On the complexity of Flavian-era literary discourse on empire, see Evans 2003.

82. E.g., Valerius Flaccus (*Argon.* 1.20) includes the Nile as one of many imperial regions whose ships will navigate according to the light of the deified Vespasian-as-star; Statius (*Silv.* 3.2.23–24) describes Alexandrian grain ships making pious libations to Minerva when approaching the Bay of Naples. On these and other Flavian literary references to Egypt, see Manolaraki 2013: 121–222.

representations of Egypt in post-Augustan literature emphasizes the degree to which later texts alter and nuance earlier representations of Egypt, frequently demonstrating "efforts to integrate the Nile more fully and less antagonistically into Roman *imperium.*"[83]

Taking 79 CE as the relevant date of analysis for Pompeian garden assemblages may, of course, appear to raise another set of problems. The houses discussed in this book were destroyed at that time, not built then. The material assemblages of most houses were multi-period palimpsests, reflecting numerous episodes of renovation, redecoration, and reuse. The inhabitants at the time of the eruption were often different from the original commissioners of the houses' frescoes, mosaics, or statuary.[84] Additionally, some household objects were imported antiques, whose production belonged to a different place as well as a different time.[85] As a result, to what degree can we actually take these assemblages as indications of Flavian-period attitudes?

It is often impossible to tease apart the different forms of human agency that contributed to assembling houses as they appeared in 79. The difficulty lies not only in assigning renovations to particular patrons; we must also deal with the different forms of agency exerted by producers, purchasers, and viewers. As Bergmann puts it, "Who, for instance, was responsible for the appearance of a renovated interior—the patron, a beneficiary, an architect, a builder, the master of a fresco or mosaic workshop, a temporary renter?"[86]

However, regardless of when they were commissioned, the interiors of these houses reflect not only the agency of their "original" owners but also that of any and all subsequent owners. Even if they did not commission all these images, later owners still made choices about retaining or replacing them.[87] Indeed, people might even purchase a house in part because they appreciated its existing interior.[88] Even more important, the new owners of a house (as well as its

83. Manolaraki 2013 (quoting 126). See also Leemreize 2014.

84. From the Augustan period on, frescoes and *tabulae pictae* were legally considered part of the *ornamenta* of a house and were typically sold with it, "treated legally just like the walls themselves" (Bergmann 1995: 99–100).

85. On Egyptian imports in Pompeian houses, see Swetnam-Burland 2002: 70–76; 2015: 32; Mol 2015.

86. Bergmann 2012: 240. Cf. also Clarke 2003: 11, fig. 3; 2007b: 9, fig. 1.

87. On later owners' purposeful conservation or restoration of earlier paintings or mosaics, see Clarke 1991: 140, 166–167, 188, 193, 215; Dickmann 1999: 253; Bergmann 2007: 99; Ehrhardt 2012.

88. Bergmann 1995: 100.

non-owning viewers, such as slaves, guests, or clients) exerted interpretive agency, as the "meanings" of images and objects altered according to their changing viewers. Even if the new owners of a house left all the original furnishings intact, they might value those furnishings for different reasons than had the original owners and interpret them in different ways.[89] Each introduction of new people, activities, and artifacts into a space would effectively transform that space into a new dynamic assemblage, generating new entanglements, affordances, and dependences.[90] Understanding those entanglements requires acknowledging the agency of objects, as well as of humans. Just as people shaped domestic space, so too did domestic space shape its inhabitants. Regardless of whether or not they had personally chosen the material culture surrounding them, the inhabitants of a house still had to live within it and engage with the interactive environments that it created.

As a result, regardless of their original production date, all the objects and images examined in this book were *also* products of 79 CE. At the time of the eruption, these Aegyptiaca participated in Flavian domestic assemblages, interacted with Flavian viewers, and collaborated in the production of meanings specific to Flavian Pompeii. As we shall see, the earlier "biographies" of many objects probably did contribute substantially to their reception by later viewers (see "Object Biographies," later in this chapter). However, the "meaning" of images, artifacts, and assemblages would have always remained in flux: not static, but emergent and dynamic.

Material Culture, Identity, and "Authenticity"

The past two decades have seen an explosion of scholarship on the cultural, economic, political, and religious networks connecting Egypt to the rest of the Roman world, as well as the functions and meanings of Aegyptiaca in Italy.[91] Within this wealth of research, three prominent developments include (1) an emphasis on "cultural," as well as "religious," interpretations of Aegyptiaca; (2) an emphasis on Aegyptiaca as "Roman" rather than (or at least in addition to) "Egyptian"; and (3) an understanding of Aegyptiaca as pointing toward not an "authentic" Egypt but, rather, a constructed "Egypt" of Roman imagination. These developments have facilitated the deconstruction of old assumptions and spurred much exciting

89. Cf. Hales 2003: 136, for a similar observation.

90. On entanglement and dependences, see Hodder 2012. On affordances, see Gibson 2015 (1986): 119-136, and chapter 2, this volume, "Multivalence and Affordances."

91. See note 16, this chapter.

research. However, they have also generated new challenges, as some of the implied binary oppositions—religious/cultural, Roman/Egyptian, constructed/authentic—do not always map smoothly onto the messier territory of ancient reality.

The early twenty-first century has seen particular critique of older assumptions that visual references to Egypt invariably evoked Isiac cult. Many early studies of Egyptian imagery in Italy relied heavily on religious explanations. In the mid-twentieth century, Karl Schefold interpreted domestic paintings of Egyptian landscapes and mythological scenes as coded references to mystery cults, especially Isis mysteries.[92] Tran tam Tinh's monumental works on Isis worship also frequently treat any Egyptian motifs as evidence that a house belonged to Isis worshipers.[93] For these and many other twentieth-century scholars, Egyptian and Egyptian-style objects were, above all, indexes of religious identity.

Much scholarship of the past twenty years has rightly reacted against the excesses of this approach, demonstrating that people might value Aegyptiaca for many reasons besides cult affiliation.[94] Current consensus reasonably holds that Egyptian and "Egyptianizing" material culture expressed a wide range of functions, goals, and values, including (among others) a desire to allude to Hellenistic luxuries; pride in imperial conquests; enjoyment of prosperity brought about by Egyptian grain; conspicuous display of imported goods as status symbols; or appreciation of the latest fashions.[95] The fashion for Aegyptiaca is often described as "Egyptomania," although some scholars have more recently critiqued this term as anachronistically loaded with eighteenth- and nineteenth-century social, political, and colonial connotations.[96]

Of course, few would deny that at least some Egyptian-looking objects do come from contexts associated with Isis worship. However, many recent publications attempt to distinguish between such "religious" artifacts and those Aegyptiaca whose significance is more generally "cultural." Malaise's influential call to "distinguer ce qui relève du cultuel et du culturel" presupposes that the "cultuel" and the "culturel" will fall into distinct and distinguishable

---

92. E.g., Schefold 1972: 61, 172–178.

93. E.g., Tran tam Tinh 1964, 1971, 1972; critiqued by Petersen 2006: 42, 246n80.

94. E.g., Versluys 2002, esp. 15–26, 329–336; Vout 2003: 189–190; Mol 2015: 31–34.

95. *Aegyptiaca or Nilotica as allusions to grain trade:* Versluys and Meyboom 2000: 126; Malaise 2005: 211. *As fashion statements:* Alfano 2001a: 286–288; Daszewski 2002: 22.

96. *Aegyptiaca as "Egyptomania":* de Vos 1980; Malaise 2005: 214–220; Merrills 2017: 118. *Critiques of this interpretation:* van Aerde 2015: 27n59; Mol 2015: 34–37; Versluys 2019. Cf. Curran's early critique of "Egyptomania" as interpretive construct (1996).

groups: "le cultuel ne se confond pas avec le culturel."[97] Malaise even proposes a new terminology reflecting this distinction, distinguishing "Aegyptiaca" from "Pharaonica" and defining the latter as Egypt-related artifacts used in Isiac sanctuaries or domestic shrines outside Egypt.[98] A similar distinction between "religious" Aegyptiaca (or "Pharaonica") and "cultural" Aegyptiaca, often freighted with similar implications of an either/or choice between the two, appears in many other recent publications.[99]

Besides pushing back against all-encompassing religious interpretations, much current scholarship makes a forceful case for the "Romanness" of Roman Aegyptiaca.[100] While Roman visual allusions to Egypt may evoke a faraway land, they also draw on Roman traditions of representation and express Roman ideas and values. Deviations from Egyptian prototypes or practices should thus not be seen simply as "misunderstandings" but, rather, as innovations that create new meanings suited to Roman cultural contexts.[101]

Accordingly, much twenty-first-century scholarship pushes back against interpretations of Aegyptiaca as necessarily expressing "authentic" Egyptian meanings.[102] Romans were not passive recipients of Egyptian culture; they actively appropriated, reinterpreted, and transformed the objects, images, concepts, and practices that they encountered (see "Rethinking Contacts," later in this chapter). Imported objects and "Egyptianizing" imagery in Italy thus need not have the same uses, or convey the same cultural significance, as similar objects or images within Egypt.[103] One of the most vocal proponents of this approach is Miguel John Versluys, who argues that the "Egypt" conjured by Roman

97. Malaise 2005 (quoting 15, 228); cf. Malaise 1972: xii; 2007: 34–38.

98. Malaise 2005: 201–214; 2007: 34–37; cf. Bragantini 2006: 161–165; 2012: 25–30. In his more recent work, Malaise's "Pharaonica" are also specifically "produites en Égypte" (2007: 35; contrast Malaise 2005: 228, where they may be "oeuvres égyptiennes ou égyptisantes"). The category thus potentially conflates two not necessarily related criteria: place of production and use-context. A third category for Malaise (2005: 210-214; 2007: 36-37) is "Nilotica," on which see Versluys 2002; Barrett 2017b. For critique of Malaise's terminology, see Bülow Clausen 2015: 40–43; Barrett 2017a.

99. E.g., Budischovsky 2004; Cibu and Rémy 2004: 142–143. On the influence of Malaise's approach, see Versluys 2010: 3–4; Bricault and Veymiers 2012: 13.

100. E.g., Vout 2003; Swetnam-Burland 2007, 2015; Davies 2011; van Aerde 2015; Bülow Clausen 2015: 34–35; Versluys 2017a.

101. Swetnam-Burland 2007; Versluys 2017a.

102. E.g., Malaise 2005; Swetnam-Burland 2007, 2015; Harrisson 2012; Versluys 2012, 2013b, 2017a; Mol 2013, 2015; Bricault and Versluys 2014a: 26–34.

103. E.g., for Italian consumers' reinterpretation of Egyptian-style ushebti figurines as representations of Osiris, see Malaise 2004: 484–485; 2005: 17–19.

Aegyptiaca is a cultural construct ultimately functioning to shape Roman identities.[104] In representing and constructing "Egypt," Romans were simultaneously representing and constructing "Romanness."

Without question, the field is much stronger for the availability of interpretive paradigms that position many Aegyptiaca as "cultural," "Roman," and "constructed," instead of (or in addition to) "religious," "Egyptian," and "authentic." Problems arise, however, when we characterize such possibilities as black-and-white alternatives, rather than positions along a spectrum. Recent years have thus seen increasing calls to move beyond the binary treatment of interpretations of Aegyptiaca as either "religious" objects (and hence "Isiac" and "authentic") or "exotica" (and hence "cultural," "secular," and testifying to "Egyptomania").[105]

Attempts to distinguish between "cultic" and "cultural" artifacts run the risk of anachronism, as Roman culture did not recognize strictly binary distinctions between "sacred" and "secular." While Latin has a rich vocabulary for relationships between people, things, and gods, that vocabulary does not map clearly onto modern constructions of "religious *versus* secular" or "sacred *versus* profane."[106] Evidence indicates that many objects might have served both religious and practical functions in antiquity. In domestic contexts, for example, oil lamps could function both as utilitarian light sources and as ritual implements in lamp-lighting rites at domestic shrines.[107] It would therefore be misleading to refer to lamps in Roman houses as either "religious" or "secular" artifacts; they are potentially either or both, depending on context and user choice, and modern distinctions between "religious" and "secular" would not have been relevant to their ancient users. Attempts to fit Aegyptiaca exclusively into one or the other category are thus similarly problematic. For example, a household statuette depicting an Egyptian deity might be simultaneously a god and an item of domestic decoration.[108] Ancient audiences appear to have been capable of viewing the "religious" and "decorative" affordances of such objects as complementary, not contradictory.

104. See esp. Versluys 2012, 2013b, 2017a.

105. E.g., Söldner 1999; Bragantini 2006: 161; 2012: 25; Mol 2013: 128; 2015: 4–5; 2017: 173; van Aerde 2015: 27–28; Versluys 2015b: 147; 2017a: 288–289.

106. Cf. Fullerton 1997: 433; Bremmer 1998; Rebillard 2012: 62, 91, 95–96; Rebillard and Sotinel 2010b. On the history of the term "religion" and its evolution from Latin *religio*, see Feil 1992; Smith 1998; Rüpke 2014: 186–193. For modern and ancient constructions of "sacred" and "profane," see Rebillard and Sotinel 2010a.

107. Bielfeldt 2014, esp. 234; Bowes 2015: 216.

108. Barrett, in press (a). On Roman approaches to *decor*, see note 20, this chapter.

Rigid oppositions of "Roman" and "Egyptian" present another false dichotomy.[109] Scholarly conceptions of Roman identities have evolved substantially in recent decades, with old paradigms of "Romanization" giving way to more nuanced approaches that acknowledge provincial actors' agency. Instead of "acculturation," individuals' adoption of Roman identities is now more often characterized as self-interested emulation, appropriation, creolization, code-switching, or globalization.[110] "Romanness" was ultimately what Wallace-Hadrill calls "a juridical category, defined by citizenship, by membership of the *populus Romanus*, or by relationship to Roman *imperium*," and it was not exclusive of other forms of cultural, ethnic, or even political identity.[111] While an awareness of one's "Romanness" was surely important in certain settings and under certain conditions, that does not mean that everyone who identified in some way, shape, or form as "Roman" necessarily foregrounded that identity at all times or viewed the world through that lens exclusively.[112] Additionally, after Egypt became part of the Roman Empire, "Roman" and "Egyptian" no longer functioned as invariably oppositional categories. Although Roman citizenship was not available to most Egyptians until 212 CE, exceptions existed for citizens of Alexandria and perhaps also army recruits.[113] A number of people of Egyptian origin lived in Rome and occupied a broad range of social and economic positions.[114]

The Roman reception of originally Egyptian gods illustrates the difficulty of drawing clear lines between "Roman" and "Egyptian." Much debate surrounds the degree to which Egyptian-*derived* deities, such as Isis and Sarapis, would have actually been *perceived* as "Egyptian" by Roman worshipers. Some argue that by the late Hellenistic period, Isis and Sarapis were so fully integrated into Greek and Roman religion as to hardly appear "Egyptian" any longer. Several recent

109. Cf. Versluys's recent critiques of "container thinking" (2015b; cf. 2013a; 2013b: 258; 2017a: 277–278), or Leemreize's (2016) demonstration that Roman literature on Egypt does not always present either wholly positive views (Egypt as respected cultural ancestor) or wholly negative views (Egypt as despised "Other").

110. *Recent overviews and reassessments of this enormous topic:* Versluys 2014, with responses; van Oyen 2017. *Globalization:* Pitts and Versluys 2015. *Relationship of "Romanization" to "acculturation":* Versluys 2015b: 144–146.

111. Wallace-Hadrill 2008: 41; cf. Versluys 2013a: 436–437.

112. On the intermittent or transitory relevance of group identities, and the danger of treating "groups" (of "Romans," "Egyptians," or anyone else) as social actors, see later in this chapter, "Material Culture, Identity, and 'Authenticity.'"

113. Plin. *Ep.* 10.6; Delia 1991: 39–47; Rowlandson and Harker 2004: 82. On Roman citizens in Egypt, see Jördens 2012: 250–252.

114. Ricci 1993; Noy 2004; Podvin 2007.

publications thus argue against referring to these deities as "Egyptian" in Roman contexts, characterizing them instead as fully "Roman" gods.[115] However, the naturalization of once-foreign cults coexisted with frequent textual and iconographic emphasis on those cults' "Egyptianness," suggesting that some conception of "Egypt"—albeit a historically contingent, culturally constructed, and potentially semi-fictionalized conception—remained central to their Roman reception.[116] Asking whether the Isis of the Roman Empire was a "Roman" or "Egyptian" goddess is thus asking the wrong question, artificially attempting to impose a binary framework onto a more complex reality. Instead of a fixed, historically stable line between "Roman" and "Egyptian," we find fluid, ongoing processes of interaction, negotiation, and reconceptualization.

Sociological research has convincingly demonstrated that identities[117] are not fixed and immutable but, rather, plural, shifting, and contextual. A person may possess many different identities, only some of which might appear relevant at a particular time or in a particular setting.[118] While such observations may sound unsurprising, they provide a necessary corrective to what Rogers Brubaker characterizes as "groupism": that is, the fallacy of treating "groups" as social actors and basic units of social analysis.[119] For Brubaker, "groups"—whether organized around ethnic, religious, cultural, or any other lines—do not constitute static entities and should not be treated as unitary actors. Instead, he sees "groupness"—that is, individuals' awareness of, and action in accordance with, collective identity—as an *event*: a transitory phenomenon that occurs only intermittently and under certain circumstances.[120] In other words, group identity

---

115. Vout 2003: 190; Beaurin 2008; Martzavou 2011: 61; Versluys 2012: 28; Gasparini 2014: 191.

116. Alvar 2008; van Andringa 2009: 164–165; Versluys 2012: 33–36; 2013b; 2015b: 148–149; Rüpke 2014: 17, 204 (arguing that Isis's Egyptian trappings made the cult recognizable transregionally). See also Mazurek 2016 on the use of Egyptian imports in Hellenistic Isiac sanctuaries.

117. Following Burke and Stets's definition of "identity" as "the set of meanings that define who one is when one is an occupant of a particular role in society, a member of a particular group, or claims particular characteristics that identify him or her as a unique person" (2009: 3).

118. Lahire 2011 characterizes individuals as "plural actors," possessed of multiple dispositions and socialized within multiple settings. Burke 2003 discusses the interaction of multiple identities within different types of social settings. On the social impact of the intersections of multiple identities, see Crenshaw 1989; Carbado et al. 2013.

119. Brubaker 2002, 2004; cf. Rebillard 2012, 2015.

120. On "groupness" as an event or variable rather than a constant, see Brubaker 2002: 168–169; 2004: 4.

is only salient in contexts that activate people's awareness of, and behavior in accordance with, that group identity.

The "groupism" fallacy can be particularly misleading when considering ancient ethnic, cultural, and religious identities.[121] While such forms of identity were undoubtedly important in the ancient Mediterranean, we should not presume that people invariably foregrounded any one identity in every social situation. For example, Rebillard has recently demonstrated that early Christians only gave salience to their Christian identity in certain contexts, seemingly treating it as less relevant for many of their other choices and interactions (including some choices and interactions where many scholars had previously assumed that ancient Christians *would* prioritize their Christianness).[122] Wallace-Hadrill emphasizes the importance of "code switching" in the Roman Empire, observing that the same individual might identify simultaneously as Roman, Greek, and a member of his or her home province.[123]

Within the study of material culture in particular, "groupism" sometimes manifests in a tendency to interpret all stylistic or iconographic content as emblematic of some form of cultural, ethnic, or religious identity. As a result, we must be careful about reading identities into Pompeian material culture. Since people had to negotiate many different forms of identity, we should not make blanket assumptions about which identities would have appeared most relevant in any given situation. The early twentieth-century tendency to interpret all visual allusions to Egypt as evidence for "Isiac" affiliation provides one manifestation of the "groupism" fallacy. However, it is just as dangerous to interpret such images and objects as straightforward indicia of cultural, ethnic, or political identities: for example, stating that domestic material culture is designed to convey the homeowner's "Romanness," or suggesting that people in Pompeii viewing paintings of festivals in Egypt would have understood themselves first and foremost as "Romans" gazing upon cultural "Others." While such identities certainly might have informed individuals' responses to material culture, they need not always have seemed as immediately relevant to ancient viewers as they do to modern scholars. The relationship between material culture and identity, as more generally between material culture and meaning, is seldom one-to-one.

Finally, while recent scholarship does well to caution against reflexively imposing "authentic" Egyptian meanings onto Roman artifacts, we should be equally wary of dismissing Egyptian meanings, uses, and values as a priori irrelevant to

121. Cf. Rebillard's 2015 cautions against looking for the "material signatures" of particular religious traditions.

122. Rebillard 2012.

123. Wallace-Hadrill 2008, esp. 3–7, 13; cf. Marvin 2008: 213–215; Versluys 2017c: 28–29.

Roman Aegyptiaca. Particularly after Egypt became a province, the empire's opportunities for individual mobility facilitated intercultural encounters. Not only did Egyptians (including intellectuals such as the Egyptian priest and Stoic philosopher Chaeremon) travel to Greece and Italy but Romans also went to Egypt as tourists, pilgrims, administrators, and soldiers.[124] Discussions of Roman travel to Egypt thus cannot simply start and end with Augustus's ban on senators' and some equestrians' access to the province, as this restriction affected only the highest elites and was in any case not completely observed in practice.[125] Egyptians, too, were active contributors to evolving Roman discourse(s) on Egypt,[126] and Roman adaptations of Egyptian objects or motifs should be understood in the broader context of those objects' or motifs' previous "cultural biographies" (see "Object Biographies," this chapter). Accordingly, we need not interpret Roman receptions of Egyptian culture as entailing *either* the museum-quality reproduction of "original" Egyptian practices *or* the invention of an imaginary fantasyland with little relationship to the real-life province of Aegyptus. In place of black-and-white alternatives, reality likely provided many shades of gray.

In pursuit of more nuanced approaches to Roman Aegyptiaca, current research in the broader fields of archaeology and art history suggests several promising directions. One important line of inquiry involves the forging of new approaches to the role of material culture in cross-cultural interactions. Additionally, the "material turn" throughout the social sciences and humanities has generated new perspectives on humans' relationship with material culture more generally. These developments are increasingly inspiring research that inserts Aegyptiaca into broader, multidisciplinary debates, with substantial implications for our understanding of the roles of "Egypt" in Roman visual and material culture.

124. *On Chaeremon,* who attained a high position in Nero's court: see van der Horst 1984. *On Egyptians in Italy*: Malaise 1972: 321–322, 324, 328; Ricci 1993; Noy 2004; Podvin 2007; Swetnam-Burland 2011: 343. *On Roman tourism and pilgrimage in Egypt*: see (from the late twentieth century onward) Bernand 1988; Foertmeyer 1989; Frankfurter 1998a; Swetnam-Burland 2002: 37–49; Klotz 2012a: 15–31. *On soldiers of Italian birth stationed in Egypt*: Alston 1995: 42–43, 66; Fischer-Bovet 2014: 112–113. One Campanian from Nucera left an inscription at Philae (Zevi 1998a: 46).

125. See, e.g., Germanicus's trip to Egypt (Tac. *Ann.* 2.59–61).

126. Moyer 2011; Versluys 2015b: 149. On Egyptian priests' "stereotype appropriation," see Frankfurter 1998b: 224–237.

Rethinking Things
Material Culture and the "Material Turn"

Since the beginning of the twenty-first century, the "material turn" in archaeological theory has emphasized that things are not just inert or passive recipients of human agency; the material world also plays an active role in shaping human behavior, cognition, and culture.[127] Within Classical art history and archaeology, this intensifying scholarly interest in "object agency" coincides with an increasing critique of logocentric approaches to ancient material culture.[128] Materiality-oriented or "thing theory"–based approaches to Roman Aegyptiaca thus hold much promise for future research.[129] Such approaches prompt questions about how Egyptian or "Egyptianizing" material culture might have influenced consumers' behaviors, opinions, and social performances, and how the material properties of Aegyptiaca helped shape the ways that people used and valued these objects.

As an example of the power of materiality in shaping viewer responses, we might consider a three-dimensional depiction of one of the most common "Nilotic" motifs: the Nile crocodile, an animal frequently presented in Roman art and literature as emblematic of Egypt.[130] The object in figure 1.9, a faience statuette of a crocodile, comes from the peristyle garden of the Casa delle Nozze d'Argento (V 2.1). This artifact served as one of four animal figurines decorating a fountain:[131] two crocodiles and two frogs, all covered with the greenish glaze typical of Egyptian faience. Chemical testing suggests that at least one was in fact imported from Egypt.[132] All four animals' open mouths originally served as water spouts.

---

127. Within the vast literature, see Hicks 2010; Hodder 2012; Knappett 2012.

128. *Objects as agents in Classical art history:* see, e.g., Whitley 2007, 2012; Collins 2008: 95–97. *Critique of logocentrism:* Squire 2009.

129. On applications of such approaches to Aegyptiaca, see Mol 2013, 2017; Bülow Clausen 2015; Versluys 2015b: 166; 2017b; Barrett 2017a; Müskens 2017; Müskens et al. 2017.

130. *For crocodiles in Roman statuary, paintings, mosaics, and coins,* see Roullet 1972: 127–128; Versluys 2002: 280–281, 283–284; Draycott 2012. *For Latin literary references to Egyptian crocodiles and crocodile gods,* see Smelik and Hemelrijk 1984: 1955–1981. According to Pliny's description of a famous painting by Nealkes, the depiction of a crocodile stalking a donkey was sufficient to locate a scene in Egypt (*HN* 35.142).

131. Ehrhardt 2004: 224 (with discussion of surviving water pipes elsewhere in the peristyle). For hydraulic facilities still preserved in the peristyle, see also Dessales 2013: 128–129, 421–423. Many lead pipes were stolen from Pompeian houses at the time of the early excavations (de Haan 2001: 43).

132. On glazed statuettes at Pompeii as Egyptian faience or imitations of Egyptian faience, see Di Gioia 2006; Mol 2015: 185, 198–204. The fabric of such Pompeian statuettes is usually

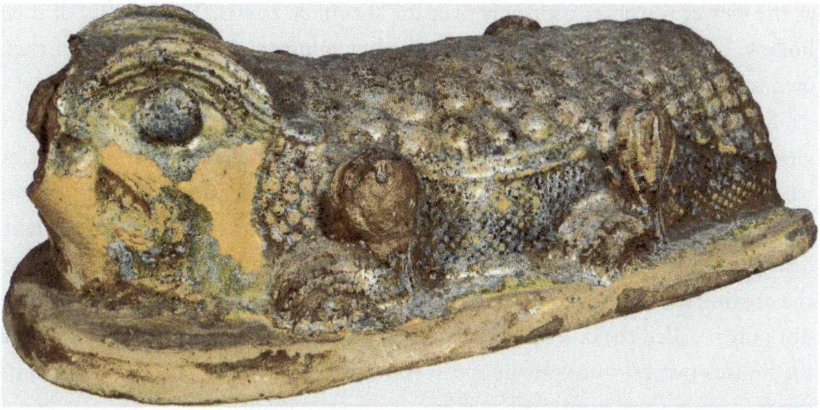

FIGURE 1.9 Egyptian faience or imitation Egyptian faience statuette of a crocodile, from the peristyle garden of the Casa delle Nozze d'Argento. Length: 26 cm (Di Gioia 2006: 124). MANN 121325.

Photograph: Soprintendenza per i Beni Archeologici di Napoli. Reproduced by permission of the Ministero dei Beni e delle Attività Culturali e del Turismo—Museo Archeologico Nazionale di Napoli.

The iconography of this figurine was clearly important to its intended impact on viewers, given its association with other figurines with similar iconography (all representations of river animals) and its location (next to flowing water that recalls crocodiles' "natural" habitat). However, the figurine's materials and physical properties would also have been crucial to its overall effect. The green glaze would have visually signaled Egyptian origins, activating associations with "Egypt" for viewers. The small scale of this 26-cm-long statuette would also have affected viewer responses. A growing body of "miniaturization theory" emphasizes the ability of miniatures to make viewers feel large, strong, and in possession of the small object.[133] Although Roman art and literature often presented crocodiles as dangerous, the small scale of these representations diminishes any implied threat,

siliceous and made with crushed quartz, like Egyptian faience, and both Di Gioia and Mangone et al. hypothesize that the majority were made in Egypt (Di Gioia 2006: 133–140; Mangone et al. 2011; *pace* Mol 2015: 185n484). Chemical testing of one of the two crocodile statuettes from this assemblage (Museo Archeologico Nazionale di Napoli, henceforth MANN, inv. no. 121324) revealed a siliceous, quartz-based fabric and suggested it was likely produced in Egypt (Di Gioia 2006: 138; Mangone et al. 2011). Frog-shaped fountains are also attested in Roman Egypt (e.g., Thomas and Higgs 2015: 32).

133. Bachelard 1964: 150, 161; Bailey 2005: 33; Langin-Hooper 2015. See also Stewart (1984: 37–69) for cross-cultural associations between miniatures and concepts of the fantastic, marvelous, and artificial.

as the viewer towers reassuringly over the shrunken beasts. The fragility of their hollow, breakable faience could have further reinforced viewers' sense of their own relative strength.

Such a small, delicate, and finely detailed object invites close-up, intimate viewing. In the garden water installation for which it was designed, this figurine would have attracted such attention by appealing to multiple senses: visitors to the garden would see the crocodile's glazed surface flashing in the sunlight, hear the sloshing of the water spouting from its mouth, and perhaps even feel the cooling spray of mist on their skins. This flashing, gurgling water would have not only recalled the crocodile's Nilotic habitat but also asserted the resources of the house's owner; since private access to running water was expensive, elaborate water features were among other things displays of wealth.[134]

Considerations of materials, scale, and other physical properties are thus crucial to the investigation of viewers' interaction with objects. Iconographically, this statuette closely parallels numerous other Roman depictions of crocodiles. The crocodile from the Casa delle Nozze d'Argento lies on its stomach, seemingly at rest but watchful, with its snout slightly raised (a pose still detectable despite the break at the snout). Depictions of crocodiles in the same attitude appear in varied contexts and media, from a granite crocodile statue in the sanctuary of Isis and Sarapis on the Campus Martius to the famous "Aegypto Capta" coins issued after Octavian's victory at Actium.[135] However, even though these assorted images of crocodiles display nearly identical iconography, both their settings and their material properties encourage different types of encounters and prompt different responses from viewers.

Object Biographies

In addition to exploring the ways that the material properties of things affect their users, much recent scholarship investigates the "social lives" of things—that is, the ways in which they participate in social relationships with people and mediate human relationships.[136] One line of research in this vein, focusing on "cultural biographies" or "object biographies," has been particularly influential among archaeologists. Coined by Igor Kopytoff, the term "cultural biography" employs the metaphor of a human life span to characterize a thing's changing

134. On leasing aqueduct water, see Richardson 1988a: 55, 62.

135. Barrett 2017a.

136. E.g., Appadurai 1986; Latour 2005; Knappett 2012: 190–192.

roles within society: the "biography of a thing," from manufacture to disuse.[137] Attention to such "biographies" helps elucidate the ways that the uses, values, and perceived meanings of originally Egyptian artifacts might change when passing from one context to another.[138]

For example, recent work by Parker, Swetnam-Burland, and Versluys applies a "biographical" approach to Romano-Egyptian obelisks.[139] In Pharaonic Egypt, obelisks were solar monuments whose shape alluded to the *benben*-stone, an object associated with the sun god's cult at Heliopolis.[140] The monumental size of obelisks also made them effective embodiments of state power, well suited for royal dedications at sanctuaries. After the conquest of Egypt, Roman emperors used some of these monuments for their own ends, exporting a series of obelisks from Egypt to Rome.[141] One of the earliest and most famous examples of such appropriation is an obelisk whose "biography" Swetnam-Burland has explored in two recent publications.[142] The sixth century BCE pharaoh Psammetichus II originally constructed the obelisk for a temple at Heliopolis. After conquering Egypt, Augustus shipped this monument to Rome, where he made it the centerpiece of a colossal sundial or solar meridian and dedicated it to the sun god Sol.[143] Part of a larger construction assemblage that included the Ara Pacis and Augustus's own mortuary complex, the appropriated obelisk presented the *princeps* as conqueror of Egypt and restorer of peace and stability.

The "biography" of this obelisk thus demonstrates numerous episodes of dramatic change, appropriation, and reconceptualization, but it also shows that the generation of new meanings need not consign the earlier history of an object

137. Kopytoff 1986 (quoting 66). Archaeological approaches to object biographies: Mills and Walker 2008: 10–16.

138. For applications of this approach to Aegyptiaca, see Parker 2003; Swetnam-Burland 2010, 2015; Versluys 2012: 25; 2013a: 433–434; 2015a; 2015b: 158, 166–167; van Aerde 2015: 56–57; Colburn 2016. Some scholars further broaden the concept of "cultural biography" to describe motifs (e.g., the figure of Isis; Bommas 2012: 177) or stylistic elements (Versluys 2017c: 31, 187).

139. Parker 2003; Swetnam-Burland 2010; 2015: 65–104; Versluys 2015a: 139–146.

140. Baines 1970; Habachi 1977: 3–14.

141. *Roman appropriations of obelisks*: see most recently Parker 2003; 2007; 2018; Swetnam-Burland 2010; 2015: 65–104; Davies 2011: 360–367; Leemreize 2014: 76–78; van Aerde 2015: 222–260; Versluys 2015a: 139–146; 2017b. *For obelisks erected in honor of (rather than by) the emperor*, see Swetnam-Burland 2015: 43.

142. Swetnam-Burland 2010; 2015: 65–104.

143. Rather than tracking the movements of the sun through the day, Augustus's solar timepiece may have tracked the length of the sun's shadow throughout an entire year (Swetnam-Burland 2010: 135). However, the functioning of the monument has been heavily disputed (Haselberger 2011, 2014). On the dedication to Sol, see Swetnam-Burland 2010: 146; 2015: 65.

to irrelevance. Augustus and subsequent emperors constructed Roman meanings for Egyptian obelisks by drawing selectively on their earlier functions and valuations. Roman valuations of obelisks were in many ways very different from the values Egyptian audiences had assigned to them—and yet those Egyptian valuations made up part of the information from which Roman audiences drew in assigning new values, as suggested by several Latin authors' explicit recognition of the Egyptian significance of obelisks as solar symbols.[144] Accordingly, when Augustus incorporated an Egyptian obelisk into his grand solar timepiece and dedicated it to Sol, he was not only displaying his control over the new province of Aegyptus; he was also adapting an Egyptian symbol to a Roman setting in a way that built on the "original" meaning of obelisks while also generating new, situationally appropriate meanings of its own.[145] Indeed, even the use of obelisks as markers of imperial power and conquest may itself draw on Egyptian precedents; the original commissioner of this obelisk, Psammetichus II, was among other things demonstrating his own military power over foreign lands in erecting this monument.[146]

The "biographies" of this and other Romano-Egyptian obelisks thus testify to the purposeful selection and adaptation of elements from Egyptian visual culture to create new meanings that build on, but also transform, their predecessors. Such examples suggest a need to nuance Versluys's assertion that it is "mainly the direct previous stage in the 'life history' of [a] theme that determines the next; other previous stages perhaps played a role; original ones most probably not."[147] Versluys's cautions provide a useful warning against assuming that objects necessarily retain their "original" functions in new contexts, but we should also be careful not to assume that cultural memory is necessarily short-lived. As Swetnam-Burland points out, there "*may* or *may not* be a relationship between the lives and afterlives of any given object, depending on the knowledge, understanding, and interpretive practices of the people involved."[148] Certainly, the "globalization" of originally Egyptian material culture often transformed its uses and meanings. However, these changes frequently occurred not through the creation of new uses and meanings "from scratch" but through more complex processes of adaptation and emulation.

144. Plin. *HN* 36.14.64; Tert. *De spect.* 8; Amm. Marc. 17.4.7.

145. On Egypto-Roman obelisks as adaptation and transformation of Egyptian solar symbolism, see Davies 2011: 364–367; van Aerde 2015: 222–260; Versluys 2015a: 141.

146. Swetnam-Burland 2010, followed cautiously by van Aerde 2015: 237–238, 241.

147. Versluys 2015b: 158.

148. Swetnam-Burland 2015: 11.

Rethinking Contacts
Connectivity, Hybridization, and Entanglement

Instead of explaining culture change through unidirectional processes of diffusion or acculturation, archaeologists now typically emphasize more active processes of appropriation, emulation, or hybridization.[149] This theoretical move has also impacted scholarship on Roman Aegyptiaca. Rather than treating Egyptian motifs in Roman art as examples of diffusion, in which Romans were essentially passive recipients of Egyptian cultural "influence," recent work typically stresses Romans' active transformation and revaluation of Egyptian images and objects (see earlier in this chapter, "Material Culture, Identity, and 'Authenticity' ").

However, the ongoing reevaluation of the interpretive construct of "hybridity" demonstrates that we still need to think critically about the ways we characterize cross-cultural interactions. For many contemporary scholars, the concept of "cultural hybridity" assumes a fundamentally flawed premise: the presupposition of some earlier state of cultural purity.[150] Since almost all societies have a long history of interactions with "other" groups, it is difficult if not impossible to point to any moment when a group was totally devoid of foreign influence. Accordingly, if all culture is thus in some sense "hybrid," it is unclear how much analytical weight this term can hold.

One possible way forward may come from recent attempts to construct specifically *archaeological* approaches to cross-cultural interactions. Philipp Stockhammer's concept of "cultural entanglement" explicitly addresses the changing valuation of *artifacts and images* in new cultural contexts. In order to jettison the baggage of contentious terms like "hybridity," Stockhammer prefers to speak of "entanglement," which he defines as a broad range of "phenomena that are the result of the creative processes triggered by intercultural encounters."[151] Within these phenomena, he distinguishes two subtypes: (1) "relational

---

149. *On hybridity, hybridization, and the applicability of these concepts to Aegyptiaca,* see Barrett 2011: 33–34. Most post-1990s archaeological literature on hybridity draws directly or indirectly on Bhabha 1994. *For critiques of diffusion-based or acculturation-based paradigms for Romano-Egyptian relations,* see Versluys 2012: 28; 2015b (esp. 144–146); 2017a: 278–279. *On appropriation within the Roman Empire,* see Loar et al. 2018.

150. See recently Stockhammer 2013; Silliman 2015.

151. Quoting Stockhammer 2013: 16; see also Stockhammer 2012a, 2012b, 2012c, 2013; and cf. Thomas 1991; Silliman 2016. Hodder 2012 employs "entanglement" differently, focusing on human–thing relations. Versluys has recently used the term "material entanglement" to describe Roman encounters with Egyptian objects (2015a: 150; see also his use of Hodder's "entanglement," Versluys 2017b), and Barrett 2017a applies Stockhammer's analytical framework to images of Egyptian landscapes in Roman gardens.

entanglement," in which people appropriate a foreign object and reinterpret it according to their own cultural conceptions, but do not substantially change its material form; and (2) "material entanglement," which involves the creation of new material forms drawing on multiple cultural traditions.[152] In other words, "relational entanglement" involves assigning new meanings to existing objects or object types, while "material entanglement" involves creating new object types that, in turn, generate new meanings.

Stockhammer's analytical framework may help cut through some of the loaded language often used for Roman Aegyptiaca. The concepts of relational and material entanglement provide an alternative way to describe Italian uses of imported "Egyptian" and locally made "Egyptianizing" objects, acknowledging the possibility of differences between the two but avoiding value-laden assumptions about function, quality, or relative "authenticity." When imported Egyptian objects appear in Italian contexts, the concept of relational entanglement may provide a value-neutral way to describe the ensuing changes in the objects' functions and valuations. Even if the journey to Rome produced relatively little material change in the physical form of, for example, a Pharaonic obelisk, that obelisk still experienced extensive relational entanglement; people used it in new ways and assigned it new meanings. The appropriation and revaluation of imported goods ultimately contributes to "material entanglement" through the production of entirely new material forms: for example, the new obelisk, complete with personalized hieroglyphic inscriptions, that Domitian commissioned in Italy.[153] These new objects then communicate another set of new meanings specific to their own cultural context. Domitian's obelisk, for example, conveys his global authority while also alluding to the historical association of his dynasty with Egypt, where his father was first acclaimed emperor.[154]

Roman Emulation of Earlier Visual Cultures

In addition to participating in archaeological debates about the role of material culture in cross-cultural exchange, research on Aegyptiaca can also contribute to art-historical debates about the roles of adaptation and allusion in Roman art. In contrast to older perceptions of Roman art as "copying" earlier visual cultures

---

152. Stockhammer 2012a: 49–51; 2013: 16–17.

153. Quack 2003: 64; Parker 2007: 212–213. For the text, see no. 501/0124 in *Recueil des inscriptions concernant les cultes isiaques* (Bricault 2005), henceforth cited as *RICIS*.

154. On Egypt in Flavian dynastic ideology, see note 78, this chapter.

(whether Greek, Egyptian, or Italian), much evidence now demonstrates that Roman artists and craftsmen also engaged with older source material in more active, innovative ways. Accordingly, rather than asking whether Aegyptiaca are "Egyptian" or "Roman," we may do better to ask how Roman artists' eclectic recombinations of Egyptian motifs and styles relate to broader patterns of emulation of older visual cultures.[155]

In art history and archaeology, much exciting work has reassessed Roman emulations and adaptations of earlier, especially Greek, material culture.[156] Such research questions the existence of a clear division between "originals" and "copies" and problematizes assumptions that Roman artists always strove to replicate earlier objects precisely. Instead, Roman artists engaged dynamically with earlier visual cultures, employing many forms of emulation, allusion, selection, and recombination. Roman "retrospective art"[157] might combine originally disparate motifs or styles, adapt older works to new functions and contexts, or otherwise engage in purposeful innovation that allowed artists to allude to the past while still generating new meanings for the present. In other words, the existence of a broad spectrum between "authentic" and "mirage," or reproduction and innovation, not only characterizes Roman adaptations of *Egyptian* visual culture but also is generally typical of *Roman approaches to visual culture*. Anachronistic modern constructs of "authenticity" have little place in this ancient aesthetics of creative emulation.

Indeed, when Roman material culture draws on iconographic, formal, or stylistic features of Egyptian art, such adaptations are often just one aspect of a highly eclectic mix of cultural and retrospective allusions. For example, the garden of the Casa degli Amorini Dorati at Pompeii contained not only Egyptian imagery (a shrine with Isiac paintings, a probably Egyptian-made statuette of Horus, and an oil lamp depicting Egyptian deities) but also a shrine to the Capitoline triad and a wide range of Hellenizing material culture (e.g., Bacchic reliefs and herms and an early Hellenistic votive relief).[158]

---

155. For previous calls to situate Aegyptiaca within the larger phenomenon of Roman emulative art, see Ridgway 1984: 108n54; Elsner 2006; Swetnam-Burland 2007: 116; Bülow Clausen 2012: 94; Tronchin 2012: 264; Versluys 2010: 17–18; 2013a: 430; 2015b; Davies 2011; Leemreize 2014: 57; van Aerde 2015, esp. 19, 52–53.

156. Within the vast literature on Roman emulative art, see recently Hölscher 2004; Perry 2005; Kousser 2008; Marvin 2008; Squire 2012.

157. I follow Fullerton (2003) in employing "retrospective" here rather than "archaizing" or "classicizing," as many Roman works allude to both Archaic and Classical (as well as Hellenistic) styles.

158. Petersen 2012; Mol 2015: 332–391; Neudecker 2015: 135–136.

Another type of eclecticism appears in paintings of Egyptian landscapes that employ a style of brushwork derived from Hellenistic and Roman art. The frescoes from the garden of the Casa dell'Efebo (figures 1.3–1.8) provide an example. These "Nilotic scenes" depict a riverine landscape filled with sanctuaries, shrines, and divine statues. Certain geographically or culturally distinctive features—crocodiles, a statue of the Apis bull, an animal-headed boat made of reeds, and human figures who resemble pygmies—locate the scene in Egypt, and many of these motifs have antecedents in Egyptian art (see chapters 2 and 3). Yet despite these Egyptian allusions, other stylistic, formal, and iconographic aspects of the paintings draw on Hellenistic and Roman precedents. The loose, "impressionistic" brushwork and the depiction of rural, shrine-filled landscapes have their closest parallels in "sacral-idyllic" Roman landscape painting.[159] Instead of a straightforward depiction of Egypt or attempt to duplicate Egyptian visual culture, the scene presents multiple layers of cultural and visual references. Such evidence suggests that Roman visual references to Egypt do not constitute an anomalous "special category" within Roman art but, rather, display the same tendencies toward eclecticism, recombination, and reinterpretation that characterize Roman art more generally.

Foreign Familiarity, Familiar Foreignness

Through a contextual analysis of Pompeian houses and gardens as interactive, multimedia ensembles, this book seeks to advance beyond binary interpretations of Aegyptiaca and to employ a more nuanced, theoretically informed approach to cultural exchange, identity, and visual culture. The uses and valuations of Italian Aegyptiaca are complex enough to resist reduction to "either/or" interpretations as Egyptian *or* Roman, objects of religious worship *or* "exotica," testifying to either "authentic" Egyptian culture *or* a culturally constructed Roman "mirage" of Egypt, and owned and used by people who either *were* or *were not* devotees of "Egyptian" gods. On close examination, such oppositions prove to be false dichotomies. This book thus adopts a new approach to the Roman adaptation of Egyptian visual culture that (1) addresses the plural nature of "identity" and the complex relationship between identities and objects; (2) situates Aegyptiaca within broader frameworks of emulation and adaptation in Roman art; (3) considers the agency of objects as well as humans in shaping behavior and creating meaning; and (4) contextualizes Aegyptiaca

159. On brushwork in Roman "sacral-idyllic" paintings, see Ling 1977: 13–14; L'Enfant 1999: 93–94. The absence of an overtly "Pharaonizing" style causes some scholars to classify "Nilotica" separately from other Aegyptiaca; see discussion in chapter 6, this volume.

within larger, interactive assemblages in order to analyze how individual objects collaborated with such assemblages, as well as with their viewers, to generate shifting meanings. Together with the larger household assemblages of which they formed part, domestic representations of Egyptian landscapes structured human movement through household space and influenced the experiences of their viewers. Such interactions would have produced kaleidoscopic, varied meanings, constantly reconfigured in response to changes in the environment, contents, and inhabitants of the space.

As the following chapters will demonstrate, the interactions of objects, assemblage, and viewers frequently generate a spectrum of possible meanings, rather than collapsing all possibilities into a single "correct" interpretation. These domestic installations thus do something more complex than signaling an Isiac religious identity, creating an exotic ambience, or mocking the supposed failings of cultural "Others." Instead of prescribing a single attitude toward Egypt, Isis, the Roman Empire, or anything else, domestic representations of Egyptian landscapes create possibilities for exploration and creative engagement. As in the Casa dell'Efebo garden, such engagement often includes *performance* of the "Other": problematizing and temporarily dissolving distinctions between "Roman" and "Egyptian," rather than (or even at the same time as!) reinforcing them. While material culture can construct distinctions between the "foreign" and the "familiar," it can also elide and deconstruct those distinctions. That which was once alien can become familiar, and vice versa—the familiar can come to seem deeply strange.

Nilotic imagery contains many motifs that strike modern viewers as unusual. The protagonists of these scenes are often pygmies or dwarfs who engage in seemingly bizarre behaviors, such as doing battle with hippopotami and crocodiles or engaging in public sex. As a result, scholars sometimes consider such scenes to be outrageous parodies expressing derogatory Roman stereotypes of Egyptians. The present study reconsiders such interpretive frameworks, arguing that the architectural settings and archaeological contexts of many Nilotic scenes advocate for more nuanced possibilities. Even as the Egyptian landscapes introduce a seemingly exotic element into domestic space, other aspects of the garden setting simultaneously familiarize and domesticate the imagery. While it is possible to interpret certain aspects of Nilotic imagery as evoking either Roman interest in Egyptian cults or Roman stereotypes about Egyptian culture, the setting of the images ultimately acts to destabilize both these readings and to open up additional possibilities for meaning. These representations of Egyptian landscapes offer no easy resolution of tensions between "Self" and "Other," but they provide viewers with the opportunity to explore those tensions and contradictions in open-ended ways.

Chapter 2 establishes and analyzes general patterns in the iconography and contexts of Nilotic imagery at Pompeii, laying the groundwork for the more specific case studies in chapters 3 through 6. A discussion of the multiple culturally recognized affordances of Nilotic iconography ("Adaptation, Emulation, and Eclecticism in Nilotic Imagery" addresses the "cultural biography" of Nilotic imagery, positioning these frescoes and mosaics within the broader context of Roman artistic emulation of earlier visual cultures. While Roman emulations of Greek material culture have traditionally attracted the most intense archaeological and art-historical attention, scholars increasingly frame "archaizing" or "classicizing" art as part of a larger "retrospective" phenomenon that also includes, among other things, Roman emulations and adaptations of Egyptian material culture. Accordingly, rather than seeking to label Nilotic landscapes as "Egyptian" or "Roman"—casting these images as either "authentic" copies of Egyptian religious iconography or as new Roman creations *ex nihilo*—we may make progress by relating Nilotic imagery to this wider phenomenon.

Chapter 2 also provides a synthetic contextual analysis of Nilotic scenes at Pompeii, analyzing the probable reasons for their frequent association with domestic gardens. Within and adjacent to gardens, the scenes are often associated with water features and installations for dining and drinking, producing an environment that in many ways mirrors that depicted in the scenes themselves. The activities associated with these settings, including dining, drinking, and the enjoyment of various forms of entertainment, create additional parallels between viewers and viewed. The context of the imagery thus provokes viewers to compare themselves to the figures it depicts, framing viewers' subsequent behavior as a response to—and collaboration with—the images.

The case studies in chapters 3 through 6 investigate the roles of Egyptian landscapes within domestic assemblages through the contextual analysis of specific installations. Chapter 3 returns to the Casa dell'Efebo, with which the present chapter began. Interrogating the relationships between the various features, images, and artifacts that make up this eclectic garden assemblage, chapter 3 argues that instead of extracting Nilotic landscapes from the rest of Roman art as an isolated, *sui generis* category, we should understand them as embedded within broader phenomena of domestic cosmopolitanism. Ulf Hannerz defines "the cosmopolitan" as someone whose demonstration of transnational cultural competence may also provide a source of cultural and social capital, even a "form of mastery," at home.[160] I contend that such demonstration was central to many of the garden assemblages presented in this book. By evoking diverse landscapes

160. Hannerz 1990 (quoting 240).

both near and faraway and emulating earlier visual cultures, householders could turn their houses into microcosms of empire and assert their own credentials as worldly cultural sophisticates. The appropriation and adaptation of Egyptian imagery is thus deeply intertwined with Roman appropriations and adaptations of other earlier visual cultures, including Archaic, Classical, Hellenistic, and Italic allusions. As model landscapes existing simultaneously within and beyond domestic space, gardens were appropriate places to depict other landscapes along the spectrum from foreign to familiar. In such contexts, representations of Egyptian landscapes exist alongside—and derive much of their impact from interaction with—evocations of other, "Greco-Roman" geographical and cultural settings.

Within this general framework, however, different domestic contexts might structure viewers' encounters with Nilotic imagery in quite different ways. Chapter 4 investigates this variability through an examination of another case study: the Casa del Medico. As at the Casa dell'Efebo, this house constructs close contextual associations between Nilotic frescoes, garden space, watery settings, and dining. Here, however, the framing and content of the images encourage different encounters with viewers and create different forms of meaning. Among other things, the Casa del Medico assemblage places much more emphasis on the "grotesque" potential of pygmy scenes. Although scholars have often taken the violent imagery of the Casa del Medico frescoes as paradigmatic for all Nilotic scenes, a close examination of this garden assemblage points instead to the diversity and thematic multivalence of domestic Nilotica. This case study thus emphasizes the open-ended possibilities generated by interactions between imagery, setting, and viewers. Additionally, since the Casa del Medico is much smaller and less luxuriously furnished than the Casa dell'Efebo, the assemblages from these houses enable us to compare the uses of Nilotic imagery at different positions along the socioeconomic spectrum of first-century Pompeii.

At both the Casa dell'Efebo and Casa del Medico, Nilotic imagery plays a central role in garden ensembles. However, in many houses, Nilotica served less as centerpieces than as marginalia: peripheral elements in larger compositions. Such is the case at the Casa del Fauno, where a Nilotic mosaic occupies the literally liminal position of threshold between a peristyle garden and an exedra showcasing the famous "Alexander mosaic." Through this case study, chapter 5 investigates the relationships between Egyptian landscapes and marginal spaces within Pompeian homes. Even when Nilotic scenes occupy more prominent visual positions, their frequent association with gardens still locates them on the margins of household space, simultaneously within and outside the *domus*. However, the example of the Casa del Fauno also demonstrates the ability of such "peripheral" visual culture to frame and recast viewers' encounters with larger images. Marginality can thus become a paradoxical source of power, enabling "ornamental" elements to shape the ways that viewers experience the images—and the forms of domestic space—that they enclose.

Finally, chapter 6 expands current constructions of "Nilotica" through an examination of the Casa di Acceptus e Euhodia, whose garden contained not frescoes or mosaics but, instead, a water feature accompanied by faience and marble statuettes of Egyptian dwarf-gods, animals, and a pygmy. The result is effectively a three-dimensional instantiation of Nilotic paintings and mosaics: the garden itself becomes the marvelous landscape depicted in Nilotic scenes, roamed by small, dwarflike beings and aquatic animals. As three-dimensional garden furnishings and "Nilotica" belong to different media and frequently employ different stylistic allusions, they are typically studied separately.[161] However, a contextually oriented examination helps bridge the gaps caused by dividing Roman art into media-based or stylistic categories. Both two-dimensional and three-dimensional "Nilotica" appear to have been treated as especially (although not exclusively) appropriate to gardens, and both create a model of the foreign within the familiar: collapsing geographical and cultural divisions to transform a Pompeian garden into a Nilotic riverbank, and in so doing, domesticating and familiarizing Egyptian imagery.

The most-discussed example of "Egyptianizing" statuary in a Pompeian garden comes from the Casa di Octavius Quartio, where much debate has addressed the ultimately inconclusive question of whether or not the house owner was an "Isis worshipper." Both at that house and the Casa di Acceptus e Euhodia, I argue that it is misleading to ask whether or not the purpose of such statuettes was "religious," as such questions frequently presuppose anachronistic definitions of religion and religious identities. Instead of debating whether garden statuary signified "religion" or "Egyptomania," I examine how these artifacts worked together with the other garden contents to create an interactive model landscape within which visitors' own actions constituted performative responses to, and interpretations of, the visual culture surrounding them.

Some evidence from the Casa di Acceptus e Euhodia suggests that its inhabitants may, in fact, have directed domestic cult toward a form of Isis. However, I argue that this evidence for "Isis worship" should not overdetermine our understanding of the garden assemblage. Previous scholarship often constructs the religious identities of Isis worshipers in exclusive, binary terms ("Isiacs" vs. "non-Isiacs") and presumes that these religious identities will correspond to distinct, identifiable material profiles within domestic contexts. However, the Casa

161. Cf. scholarly discourse on "Pharaonica" versus "Nilotica": Malaise 2005: 201–204; 2007: 34; Bragantini 2006: 161–165; 2012: 25–30.

di Acceptus e Euhodia provides an opportunity to develop more sophisticated approaches to the relationship between religious identity and domestic material culture. In place of a binary divide between "Isiacs" and "non-Isiacs," evidence from Pompeii suggests a broad spectrum of ways in which individuals might choose to engage with Isis or other Egyptian-derived deities. Additionally, the Casa di Acceptus e Euhodia suggests commonality, not difference, between the domestic assemblages of "Isis worshipers" and their neighbors. Regardless of the degree to which they interacted with Egyptian deities, the inhabitants of first-century Pompeii appear to have employed domestic Aegyptiaca in closely analogous ways.

Although interaction with Egyptian landscapes in the household entails interaction with an imagined "Egypt," explicit commentary on the role of Egypt within the empire need not have been (and was probably not) the primary goal of many patrons, artists, or viewers at Pompeii. Different visitors to a house might enjoy representations of Egyptian landscapes and figures for very different reasons, depending on their own backgrounds and/or their degree of knowledge about Egyptian culture.[162] While individual viewers would undoubtedly have understood Nilotic imagery through the perspective of their own various social identities, we should not assume that they would all, at all times, have prioritized any *single* type of identity, whether ethnic, political, religious, or anything else. As in any other human society, Pompeian viewers' identities were not fixed and monolithic but, rather, plural, contextual, and shifting; any one person might possess many different social identities, only some of which might appear particularly relevant at a given time.[163] While representations of Egyptian landscapes could have prompted some viewers to think about "Romanness" vis-à-vis "Egyptianness," they could have activated other identities and interests in other viewers: professional, religious, economic, social, and the like.

However, whether or not people *valued* Nilotic imagery primarily for its implicit statements about the relationship of Egypt to the rest of the empire, the incorporation of Egyptian imagery into the daily life of the household still has *consequences* for our understanding of that relationship. No matter what interpretations people chose to find in domestic material culture, they still had to exist and interact within the environments that it created. Whether or not

162. Cf. the remarks of Versluys and Meyboom 2000: 126; Malaise 2005: 211–212; Meyboom and Versluys 2007: 208; Versluys 2007: 14.

163. Cf. Brubaker 2002, 2004 (on ethnic "groupness" as event, not immutable given); Burke and Stets 2009, esp. 112–154; Lahire 2011; Rebillard 2012, esp. 1–8.

homeowners' goals explicitly included the domestication and familiarization of once-foreign themes, this book will argue that may have been one of the results of the household ensembles they constructed. During the period of profound political and social change that was the early empire, domestic cosmopolitanism provided a means to engage, interact, and experiment with changing constructions of local and imperial identities. Domestic material culture enabled individuals to transform the *domus* into a microcosm of the *orbis* and to participate through everyday practice with an increasingly interconnected, interdependent world.

2

Imagining the Nile

AFFORDANCES AND CONTEXTS

Introduction
Navigating "Nilotic Scenes"

Scholars typically apply the term "Nilotic scenes" to Roman artistic representations of the Nile river in flood.[1] Besides the river itself, with its distinctive flora and fauna (e.g., crocodiles, hippopotami, lotuses, etc.: figures 2.1 and 2.2), such images often depict human figures engaging in a range of behaviors that include fighting with riverine beasts, engaging in outdoor sex, drinking alcohol, playing music, fishing and fowling, and performing acts of rural cult. While some of these motifs also appear in Roman representations of people and places outside Egypt, others—for example, depictions of characteristically Egyptian wildlife, or of animal worship—were consistently associated with Egypt in Roman literary and visual sources.[2] The presence of such elements suggests that Roman viewers would almost certainly have recognized these scenes as taking place in Egypt.

The physical features of many of the people in Nilotic scenes resemble those of achondroplastic dwarfs, with large heads, short limbs, and stocky bodies.[3] Depictions of a faraway landscape populated by dwarfs evoke Greek and Roman legends of *pygmaioi* or *pygmaei*: short-statured, dwarflike tribes about

1. On the indications that the river is flooding, see Meyboom 1995: 43–79; Assmann 2000: 75–76, 78–79; Versluys and Meyboom 2000: 112–114; Versluys 2002: 40–41; Malaise 2003: 314.

2. For Roman tropes concerning Egypt, see note 10, chapter 1, this volume.

3. Versluys 2002: 276; Malaise 2003: 314–315; Clarke 2007b: 75–76. Characters in Nilotic scenes sometimes, but not always, also have curved spines and prominent buttocks (e.g., Versluys 2002: no. 60).

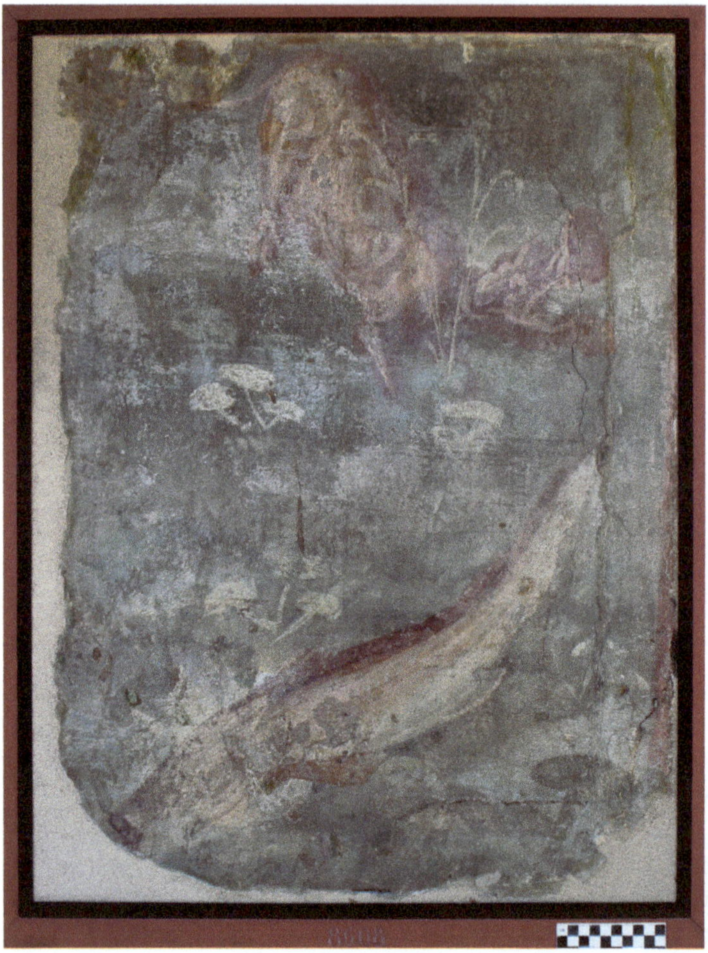

FIGURE 2.1 Nilotic wall painting with hippopotamus, fish, and lotuses. House II 4.2 (*praedia* di Julia Felix), triclinium. MANN 8608.

Photo: C.E. Barrett. Reproduced by permission of the Ministero dei Beni e delle Attività Culturali e del Turismo—Museo Archeologico Nazionale di Napoli.

whose actual existence there was much disagreement, but who were sometimes thought to live on the fringes of the inhabited world.[4] While such tribes were imagined to live in various distant lands, Romans associated them with Egypt

4. Arist. *Hist. an.* 8.12.597a; Strabo 1.2.30, 1.2.35, 2.1.9, 15.1.57, 17.2.1; Pompon. Mela 3.81; Dasen 1993: 178–179; Tybout 2003: 515; Harari 2004; Swetnam-Burland 2015: 162–164. Strabo expresses skepticism about the actual existence of pygmies, and Aristotle, while asserting their reality, sounds a bit defensive. For Pomponius Mela, pygmies existed in former days but are now extinct.

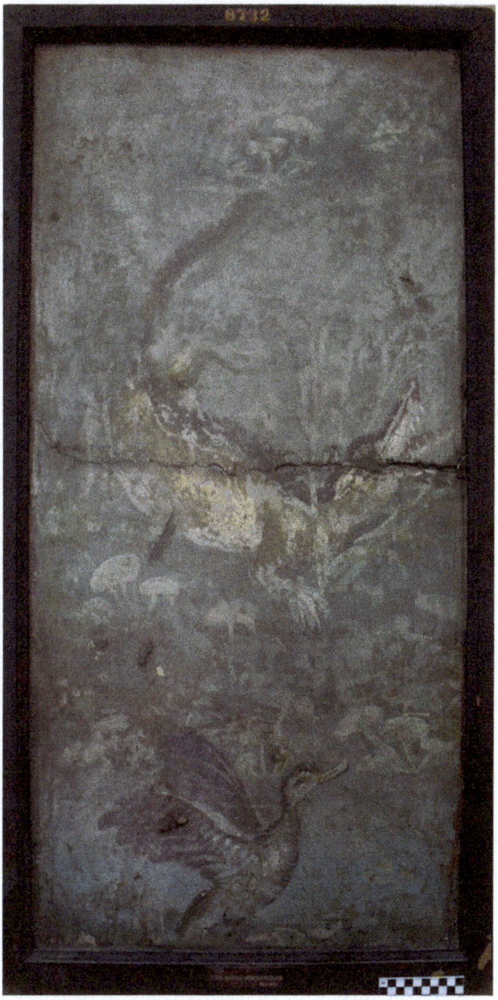

FIGURE 2.2 Nilotic wall painting with duck, crocodile, and lotuses. House II 4.2 (*praedia* di Julia Felix), triclinium. MANN 8732.

Photo: C.E. Barrett. Reproduced by permission of the Ministero dei Beni e delle Attività Culturali e del Turismo—Museo Archeologico Nazionale di Napoli.

in particular, probably partly because of the actual importance of pygmies and dwarfs in Egyptian art and religion.[5] Unlike the clear distinction that English draws between pygmies (normally proportioned people from short-statured

5. *Pygmies and dwarfs in Egypt:* Dasen 1993: 25–162; 2013: 265–267; Meyboom and Versluys 2007. *Greco-Roman associations of pygmies with Africa and Egypt*: Dasen 1993: 176 (and add to her references Philodemus, *De Signis* 4, on pygmies displayed in the Egyptian town of

ethnic groups) and dwarfs (people with a congenital condition that restricts their stature), Classical texts often conflate the two. The Greek *pygmaios* can describe both members of short-statured populations and individuals with dwarfism.[6] While the Latin cognate *pygmaeus* generally describes whole tribes of short-statured people, such *pygmaei* were imagined to physically resemble dwarfs.

Artistic representations of geranomachies (pygmy–crane battles) provide a way to track the changing images of *pygmaioi/pygmaei* in Greco-Roman culture, as textual descriptions of these battles characterize the pygmies in question as an ethnic group rather than as people with a pathological condition.[7] Although Archaic Greek vase paintings depict pygmy tribes as short but normally proportioned, artists of the fifth century BCE and later typically depict them with features of achondroplastic dwarfism, as well as oversized phalli.[8] Dwarflike features subsequently also characterize pygmies fighting cranes in Roman art.[9] Greek and Latin textual references to tribes of pygmies along the Nile[10] further suggest that the typical protagonists of Nilotic scenes should be understood as members of a short-statured ethnic group, rather than as individuals affected by congenital dwarfism. In following Roman art historians' common practice of referring to such figures as "pygmies," I am explicitly using the term in its ancient, not modern, sense.[11] While some short-statured peoples

---

Akoris). *Pygmies in multiple distant lands:* Romm 1992: 96–97, 102–103; Swetnam-Burland 2015: 162–163.

6. Dasen 1993: 163–164 (with discussion of the interchangeability of *pygmaios* and *nanos*); Cappel 1994: 100; Garmaise 1996: 18, 22, 24. The Latin *pumilio* similarly characterizes people, plants, or animals smaller than usual, without regard to cause; it describes not only individuals with dwarfism but also dwarf varieties of plants and animals (*OLD*, s.v. *pumilio*).

7. Discussions of ancient textual sources on pygmy–crane battles (starting with Hom. *Il.* 3.2–6) and pygmy tribes: Ballabriga 1981; Dasen 1993: 175–176.

8. Sparkes 2000: 90; Harari 2004: 168–169. Compare Ctesias's description of *pygmaioi* as snub-nosed, ugly, and macrophallic (*FGrH* 688 F45fα).

9. See examples in Dasen 2009. Pygmy–crane scenes at Pompeii appear in houses II 9.2 (Versluys 2002: 112; examination in situ indicates a crane rather than an ibis, *pace* Versluys), VIII 7.24 (Maiuri 1956: pl. VI; Dasen 1993: no. 44; Clarke 2007b: 93), and in the Terme del Sarno (Cappel 1994: no. W20, identifying the bird as an ibis).

10. Arist. *Hist. an.* 8.12.2 (cf., on pygmies and cranes, Hom. *Il.* 3.2–6; Lucian, *Rhetorum praeceptor* 6; Philostr. *Imag.* 1.5; Plin. *HN* 6.188, 8.92–93; Pompon. Mela 3.81); Dasen 1993: 595–597. Herodotus (2.32.6, 4.43.5) describes a short-statured ethnic group in Africa (2.32.6, 4.43.5).

11. Cf. the similar position of Clarke 2007b: 75–76.

do live in central Africa, they were probably unknown to the Romans.[12] For Greco-Roman audiences, *pygmaioi/pygmaei* were largely legendary beings.

The Casa dell'Efebo frescoes, introduced in chapter 1, provide numerous examples of figures with seemingly achondroplastic features (figures 1.3–1.8).[13] A few figures in these frescoes have more idealized proportions, with longer limbs, taller and slimmer bodies, and smaller heads (e.g., figure 1.4, A; figure 1.8, A, B, C; and the painted statue of Isis-Fortuna in figure 1.3, A). However, Roman viewers likely interpreted most of the other figures as pygmies or dwarfs, and the Nilotic setting would have suggested pygmies. Many Nilotic scenes are populated entirely with pygmy-like figures, although some other scenes have more somatically "normal" populations.[14] Several scenes, like those in figures 1.3–1.8, depict both pygmies and non-pygmies within the same landscape.[15]

The depiction of the figures' skin tone may provide another indication of setting. Greek and Roman authors often described Egyptians as having darker skin than other Mediterranean peoples, among other perceived somatic differences.[16] Many Nilotic scenes depict their protagonists with dark brown skin, as opposed to the lighter skin colors (tan for males, pale beige or white for females) that typically characterize Mediterranean populations in Roman painting.[17] Other Nilescapes include both dark-skinned and lighter-skinned figures, with the distinction sometimes seemingly conveying geographical origins. In the Palestrina mosaic, the inhabitants of the Delta have beige skin while those of Upper Egypt and Nubia are darker brown.[18] In the Casa dell'Efebo frescoes, both darker-skinned figures (e.g., figure 1.4, F and G; figure 1.6b figure 1.7, D and G; figure 1.8, C) and lighter-skinned figures (e.g., figure 1.3, C; figure 1.4, A; figure 1.7,

12. Cf. Versluys 2002: 276; Malaise 2003: 314–315; Boissel 2007: 331–332. On Egyptian conceptions of pygmies vs. dwarfs, and the possibility of Egyptian interactions with actual short-statured peoples, see Silverman 1969; Romano 1989: 104–107; Dasen 1993: 25–33.

13. The largeness of the figures in comparison to the architecture is common in Roman landscape painting (Ling 1977: 13) and indicates an emphasis on human activities, not a literal rendering of relative size.

14. Somatically "normal" figures: see, e.g., the Palestrina mosaic (Meyboom 1995).

15. Pygmies and somatically "normal" figures together: cf., e.g., Versluys 2002: nos. 50, 60 (Casa delle Quadrighe, Casa dello Scultore).

16. Aesch. *Supp.* 70–71, 154–155, 719–720; Arist. [*Phgn.*] 812a; Amm. Marc. 22.16.23; Hdt. 2.57, 2.104. On ethnicity and ethnic diversity in Greco-Roman Egypt, see sources in Barrett 2015b, s.v. "Ethnicity and Ethnic Identity."

17. On the variations in Nilotic pygmies' skin color, see Boissel 2007: 202–204.

18. On the Palestrina mosaic, see Meyboom 1995.

B and C, and the figures near 1.7, L and M; figure 1.8, A, B, D, E, and F) occupy the landscape. Both the darker-skinned and lighter-skinned groups include figures with dwarflike proportions, as well as more somatically typical individuals; contrast, for example, the large heads and short limbs of the dark-skinned figures in figure 1.7, D with the more idealized dark-skinned figure in figure 1.8, C.

Modern interpretations of Nilotic iconography are varied. As with many other forms of Aegyptiaca, scholars once frequently interpreted such images as indications of house owners' affiliation with Isiac cult.[19] However, as subsequent critiques noted,[20] such interpretations generally failed to explain what, exactly, was "Isiac" about these scenes. Furthermore, only a minority come from houses with other evidence of cults for Egyptian deities.[21] Accordingly, much scholarship of the late twentieth and early twenty-first centuries reacted against the automatic labeling of all Egyptianizing imagery as "Isiac." Instead, it became common to interpret Nilotic scenes as outrageous parodies expressing derogatory Roman stereotypes of Egyptians for comic and/or apotropaic purposes.[22]

However, the interpretation of Nilotic scenes as caricature also encounters a number of problems. For one thing, some of the earliest Nilotic mosaics come from Ptolemaic Egypt[23] (e.g., figure 2.3). Outside Egypt, Nilotic scenes appear

19. E.g., Alföldi 1965–66: 60–61; Schefold 1972: 61; Roullet 1972: 46; Leclant 1984: 440–441.

20. E.g., Meyboom 1995: 83–84; Versluys and Meyboom 2000; Versluys 2002: 32–33.

21. Versluys and Meyboom 2000; Versluys 2002. On what constitutes such evidence, see note 208, this chapter.

22. E.g., Cappel 1994: 100–102; Versluys 2002: 436–437; 2013b: 245; Walker 2003: 191–202; Clarke 2007b: 87–107; 2007c; 2014: 653; Merrills 2017: 106–149. *Pygmy scenes as apotropaia:* see note 186, this chapter.

23. Versluys 2002: nos. 99 (c. 100–75 BCE), 101 (c. 80 BCE). Both include at least one pygmy (*pace* Pearson, who argues that pygmies do not appear in pre-first century CE Nilotica: 2015: 4), and the pygmy in Versluys's no. 99 is macrophallic (Daszewski 1985: 139, no. 31). The Palestrina mosaic is often dated to the late second century BCE (Meyboom 1995: 16–19). However, Fuentes (1987: 51), Schrijvers (2007: 233) and Rose (2013: 548–549) have argued persuasively for redating it to the early imperial period. Particularly compelling arguments for this redating come from archaeological context (Rose 2013: 548–549) and the representation of a Roman shield type otherwise unattested before the late first century BCE (Fuentes 1987: 51, 65n44; on the shield type, see Bishop and Coulston 1993: 81). If this redating is accepted, then the earliest securely provenienced Nilotic mosaics come from Egypt, Delos, and Pompeii; see Versluys 2002: no. 121 (Delos, c. 100 BCE), no. 47 (Pompeii, c. 90–80 BCE), and no. 99 (Abu Qir, Egypt, c. 100–75 BCE; this example, unlike Versluys's no. 101, comes from an archaeological excavation). Another set of Nilotic mosaic fragments in Madrid resemble the Palestrina mosaic in their high-quality, naturalistic depictions of normally proportioned human figures (Versluys 2000, 2001: no. 25), and Versluys dates them to c. 100 BCE based on the traditional date of the Palestrina mosaic (2000: 249–250). However, the Madrid fragments lack clear provenience, and as with the Palestrina mosaic itself, their stylistic date is controversial. Helen Whitehouse considers them potentially "Hadrianic or later," rather than Republican (2001: 179–185,

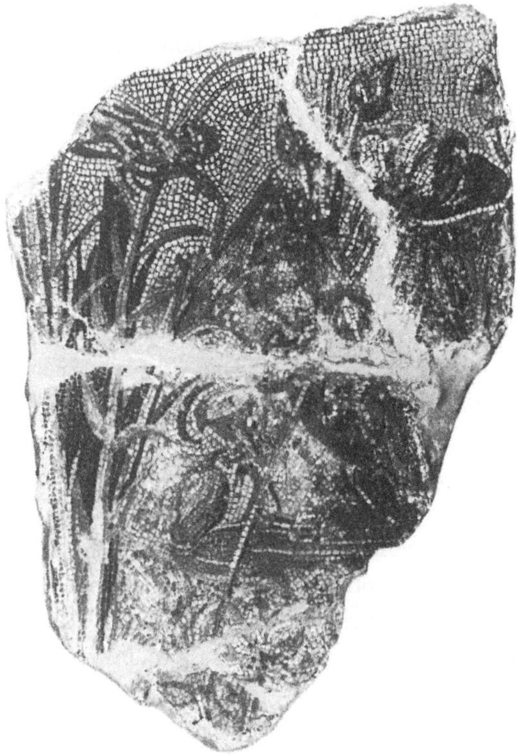

FIGURE 2.3 Nilotic mosaic from Egypt, context unknown, c. 80 BCE (Cairo Agricultural Museum 477). A pygmy or dwarf poles a reed boat through marsh plants.

Source: After Parlasca 1975: pl. 175, no. 2 (republished as Versluys 2002: 199, fig. 127). Reproduced by permission of Klaus Parlasca and Miguel John Versluys.

occasionally in Isis temples or houses containing Isis shrines,[24] suggesting that people did not necessarily see this imagery as demeaning to Egypt or Egyptian culture. Nilotic landscapes and pygmy images even appear in the portico of the Isis temple at Pompeii.[25] Additionally, pygmy scenes do not appear to have become widely popular in Italy until the Flavian period,[26] when Augustan-era

quoting p. 184). On potential parallels between the Palestrina mosaic and the Nile mosaic of the Casa del Fauno, see note 24, chapter 5, this volume.

24. See later this chapter, "Distribution Within the City."

25. Versluys 2002: no. 143.

26. While earlier examples certainly exist (e.g., Merrills 2017: 133–135), Flavian examples are much more numerous (Tybout 2003: 510–512; cf. Mol 2015: 270–271, 290–291; *pace* Merrills 2017: 117).

propaganda about the supposed moral failings of Egyptians generally, and Cleopatra VII specifically, was no longer as politically relevant. More broadly, recent studies of Greek and Roman literary and visual references to Egypt are increasingly problematizing exclusive emphases on "Othering" strategies and finding evidence for more complex, nuanced perspectives.[27]

Furthermore, taking these scenes' pygmy protagonists as stand-ins for all Egyptians is problematic.[28] Although scholars often interpret these figures as caricatures of Egyptians, there is no evidence that dwarfism was any more common in Egypt than elsewhere,[29] and no reason to think Roman audiences—whatever other prejudices they held—imagined that average Egyptians actually looked like this. Actual Egyptians might not have been such unfamiliar sights in Campania; merchants and sailors from Egypt regularly came to Puteoli on grain ships, and some Egyptian expatriates may have lived on the Bay of Naples.[30] Pygmies, however, were quasi-mythical figures of uncertain reality.

Accordingly, a number of scholars have recently attempted to moderate between the more extreme positions of earlier literature, arguing that Nilotic scenes recall some elements of Egyptian culture in addition to whatever humorous or apotropaic functions they might also have performed.[31] Nilotic imagery is also frequently explained in more general terms as evoking concepts of fertility and *tryphe* (luxury).[32]

Much previous research on Roman "Nilotica" thus proposes, or presupposes, conflicting narratives concerning the "cultural biography" of Nilotic imagery. Behind the debate is a (sometimes explicit, sometimes implicit) question: To what degree do these scenes draw on elements of "authentic" Egyptian religious

---

27. E.g., Moyer 2011; Manolaraki 2013; Leemreize 2014; Versluys 2013: 253; 2015b. Reconsiderations of Augustus's supposed antipathy to Egyptian cults: Orlin 2008; Bommas 2012: 180; 2013a: 91–92. Gruen (2011) more generally questions scholarly emphases on "Othering."

28. Tybout 2003: 515; Swetnam-Burland 2015: 161–164; *pace,* most recently, Merrills 2017: 146.

29. Dasen 1988: 253; *pace* Meyboom 1995: 150; Versluys 2002: 276; Meyboom and Versluys 2007: 182.

30. *On trade between Egypt and Campania,* see note 55, chapter 1, this volume. *On evidence for Egyptians visiting or dwelling in Campania,* see Swetnam-Burland 2011: 343. *On Egyptians in Italy more generally,* see note 124, chapter 1, this volume.

31. Whitehouse 1977: 64; Meyboom 1995: 99–107; Assmann 2000: 72–82; Swetnam-Burland 2002: 150–153; Meyboom and Versluys 2007; Barrett 2011: 313–315.

32. *Evocation of luxury:* Amedick 1993: 181–182; Dunbabin 2003: 61. *Fertility and* tryphe: e.g., McKay 1985 (specifically addressing pygmy landscapes); Meyboom 1995: 153–154; Söldner 1999: 99; Versluys 2002, esp. 295; Meyboom and Versluys 2007; Merrills 2017: 143.

iconography, and to what degree are they Roman creations, whose only relationship to Egyptian models might be at most a parodic one? However, the extensive recent rethinking of old narratives about "originals," "copies," and "authenticity" in Roman art[33] advocates for reframing such questions. Rather than asking whether—or to what degree—Nilotic scenes are "Egyptian" or "Roman," we may do better to ask how this imagery relates to the broader phenomenon of Roman emulation of earlier visual cultures. A close examination of Nilotic frescoes and mosaics suggests that they do in fact engage extensively with older Egyptian antecedents, but they do so in an eclectic, selective fashion that recombines and transforms those models.

Multivalence and Affordances

As this chapter will argue, substantial evidence suggests that many Nilotic motifs and themes draw on Egyptian sources. After examining this evidence, we will explore the ways that Roman craftsmen adapted and reshaped these foreign elements. However, the goal is not to suggest that if Roman craftsmen employed motifs derived from Egyptian art, then the resulting images must necessarily have conveyed the same meaning(s) as their Egyptian prototypes. Instead, this discussion aims to position the "cultural biography" of Nilotic imagery within the broader framework of Roman emulation and eclecticism.

I argue that as a result of this complex cultural biography, Nilotic imagery could potentially convey a wide range of possible "meanings": some derived from Egyptian representational traditions and Greco-Roman receptions of Egyptian culture, and some derived from Greek and Roman representations and practices. James Gibson's concept of "affordances" provides a helpful way to think about such multivalence. Gibson defines the "affordances" of an object as its potentials for facilitating certain outcomes, actions, or behaviors.[34] While Gibson's formulation of "affordances" encompasses an extremely broad range of possible interactions between people and objects, I will focus here on "affordances" in a narrower sense: those affordances that ancient Pompeians themselves would have been culturally habituated to see as salient.[35] In other words, I seek to determine the range of *emically recognizable* functions, values, and meanings that ancient Pompeians could potentially have associated with Nilotic imagery.

33. See chapter 1, this volume, "Material Culture and the 'Material Turn.'"

34. Gibson 2015 (1986): 119–136.

35. Cf. the emphasis on "culturally specific" affordances in design theory (Swift 2017: 8, 196–197).

This perspective provides a way to integrate viewer responses with the agency of objects and images themselves. While Nilotic imagery presented its viewers with a range of affordances for making meaning, differences in individual viewers' backgrounds and interests might cause them to experience some of those affordances as more available than others. Additionally, considerations of context and framing could have primed viewers to perceive some of the many possible associations of Nilotic imagery as more relevant than others. To understand how people made meaning with these images, we therefore need to put iconography in dialogue with context.

The second half of this chapter ("Contextualizing Nilotic Imagery at Pompeii") thus establishes and analyzes general patterns in the contextualization of Nilotic imagery at Pompeii. Nilotic scenes are frequently associated with certain specific settings and features—in particular, gardens, water features, and dining installations. Within the many possible associations and significations of Nilotic imagery, these domestic features activate specific associations for the images and prompt viewers to relate Nilotic iconography to their own activities and social performances. This general investigation of the contexts of Nilotic scenes lays the groundwork for the more detailed case studies in the subsequent chapters 3–6, which explore the ways that this confrontation between images and viewers played out in practice.

Adaptation, Emulation, and Eclecticism in Nilotic Imagery

Egyptian Antecedents and Affordances

Egyptian visual culture includes numerous depictions of the Nile, its yearly inundation, and the festivals and activities associated with its flooding. As an agricultural necessity without which crops could not grow and permanent settlements would be impossible, the Nile was a central feature of Egyptian life. The river structured both space and time, dividing the landscape into "black land" (fertile floodplain) and "red land" (desert) and dividing the year into three seasons based on riverine agriculture (inundation season, growing season, and harvest season). Unsurprisingly, therefore, the inundation was the focus of intense religious and cultural interest, and people performed numerous rituals to ensure its success. The yearly return of the floodwaters coincided with festivals for a deity identified in different regions with different goddesses, including Isis, Hathor, Tefnut, Mut, Bastet, and others.[36] This goddess, sometimes characterized as the personified

36. Junker 1911, 1917; Sethe 1964 [1912]; Smith 1984; Desroches Noblecourt 1995; Quack 2005: 128–139; Richter 2010.

Eye of the sun-god Re, traveled south to Nubia every winter, and when the river rose in summer, she returned to Egypt with a jubilating Nubian retinue. Egyptian festivals celebrating the goddess's return at the time of flood frequently involve themes of inebriation and sexual imagery, echoing the agricultural fertility brought by the floodwaters.[37] By the Greco-Roman period, these sexual themes were also increasingly associated with the union of Isis and Osiris.[38]

Nilotic scenes in Roman art may contain a range of motifs, including Egyptian animals and plants; pygmies; cult buildings or divine images; sexual imagery; combats with Nilotic beasts; dancing and musical performance; and drinking and inebriation (appendix B).[39] Many of these motifs have parallels in Egyptian religious texts and iconography. The pygmies may recall not only the possible apotropaic function of dwarfs in Roman iconography but also their Egyptian associations with cultic activity.[40] The cults of dwarf-gods like Bes were closely linked to that of Hathor, and Bes-gods played a role in the myth of the wandering goddess, dancing and playing music to pacify the goddess and encourage her to return to Egypt.[41] Real-life dwarfs and pygmies were also highly valued as temple performers and ritual dancers[42]—a connection that may help explain the Nilotic pygmies' frequent engagement in dancing or musical performance. The Nilotic pygmies' reed huts resemble Egyptian depictions of reed huts in Punt, south of Egypt.[43] The frequent depiction of their skin as dark brown is

---

37. Brunner 1954; Gutbub 1961; Sternberg-El Hotabi 1992: 101–114; Perpillou-Thomas 1993: 116–117; Darnell 1994; 1995; 2010b; 2010c: 44–45n34; von Lieven 2003; DePauw and Smith 2004; Graves-Brown 2010: 168–170; Richter 2010; Jasnow and Smith 2010–11; Barrett 2011: 124–127; Bryan 2014. Meyboom (1995: 71–75) discusses issues caused by the rotating Egyptian civil calendar and implications for the timing of inundation-related festivals.

38. Bonneau 1964: 247–248; Merkelbach 1995: 6–7, 13–17, 70; Versluys 2002: 281–282; Meyboom and Versluys 2007; Kleibl 2009: 141; Manolaraki 2013: 125; and see further note 142, this chapter.

39. Cf. the inventory of Versluys 2002: 261–285.

40. *Dwarfs and pygmies as apotropaic in Roman art:* see note 186, this chapter. *Nilotic pygmies' associations with Egyptian religion:* McKay 1985: 8–25 (somewhat overemphasizing afterlife themes); Meyboom and Versluys 2007; Barrett 2011: 313–315.

41. *Bes and Hathor:* Derchain 1981: 168; Quaegebeur and Cherpion 1999, esp. 129–131; Malaise 2004: 268–269, 286–287; Volokhine 2010: 237–248; Barrett 2011: 274–275. *On Bes's later associations with Isis:* Malaise 2004; 2005: 79–80. *Bes-gods in the myth of the wandering goddess:* Junker 1911: 5, 86.

42. Silverman 1969: 59; Dasen 1993: 131–133, 149–155; Barrett 2011: 265–270. *Egyptian dwarfs' Hathoric and erotic associations:* Derchain 1981: 168–169; O'Connor 1996: 629–630.

43. Phillips 1996; cf. Strabo's description of huts in Ethiopia made of woven palm wood (17.2.2). The straw huts contrast with Pliny's characterization of pygmies' houses as made of mud,

compatible not only with some Roman perceptions of Egyptians' appearance but also with Egyptian depictions of the Nubian retinue that accompanied the wandering goddess as she returned from the south.[44]

Some of the pygmies' more striking behaviors, such as fighting with river beasts (figure 2.4), drinking and carousing (figure 2.5), and engaging in public sexual activity (figures 1.8, A, and 2.5), also have Egyptian antecedents. Their battles with hippopotami and crocodiles recall Egyptian temple and tomb depictions of hippopotamus and crocodile hunts, which in Egypt represented the struggle of order over chaos (figures 2.6, 2.7).[45] The pygmies' consumption of alcohol also evokes the theme of inebriation in Egyptian inundation festivals.[46] The flooding Nile was identified with wine, and alcohol was thought to appease the returning goddess.[47] From the New Kingdom through the Greco-Roman period, one of the major cultic events in the first month of the inundation season was the Festival of Inebriation (*tḫy*).[48] In Demotic hymns and ritual texts from Ptolemaic and Roman Egypt, drunkenness induces divine epiphanies, and Herodotus's fifth-century description of the Bubastis festival emphasizes heavy wine consumption.[49] Depictions of local drinking festivals appear in a number of New Kingdom tombs from Thebes, where reliefs depict banqueters drinking heavily, stating their desire to become inebriated as fast as possible, and even vomiting[50]

feathers, and eggshells (*HN* 7.26), but Roman prejudices likely considered both house types similarly "primitive." Nilotic huts may alternatively recall the temporary festival structures of branches or reeds that appear in some Egyptian texts as sites for drinking on the riverbank (e.g., Darnell 1995: 59).

44. Egyptian representations of this retinue: Darnell 1995.

45. *On Nilotic scenes and Egyptian depictions of hippopotamus hunts:* de Vos 1983: 71; Meyboom 1995: 99, 148, 255–256n114, 325n169; Boissel 2007: 288–289; Meyboom and Versluys 2007: 183; Leclant 2010 [1976]: 280). *On the religious significance of Egyptian hippopotamus hunting:* Säve-Söderbergh 1953; Capriotti Vittozzi 2002b: 113–114. *On ritual crocodile hunts at Dendera and Edfu in Greco-Roman Egypt:* Wilson 1997. Such practices likely inform Greek and Latin descriptions of battles between crocodiles and the people of Tentyra (Dendera): Plin. *HN* 8.92–93; Strabo 17.1.44; Sen. *QNat.* 4a.2.15; Ael. *NA* 10.21; cf. Hdt. 2.69 on crocodile-eating at Elephantine.

46. Cf. Assmann 2000: 78.

47. *Nile flood as wine:* Merkelbach 1963: 17; Barrett 2011: 290n1120. *Alcohol appeasing the goddess:* Barrett 2011n396.

48. Sternberg-El Hotabi 1992: 101–114; Spalinger 1993; Jasnow and Smith 2010–11; Bryan 2014.

49. Hdt. 2.60. *Demotic sources:* DePauw and Smith 2004; Hoffmann and Quack 2007: 305–311, 370; Jasnow and Smith 2010–2011; see also note 53, this chapter.

50. Bryan 2014.

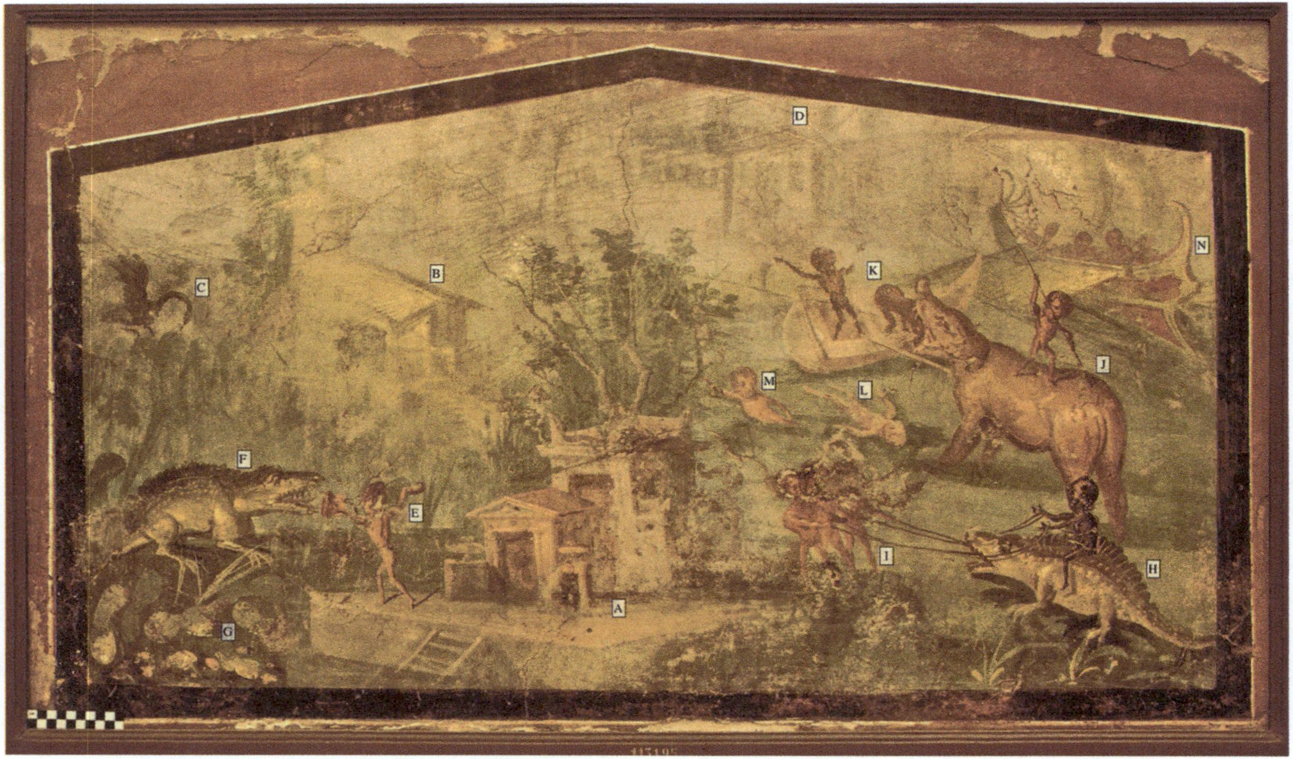

FIGURE 2.4 Wall painting depicting pygmies fighting Nilotic beasts (hippopotamus and crocodile). Pompeii, Casa del Medico (VII 5.24), pseudo-peristyle. MANN 113195. Letters have been added to show the location of particular features discussed in the main text (Chapter 4).

Photo: C.E. Barrett. Reproduced by permission of the Ministero dei Beni e delle Attività Culturali e del Turismo—Museo Archeologico Nazionale di Napoli.

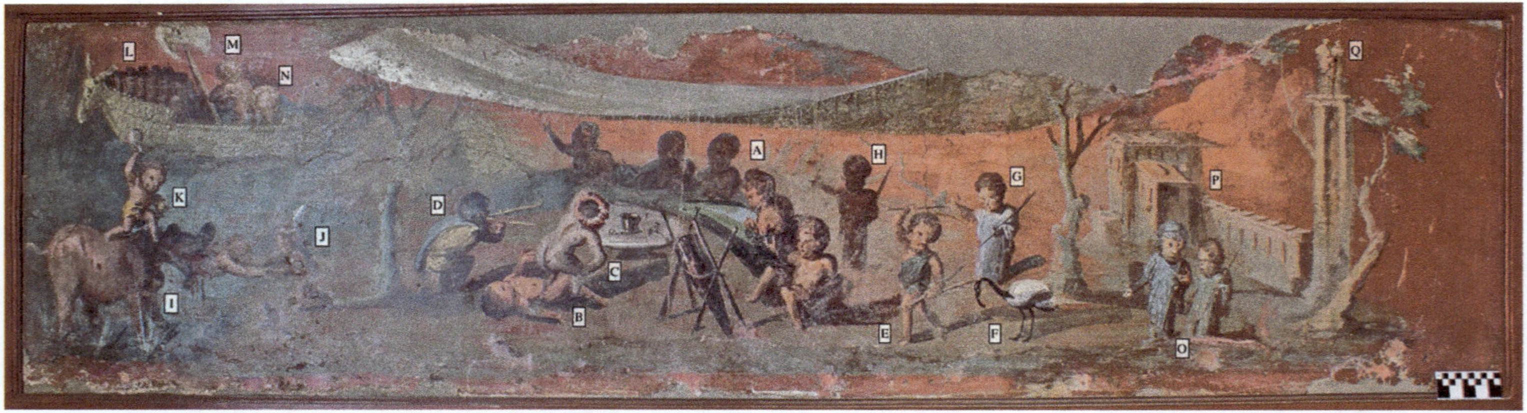

FIGURE 2.5 Wall painting depicting an outdoor symposium with music, dance, and sexual encounter, as well as a pygmy-hippopotamus battle. Pompeii, Casa del Medico (VIII 5.24), pseudo-peristyle. MANN 113196. Letters have been added to show the location of particular features discussed in the main text (Chapter 4).

Photo: C.E. Barrett. Reproduced by permission of the Ministero dei Beni e delle Attività Culturali e del Turismo—Museo Archeologico Nazionale di Napoli.

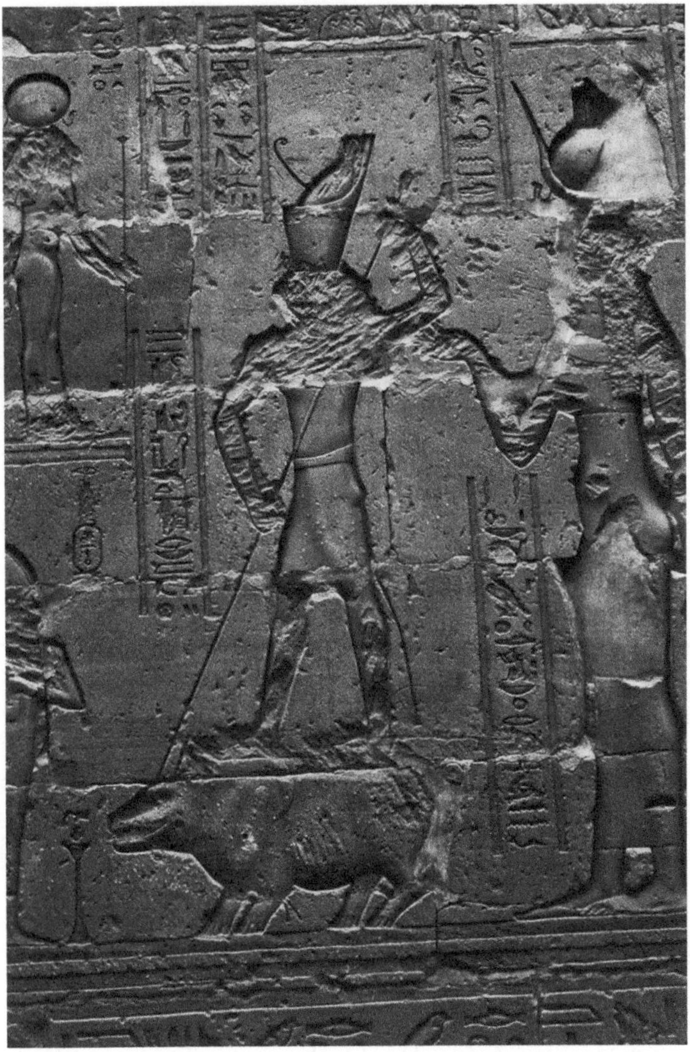

FIGURE 2.6 Horus spearing Seth in the form of a hippo, Edfu Temple, interior of west enclosure wall.
Photo: Arielle Kozloff Brodkey; reproduced with permission of Arielle Kozloff Brodkey.

(figure 2.8). At one tomb from Elkab, speech captions next to female banqueters include statements such as "For your *ka*: drink until you are drunk. Make festival!" and "Give me 18 jars of wine. To be drunk is what I continually desire."[51]

51. Trans. Bryan 2014: 117.

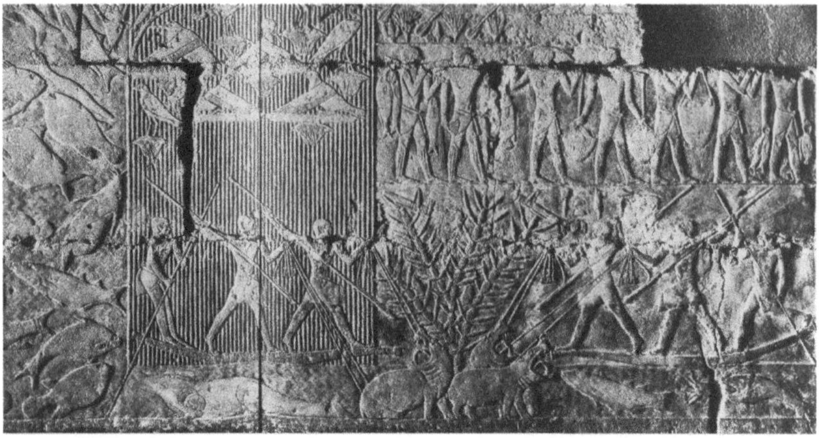

FIGURE 2.7. Depiction of a hippopotamus hunt in the mastaba of Mereruka, Saqqara, 6th Dynasty.

Photo: After Sakkarah Expedition 1938: pl. 10, detail. Courtesy of the Oriental Institute of the University of Chicago.

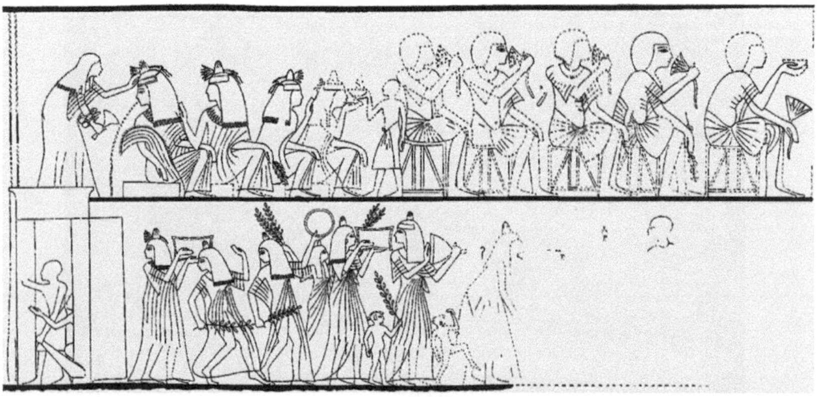

FIGURE 2.8 Depiction of a festival involving banqueting, drinking, and dancing in the 18th Dynasty tomb of Neferhotep, Thebes, Egypt (TT 49). The seated woman at the far left of the upper register is vomiting from drinking too much alcohol.

Source: After Davies 1933: pl. 18; Bryan 2014: fig. 6.31.

Depictions of pygmies engaging in sex on boats or the riverbank further recall the prominence of sexual themes in these *tḫy*-festivals.[52] Ptolemaic ostraca and Roman-period Demotic texts associate sexual activity and alcohol consumption

52. Cf. Meyboom 1995: 329–330n180; Meyboom and Versluys 2007: 182–202.

with festivals of the returning goddess.[53] Herodotus's and Strabo's comments on supposedly licentious festivals at Bubastis and Canopus likely indicate some Greco-Roman awareness of these traditions.[54]

Most of the sexual aspects of Egyptian flood myths pertain to divine couplings rather than human, but there is evidence that couples interpreted their own sexual activity at this time as imitating divine unions.[55] In a recent examination of the evidence for ritual sex at *tḫy*-festivals, Jasnow and Smith conclude that the practice is indeed attested in some Ptolemaic and Roman-period sources.[56] Indeed, one 20th Dynasty stele dedicated to Hathor may actually claim that the dedicant was conceived *inside* Hathor's chapel at Deir el-Bahri, although the text is not wholly clear.[57] Bryan and Ryholt consider some form of sexual behavior to have been a common feature of *tḫy*-festivals in both New Kingdom and Greco-Roman Egypt, and Bryan associates erotic images from some 18th Dynasty tombs (e.g., figure 2.9) with these festivals.[58]

It is unclear how much of this activity took place in public as opposed to private. Jasnow and Smith suggest that in the Greco-Roman period at least, actual public intercourse was probably "not widespread" and may have characterized only certain cults.[59] They propose that acts of "ritualized ribaldry," such as self-exposure and obscene speech, may have been more common in public settings.[60]

Nonetheless, there is evidence that Egyptian couples considered the inundation a particularly auspicious time for sexual activity (whether conducted in private or otherwise). Bioarchaeological study of remains from a late Roman cemetery in Dakhleh Oasis has shown that the peak months for birth at this site

---

53. DePauw and Smith 2004; Quack 2005: 87; Hoffmann and Quack 2007: 305–311, 370; Kockelmann 2008: 84; Dieleman and Moyer 2010: 440; Barrett 2011: 199–202; Jasnow and Smith 2010–2011; Quack 2013: 76–79; 2016: 70–74. For a 20th Dynasty hieratic precedent, see Brose 2009.

54. Herodotus 2.60; Strabo 17.1.16–17.

55. Meyboom and Versluys 2007: 191–192; Darnell 2016.

56. Jasnow and Smith 2010–11: 31–53.

57. The word Bryan (2014: 119) reads as "conceived" can alternatively mean "born" (Darnell 2016: 54n108). In most contexts, intercourse in temples was forbidden (Manniche 1977: 12–13). However, one unpublished Demotic papyrus may situate a sexual episode within a temple of Mut (Quack 2016: 73–74), providing a potential parallel to Bryan's interpretation.

58. Ryholt 2010: 729; Bryan 2014: 105, 107, 112, 118–119; see also Darnell (2016: 54) on the Greco-Roman period.

59. Jasnow and Smith 2010–11: 31–53, quoting 50.

60. Jasnow and Smith 2010–11: 48; Dunand 2013: 212–215; Thomas 2015a: 62–63.

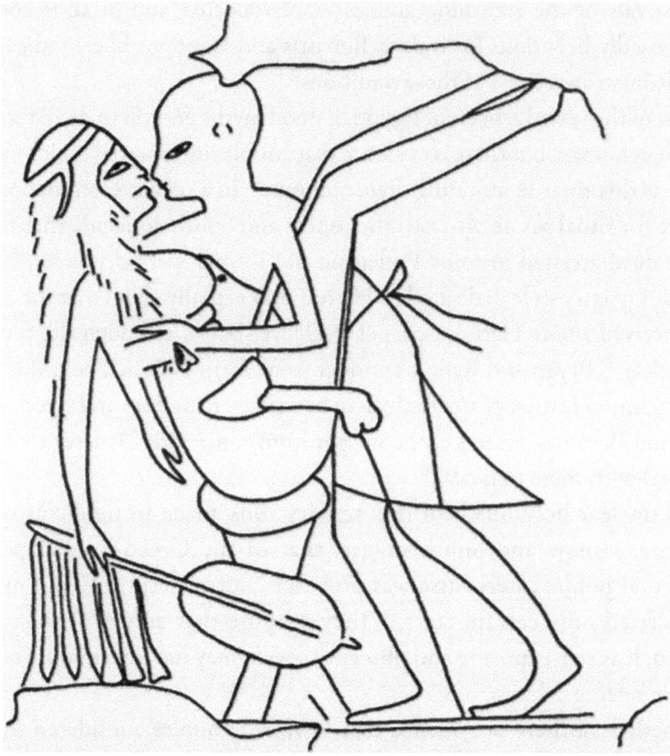

FIGURE 2.9 A sexual encounter between a man with a bald or shaved head (possibly a priest) and a female musician. Drawing on wood (identified as a piece of a coffin), now lost, from the tomb of Puiemra, Thebes, Egypt (TT 39), 18th Dynasty. A similar image, whose male protagonist also has a shaved head, appears on a New Kingdom ostracon in the Cairo Museum (Manniche 1977: fig. 2).

Sources: Drawing after Manniche 1977: fig. 1; Bryan 2014: 119.

were March and April, meaning the peak period for conception was in July and August—that is, the height of the flood.[61] This seasonal pattern differs substantially from that of other Mediterranean peoples, suggesting some cultural or environmental explanation particular to Egypt. The researchers point to Egyptian beliefs associating the inundation with human fertility, proposing that couples may have increased their sexual activity during this favorable season. Literary evidence from earlier periods also eroticizes the Nile and its flood. For example, New Kingdom love poems frequently take place in Nilotic settings and may often

61. Williams et al. 2012; Dupras et al. 2015: 59; 2016: 291–292.

be set during the flood season.[62] On a practical level, the cessation of agricultural activity during the high waters may have given farmers more leisure time,[63] creating more opportunities to spend time with their wives.

The mythical narratives associated with the Nile inundation are best known from Ptolemaic and Roman-period textual sources: hieroglyphic temple inscriptions, Demotic narrative and ritual texts, and Greek papyri.[64] However, the settings, protagonists, and festivals associated with those narratives also appear frequently in material culture from Ptolemaic and Roman Egypt. These visual representations of the Nile and its associated festivals and deities do not simply serve as illustrations to textual narratives but also make statements of their own about worshipers' relationship to the riverine landscape and its deities.

On temple and tomb walls, visual representations typically adhered to stricter rules of decorum than in some other media.[65] In Ptolemaic and Roman Egypt, temple reliefs were still expected to employ traditional conventions such as two-dimensionality, nonnaturalistic perspective, and a prescribed canon of proportions.[66] From the Pharaonic through Roman periods, such reliefs usually depict sexual intercourse in allusive rather than overt ways—e.g., indicating a divine sexual union by depicting the protagonists sitting together on a bed with their legs overlapping (figure 2.10)—although depictions in tombs are occasionally more explicit.[67] Temple reliefs also usually suggest inebriation in

62. Darnell 2010a, 2016; though see the comments of Quack (2016: 81–82) on seasonality.

63. With the caveat that their corvée duties increased: see sources in Barrett 2015, s.v. "Taxation, Corvée, and Liturgies."

64. References to the myth and associated festivals appear in at least 22 Ptolemaic temples (Richter 2010: 155). See especially Philae (Žabkar 1988; Inconnu-Boquillon 2001) and Medamud (Darnell 1995). *Demotic myth of the returning solar eye goddess:* Bresciani 1992; Spiegelberg 1994 [1917]; De Cenival 1988; Quack 2005: 148–160; Stadler 2012: 60–68. *Greek translation:* West 1969; Signoretti 2010. *Pre-Ptolemaic textual and iconographic references to this myth:* Verhoeven and Derchain 1985; Darnell 2010a; Barrett 2011: 122–124.

65. *Egyptian artistic "decorum":* Baines 1985: 277–286; 1990: 20–23.

66. Robins 1997: 235–239.

67. Pharaonic tomb scenes often allude to sexual activity in coded ways, with couples holding hands on a bed or fishing and fowling in the marshes (Derchain 1975; cf. O'Connor 1996: 629–632). However, there are also multiple depictions of explicit sex in tomb contexts, *pace* Venit (2016: 191). The famous "Dirty Grotto" New Kingdom graffito from a Theban tomb depicts a couple engaged in intercourse (Wente 1984). An unusual hieroglyph in an early Middle Kingdom tomb from Beni Hasan illustrates sexual intercourse (Collombert and Volokhine 2005: 49–50). A New Kingdom Theban tomb contained a now-lost depiction of intercourse (fig. 2.9). A Roman-period tomb painting from Akhmim depicts a sexual encounter between Anubis and a female figure (Klotz 2012b; contra Venit 2010: 113–119; 2016: 191–192, 236n1237; Venit's attempted rebuttal of Klotz's argument does not address the fundamental problem

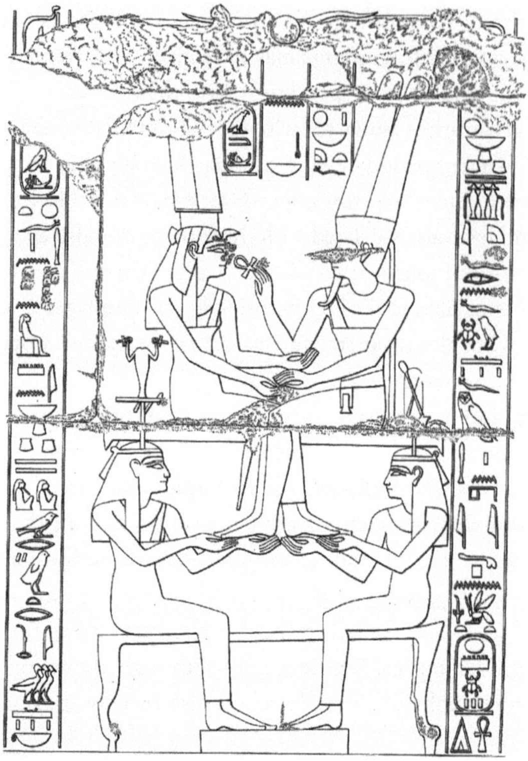

FIGURE 2.10 The god Amun impregnates Queen Mutemwia with the future king Amenhotep III. Relief from the Divine Birth cycle at Luxor Temple.

Source: After Kemp 2006: 263, fig. 96; drawn by Barbara Garfi after Brunner 1964: pl. 4. Reproduced by permission of Barry Kemp.

discreet ways: for example, showing the king offering wine to Hathor without illustrating the state of drunkenness she was expected to attain and facilitate, even when textual inscriptions on temple walls are more frank.[68] Depictions

that the feature she interprets as a coil of intestines is probably not part of the painting, but a gap where some plaster fell off the wall—see Klotz 2012b: fig. 2). A Roman-period tomb at Hermopolis depicts an unusual image of sexual intercourse between a woman and an amphibian-headed ithyphallic figure (Gabra et al. 1941: 102–103, pl. 48). *Pace* the speculation of Gabra et al. that this last image memorializes a woman who succumbed to a uterine ailment "causé, pensait-on, par un succube" (1941: 102), the scene may more likely recall other Egyptian representations of sexual activity in watery settings (as evoked by the amphibian, perhaps frog, head), as well as representations of deities such as the Ogdoad as amphibian-headed.

68. *Depictions of wine offerings in temples:* Poo 2009 [1995]. *Drink offerings for Hathor:* Sternberg-El Hotabi 1992. *On the relative frankness of texts*, see ancient textual descriptions of Hathor as the "lady of drunkenness": e.g., note 80, this chapter.

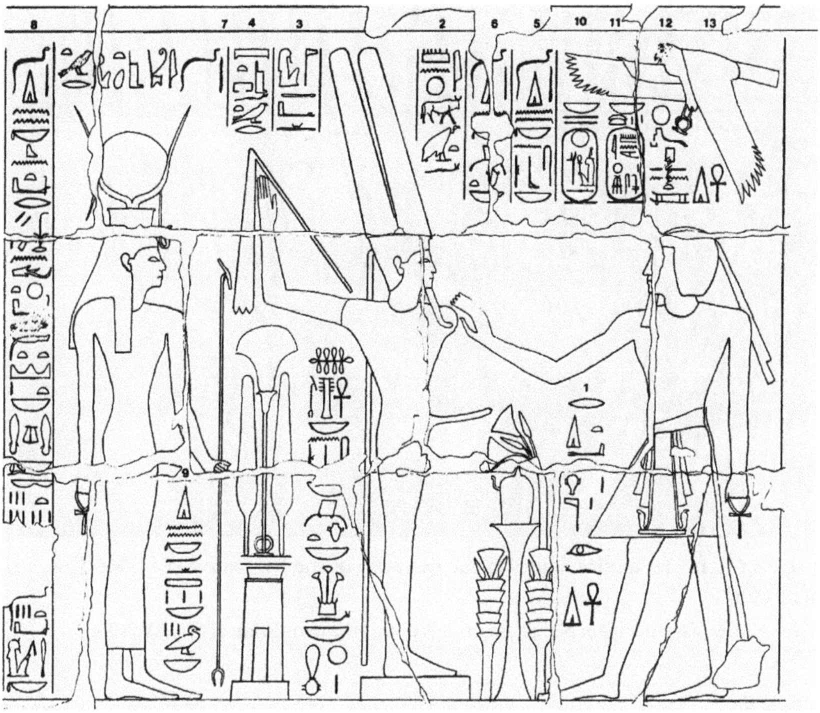

FIGURE 2.11. Ramses II steadies a statue of ithyphallic Amun-Re Kamutef, with Isis standing at left. Relief from the Great Hypostyle Hall of Karnak Temple (west wall, south half, south part, second register), 19th Dynasty.

Source: After Nelson and Murnane 1981: pl. 18. Courtesy of the Oriental Institute of the University of Chicago.

of ithyphallic deities were perfectly permissible in temple reliefs and statuary, however (figure 2.11), and ithyphallism is actually a frequent identifying feature of creator deities.[69] The Nile flood appears in temples in several forms. The inundation itself was personified as the god Hapy, and reliefs frequently depict Hapy and related fecundity figures taking part in processions (figure 2.12), often accompanying goddesses who personify agricultural fields.[70] Other images depict the deities responsible for the inundation. For example, a relief at Karnak—carved in the Ramesside period, but visible still in Greco-Roman times—depicts

---

69. Particularly recalling the masturbation or autofellatio of Atum in the Heliopolitan cosmogony (Allen 1988: 13–35; Robins 2007: 208–209).

70. Baines 1985, and see also note 75, this chapter. A Ptolemaic mosaic from Dendera shows the king and queen in a New Year procession with personifications of the flood and fields (Meyboom 1995: 78–79).

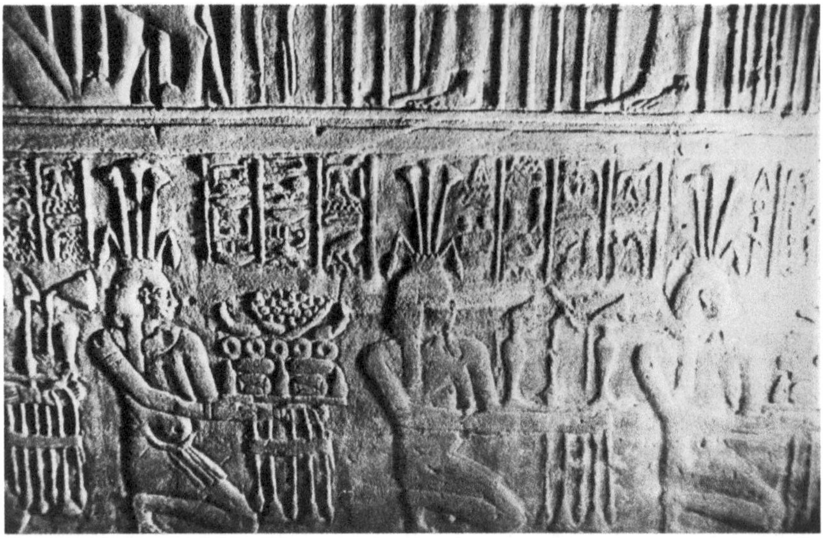

FIGURE 2.12 Fecundity figures in the *mammisi* of the Isis temple at Philae (Room II, west wall).
Source: After Vassilika 1989: pl. 14c. Reproduced with the permission of Eleni Vassilika.

that temple's chief deity Amun sitting atop the floodwaters.[71] Also suitable for temple decoration were images of riverine festivals, including processions of divine images, boats, musicians, singers, and Nubian and Libyan dancers[72] (e.g., figure 2.13). On temple walls at Edfu and Dendera, scenes of crocodile and hippopotamus hunts represent the battle of order against chaos, characterizing the hostile beasts as Sethian adversaries of Horus and the ruling Horus-king (e.g., figure 2.6).[73]

Riverine imagery was particularly common in certain parts of temples. Outer courtyards, the temple area most accessible to the general populace,[74]

71. Gabolde 1995. Cf. depictions of Hathor and Isis seated atop sacred waters (16 vases under Hathor; an *isheru*-lake under Isis) at the Per-wer sanctuary of the Hathor temple at Dendera (Richter 2010: 173–174, 185 fig. 20).

72. See, e.g., the representations of the Opet festival at Luxor Temple (Epigraphic Survey 1994; Darnell 2010b); cf. textual descriptions at the Greco-Roman-era temple of Medamud (Darnell 1995). The Luxor reliefs date to the New Kingdom but would have been visible into the Greco-Roman period, when the Opet festival was likely still practiced (it is mentioned in Roman-period mortuary papyri from Thebes: Klotz 2012a: 386–388).

73. Säve-Söderbergh 1953; Finnestad 1997: 225; Wilson 1997.

74. Bell 1985: 270–271, 275–276; 1997; Barrett 2011: n1393, n1724. The still-unpublished "Book of the Temple"—a Roman-period manual for the ideal Egyptian temple, known from

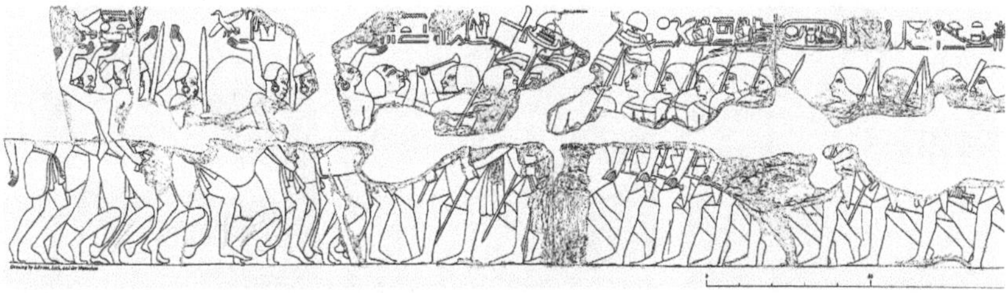

FIGURE 2.13 Dancing Nubians, trumpet player, and soldiers accompanying the procession for the Opet festival. 18th Dynasty relief in the colonnade hall of Luxor Temple. *Source:* After Epigraphic Survey 1994: pl. 94. Courtesy of the Oriental Institute of the University of Chicago.

were typically decorated with images of riverine vegetation. Such images often included wall reliefs of water plants and fecundity figures personifying the Nile flood, as well as vegetal columns.[75] Wall bases depicted low-growing plants, while the columns represented "palms, papyrus, lotus, and reeds, vegetation growing from the inundated earth"[76] (figure 2.14). This aquatic imagery evokes creation mythology. The innermost part of Egyptian temples represents the "mound of creation" on which the sun-god first emerged from the primordial Nun-waters, while the outer courtyards represent the marshy regions at the edge of the emerging land.[77] Nilotic imagery was particularly appropriate here because the flood was theologically equated to the primordial waters of creation.[78] Additionally, some temple inscriptions suggest that columned courtyards could serve as sites for the drinking and celebration associated with *tḫy*-festivals.[79] In the columned forecourt of the small temple of Hathor at Philae, Augustan-era reliefs depict scenes appropriate to such festivals, with human, divine, and animal musicians and dancers màking merry. On the columns, Bes-like figures dance and play music for Hathor, while an accompanying text describes Hathor as the "mistress of the dance, great of love . . . lady of drunkenness, with many

numerous copies in hieratic, Demotic, and (in one case) Greek— discusses the relative accessibility of different areas of temples (Quack 2004, 2007a).

75. Darnell 1995: 52. *Contexts of fecundity figures:* Baines 1985: 277–305, 330–343. *Vegetal columns:* Arnold 1999: 292–302.

76. Quoting Finnestad 1997: 204; cf. Robins 1997: 235.

77. Finnestad 1985; 1997: 203–204, 211.

78. Finnestad 1985: 42–46.

79. Darnell 1995: 52; DePauw and Smith 2004: 88–89; Bryan 2014: 103–106. In other contexts, drunkenness inside temples was prohibited (Depauw and Smith 2004: 82–83).

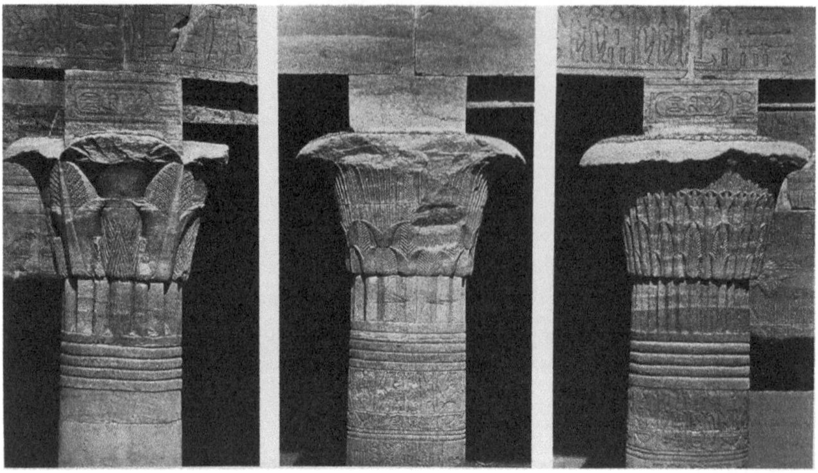

FIGURE 2.14 Vegetal columns. Examples of palmette capitals of three different types (single, double, multi-story) from the Augustan court at Philae.

Source: Photo by D. Arnold; after Arnold 1999: fig. 256. Reproduced with the permission of Dieter Arnold.

festivals" and the one for whose *ka* "the Hity-gods [dwarf deities similar to Bes] strike the tambourine"[80] (figure 2.15). In real-life festival processions, dancers and musicians sometimes wore masks to impersonate these dancing, music-playing dwarf-gods.[81]

Representations of dwarf-deities, marshes, riverine vegetation, and Hathoric imagery are particularly common in (1) kiosks located near the entrance of temples, designed to hold the barques of visiting divine images and sometimes incorporated into celebrations for the return of the "wandering goddess"; and (2) *mammisis*: shrines of a type common in Late Period through Roman-period temples, celebrating the birth of a divine child who was identified simultaneously with the reborn sun and the reigning king[82] (figure 2.16; for an earlier

80. Translation adapted from Daumas 1968: 11; see 10 for hieroglyphs. Similarities between Hity and Bes: Volokhine 2010: 234–235; Derchain 1981: 168.

81. As indicated by statues, paintings, and excavated masks (Volokhine 2010: 248–249; Barrett 2011: 271).

82. *Association with temple entrances:* Festival processions for the returning goddess could conclude at a "place of inebriation" at the temple entrance (Darnell 1995: 59–61; Bryan 2014). *Representations of dwarf-gods in kiosks and at temple entrances:* Labrique 2011: 189–190. *Association with* mammisis: Daumas 1968; Darnell 1995: 47–48; Arnold 1999: 302, fig. 263; Volokhine 2010. *On* mammisis, see generally Daumas 1958; Arnold 1999: 285–288.

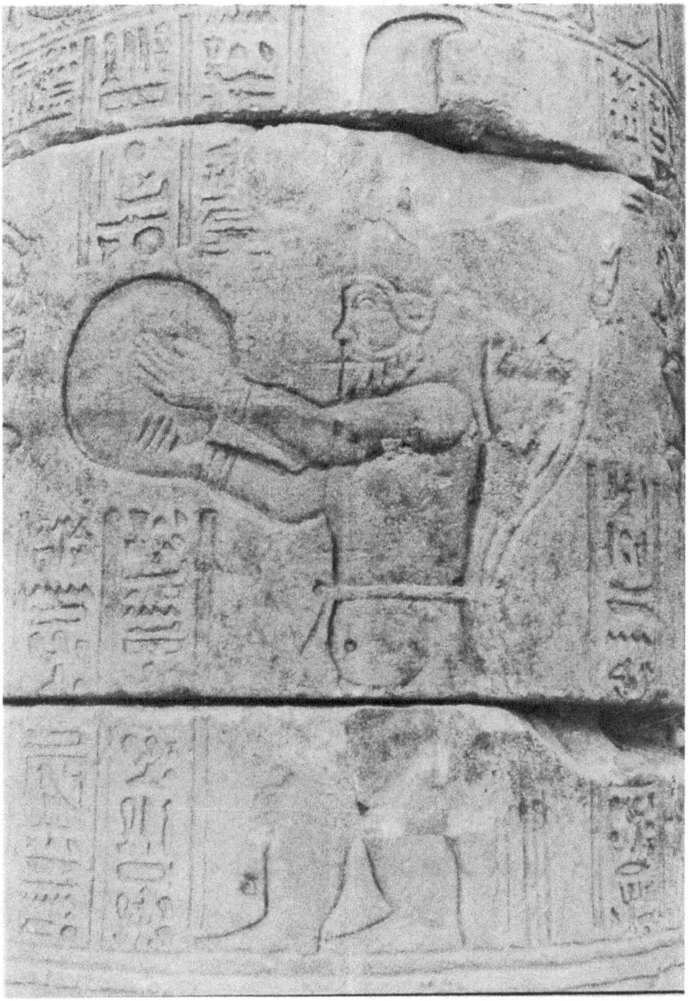

FIGURE 2.15 Dancing, tambourine-playing dwarf-god. Relief on a column at the small temple of Hathor at Philae.

Source: Photo by A. Tuor; after Dasen 1993: pl. 9, 1a. Reproduced with the permission of Véronique Dasen.

representation of the "divine birth" cycle, see figure 2.10). This solar child was conceived at the time of the flood and born in the Delta marshes.[83] In these temple areas, Bes and related dwarf-deities appear as attendants of the female

83. Daumas 1958: 297; Arnold 1999: 287, 300–301; Barrett 2011: 127; Klotz 2012a: 91–92. Cf. Hdt. 2.156 on the birth of Horus.

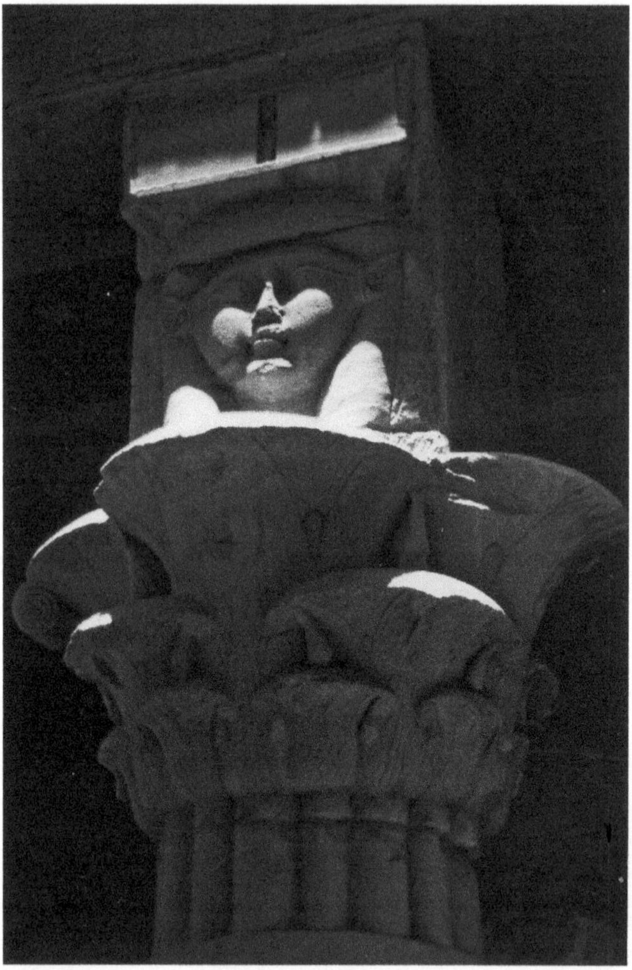

FIGURE 2.16 Hathor-headed vegetal column from the pronaos of the Philae *mammisi*.
Source: After Vassilika 1989: pl. 42d. Reproduced with the permission of Eleni Vassilika.

deity who serves as the local form of the Eye of Re.[84] These dwarf-gods[85] first dance to celebrate the reappearance of the goddess (figure 2.15), and then serve as companions and protectors for her and her son.

84. Egyptian temples are often dedicated to a family triad of gods: a male solar creator deity, his wife, and their child. The precise identification of the deities in question varied regionally. On triads, see te Velde 1971.

85. On Bes, similar-looking dwarf-gods (e.g., Aha or Hity), and their iconography, see Altenmüller 1975; Romano 1989; Dasen 1993: 55–83; Volokhine 2010: 234–235.

Monumental images of Bes are also attested. These appear in a number of shrines or temples that may be dedicated either to the dwarf-god himself or to the wandering goddess for whom he danced. Large-scale stucco reliefs at a Ptolemaic shrine at Saqqara depict dwarf-deities, 1.5 meters tall, in the company of nude Hathoric females[86] (figure 2.17). Two temples in the Bahariya Oasis feature Bes prominently: one 26th Dynasty temple with gigantic painted reliefs of Bes-gods accompanying the seven Hathors, and one Ptolemaic temple containing a 1.22-meter sandstone statue of a nude, squatting Bes.[87] Although the painted reliefs from the earlier temple are now mostly destroyed, they once covered at least two registers of the wall, and a single foot from one Bes figure is approximately 60 centimeters long.[88] While some scholars see these temples as dedicated to Bes himself, Labrique argues persuasively that they were more likely dedicated to female divinities for whom Bes served as attendant, dancer, and appeaser. This is certainly the case at Gebel Barkal, where a (now-destroyed) 25th Dynasty rock-cut temple celebrated Mut as the wandering solar eye goddess; the columns in the forecourt of this temple forecourt took the form of gigantic Bes statues.[89] Similar statue-columns of Bes appear in the temple of the deified 19th Dynasty queen Nefertari at Abu Simbel and in Meroitic temples at Wad ban Naqa, Musawwarat, and Meroe.[90]

Moving from temple to mortuary contexts, depictions of the Nile appear frequently in reliefs and paintings on Pharaonic through Roman-period Egyptian tomb walls. In funerary settings, Nilotic imagery expresses themes of regeneration, fertility, and the triumph of order over chaos. Scenes of boatmen hunting hippopotami echo the deceased person's own hoped-for triumph over chaotic forces (figure 2.7).[91] Similarly, depictions of riverine resources or fecundity figures (figure 2.18) suggest new growth, fertility, and the continued provision of the deceased.[92] Hathoric and riverine imagery—and, in some New Kingdom cases, representations of *tḫy*-festivals (figure 2.8 and possibly figure 2.9)—take

86. Quibell 1907: 12–14; Klotz 2012b: 394–395. The "Bes chambers" stood within a larger sanctuary of Anubis (Labrique 2011: 187; Klotz 2012b: 395). On their possible relationship to *mammisis*, see Thomas 2015a: 56–57.

87. Hawass 2000: 169–173; Labrique 2011.

88. Labrique 2011: 185.

89. Labrique 2011. *Mut as wandering solar eye goddess at Gebel Barkal*: Robisek 1989: 78.

90. Shinnie and Anderson 2004: 59–60, with n31.

91. See note 45, this chapter.

92. Desroches Noblecourt 1995: 179–185.

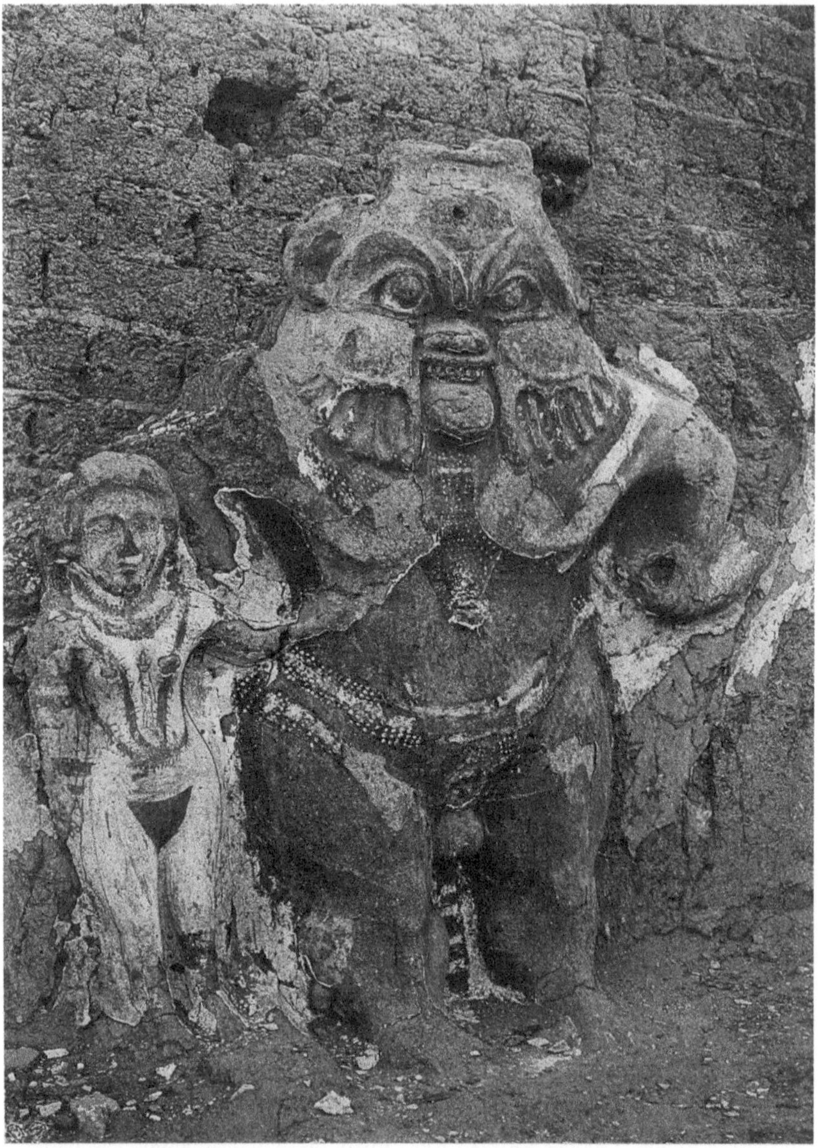

FIGURE 2.17 Painted stucco relief of Bes (1.5 m tall: see Quibell 1907: 12–14) and a nude female companion from the "Bes Chambers," Saqqara.

Source: After Quibell 1907: pl. 27.

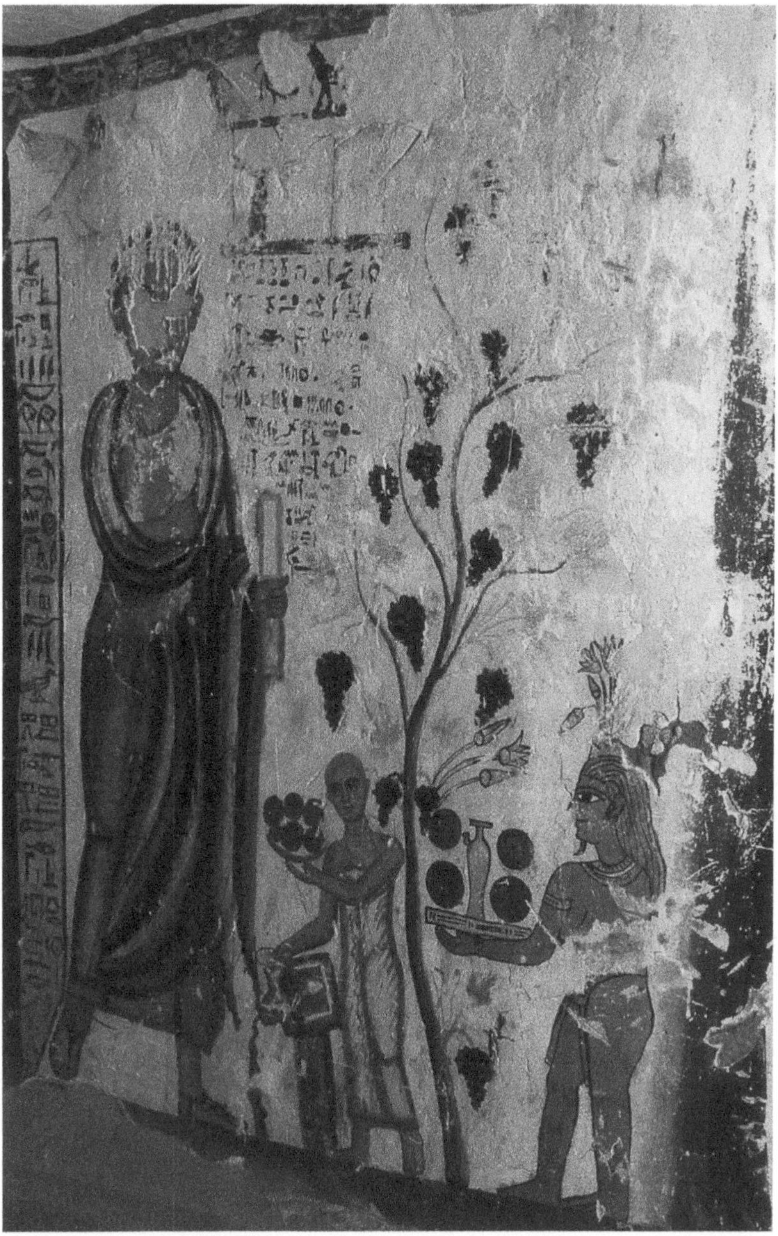

FIGURE 2.18 Painting from the tomb of Petosiris, Qaret el-Muzawaqqa, Dakhla Oasis, first century CE. The tomb owner stands at left; to the viewer's right, a priest (right) and a fecundity figure (far right) approach him with offerings. The tomb owner and priest are depicted in a "Hellenizing" style, while the rendering of the fecundity figure follows more traditional Egyptian conventions.

Source: Photograph by Helen Whitehouse (previously published as Riggs 2005: fig. 76). Reproduced with permission of Helen Whitehouse and Christina Riggs.

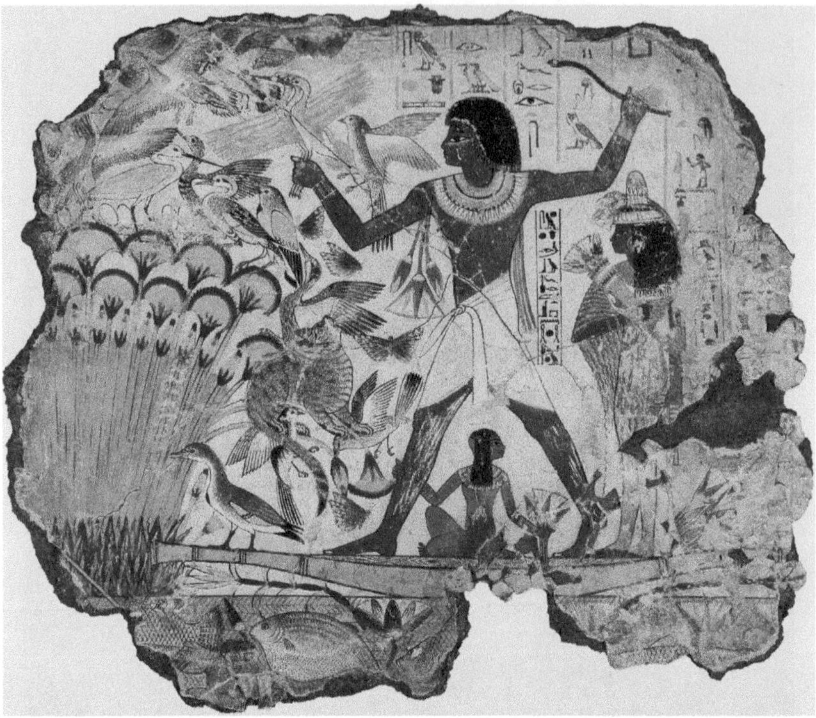

FIGURE 2.19 Painting from the tomb of Nebamun, Thebes, Egypt, 18th Dynasty. Nebamun and his family hunt river birds in the marshes of the Nile.

Source: Image courtesy of the British Museum (EA 27977, AN). © Trustees of the British Museum.

on eschatological significance in tombs, expressing the deceased's hope for personal regeneration.[93] Representations of couples fishing and fowling on riverboats while wearing elaborate festival garments (figure 2.19) not only suggest a paradisiacal afterlife but also carry erotic implications.[94] "Traversing the marshes" (*sꜣb sš*) was, among other things, a euphemism for sexual activity.[95]

While certain forms of decorum appear to have characterized temple reliefs and (to a somewhat lesser degree) tomb paintings, other media—attested in both temple and domestic contexts—were not as constrained.[96] Numerous

93. See, e.g., funerary depictions of Hathor as a cow emerging from the marshes (Pinch 1993: 175–183). Possible tomb images of *tḫy*-festivals: Bryan 2014: 118.

94. Derchain 1975: 61–64; Manniche 2003; Parkinson 2008: 122–132; Darnell 2010a.

95. DePauw and Smith 2004: 81–82.

96. Derchain states that allusions to sex in Pharaonic-period religious contexts were "toujours . . . discrète" (1981: 169), but Pinch points out that even in the New Kingdom, this

Ptolemaic and Roman-period statuettes, stelae, and amulets of various media (stone, terracotta, faience, and bronze) depict figures, often dwarflike or pygmy-like, whose activities closely resemble those of the protagonists of Roman Nilotic scenes.[97] Although such figurines are attested all over Egypt, they appear to be particularly common at Lower Egyptian sites.[98] The activities of these figures include, among other things, participating in explicit sex or exposing their (often exaggerated) genitals (figures 2.20–25, 2.28, 2.32, and possibly 2.26), playing music and/or dancing (figures 2.24–25 and perhaps 2.26), contending with hostile river beasts (figure 2.27), sailing on papyrus skiffs (figure 2.28), carrying divine images or shrines in religious processions (figures 2.29–32), drinking (figure 2.33), or carrying wine vessels and/or offerings (figure 2.34), which they sometimes bring to altars (figure 2.35).[99]

While figurines often feature more explicit imagery than that which appears on temple walls, this difference does not seem to reflect a dichotomy between

---

was not the case with phallic votive figurines (1993: 245). Sexual imagery also appears in some Pharaonic-period tombs (see note 67, this chapter).

97. *Previous comparisons of Nilotic scenes to bronze and terracotta figurines:* Cappel 1994: 101; Meyboom 1995: 100; Bragantini 2006: 162–163; 2012: 27; Barrett 2011: 313–315. *On the dating of the terracotta parallels*, note archaeological evidence for the production of such figurines in Egypt as early as the Ptolemaic period; Szymańska 2005 (esp. pls. XIV–XIX for dwarfs and Bes figures); Mathieu 2001: 548 (Bes striking a dance pose paralleled in Nilotic scenes, e.g., Figure 2.5, E); Thomas 2015a, 2015d. Dunbabin interprets some such figurines as images of mimes, suggesting that their "bald head, large ears, and stupid expression" recall the features of the *stupidus* (2016: 124). However, the bald heads and exaggerated facial expressions of the Egyptian figurines more closely resemble the iconography of the so-called "Pataikos," an Egyptian figure with features of dwarfism (see further discussion later in this chapter, and see especially note 114, this chapter).

98. Thomas 2015a, esp. 65–72.

99. *Explicit sex:* E.g., Bianchi 1988a (paralleled by two limestone figurines from Ptolemaic Memphis: Petrie Museum UC 38350, 35954); Ewigleben and von Grumbkow 1991: no. 99; Fischer 1994: nos. 519–521; Török 1995: no. 212 (though the figures in this example are not dwarfs); Bailey 2008: 3170–3171 (*pace* Bailey's identification of both figures as female); Sandri 2012: 638; Griesbach 2013: no. 33. *Ithyphallic or macrophallic males:* Fischer 1998, 1999; Barrett 2011: 202–208; 2015a. *Women exposing their genitals:* Barrett 2011: 208–211. *Playing music or dancing:* Perdrizet 1921: nos. 140, 142; Philipp 1972: figs. 36, 40–41; Ewigleben and von Grumbkow 1991: nos. 30, 105; Bailey 2008: 58; Barrett 2011: 265–268; Griesbach 2013: no. 66; Vendries 2013: 208; Dasen 2013. *Bes or other dwarflike figures standing atop crocodiles*: Berlandini 1995: pl. IV; Frankfurter 1998b: 125–126, pl. 12. Such images recall New Kingdom through Greco-Roman "magical" stelae depicting Horus or Harpocrates—often surmounted by Bes's head—standing on crocodiles (Ritner 1989; Barrett 2011: 282n1075). Isis takes a similar pose in a painting at the Pompeii Iseum (Swetnam-Burland 2015: 134). *Carrying divine images or shrines:* Perdrizet 1921: pl. 106; Philipp 1972: fig. 16; Fjeldhagen 1995: nos. 97–98; Bailey 2008: nos. 3147–3153. *Drinking:* Myśliwiec 1998, color plate; Szymańska 2005: no. 136. Cf. figurines of boats stacked with amphorae (e.g., Ewigleben and von Grumbkow 1991: no. 147). *Carrying wine vessels or offerings:* Perdrizet 1921: pl. 35, 104–106; van Wijngaarden 1958: no. 54, 57; Philipp 1972: figs. 6, 8, 9; Ewigleben and von Grumbkow 1991: nos. 31, 91, 104, 106–108;

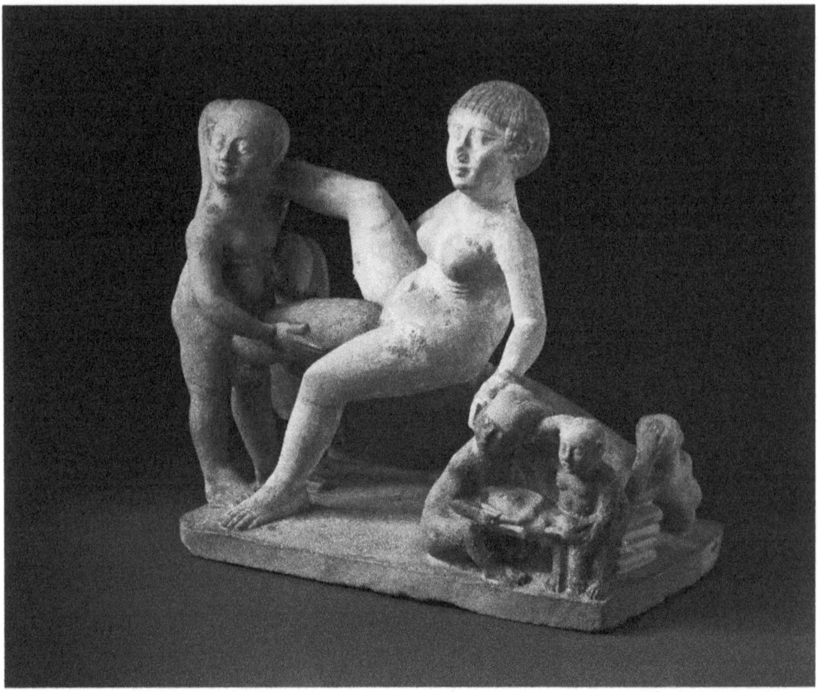

FIGURE 2.20 Painted limestone statuette group, Ptolemaic (305–30 BCE), with a naked woman and six smaller naked males (juveniles, pygmies, or dwarfs?) wearing sidelocks. One male penetrates the woman from the front, another from behind; two smaller males, not visible from this angle, lie on their backs with erect phalli; and two small males hold a bound oryx, a symbol of defeated chaotic forces (see discussion in Bianchi 1988a; Török 1995: 53). 16.5 × 9.5 × 17 cm (Brooklyn Museum). Brooklyn Museum, gift in memory of Dr. Jacob Hirsch and Charles Edwin Wilbour Fund (58.13).

Source: Image courtesy of Brooklyn Museum.

"religious" and "secular" images. Figurines of festival-making, dancing, drinking, macrophallic, or sexually active individuals—often, but not always,[100] shown as dwarfs or pygmies—are most often associated with household contexts, but the full range of their findspots is in fact quite broad. They are certainly most common

Dasen 1993: 79–80; Nenna and Seif el-Din 1994: 306–311; Fjeldhagen 1995: nos. 96, 124–126; Török 1995: no. 238; Barrett 2015a: 198–199. *Bringing offerings to an altar*: Philipp 1972: fig. 22; Sandri 2012: 640, fig. 38.5.

100. For figurines, lamps, and relief vases from Greco-Roman Egypt depicting sex between somatically typical individuals, see Montserrat 1996: pls. 21–25; Myśliwiec 1998: 64–65, and color plate. Ithyphallic imagery also need not always involve dwarfs (see, e.g., Bailey 2008: no. 3264: an otherwise normally proportioned youth with two phalli, on one of which he balances an amphora).

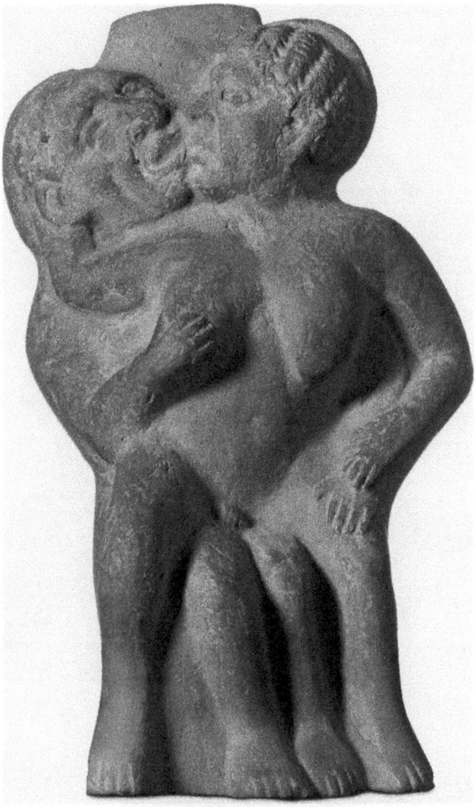

FIGURE 2.21 Terracotta *symplegma* from Egypt, Ptolemaic or Roman. Height 13.0 cm (see Ewigleben and von Grumbkow 1991: no. 99). Museum für Kunst und Gewerbe, Hamburg, inv. no. 1989.1266.

Source: Photo by Joachim Hiltmann; published by permission of the Museum für Kunst und Gewerbe, Hamburg.

in settlements, where evidence suggests they were sometimes associated with domestic shrines, but their findspots also include non-domestic ritual contexts, including a Ptolemaic shrine decorated with monumental Bes figures (figure 2.17); foundation deposits underneath private dwellings; and burials.[101] Additionally, a number of terracotta and limestone statuettes represent figures with exaggerated

---

101. *Settlements:* Allen 1985: nos. 78, 80, 104 (Karanis), no. 156 (Soknopaiou Nesos); Nenna and Seif el-Din 1994: 318 (Apollinopolis Magna and Tanis); Nachtergael 1985: 229, 230, 234 (Coptos, Edfu, Heracleopolis Magna); 1995: 280–281, 283 (Bubastis, Antinoe); Szymańska 2005: nos. 128–141 (Athribis); Thomas 2015a: 61, 77 (Naukratis). *Domestic shrines:* Barrett 2015d. *Sanctuary or other ritual contexts:* Quibell 1907: 12–14, 29 (the Saqqara "Bes chambers"); Derchain 1981 (a site of popular cults in the sacred animal necropolis of Saqqara), with pls.

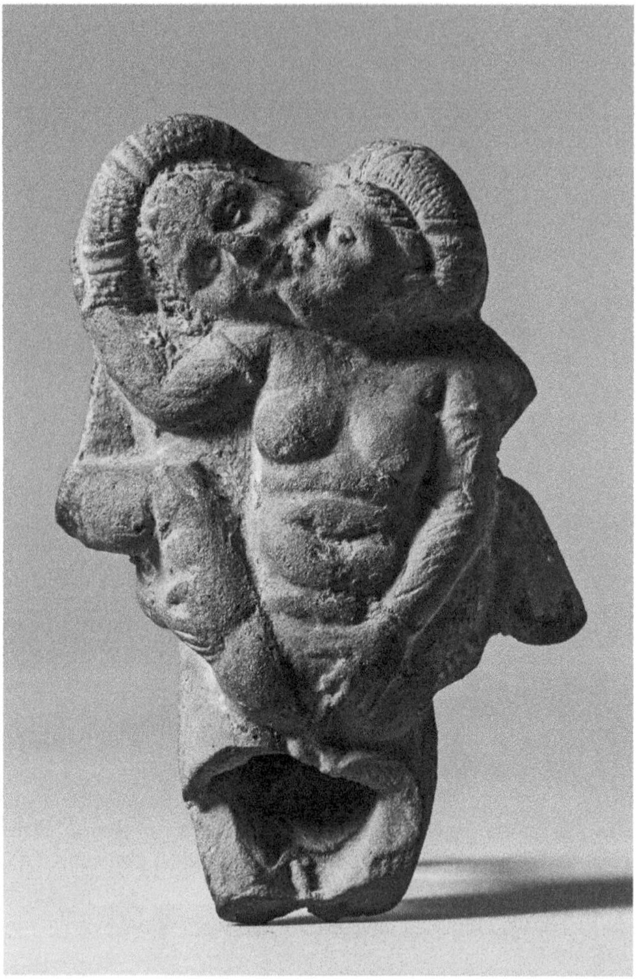

FIGURE 2.22 Terracotta *symplegma*, Egypt, Ptolemaic. Both figures wear wreaths. The woman's pose, with spread legs and hand pointed to her genitals, recalls a widespread genre of similarly posed female figurines from Greco-Roman Egypt (cf. Barrett 2011: 208–218). Height 11 cm (see Török 1995: no. 212). Szépmüveszeti Múzeum 74.24.A.

Source: photo © Lászlo Mátyus; courtesy of Szépmüveszeti Múzeum, Hungary.

23–28; Allen 1985: no. 105 (the South Temple Complex at Karanis; found in construction fill, however, rather than in situ); Adriani 1940–50 and Nenna and Seif el-Din 1994: 316 (Ptolemaic faience and terracotta figurines from Ras el-Soda—earlier in date than the Roman temple, but seemingly deposited as a ritual cache, and associated with Roman as well as Ptolemaic coins); De Maria 1999 and Dasen 2013: 263 (both describing a dancing dwarf figurine from a sanctuary at Ptolemaic Bakchias); Thomas 2015a: 12–13 (the "Hellenion" at Naukratis). *Domestic foundation deposits:* Boutantin 2014: 155–156; Barrett 2015d: 411–413 (from Ptolemaic Tebtunis). *Burials:* Nenna and Seif el-Din 1994: 317 (ithyphallic faience figurines from the Hadra necropolis).

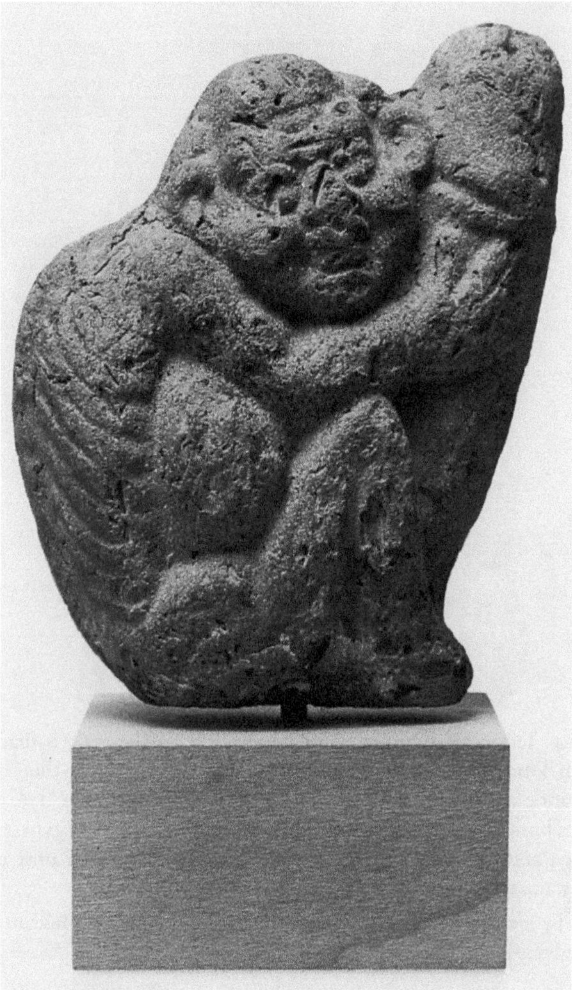

FIGURE 2.23 Terracotta figurine of an ithyphallic, hunchbacked male, from Ptolemaic or Roman Egypt. Height 6.5 cm (see Fjeldhagen 1995: no. 121). Ny Carlsberg Glyptotek AE.I.N. 1771.

Source: photo by Ole Haupt; courtesy of Ny Carlsberg Glyptotek, Copenhagen.

sexual imagery—either ithyphallic males or nude women exposing their genitals—as *cult statues* carried in procession by other figures (figures 2.30–32).[102] Egyptologists generally interpret figurines depicting processions, dancing, music,

102. Barrett 2011: 207, 211. The phallophoric procession group depicted in fig. 2.30 likely dates to the early fourth century BCE (Ross Thomas, personal communication). For parallels to this object, see Thomas and Nicholson 2013: 212; Thomas 2015a: 39.

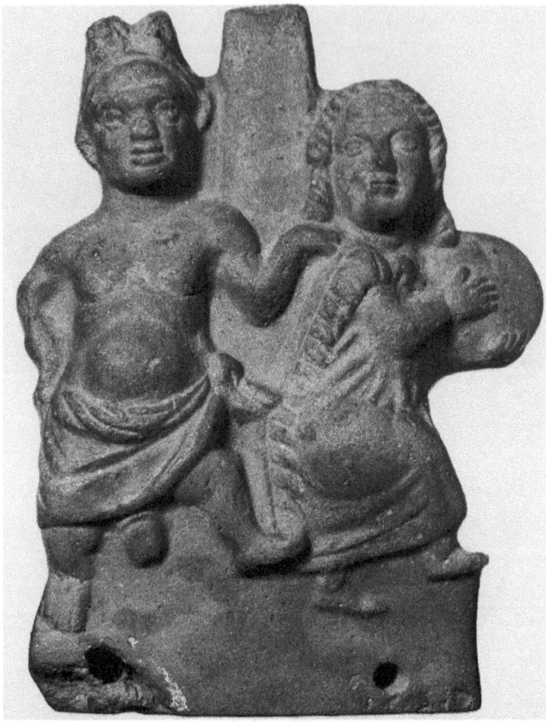

FIGURE 2.24 Terracotta figurine of a dancing couple. The male is macrophallic and crowned with lotus buds; the tambourine-playing woman wears the "Isiac" costume of fringed mantle and corkscrew curls (on which, see Barrett 2011: 136, 143–157). The proportions of both figures suggest dwarfism. Ptolemaic or Roman Egypt. Height 12.6 cm (see Ewigleben and von Grumbkow 1991: no. 30). Museum für Kunst und Gewerbe, Hamburg, inv. no. 1989.482.

Source: Photo by Joachim Hiltmann; published by permission of the Museum für Kunst und Gewerbe, Hamburg.

and/or wreath-wearing as allusions to festivals or religious rituals.[103] The conjunction of sexual imagery, alcohol, and dwarfs may be particularly suggestive of inundation festivals, although some images might also evoke other festival contexts.[104] For example, phallic imagery was also appropriate in Osirian festivals, representing the restored potency of the revived god.[105]

103. Dunand 1990: 178–209; Nachtergael 1991: 22–23; Fjeldhagen 1995: 114; Ballet 2000: 99–100; Bailey 2008: 55–58; Sandri 2012: 638, 640–641; Boutantin 2014: 143–154; Barrett 2011; 2015d: 415; Thomas 2015a: 39.

104. Barrett 2011: 199–247; Thomas 2015a: 61–63, 73–74, 79–80.

105. Derchain 1981: 167–168; Török 1995: 52–54; Meyboom and Versluys 2007: 202–204.

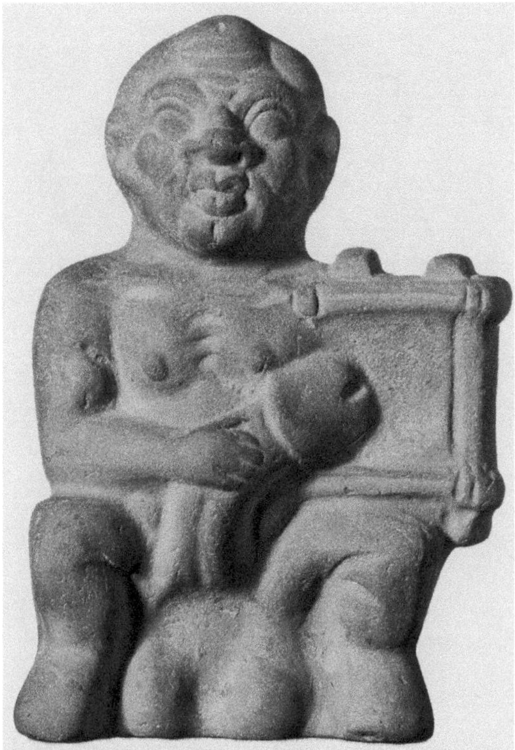

FIGURE 2.25 Dwarf or pygmy, naked and playing a kithara with his own ithyphallus. Terracotta figurine, Ptolemaic or Roman Egypt. Height 14.2 cm (see Ewigleben and von Grumbkow 1991: no. 105). Museum für Kunst und Gewerbe, Hamburg, inv. no. 1989.1437.
Source: Photo by Joachim Hiltmann; published by permission of the Museum für Kunst und Gewerbe, Hamburg.

As is frequently the case for small-scale statuettes from Ptolemaic and Roman Egypt, these images often show a mixture of "Pharaonizing" and "Hellenizing" stylistic features,[106] but much of their iconography also has much older precedents. The iconographic association of Bes and other dwarf-gods with sex, fertility, and drinking goes back at least to the New Kingdom, and representations of dwarfs in group sex scenes are attested as early as the New Kingdom "Turin Erotic

106. Cf. Barrett 2011: 91, 97–109, on the "technological style" of Egyptian terracotta figurines. There are Egyptian as well as Hellenistic precedents for the "grotesque" wrinkled faces and bodies of dwarf figurines. Some Middle Kingdom representations of dwarfs have similar features; see, e.g., Trope et al. 2005: no. 48.

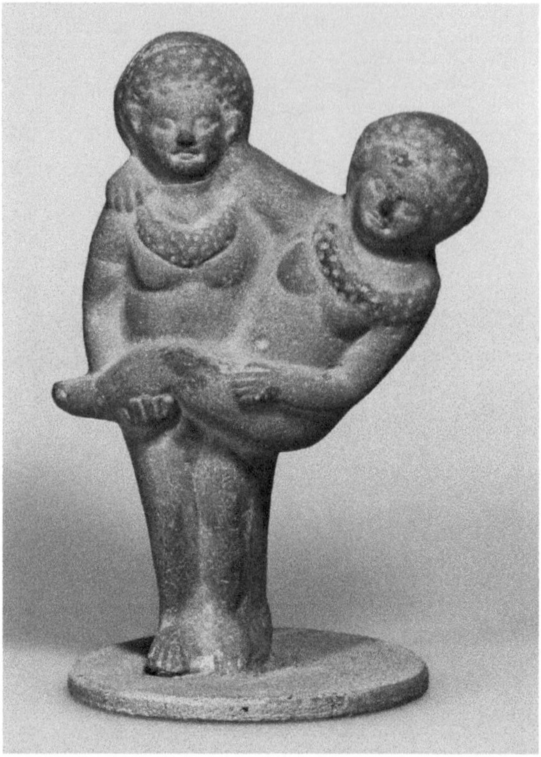

FIGURE 2.26 Terracotta figurine of a pair of figures engaged in dancing and/or sexual activity. Ptolemaic or Roman Egypt. Both figures wear thick wreaths around their heads and necks. Height 9.6 cm (see Griesbach 2013: no. 33). Martin von Wagner Museum A1176.

Source: Photo by P. Neckermann, reprinted by permission of the Antiquities Collection of the Martin von Wagner Museum. © Martin von Wagner Museum der Universität Würzburg.

Papyrus"[107] (figure 2.37). Figurines of nude females, ithyphallic males (sometimes playing instruments, participating in religious processions, or balancing amphorae or other vessels on their phalli), disembodied phalli, and erotic groups also have a very long history in Egypt.[108] In Pharaonic as well as Ptolemaic and

107. *Dwarf-gods and sex, fertility, and drinking:* Derchain 1981: 168; Bulté 1991 (Third Intermediate Period); Dasen 1993: 67–75, 79–80; Graves-Brown 2010: 112–113; Volokhine 2010: 237–238, 251; Budin 2011: 135–138. *Turin Erotic Papyrus:* Omlin 1973: pls. XVI, XVII, XIX; Dasen 1993: 144–145. *Sketches on Ramesside ostraca with similarly explicit sex scenes, but more physically idealized protagonists:* Manniche 1977: figs. 2, 3.

108. Pre-Ptolemaic (Late Period) examples include Thomas 2015a: 17, 22–23, 25–26, 39, 45, 48–49, 57–58.

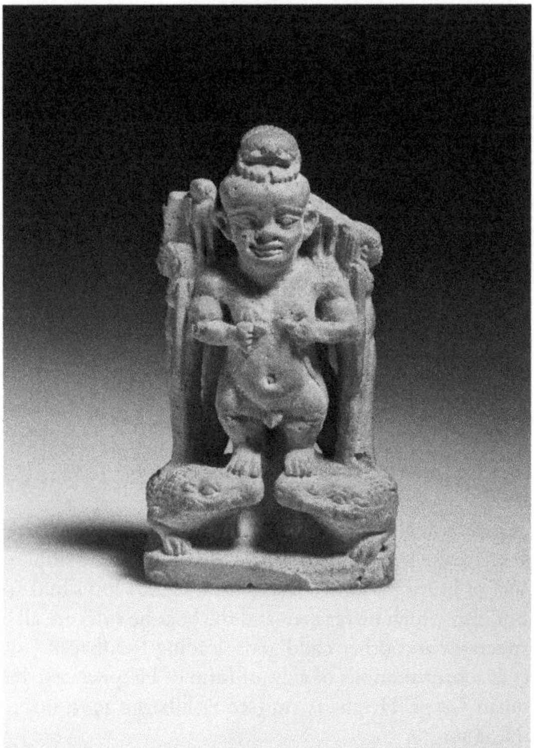

FIGURE 2.27 Faience amulet of dwarf deity (so-called Pataikos) as a solar god and vanquisher of evil. The scarab on his head identifies him as a form of the morning sun. The goddesses Isis and Nephthys stand on either side of him like the hills of the horizon, while two falcons perch on his shoulders. He stands victorious atop two crocodiles while strangling two snakes in his hands. On the back of the amulet is a *ba*-bird, representing one of the several forms of the human soul. Late Period or Ptolemaic Egypt (664–30 BCE). Dimensions: 7.5 × 4.3 × 2.5 cm (Brooklyn Museum). Brooklyn Museum, Charles Edwin Wilbour Fund 37.949E.

Source: Image courtesy of Brooklyn Museum.

Roman-period contexts, such figurines appear in houses (including ritual deposits and household shrines), sanctuaries, and burials.[109] Early excavators often avoided publishing such "obscene" materials, making them underrepresented in the scholarly literature, but more recent research has rectified this omission.[110]

109. *Greco-Roman-period examples:* see note 101, this chapter. *Pre-Ptolemaic precedents:* Barrett 2011: 164, 207, 308–309; see also numerous examples in Thomas 2015a. On iconographic continuity between Pharaonic and Greco-Roman-period ithyphallic and erotic figurines, cf. Thomas 2015d: 10.

110. As discussed by Derchain 1981: 166; Thomas 2015a: 6–7.

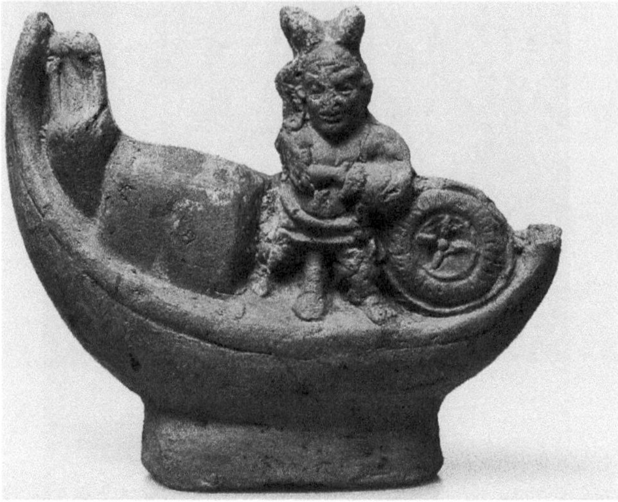

FIGURE 2.28 A macrophallic dwarf or pygmy rides a boat whose shape suggests a river-going craft made of reeds. The contents of the boat include an amphora (at left) and a round, molded loaf of bread (at right). The central figure's lotus-bud crown, sidelock of hair, the round pot into which he reaches, and the boat he rides are all best paralleled in figurines of Harpocrates and other child-gods, leading Fjeldhagen (1995: 40) to interpret these images as representations of a dwarf form of Harpocrates. Terracotta figurine, Ptolemaic or Roman Egypt. Height 13 cm (see Fjeldhagen 1995: no. 16). Ny Carlsberg Glyptotek AE.I.N. 1686.

Source: Photo by Ole Haupt; courtesy of Ny Carlsberg Glyptotek, Copenhagen.

The precise identity of these dwarflike beings is not always clear. In some cases, characteristics such as leonine features, protruding tongues, and tails identify them as Bes or related dwarf-deities[111] (e.g., figures 2.30, 2.36). Other figures lack some or all of Bes's identifying features and look more like bald-headed humans with achondroplasia. Figures on a number of Late Period limestone plaques have certain Bes-like features (e.g., the large feather crown and the characteristic squatting pose with hands on thighs) but replace Bes's leonine head with a bald human head.[112] Another large corpus of amulets and figurines from the New Kingdom

111. E.g., Perdrizet 1921: I.41–51; Dunand 1990: nos. 30–65; Ewigleben and von Grumbkow 1991: 49–52; Török 1995: nos. 10–28; Bailey 2008: 17–19. On dwarf-gods similar to Bes, see notes 80 and 85, this chapter.

112. E.g., Thomas 2015a: 24, fig. 30. Thomas sees these figures as possible child-gods, but their proportions seem more characteristic of dwarfism.

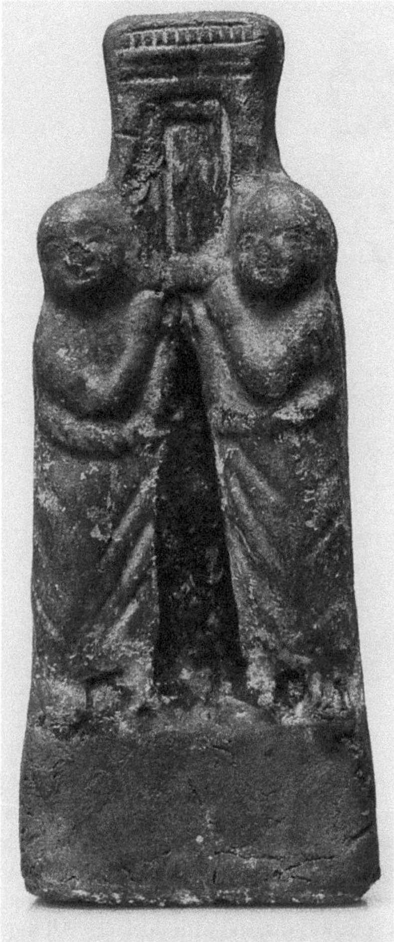

FIGURE 2.29 Two priests with shaved heads carry a shrine in procession. Their proportions may perhaps suggest dwarfism. The standing figure of a Harpocrates-like child god, with his finger at his mouth, is faintly visible within the shrine. Terracotta figurine, Ptolemaic or Roman Egypt. Height 15.5 cm (see Fjeldhagen 1995: no. 97). Ny Carlsberg Glyptotek, AE.I.N. 489.

Source: Photo by Ole Haupt; courtesy of Ny Carlsberg Glyptotek, Copenhagen.

through the Roman period, attested in numerous media but particularly common in faience,[113] represents achondroplastic dwarfs who often have various attributes of divinity. Among other things, they may carry crowned falcons on their shoulders,

113. On the materials of these amulets, see Dasen 1993: 85.

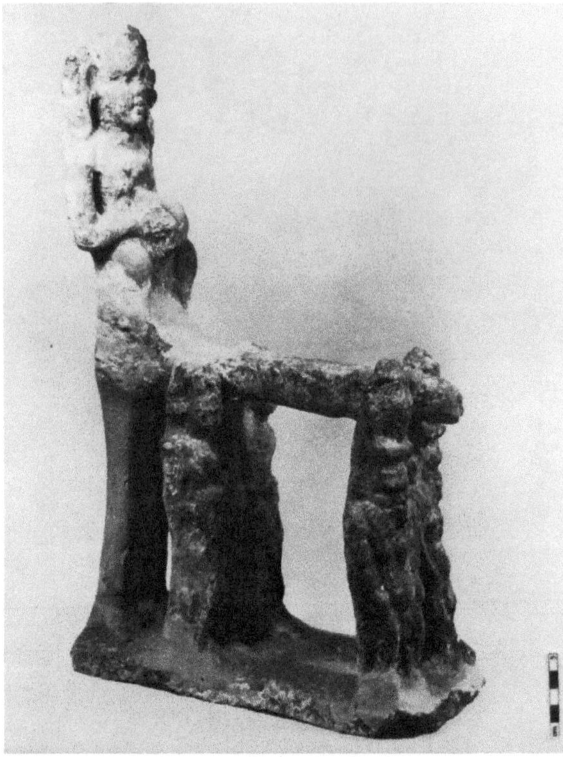

FIGURE 2.30 Terracotta votive figurine from a cult site at Saqqara, early fourth century BCE, depicting four figures carrying an ithyphallic statue of a naked child-god (often identified as Harpocrates) with a sidelock, a tambourine in his hands, and an indistinct figure, possibly either a monkey or a small female figure (Derchain 1981: 167), on his left shoulder. The two figures at the front take the form of Bes-gods, while the two at the back are naked humans (possibly dwarfs) with sidelocks. All four may perhaps represent costumed priests. On the date, see note 102, this chapter.

Source: After Derchain 1981: pl. 23. Courtesy of the Egypt Exploration Society.

stand atop crocodiles, or wear a skullcap similar to that of the god Ptah (see, e.g., figure 2.27 for an example with falcons, crocodiles, and a scarab crown).[114]

114. Cf. also an iconographically similar faience *cippus* at the Metropolitan Museum, New York (inv. no. 30.8.87). On "Pataikos" images (attested in Egypt from the New Kingdom on) and their iconographic connections with Ptah and other creator gods, see Meeks 1971: 55–56; Dasen 1993: 40–41, 84–98; 2000: 91–93; Berlandini 1995; Bulté 1998; Amenta 2002; Győry 2003; Minas-Nerpel 2013; van Erp 2014. See further chapter 6, this volume, "Dwarf Deities, Pygmies, and Sileni." For the survival of images of dwarf-gods astride crocodiles into the Roman period, see, e.g., Amenta 2011.

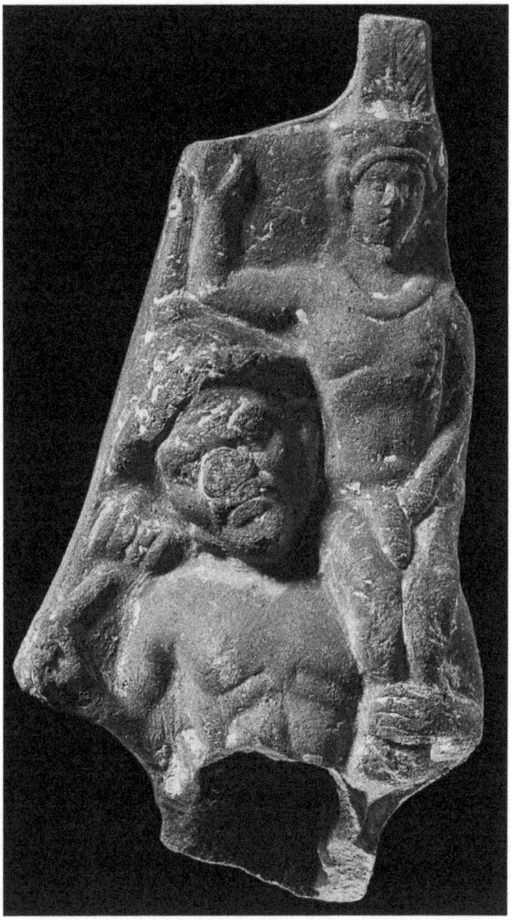

FIGURE 2.31 A cult attendant or priest with "grotesque" features and pygmy-like proportions carries a statue of Amun-Min (recognizable by the plumed crown, upraised right arm, and masturbating left arm). The cult attendant wears a conical hat and carries an object identified by Bailey as an "incense-diffuser or a bullroarer" (2008: 60) in his right hand. Terracotta figurine, Ptolemaic, acquired in the Fayum. Height 11.1 cm. For height and stylistic dating, see Bailey (2008: 60). British Museum 3153 EA.

Source: © Trustees of the British Museum; courtesy of the British Museum.

At least some of these human-headed dwarfs are probably to be understood as forms of Ptah. The skullcap resembles his headgear, and in at least one case, a hieroglyphic inscription on a statuette of this type actually labels the figure as Ptah.[115] Herodotus (3.37) also claims that a cult statue at Ptah's chief temple in Memphis

115. Berlandini 1995, esp. pp. 24, 35, pl. IV (*pace* the downplaying of possible connections to Ptah in Dasen 1993: 85; Sandri 2012: 640).

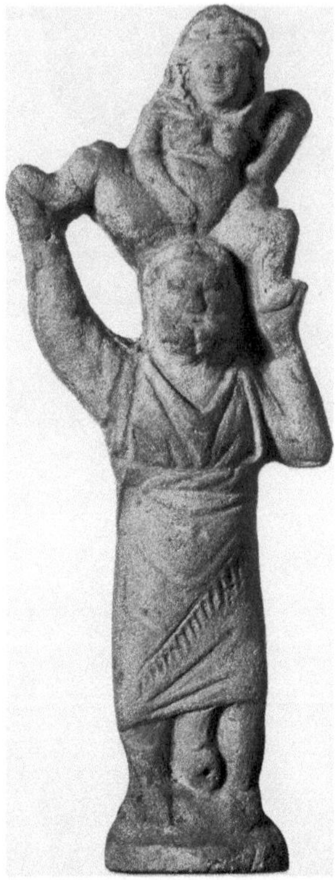

FIGURE 2.32 A macrophallic (but not dwarf-proportioned) cult attendant or priest, wearing a fringed mantle, carries a statue of a naked, wreathed female spreading her legs and pointing to her genitals. Although images like this female figure are sometimes called "pseudo-Baubo" in scholarly literature, they do not appear to be connected to the Eleusinian Baubo (Barrett 2011: 208–218). Terracotta figurine, Ptolemaic or Roman Egypt. Height 13.9 cm (see Ewigleben and von Grumbkow 1991: no. 110). Museum für Kunst und Gewerbe, Hamburg, inv. no. 1989.1490.

Source: Photo by Joachim Hiltmann; published by permission of the Museum für Kunst und Gewerbe, Hamburg.

took the form of a pygmy, "πυγμαίου ἀνδρὸς μίμησις." Scholars typically call these statuettes of achondroplastic dwarfs "Pataikoi," after the term Herodotus uses for some Phoenician dwarf-images that he thinks Ptah's statue resembles.[116] For Egyptian figurines, this term is purely conventional and should not be assumed to correspond to emic use; Egyptians may have simply referred to the figures as, for example, "Ptah," or some other divine name.

116. Hdt. 3.37; see Barrett 2011: 279–280n1062. "Pataikos" may also be attested as the personal name of an individual on Hellenistic Delos (Barrett 2011: 280).

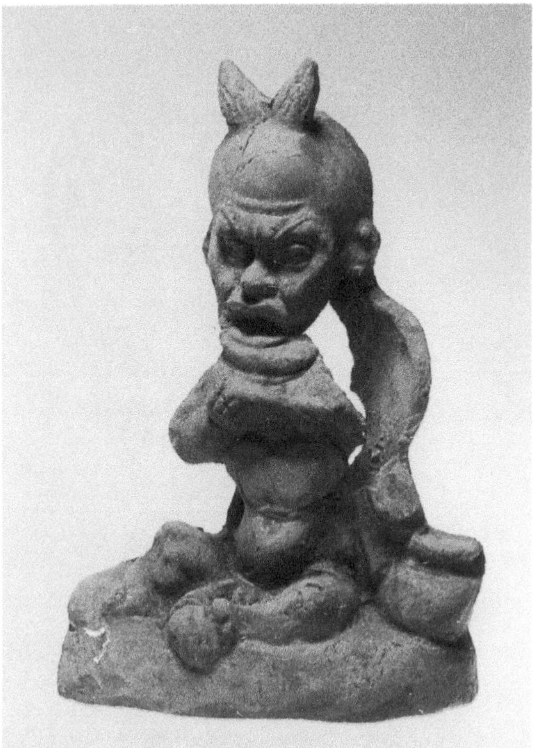

FIGURE 2.33 A pygmy or dwarf with "grotesque" facial features and a macrophallus drinks from a round pot, with a similar pot resting next to him. His bald head is crowned with two lotus buds. Terracotta figurine excavated at Athribis from a late third or early second century BCE stratum underneath the baths of Ptolemy VI (see Szymańska 2005: 207; object previously published as Szymańska 2005: pl. 16, no. 136).

Source: Photograph by Waldemar Jerke, Polish Center of Mediterranean Archaeology, University of Warsaw (reproduced by permission of Karol Myśliwiec).

However, although some figurines of "Pataikos" type do seem to have connections with Memphite cults in general and Ptah in particular, this is not always the case. Some bear inscriptions identifying them with deities other than Ptah, including the Theban god Amun.[117] Ritual papyri further testify to other solarized dwarf deities, not always specifically named.[118]

117. For a figurine labeled with names of Amun, see Van Erp 2014: 82. On one Roman-period amulet, a hieroglyphic trigram identifies a Pataikos-like figure with the morning, noon, and evening forms of the sun-god (Amenta 2011: 4).

118. For other solar dwarf deities, see Dasen 1993: 85; Barrett 2011: 283, with n1079.

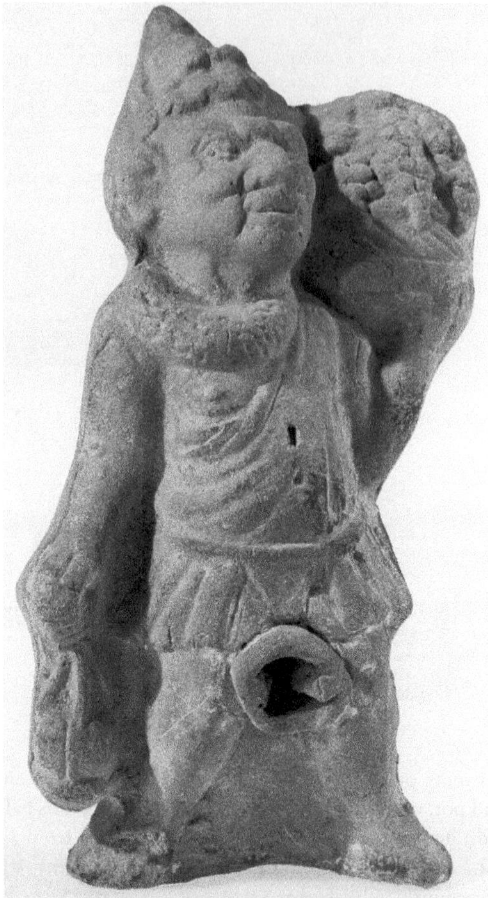

FIGURE 2.34 Dwarf or pygmy carrying a basket of grapes. A separately molded ithyphallus is now missing. The figure wears a short tunic, a conical cap, and a floral wreath around his neck. In his right hand he carries a smaller hanging basket which appears to contain a cup. Terracotta figurine, Ptolemaic or Roman Egypt. Height 17.3 cm (see van Wijngaarden 1958: pl. 12, no. 54). National Museum of Antiquities, Leiden, AED 132.
Source: Courtesy of the National Museum of Antiquities, Leiden.

In addition to numerous figurines and plaques of Bes (sometimes accompanied by a female counterpart[119]), other figurines (especially in terracotta) depict bald, achondroplastic dwarfs of a more "Pataikos"-like type. These figures often wear headdresses of lotus buds (e.g., figures 2.24, 2.28, 2.33, 2.35)—which, when

119. E.g., Perdrizet 1921: nos. 142–144, 152; Ewigleben and von Grumbkow 1991: nos. 53–55; Fjeldhagen 1995: nos. 62–63; Török 1995: 26–27; Bailey 2008: 19.

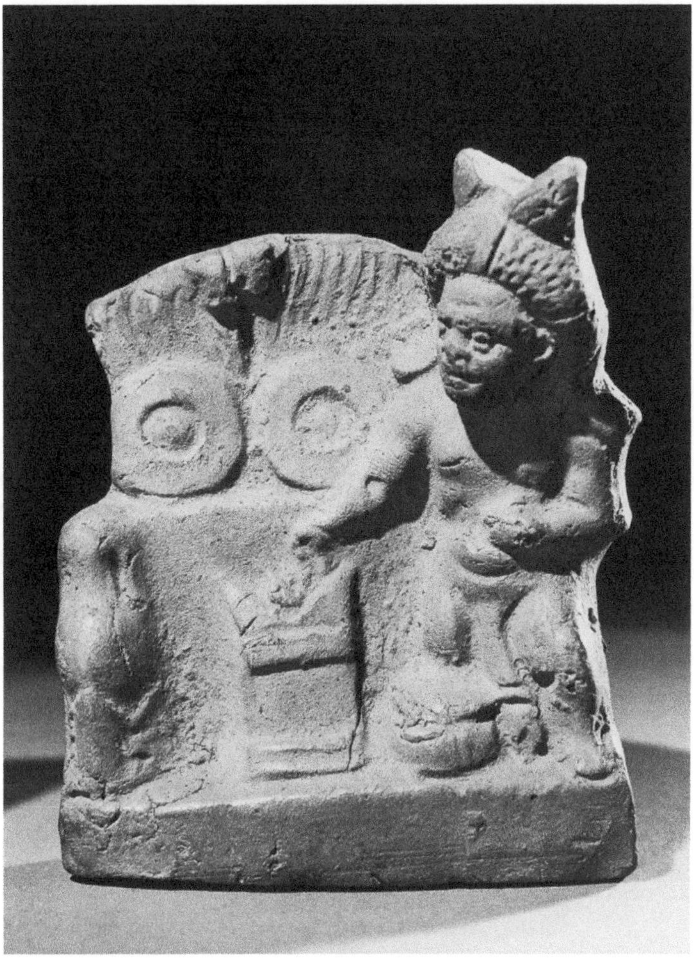

FIGURE 2.35 Terracotta figurine of a "grotesque" dwarf or pygmy making an offering at an altar. The pygmy is naked and macrophallic, his head adorned with a thick floral wreath and two lotus buds. Behind him are two *phialai* or, perhaps more likely, offering breads on a bench. His right foot stands atop what may be a small pot or jar. At the viewer's left, an amphora in a stand occupies the edge of the scene. Ptolemaic or Roman Egypt. Height 10.2 cm (see Philipp 1972: cat. no. 24). Ägyptisches Museum und Papyrussammlung, Berlin, inv. no. 13731.

Source: Photo by Jürgen Liepe; courtesy of Ägyptisches Museum Berlin.

present, characterize them either as deities or as cult attendants—and/or large wreaths characteristic of festival participants (e.g., figures 2.22, 2.26, 2.34, 2.35).[120]

120. The lotus-bud headdresses evoke Harpocrates (and other child-gods) and may recall the earlier lotus crown of the Memphite god Nefertem, as well as the birth of the solar

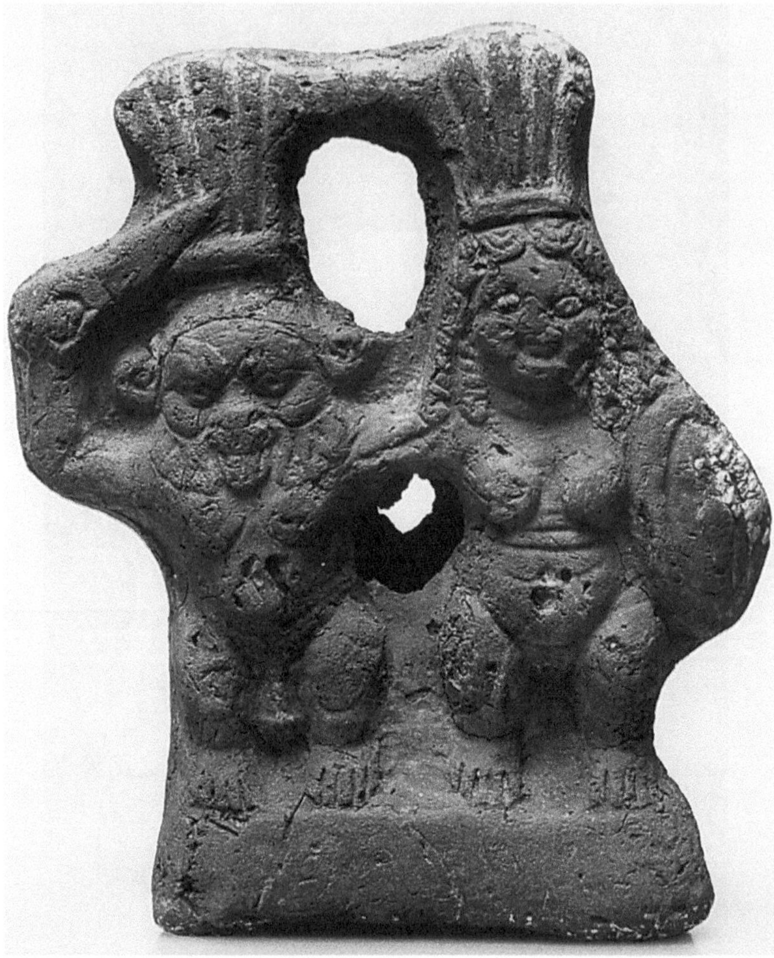

FIGURE 2.36 Terracotta figurines of Bes and his female companion, often identified as "Beset." Both are naked apart from their plumed crowns; Bes is macrophallic. Bes raises a sword above his head, while Beset carries a shield. Both figures' proportions suggest dwarfism. Egypt, probably Roman period. Height 11.5 cm (see Fjeldhagen 1995: no. 62). Ny Carlsberg Glyptotek AE.I.N. 1688.

Source: Photo by Ole Haupt; courtesy of Ny Carlsberg Glyptotek, Copenhagen.

---

deity from a lotus flower (Barrett 2011: 202, 260–261). They also resemble the iconography of Hapy, personification of the inundation, who wears a crown of river plants (Thomas 2015a: 61). On wreath-wearing figurines from Egypt as depictions of festival celebrants, see Barrett 2011: 305–307. Thomas (2015d: 7, fig. 14) notes the difficulty of distinguishing priests from "lay celebrants."

(a)

(b)

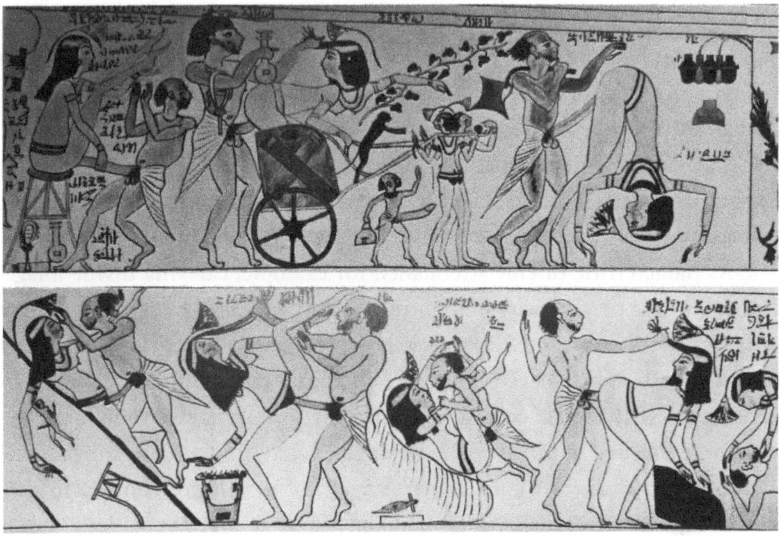

FIGURE 2.37 Two excerpts from the Ramesside "Turin Erotic Papyrus," depicting a series of sexual couplings between women and men with exaggerated ithyphalli. In the top vignette (a), an ithyphallic dwarf stands beneath a chariot on which a couple is copulating; in the bottom vignette (b), a series of sexual pairings includes one coupling between a man with dwarfism and a woman of average height.

Source: After the drawing of Omlin 1973: pl. 17.

Alternatively, they occasionally wear a conical cap similar to that sometimes worn by pygmies in Nilotic scenes (figures 2.31, 2.34; cf. the conical cap on one of the pygmies at the Casa del Medico, figure 2.5, J).[121] Their attributes sometimes assimilate them to deities. For example, the lotus buds, sidelock, round pot, and boat of the macrophallic dwarf or pygmy in figure 2.28 are best paralleled in figurines of Harpocrates and related child-gods.[122] Another figurine depicts a naked, macrophallic dwarf or pygmy standing next to Harpocrates (figure 2.38).[123] The image clearly constructs a relationship between the two figures, but the nature of that relationship is unclear: Are they both deities, or is the smaller

121. On conical caps on pygmies and dwarfs in Roman art, see Garmaise 1996: 78–80; Boissel 2007: 206–208.

122. Cf. Fjeldhagen 1995: 40. On the sidelocks on Egyptian figurines of dwarfs, "grotesques," and/or ithyphallic figures, cf. figure 2.20, this volume; Philipp 1972: figs. 5, 9; Thomas 2015a, 37–40, 44–51. Characteristic of childhood, the sidelock was worn by children, child-gods, and certain priests (Barrett 2011: 247; Teeter 2011: 24–25; and see also note 127, this chapter).

123. Ewigleben and von Grumbkow 1991: no. 20. While using the name "Harpocrates" for convenience, I acknowledge that such figurines might actually depict a range of related child-gods (Sandri 2004; 2012: 634; Thomas 2015a: 15–16).

figure a cult attendant of Harpocrates, perhaps even carrying his statue? The association with the child-god Harpocrates may recall dwarf-gods' longstanding role in Egypt as protectors of child-gods, as well as the Egyptian conceit that dwarfs, as small adults, were in a sense simultaneously adults and juveniles.[124]

While some of these dwarflike figures may be deities, many probably represent human dwarfs serving as cult attendants or ritual dancers—both occupations well-attested for people with dwarfism in Egypt.[125] The figures often perform activities characteristic of cult attendants: for example, carrying divine images or shrines (a task particularly, though not exclusively, associated with the lower-ranking priests known as *wab*-priests), or bringing offerings to altars.[126] They also frequently sport shaved heads and costumes characteristic of certain types of priests.[127] The female companions of these pygmy figures sometimes wear a costume associated with Isis and her cult attendants, including a fringed mantle, Isis knot, and corkscrew curls (figure 2.24).[128] The fringed mantle also appears on some statuettes of male cult attendants (figure 2.32), ithyphallic "grotesques" (including at least one from a popular cult site), hunchbacked dwarfs, and female musicians.[129] Some of the figures' other characteristic activities—dancing, carrying amphorae or other vessels that probably contain wine, and exposing their genitals—may also evoke a festival context.[130] In general, rather than imposing a single totalizing reading on all of these figurines, it is preferable to view them as potentially representing a range of dwarflike beings.[131] Such iconographic ambiguity may be intentional, enabling consumers to identify the statuettes differently according to regional origin, ritual needs, or personal choice.

---

124. On Harpocrates's occasional identification with dwarf-gods, and the treatment of dwarfs as simultaneously adults and children, see Amenta 2011: 2–3; Barrett 2011: 260–261; 2015a.

125. See note 42, this chapter.

126. *On carrying offerings to altars*, see note 99, this chapter. *On* wab-*priests and the carrying of divine images and shrines*, see Kruchten 1989: 251–252. Higher-ranking clergy could sometimes also carry statues in processions: Degardin 2010: 230–231.

127. Derchain 1981: 169–170; Török 1995: 53 (identifying sidelocked adult males as *sem*-priests).

128. On this costume, see Barrett 2011: 136, 143–157.

129. *Ithyphallic "grotesque" from a site of popular cult in the sacred animal necropolis of Saqqara:* Derchain 1981: pl. 25, no. 1210. *Hunchbacked dwarf:* Philipp 1972: fig. 4. *Female musicians:* e.g., Ewigleben and von Grumbkow 1991: fig. 42 (figure at left), also wearing an Isis-knot in her drapery. *Dwarf musicians:* Garmaise 1996: 73–74, 140–141; Dasen 2013: 262.

130. See discussion earlier in this chapter.

131. Fischer 1998: 348–350; Sandri 2012: 640–641.

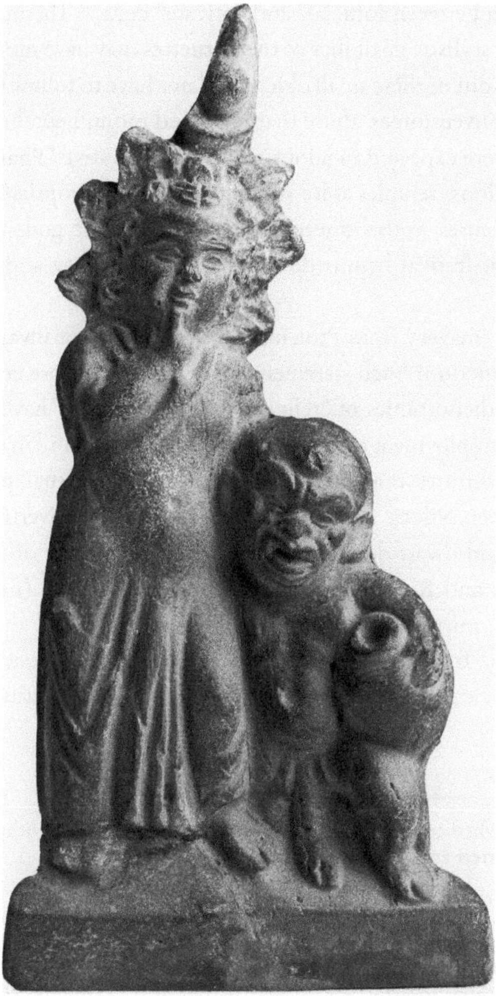

FIGURE 2.38 Terracotta figurine depicting Harpocrates (or a related solar child-god) together with a naked, macrophallic dwarf or pygmy. The smaller figure wraps his right arm around Harpocrates and carries a round vessel in his left hand. Harpocrates wears a double crown, a wreath of ivy leaves, and a long diaphanous garment. He holds his right finger to his mouth and embraces the pygmy with his left hand. Ptolemaic or Roman Egypt. Height 13.8 cm (see Ewigleben and von Grumbkow 1991: no. 20). Museum für Kunst und Gewerbe, Hamburg, inv. no. 1989.629.

Source: Photo by Maria Thrun; published by permission of the Museum für Kunst und Gewerbe, Hamburg.

Because most of these figurine types appear in both temple and domestic contexts, the differences between this imagery and that found on temple and tomb walls do not seem to reflect a division between "religious" and "decorative" imagery, or even between "official" and "private" cult.[132] The more explicit content and greater stylistic flexibility of the statuettes may have more to do with the difference in medium; these small objects did not have to follow the same rules of decorum and convention as those that governed monumental reliefs. Although temple reliefs were expected to adopt a more conservative "Pharaonizing" set of artistic conventions, temples were nonetheless still appropriate sites for reliefs of ithyphallic deities, votive statuettes depicting phallic gods and dwarfs, and the recitation of festival hymns describing sexual activity and drunkenness at *tḫy*-festivals.[133]

Was Nilotic imagery from Ptolemaic or Roman Egypt invariably "religious" in intent and function? Such statements oversimplify a more complex reality in which modern dichotomies of "religious" versus "secular" have little place (see chapter 1). The employment of figurines as temple votives and in domestic shrines certainly demonstrates that they could play roles in ritual practice. Beyond figurines, however, Nilotic imagery—including images of riverine plants and animals, dwarfs, and dwarf-deities—appears on a wide range of consumer goods from Ptolemaic and Roman Egypt. These include mosaics (figure 2.3), lamps, amulets, vessels, mirrors, and textiles.[134] Although many of these artifacts unfortunately come from unprovenienced museum collections, at least some were used in domestic contexts and need not have functioned in ways that we would

---

132. It is unclear whether or not figurine workshops were affiliated with temples in Greco-Roman Egypt (Barrett 2011: 106; Boutantin 2014: 91–94).

133. Barrett 2011: 200; see also discussion earlier in this chapter.

134. *Mosaics:* Daszewski 1985: nos. 28–37, 45; Versluys 2002: nos. 99–101. *Lamps:* Dasen 1994: 597; Versluys 2002: 451–452; Thomas 2015b: figs. 4, 32; Vucetic 2014, esp. 148, 154. *Amulets and gems depicting dwarf-gods:* Barrett 2011: 283n1078, 310–311; Frankfurter 1998b: 125n98. Amulets from a Roman-period tomb at Tuna el-Gebel included (among other subjects) multiple motifs common in Nilotic imagery, including dwarf-gods, ithyphallic dwarfs, phalli, women exposing their genitals, amphorae, and musicians (Kessler and Brose 2008: 40–75). *Glass vessels:* e.g., D. Whitehouse 2001: no. 854; 2003: 1099. *Faience vessels* may depict Nilotic flora and fauna (Friedman 1998: nos. 136, 140, 141, 170; Nenna and Seif el-Din 2000, *passim*) or dwarf-gods (Bianchi 1988b; Friedman 1998: no. 139; Nenna and Seif el-Din 2000: nos. 280, 293, 307, 424; Walker and Higgs 2001: no. 121). Earlier precedents include New Kingdom faience bowls with Hathoric motifs (e.g., Friedman 1998: nos. 76–79, and see the description of no. 170). "Pilgrim flasks" often have vegetal imagery (Seif el-Din 2006, *passim*) and sometimes Bes-related and/or erotic imagery (Seif el-Din 2006: pls. 3, 29–31, 34–40). One moldmade kantharos shows a dancing Bes; see Walker and Higgs 2001: no. 120. A plaster mold for a cup from late Hellenistic Athribis depicts a religious procession with some parallels to Roman Nilotic scenes (Meyboom 1995: 308–309n123). *Mirrors:* e.g., a Bes-shaped mirror handle, Late Period or Ptolemaic (Walters Art Museum 48.1537). Mirrors from many periods feature Hathoric motifs (e.g., Capel and Markoe 1996: 76–77). *Textiles from (later) Roman Egypt:* Versluys 2002: 470–471.

recognize today as "religious rituals."[135] Other items, like the lamps, could function both as practical tools and as ritual paraphernalia, depending on the needs of the moment.[136] The absence of a clear distinction between "religious" and "decorative" imagery also characterizes many other forms of Egyptian domestic material culture. Images of deities commonly appear on objects of domestic furniture, decoration, and personal adornment from the Pharaonic through Greco-Roman periods.[137] Although such objects do not come from exclusively "religious" contexts or serve exclusively "religious" purposes, describing their imagery as purely "decorative" or "secular" would be equally misleading, as they recognizably refer to divinities who formed part of their users' experience and conception of the universe. The deities encountered in domestic settings seem to be chosen, at least in part, for their appropriateness to household concerns. From the New Kingdom on, much imagery in domestic wall painting, furnishings, and personal jewelry relates to deities such as Hathor, Bes, and Taweret, whose spheres all include ensuring fertility, overseeing sexuality, and protecting women and children.[138] Depictions of dwarf-deities and Nilotic flora and fauna evoke a similar set of concepts. As with Roman material culture, binary oppositions of "religious" versus "decorative" imagery fail to map neatly onto Egyptian domestic contexts. While representations of the Nile and its associated festivals would have evoked positive associations with, and been particularly appropriate to, certain times of the year (e.g., the inundation season), the employment of such images was in no way *restricted* to those times of year or to temple settings. Nilotic and related imagery recalled a variety of interrelated concepts and appeared in a broad range of contexts in Egypt, including private houses, as well as temples and tombs.

135. *On figurines from excavated contexts in Ptolemaic and Roman Egypt*, see Boutantin 2014: 97–129; Barrett 2015d: 405n34. *From settlement contexts*, see note 101, this chapter. A Ptolemaic Nilotic mosaic excavated at Abu Qir may have come from a dwelling, though the interpretation of the context has been questioned (Daszewski 1985: 137).

136. Cf. Barrett (2015d: 406–407) on domestic lamp-lighting rituals.

137. Kemp 1979; Derchain 1981: 168; Meskell 1998; Gahlin 2007: 328–333; Koltsida 2007; Ritner 2008: 179–186; Budin 2011: 135–147; Thomas 2015b: figs. 3, 26, 27, 31. A wall painting of Bes figures decorates a bedroom at Amenhotep III's palace at Malqata (Stevenson Smith 1998: 164–166).

138. See note 137, this chapter. However, the spheres of these deities were not exclusively limited to such functions; e.g., Bes could also appear as a creator god (Barrett 2011: n720, n1074) and a deliverer of oracles both in temple contexts and in private magical rituals (*PGM* VII.222–249, VIII.64–110; Dunand 1997; Frankfurter 1998b: 169–174). Other deities with public, state-sponsored, and institutional cults, such as Amun and Ptah, also appear in Pharaonic household rituals (Ritner 2008: 185), further blurring the borders of "official" and "popular" religion.

Encounter, Communication, and Consumption

There are several ways that Roman audiences could have encountered imagery or themes associated with Egyptian representations of the Nile and its yearly flood, as well as Egyptian representations of dwarfs or pygmies. The Nile inundation was a matter of frequent comment; Strabo lists it as one of the features of Egypt that is "most widely known, and manifest to everybody."[139] The existence of cults of the Nile in Greco-Roman Egypt also seems to have been fairly well known.[140] Historical, geographical, and literary accounts of Egypt and its people provide one way that (literate) Romans could have formulated impressions of Egyptian culture. Greek translations of some Egyptian literary and religious texts also existed, including the myth of the wandering goddess's return to Egypt.[141] In addition to Herodotus's and Strabo's references to debaucherous riverine celebrations, several imperial-period authors refer to the Nile flood as the sexual union of Isis and Osiris,[142] suggesting some awareness of the sexual themes associated with Egyptian Nile festivals. Statius describes the flood itself in eroticized terms as "the fertile license (*fecunda licentia*) of marshy Nile"; the word *licentia*, here describing the unrestrained overflow of the water, also inevitably evokes Egypt's reputation for sexual licentiousness.[143]

Individual mobility provides another means of encounter with Egyptian images and practices. Temples and tombs (both Pharaonic and Ptolemaic) were prominent on the standard tourist routes for visitors to Egypt.[144] Trade facilitated further connections, particularly in Campania; Alexandrian merchants and sailors regularly came to Puteoli on the grain ships, which also transported Italians traveling to

139. Strabo 1.2.29; trans. Jones 1960: 131.

140. See Bonneau 1991: 27–31. Worship of the river itself was a "Hellenizing" innovation. Pharaonic-era Egyptians worshiped the inundation, not the river, as Hapy (Bonneau 1964: 219–242; 1991: 25; Baines 1985, esp. 112–116).

141. The extant manuscript of the Greek version of this myth dates to the third century CE (West 1969); it is unclear when it was first translated, or whether it ever circulated outside Egypt. See also note 64, this chapter.

142. Heliod. *Aethiopica* 9.9.2–5; Plut. *De Is. et Os.* 32 = *Mor.* 363d, 366a; Porph. *De cultu simulacrorum* fr. 10.

143. Stat. *Silv.* 3.2.108, trans. Shackleton Bailey 2003: 197. Statius further alludes to Egyptian licentiousness two lines later in the poem, in reference to the famously debaucherous resort of Canopus: a young Roman visiting Egypt will learn "curve Therapnaei lasciviat ora Canopi" (3.2.111).

144. Malaise 1972: 329–330; Smelik and Hemelrijk 1984: 1938–45; Dieleman 2005: 252–253.

Egypt.[145] Some Pompeians may themselves have visited Egypt.[146] Additionally, at least a few Egyptian expatriates appear to have lived on the Bay of Naples.[147] Earlier, in the Republican period, Campanian and Egyptian merchants had interacted at other trading ports: for example, Delos, which also features one of the earliest known Nilotic mosaics.[148] Some painters and mosaicists may also have been mobile.[149] Mobility of artists is better documented than that of supposed pattern books for paintings or mosaics; although scholars often assume that such pattern books circulated, their existence is controversial and lacks direct supporting evidence.[150]

Aside from direct encounters with Egypt or Egyptians, Roman audiences could also have encountered Nilotic or inundation-related imagery and activities through other means. Those with a personal interest in the cults of Egyptian deities may have participated in rituals in Italy that involved sacred "Nile" water. Although one might expect the absence of the actual Nile to pose a problem here, Hellenistic and Roman sanctuaries of Egyptian gods were often equipped with water features assimilated to "Nile" water for ritual use.[151] This practice probably underlies Juvenal's reference to devotees sprinkling water inside a temple of Isis.[152]

145. Stat. *Silv.* 3.2. On the grain ships, see note 55, chapter 1.

146. It is just possible that a graffito from the Pompeii Iseum may commemorate a visit to Egypt; someone scratched *veni hoc* ("I came here") into a Nilotic fresco in the portico (Sampaolo 1992: 40, no. 1.2, pl. VI). However, the graffito lends itself to many other possible interpretations besides recollection of a past voyage: Swetnam-Burland 2009: 451; 2015: 145–146.

147. See note 30, this chapter.

148. See note 62, chapter 1, and note 23, this chapter.

149. Some Alexandrian artists may have traveled to Rome (Malaise 1972: 324–325; Carettoni 1983a: 417; 1983b: 91–92; Ling 1991: 212). Much ink has been spilled over the shadowy figure of one Demetrios, whom Diodorus (31.18.2) and Valerius Maximus (5.1.1) describe as an Alexandrian artist living in Rome in the mid-second century BCE. Since Diodorus calls Demetrios a *topographos*, scholars sometimes connect him with the development of Roman Nilotica (e.g., Meyboom 1995: 100–101, 189; Daszewski 2002: 25). Malaise (1972: 324) similarly interprets Nilotica as evidence of Alexandrian artists in Rome. Unfortunately, we have no real evidence for the elusive Demetrios's actual output. Similarly mysterious is the "Serapion" whom Pliny praises for his ability in painting *scaenas* (*HN* 35.113; see Plantzos 2012: 182–184). Although it is tempting to associate his theophorous name with Egyptian origins (e.g., Meyboom 1995: 86), the name alone cannot provide solid evidence for his origins or ethnicity; see note 205, chapter 6, in this volume.

150. Roller 2006: 48; Squire 2009: 122–139; 2011: 129–148.

151. Wild 1981; Merkelbach 1995: 152; Capriotti Vittozzi 2002a: 26–27; Alvar 2008: 314–317; Bommas 2013b: 60, 66n25–27.

152. Juv. *Sat.* 6.526–529. Juvenal claims that if inspired by a dream, a devotee would travel to Meroe itself to get water for such purposes—i.e., that she would get *real* Nile water, from as close as possible to the river's mysterious sources. Presumably she would otherwise use water from the temple crypt, as appears to have been the general practice (see note 151, this chapter).

It is uncertain whether the Egyptian New Year festival, which took place on the first day of the inundation season, was ever celebrated in Italian temples of Isis or Sarapis.[153] However, the *Isia* or *inventio Osiridis* festival—which celebrated Osiris's resurrection and reunion with Isis, and drew on many features of the ancient Khoiak festivals—was extremely popular outside Egypt and expressed many of the same themes of renewal and regeneration that also characterized inundation festivals.[154]

As discussed earlier, imagery related to that of Nilotic scenes appears on numerous portable consumer goods made in Ptolemaic and Roman Egypt, such as bronze and terracotta figurines, lamps, amulets, glass and faience vessels, and textiles. Many of these were also popular export items.[155] In some cases, craftsmen outside Egypt clearly drew on such objects as models. For example, coroplasts on Hellenistic Delos manufactured a large number of terracotta figurines with close parallels in Egyptian coroplastic production. Many of these Delian products display themes also common in Nilotic scenes, including (among others) dwarf or pygmy figures, sexual imagery, and alcohol consumption.[156] Objects of portable material culture depicting Nilotic motifs came to be consumed and manufactured throughout the Roman world, in multiple media.[157]

Romans could thus have encountered Egyptian versions of Nilotic motifs in numerous different contexts, with widely varying implications. Some people would have firsthand knowledge of Nilotic festivals, either through personal involvement in Egyptian-derived cults or through visits to Egypt. Others would have heard about such events secondhand or through literary sources. Still others would have encountered Nilotic motifs through consumer goods, whose places of production, consumption, and use were potentially numerous. Consumers' reasons for desiring such products might differ widely, as might their

153. Kleibl 2009: 139–140. Alvar is more skeptical (2008: 296n354). On the festival date, see Barrett 2011: 124–125n391.

154. On the *Inventio Osiridis* festival, see Merkelbach 1963: 32–36; 1995: 153–156; Malaise 1972: 221–228; Quack 2003: 61; Alvar 2008: 300–305; Kleibl 2009: 140–141; Manolaraki 2013: 125. On Nilotic and Osirian festivals in Khoiak, see Bonneau 1964: 368–372; Eaton 2006. On *Isia* festivals in Greco-Roman Egypt, see Perpillou-Thomas 1993: 94–100.

155. Exports from Egypt: see note 55, chapter 1, this volume.

156. Barrett 2011. For similar iconography on appliqué relief vessels from Delos, see Bruneau 1970: 244, pl. 42, nos. D37, D38.

157. Dasen 1994: 597, 599–600; Pfisterer-Haas 1994; Rauch 1999; Versluys 2002: 166, 169–170, 449–453, 455–456; Meyboom and Versluys 2007: figs. 12, 13; Barrett 2013. Some Nilotic glass vessels may have been made in Augustan Rome (van Aerde 2015: 173–178, 193–194, 208, 287). Nilotic imagery also appears on engraved gems of unknown provenance: Dasen 1994: 596–597.

interpretations of the imagery and their knowledge (or lack thereof) of its original significance. The process is thus replete with opportunities not only for the intercultural exchange of objects, concepts, and practices but also for the reinterpretation, recombination, and transformation of older motifs and the generation of new meanings. Older and newer associations could have adhered simultaneously to the same objects, accessible to different viewers or even simultaneously accessible to the same viewer.

Additionally, while there are indeed Egyptian *antecedents* for many of the features of Nilotic scenes, the Roman frescoes and mosaics do not appear to emulate any *single* prototype. Egyptian representations of inundation-related imagery cover an extremely wide range of media, from temple reliefs to terracotta figurines; and while numerous Egyptian images depict some (or even many) motifs and themes that also appear in Roman Nilotic scenes, they seldom feature all of them. Additionally, no two Nilotic scenes from Pompeii are identical[158]—suggesting that even if artists were drawing on shared models, they did not duplicate those models precisely.

As a result, Roman painters' and mosaicists' engagement with Egyptian sources parallels some of the ways in which Roman artists also engaged with Greek sources. Rather than exclusively producing exact copies of older Greek works,[159] it was common to excerpt various elements from different sources to produce new, eclectic compositions. A classic example is the *pinacotheca* at the Villa della Farnesina, whose painters purposefully alluded to artistic styles from numerous time periods and geographical regions without precisely duplicating single prototypes. In Bergmann's words, they produced not "transparent copies or reflections, but . . . creative paraphrases and simulations."[160] We might also compare the numerous variations in Roman depictions of Greco-Roman myths: for example, the many domestic wall paintings depicting Narcissus. These images play upon a common theme, but do not appear to derive from a single pictorial prototype.[161] Nilotic scenes similarly do not seem designed to recall, for example, a particular relief from Memphis, painting from Alexandria, or figurine from the Fayum but, rather, to produce a general ambience evocative of the Nile

158. Cappel 1994: 72. Roman painters usually produced "variations upon a theme" rather than identical duplicates (quoting Bergmann 2007: 82; cf. Trimble 2002: 246–247).

159. Roman writings on art refer to both exact copying and creative emulation (Perry 2005: 90–96). Quintilian sees precise copying as a lesser form of painting (*Inst.* 10.11.6–7), but Kousser reminds us that copying was not always devalued (2008: 9, 149–150; cf. Marvin 2008: 199).

160. Bergmann 1995: 106–107; cf. Wyler 2006; Squire 2012: 608.

161. On variations on the Narcissus theme in Roman painting, see Platt 2002: 91–94; Elsner 2007: 132–176; Hodske 2007: 166–171; Lorenz 2008: 187–193; Marvin 2008: 199–200.

and its festivals.[162] Their composition thus recalls the same principles of *decorum* that shaped other aspects of Roman visual culture. When creating images of the flooding Nile, Roman artists compiled motifs whose original models were diverse, but generally potentially *appropriate* to Egyptian depictions of the Nile or the celebrations associated with its inundation.

Antecedents and Affordances in Greek and Roman Visual Culture

In addition to these Egyptian and Greco-Egyptian allusions, Roman Nilotic scenes clearly reference many other cultural and visual antecedents. Stylistically, the handling of the figures and the landscape draws on Hellenistic and Roman precedents.[163] As examples in both statuary and painting demonstrate, Roman artists could emulate "Pharaonic" styles when they wished to do so. Many Roman depictions of Egyptian figures employ stylized, hieratic forms that clearly evoke the conventions, proportions, and poses of Pharaonic art.[164] For Nilotic scenes, however, artists almost entirely avoided such "Pharaonizing."[165]

Instead, painted Nilotic scenes display much stronger stylistic, formal, and even thematic similarities to so-called sacral-idyllic Roman paintings.[166] Images of shrines, divine statues, and worshipers making offerings are among the most common themes in Nilotic scenes. Such representations of religious structures in a rural setting closely resemble those in "sacral-idyllic" scenes, which similarly depict pastoral landscapes dotted with cultic monuments, worshipers, and the occasional passing

---

162. Speculation that the Palestrina mosaic copies a model from Egypt (Dunbabin 1999: 51; Meyboom 1995: 96–107) remains unproven.

163. Malaise 2003: 324; 2005: 211; Mol 2015: 283–287.

164. De Vos 1980: 1–60; Söldner 1999: 112; Swetnam-Burland 2002: 77–78; 2007: 119; Malaise 2005: 209–210; Bragantini 2006: 163–164; 2012: 25; Versluys 2010: 18–20; Pearson 2015. See further chapter 6, this volume, "Style, Function, and Value: 'Pharaonizing' in the Garden."

165. As also noted by Mol 2015: 254; Pearson 2015: 4. One exception is the Villa dei Misteri "black tablinum" (Versluys 2002: no. 67), where miniature Nilotic motifs (lotuses and waterbirds) accompany images of Egyptian deities that follow "Pharaonic" conventions of representation (e.g., figures partially frontal and partially profile). The Nilotic mosaic in the Casa del Bracciale d'Oro (Versluys 2002: no. 48) is not itself Pharaonizing, but it shares a room with Pharaonizing garden paintings (see chapter 6, this volume, "Approaching 'Pharaonizing': An Example from the Casa del Bracciale d'Oro").

166. Previous comparisons of Nilotic and "sacral-idyllic" paintings: Foss 1997: 214; Mol 2015: 284–285; Barrett 2017b: 313. Clarke interprets pygmy landscapes as a parodic "riff on the familiar genre of the familiar sacral-idyllic landscape" (2007b: 102; see also 98, 103).

shepherd or traveler. As in such images, representations of religious architecture and rituals in Nilotic scenes appear to evoke an atmosphere of rural piety. Both "Nilotic" and "sacral-idyllic" landscapes also share additional stylistic and formal features, such as loose brushwork and (at least in Third and Fourth Style paintings) miniature scale.[167] Such observations offer further support for Bergmann's argument that the "traditional categories of the 'sacral-idyllic,' 'villascape,' and 'mythological landscape' are misleading."[168] The blurry lines between "Nilotic" and "sacral-idyllic" landscapes provide another attestation of the limits of such rigid typologies.

Other elements of Roman Nilotic iconography, especially the representation of the pygmy figures with achondroplastic features and sometimes exaggerated phalli, have parallels not only in Egyptian art but also in Greco-Roman traditions of representation that go back ultimately to fifth century BCE vase painting. In contrast to Archaic art depicting pygmies with the same proportions as taller people, fifth-century images of pygmies often have achondroplastic features and large phalli.[169] Harari interprets the pygmies' phalli and facial features as assimilating them to satyrs, and thus incorporating them within the Dionysiac sphere.[170] Beliefs in the unusual sexual endowments of dwarfs[171] may also have contributed to both their macrophallic depictions and their association with satyrs.

Within the specific context of Roman gardens, sexualized imagery has many additional parallels beyond depictions of pygmies per se. The ithyphallic god Priapus protected domestic gardens, and ithyphallic herms, their form and appearance alluding to earlier Greek statuary, often appeared in Roman garden contexts.[172] The popularity of garden imagery recalling Venus or the Bacchic sphere would also have facilitated the adoption of Nilotic imagery in similar

---

167. On brushwork in "sacral-idyllic" paintings, see note 159, chapter 1, this volume. Two earlier (Second Style) Nilotic scenes are larger in scale, in the Casa dello Scultore (VIII 7.24) and the Casa di Maius Castricius (VII [*Ins. Occ.*] 16.17). On their dates, see Maiuri 1956; Bragantini 1997; Sampaolo 1998b; Clarke 2003: 192; Moret 2012: 157n251; *pace* Versluys (2002: 132–134, 140–142), who sees both paintings as Fourth Style.

168. Bergmann 1991; 1992: 43n 7. For further critique of the "sacral-idyllic" category, see Allison 2002: 76–77.

169. See note 8, this chapter.

170. Harari 2004: 169. On the relationship between pygmies, satyrs, and the Dionysiac *thiasos*, see further chapter 6, this volume, "Dwarf Deities, Pygmies, and Sileni."

171. Arist. *Hist. an.* 6.24.577b; Dasen 1993: 218–219; Garmaise 1996: 26–27; Versluys 2002: 276.

172. *Priapus in gardens:* Grimal 1969: 43, 46–49; Stewart 1997. *Herms in gardens:* see chapter 3, this volume, "Interactive Statuary: Social Performance and Service."

contexts, where the pygmies' activities might evoke similar themes of eroticism and *tryphe*.[173]

The pygmies' battles with animals also draw on Archaic and Classical precedents. Rarely but occasionally, pygmies in Roman Nilescapes fight not with crocodiles or hippopotami but with cranes.[174] This motif derives from a Greek legend first attested in Homer.[175] Such "geranomachy" scenes have a long history in Archaic, Classical, and Etruscan art.[176] Besides geranomachies, representations of battles with crocodiles have precedents not only in Egyptian temples and tombs but also in Classical Greek adaptations. A number of Greek plastic vases from the fifth and fourth centuries BCE depict a crocodile attacking a young man with African features.[177] Such images may themselves owe a debt to Egyptian models,[178] but first century CE audiences need not have recognized that debt, instead registering such older Greek traditions (if they were aware of them) as providing yet another source for representations of combats between crocodiles and people.

Numerous models for the depiction of unusual-looking people also come from Hellenistic art.[179] In painting, Pliny the Elder famously traces back this trend to one Antiphilus, who painted a figure of "ridiculous appearance" (*deridiculi habitus*) and supposedly gave rise to a great vogue for such pictures.[180] Images that scholars today label with the (problematically overgeneralizing) term "grotesque" come from all over the Hellenistic world, including Egypt.[181] Indeed, Pliny's Antiphilus was himself *in Aegypto natus* (*HN* 35.114). This extensive *koine* of visual culture blurs older cultural boundaries, breaking down distinctions between "Egyptian" and "Greek" images and identities. Visual *koine* thus forms a material counterpart to the increasing

173. *On Nilotic scenes as evocations of* tryphe, see note 32, chapter 2, this volume. *On Venus and Bacchic imagery in gardens*, see notes 91 and 112, chapter 3, this volume.

174. See note 9, this chapter.

175. Hom. *Il.* 3.2-6; see further notes 7 and 10, this chapter.

176. McKay 1985: 26–55; Dasen 1993: 182–184; Lissarrague 1995: 7; Harari 2004; Williams 2004: 95–103.

177. Lissarrague 1995: 7–8; Leclant 2010 [1976]: 277; Williams 2004: 102–103.

178. Ritner 1993: 125n573.

179. See, e.g., McKay 1985: 56–61; Pfisterer-Haas 1994: 495–500; Mitchell 2013.

180. Plin. *HN* 35.114. Meyboom (1995: 372n20) relates such *grylli* to Nilotic scenes.

181. Claims that "grotesque" art is somehow particularly "Alexandrian" are hence reductive. For reassessments of older debates on a supposed "Alexandrian style," see Stewart 1996; Hardiman 2013.

porousness of ethnic identities during the Hellenistic period. In Ptolemaic Egypt, "Egyptian" and "Hellene" were no longer conceptualized as exclusive categories, and many of the inhabitants of Egypt would in fact have identified as both.[182]

Finally, Roman visual culture also provides many depictions of non-Nilotic dwarfs or pygmies. At Pompeii, for example, paintings from the baths at the Casa del Menandro depict Greek gods and heroes (accompanied by Greek name labels) in pygmy form, and paintings from the atrium of the Casa del Toro depict a pygmy market scene.[183] Scholars often interpret such images as comical, either satirizing social institutions or lampooning the pretensions of more "serious" art.[184] Ling suggests that when pygmies replace more idealized figures in mythological scenes, they function as, effectively, "cartoon characters" whose presence marks the scene as playful and humorous.[185] Pygmies are also frequently interpreted as serving apotropaic functions in Roman art, although Jewell has recently raised questions about this interpretation.[186] While there is good evidence that Romans thought that phallic imagery and possibly also hunchbacks could ward off the evil eye, it is less certain that dwarfism alone had such powers. However, at least in those cases when dwarfs or pygmies appear as macrophallic, ithyphallic, or hunchbacked,[187] people may well have valued their images for apotropaic and other reasons.

182. On the legal and fiscal ramifications of the ethnic status of "Hellene" in Ptolemaic Egypt, and the evidence that this status depended less on ancestry or geographical origin than linguistic ability and/or occupation, see Barrett 2011: 16–22; 2015a, s.v. "Ethnicity and Ethnic Identity."

183. *Casa del Menandro:* Clarke 2007b: 133–146; Jewell 2013, and forthcoming. *Casa del Toro:* Clarke 2007b: 96–98; Jewell 2013: 21–22. I follow Versluys (2002: 463) in seeing no evidence for Nilotic landscape elements in the Casa del Toro. The blue background might conceivably remind viewers of the Nile, as Clarke (2007b: 96) suggests, but blue backgrounds appear in many frescoes and need not necessarily impose Nilotic interpretations.

184. E.g., Garmaise 1996: 146; Clarke 2007b: 87–108, 133–146.

185. Ling 1991: 166.

186. *For the standard view*, see Garmaise 1996: 34–35, 135–139; Clarke 2007b: 87–107; 2007c. *Critique of the apotropaic interpretation*: McKay 1985: 5–6; Jewell 2013, and forthcoming. Beard casts doubt on the supposed Roman belief in apotropaic laughter (2014: 58, 146, 234n25).

187. E.g., in the famous mosaic from second century CE Antioch (Levi 1941; Clarke 2007b: 64–67; Trentin 2015: 54–59; 2017: 236–237). *Apotropaic powers of phalli and hunchbacks:* Slane and Dickie 1993: 486–494; Masséglia 2015: 295–296; Trentin 2015: 51–64; 2017 (though Trentin 2017 stresses the limits of interpretations that focus solely on apotropaic value); Mitchell 2017: 187, 192.

Images of dwarfs or pygmies may also have reminded viewers of the roles of actual people with dwarfism in Roman society. Slaves with dwarfism and other visible congenital conditions were valuable status symbols, treated as curiosities and *deliciae*, and dwarf performers sometimes danced at banquets.[188] Many Roman images of dwarfs, like most Roman individuals with dwarfism, had no direct connection to Egypt or the Nile. Accordingly, we should certainly not categorize all dwarfs as "Nilotic." However, encounters with other Roman images of dwarfism would have formed part of the conceptual and experiential background that Roman viewers brought to their encounters with Nilotic images. Images of non-Nilotic dwarfs could thus have also shaped Roman viewers' receptions of Nilotic pygmy scenes.

Cultural and Visual Eclecticism

Nilotic scenes thus present complex creative syntheses of sources Pharaonic, Archaic, Classical, Hellenistic (including Ptolemaic), and Roman. Furthermore, many of these sources—especially the Ptolemaic material—are themselves already products of extensive negotiation between multiple cultural and visual traditions.[189] Neither wholesale importations of foreign iconography nor completely new exoticizing fantasies *ex nihilo*, Nilotic scenes thus employ the same general principles of emulation, adaptation, and eclecticism that characterize Roman art as a whole. Yet such "eclecticism" was not just a matter of throwing randomly chosen motifs at a wall and seeing what sticks. General principles of thematic, formal, and stylistic appropriateness appear to have shaped the selection and recombination of images. Insofar as Nilotic scenes draw on Pharaonic and Ptolemaic models, they tend to employ imagery appropriate to Nile festivals; and insofar as they draw on other Greco-Roman models, they tend to employ imagery appropriate to pastoral landscapes, faraway lands, and the exotic figures who might inhabit such lands.

Yet such conclusions, on their own, provide only partial evidence for the range of meanings these images may have conveyed to their users. Nilotic and pygmy imagery possessed multiple affordances, derived from both Egyptian and Greco-Roman traditions, for meaning-making. Iconographic analysis alone cannot determine whether viewers would have interpreted a scene of merrymaking Nilotic

---

188. Garland 1995: 46–48; Jewell 2013: 11–15. On dwarf dancers, see chapter 3, this volume, "Interactive Statuary: Social Performance and Service."

189. See, e.g., Fischer 1998, 1999 on both Greek and Egyptian sources for Ptolemaic figurines of dwarfs.

pygmies as a depiction of Egyptian inundation festivals, a reference to dwarf entertainers in contemporary Roman culture, a geographically specific variant of "sacral-idyllic" imagery, or all of the above. Because of the complex cultural biography of Nilotic imagery, the potential for multiple meanings is inherent in such iconography.

Choosing within the range of available cultural associations would thus have depended in part on individual viewers' background, knowledge, and interests. However, the activation of one particular subset of possible meanings would also have depended on context. Roman viewers did not encounter Nilotic scenes in a vacuum but, rather, in architectural spaces within which Nilotica appeared juxtaposed with many other forms of material and visual culture. These larger assemblages would have framed viewers' encounters with Nilotic imagery, potentially activating some of the available associations of these images while downplaying others. To bridge the gap between iconography and meaning, we need to turn to context.

Contextualizing Nilotic Imagery at Pompeii

Goals and Methods: Context Beyond Room Labels

The remainder of this chapter takes a macro-scale approach to the contextual analysis of Nilotic scenes from Pompeii, analyzing distribution patterns across the city and relating them to broader questions about the use of household space and the creation of domestic environments. After these basic patterns are established, the case studies in chapters 3–6 will paint with a finer brush, interrogating household-specific variations on (and deviations from) general trends at Pompeii. By positioning Nilotic imagery within specific household ensembles, these case studies will enable us to move beyond domestic architecture in order to situate Egyptian landscapes within more comprehensive multimedia assemblages.

While some studies have taken valuable initial steps toward contextualizing domestic Nilotica, these are largely confined to general observations about architectural room types. Versluys, Meyboom, Clarke, and Mol have demonstrated that the most common domestic locations of these scenes are gardens and/or dining rooms.[190] In 2002, Versluys attributed 40 percent of Pompeian Nilotic scenes to gardens, 20 percent to dining rooms (triclinia or oeci), and 14.3 percent to peristyles.[191] His earlier co-authored study with Meyboom produced figures

---

190. See note 41, chapter 1, this volume. On the more general popularity of landscape scenes in gardens, see Ling 1991: 148–149.

191. Versluys 2002: 257, diagram 10.

differing by only a few percentage points.[192] Mol's recent dissertation produces comparable results for Pompeian Nilotica but subdivides her room types into a greater number of categories: for example, treating "triclinia" (6%) separately from "summer triclinia" (16%), and treating "gardens" (12%), "viridaria" (12%), and "peristyles" (12%) as separate.[193] Clarke identifies a similar pattern for pygmy scenes (which are often, but not always, set in Egyptian landscapes). He attributes 33 percent of Pompeian pygmy scenes, Nilotic or otherwise, to gardens and/or peristyles.[194]

These findings have generated useful initial data on which it is now possible to build. The assignment of images to room types provides a valuable starting point for contextual analysis, but should not be taken as its sole end goal. For one thing, unreflective application of generic room labels runs the risk of obscuring the potential multifunctionality of those rooms. Analyses of artifact distribution and room use have demonstrated the danger of overreliance on Vitruvius's idealized architectural classifications, demonstrating that the actual uses of rooms were often much more varied.[195] Additionally, the assignment of rooms to exclusive typological categories can conceal functional similarities between some room "types" while glossing over meaningful differences within others. For example, the creation of separate categories for "gardens," "viridaria," and "peristyles" obscures the functional similarities between these spaces, concealing the role of peristyles and partial peristyles as (colonnaded and architecturally "Hellenizing") forms of garden space and reifying the undefined category of "viridaria," a term often used inconsistently in modern scholarship.[196] Similarly, the treatment of "gardens" and "triclinia" as exclusive categories obscures the frequent occurrence of dining in gardens, often at outdoor triclinia made of wood or masonry.[197]

---

192. Versluys and Meyboom 2000: 124, table 3.

193. Mol 2015: 289, fig. 4.34b.

194. Clarke 2007b: 95, fig. 41.

195. See note 42, chapter 1.

196. *Peristyles as garden space:* see note 49, chapter 1, this volume. For a critique of attempts to identify peristyle and non-peristyle gardens with, respectively, decorative and practical/economic functions, see von Stackelberg 2009: 22–23. *Problems with the category of* viridaria: Purcell 1996: 151n29. Allison sees the term *viridarium* as inappropriate for Pompeian houses, as ancient sources usually use it to describe "a fairly extensive garden or even a park" rather than a small domestic planted space (2004a: 169). See also note 47, chapter 1, this volume.

197. See later this chapter, "Parallel Landscapes in the Garden."

Another, subtler problem concerns the functional and visual interdependence of rooms. Rather than existing as wholly self-contained, hermetically sealed environments unto themselves, rooms in Roman houses were frequently functionally connected and/or intervisible. Applying methodologies derived from space syntax theory and access analysis, much recent research has examined access, movement, and visibility within Roman houses at Pompeii and beyond.[198] While such considerations are important throughout houses, there are also particular reasons to stress their significance for the types of space in which Nilotic imagery is most common: gardens (including peristyles) and dining rooms. Many triclinia and gardens were explicitly designed as interdependent, with dining areas opening on to gardens in such a way that their doorways became frames for garden views.[199] Vitruvius emphasizes the desirability of garden views while dining, and Zaccaria Ruggiu's study of the placement of triclinia in Pompeian houses suggests that almost half (46% of her sample) open onto peristyles or other forms of planted space.[200] The treatment of gardens and triclinia as mutually exclusive, either/or locations for Nilotic imagery thus flattens the frequent interrelationships between such spaces, in which open-faced dining rooms functioned at least in part as shaded, reclining-friendly extensions of the garden space they adjoined. For example, previous studies state that the Nilotic mosaic from the Casa del Bracciale d'Oro comes from a triclinium.[201] While this attribution is certainly accurate, it also arguably oversimplifies the context of this mosaic, as the triclinium in question is located in a niche opening off the house's garden (figure 2.39). The treatment of contextual attributions as exclusive, single-room labels (e.g., "garden" versus "triclinium") thus obscures the complex interplay of interdependent contextual associations that shaped the settings of these scenes. By moving beyond the simple labeling of rooms to integrate broader considerations of access, sightlines, and visibility, we can obtain a more nuanced understanding of how Nilotic imagery actually participated within broader domestic visual landscapes and activity areas.

198. The fundamental work on space syntax theory and access analysis (also called gamma analysis) is Hillier and Hanson 1984, with further refinement in Hillier 1996; 2014.

199. Bek 1980: 189–194; 1983: 82–88; Clarke 1991: 17; Fredrick 1995: 278–279; Ellis 1997: 42–44; Dickmann 1999: 146–151; Roller 2006: 53–56; Dwyer 2012: fig. 7; Dessales 2013: 363–367.

200. Forty-one percent open onto peristyles, 5 percent onto "viridaria" (Zaccaria Ruggiu 1995: 144). Vitruvius characterizes Cyzicene oeci as dining rooms planned around garden views so that "de tectis . . . viridia prospiciantur" (Vitr. 6.3.10).

201. E.g., Versluys 2002: 123; Mol 2015: 278.

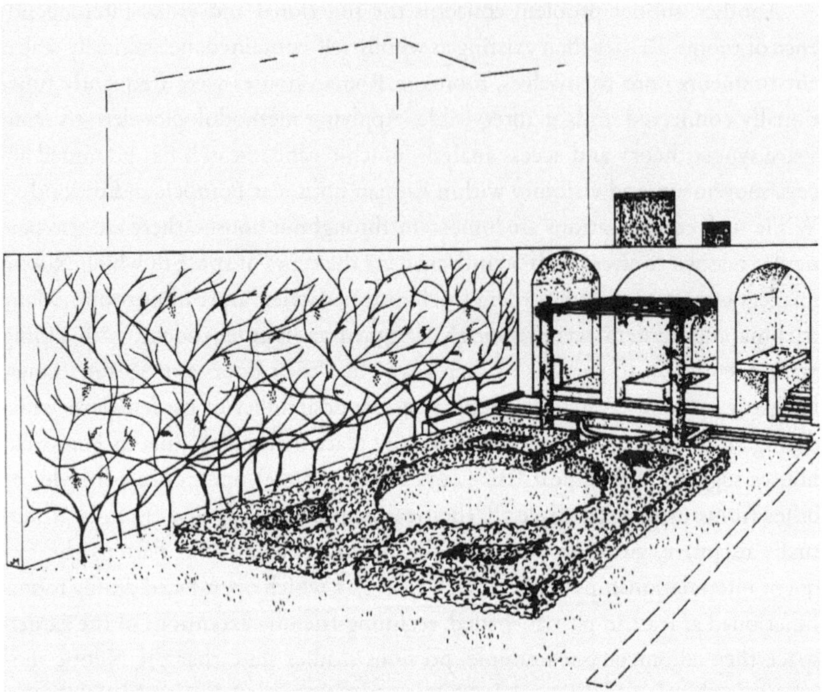

FIGURE 2.39 Reconstruction drawing illustrating the visual and spatial interrelationships between a garden space, and an adjacent water-triclinium located in a niche that opens off the garden. Casa del Bracciale d'Oro (VI 17 [Ins. Occ.] 42).
Source: After Jashemski 1979–93: 2:fig. 201.

Similarly, a recent statement that only three Pompeian Nilotic scenes derive from "water contexts" is limited by the author's restriction of "water contexts" solely to images that actually appear *on* nymphaea or fountains,[202] obscuring the wide variety of alternative ways in which water features and Nilotic imagery frequently participate in a shared visual landscape. In the Casa della Caccia Antica, for example, the Nilotic wall paintings are not directly attached to the large pool in the center of the pseudo-peristyle (figure 2.40). However, a viewer gazing at the paintings could hardly have failed to be aware of that pool—or of the crocodile statue that once stood on its bank, bringing to (potentially ominous) life the pygmy/crocodile battle painted on the wall above![203] In order to understand how

202. Mol 2015: 288–289.

203. The evidence for the presence of this crocodile (now lost) is somewhat complicated. Neither Allison and Sear's publication of the house (2002) nor Appleton's study of animal

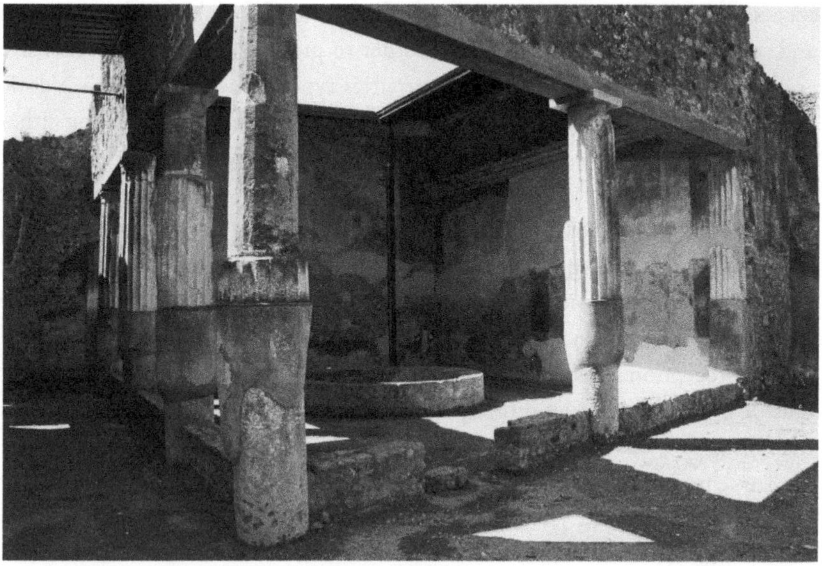

FIGURE 2.40 Pseudo-peristyle of the Casa della Caccia Antica (VII 4.48), facing southwest. A Nilotic scene (now mostly lost), with pygmies fighting crocodiles, originally covered the upper zone of the west wall. The central panels of the wall featured "sacral-idyllic" and mythological landscapes. A large pool forms the center of the courtyard. A crocodile statue of black stone (now lost) originally stood inside the pseudo-peristyle, extending the Nilotic landscape into three dimensions.

Photo: C.E. Barrett.

ancient Pompeians related to their domestic visual environments, we need to go beyond the assignation of room labels to examine the whole spectrum of objects, images, and activities that made up those visual environments.

This chapter therefore seeks to articulate a broader framework for the contextual study of domestic imagery. In addition to identifying the architectural

---

sculpture from Pompeii (1987) refers to a crocodile from this house. However, Versluys (2002: 129) attributes a "statue of a crocodile" to the pseudo-peristyle, citing Whitehouse (1977: 64n49), who attributes a pygmy–crocodile scene and a crocodile statue to the garden of a "small house in reg. VIII, 4." There are no Nilotic scenes today attested within the insula of VIII 4. However, Whitehouse's own source, Bechi (1833: 2–3), describes the house, its location, and its pygmy scenes in a way that can only apply to the Casa della Caccia Antica (which is located in VII 4, as opposed to VIII 4), and Allison (1997: 8) confirms the identification of Bechi's description with the Casa della Caccia Antica. Furthermore, other houses with pygmy–crocodile paintings (appendix B, this volume) were all either excavated too late or have iconography or locations incompatible with Bechi's description. Versluys is therefore certainly correct to attribute the crocodile statue, which Bechi characterizes as "un coccodrillo scolpito in nero di paragone" (1833: 3), to this house. On other crocodile statues from Roman gardens, see Appleton 1987: nos. 19–20; Zarmakoupi 2014: 159.

spaces within which Nilotic imagery is often situated, I also consider the visual and spatial settings of those rooms in order to produce a more comprehensive account of the domestic spaces and activities typically associated with Nilotic imagery. For present purposes, I define such "association" as comprising either *location within the same room* or *visibility from the vantage point of the Nilotic fresco or mosaic*, so that objects or features "associated" with Nilotic scenes are those that form part of the visual landscape framing viewers' encounter with those scenes. Following Fisher, I define visibility as the ability "to see at least 30% of the room's interior" from a vantage point within another room.[204] Since this project focuses specifically on the visual setting that would have informed viewers' reception of Nilotic imagery, the relevant vantage point in this case is that of a person standing directly in front of the Nilotic fresco or mosaic in the room.

This chapter thus situates Nilotic landscapes not only within individual rooms but also within broader visual and functional landscapes. Appendix A presents a full list of the data, with contextual information for all Nilotic scenes from known domestic findspots at Pompeii. On the most basic level, the results confirm and strengthen previous scholars' observation of associations between Nilescapes, gardens, and dining, while also reinforcing the evidence for a frequent association with water features. Following the definition of "association" just proposed, almost all Nilotic frescoes or mosaics from known domestic contexts at Pompeii are associated with *either* gardens, dining, or water features, and approximately half are associated with all three. Beyond simply affirming this pattern, however, the analysis of these images' broader visual, spatial, and behavioral settings also breaks newer ground by helping us understand the reasons for those associations and the ways in which those contexts would have framed viewer responses. Many Nilotic scenes appear in domestic settings whose physical environment and associated human activities construct various parallels between images and "real life." Such contexts frame both visual culture and human behavior as commentaries on each other, but leave the content of that commentary up to the viewer to determine. The case studies in chapters 3–6 suggest some of the variety of forms that viewer responses might take.

Distribution Within the City

The vast majority of two-dimensional Nilotic scenes from known contexts at Pompeii—39 of 47, or 83 percent—come from domestic contexts

204. Fisher 2009: 449.

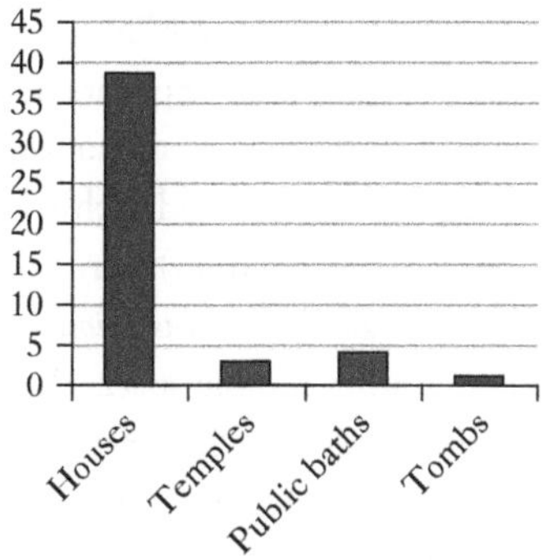

FIGURE 2.41 Graph showing archaeological contexts of Pompeian Nilotic scenes. The *y*-axis shows the number of Nilotic scenes from each building type. Out of 47 Nilotic scenes from known contexts at Pompeii, 39 (83%) come from houses.

(figure 2.41).[205] The houses that produced Nilotic scenes were widely distributed throughout the city.[206] They do not appear bound to any particular socioeconomic level. Such scenes appear in some of the largest and most sumptuous, as well as the smallest and least elaborate, houses at Pompeii.[207]

Of 39 houses and villas containing Nilotic imagery, only 7 can be securely shown to have produced shrines, statuary, or ritual equipment potentially depicting, or indicating the practice of a cult for, Egyptian gods.[208] Accordingly—and in keeping with Versluys and Meyboom's conclusions—archaeological context provides no

205. Note, however, that some of these houses may have been modified for commercial use by 79 CE, through whole or partial conversion into *cauponae, hospitia,* or rental establishments. At least two houses (I 2.24 and VI 2.4) may have been converted into *cauponae* or *hospitia* (Jashemski 1964: 339–340; M. de Vos 1990: 53–54; Sampaolo 1993: 87–88; DeFelice 2007: 476–477; Calabrò 2012: 91. For dissenting opinions on VI 2.4, see Wallace-Hadrill 1994: 223n24; Dwyer 2012: 316–317. On II 2.2 (Casa di Octavius Quartio), see chapter 6, this volume, "Garden Luxury: The Casa di Octavius Quartio". On II 4.2 (Praedia di Julia Felix), see Parslow 1989. On the coexistence of commercial and domestic activity within the Casa delle Quadrighe, see Calabrò 2012: 77.

206. See appendix A, this volume, and cf. the map in Mol 2015: 480.

207. Mol 2015: 288–289, table 4.24.

208. Such evidence is here taken to consist *not* of any and all "Aegyptiaca," but of three-dimensional objects that depict Egyptian deities or have identifiable links to ritual practice associated with such deities: for example, shrines containing images of Egyptian gods; statuettes of Isis or other Egyptian deities known to have received worship at Pompeii; or artifacts, such as sistra, whose sole known function was as ritual implements in Egyptian-derived cults. Houses containing such materials *in addition to* Nilotic imagery include the following: I 2.24 (terracotta bust of Isis; Kekulé 1880: 41); Praedia di Julia Felix (garden shrine with representations of Egyptian deities; Parslow 2013); Casa di Acceptus e Euhodia (see chapter 6, this volume); Casa del Centenario (bronze Isis statuette and sistrum; Tran Tam Tinh 1964: 183; Versluys 2002: 152); Casa delle Amazzoni (garden shrine with painting of Egyptian deities; Boyce 1937: 44, no. 141;

evidence for a one-to-one correlation between Nilotic scenes and Isiac religious affiliation.[209] However, Nilotic decoration also appears not inherently *incompatible* with the worship of Egyptian deities, since some houses do provide evidence for both.[210] Similarly, Nilotic imagery also appeared within the Isis temple at Pompeii. Painted Nilescapes (one populated by pygmies) were present in two rooms of the temple – the portico and the so-called *ekklesiasterion* – and another pygmy decorated one of the pilasters of another temple space, the so-called *purgatorium* or "Nilometer."[211] The distribution of Nilotic imagery thus appears to support neither exclusively Isiac nor exclusively caricatural interpretations, as sites of Isis cult (either temple or domestic) seem unlikely venues for anti-Egyptian sentiment.[212]

Distribution Within Domestic Contexts

Of the 39 Nilotic scenes from domestic contexts at Pompeii, 36 are sufficiently well-documented that we know not only what house they came from, but also what room.[213] A strong association with gardens is instantly apparent (figure 2.42). Almost half of the scenes—18, or 50 percent—come from gardens.

Bragantini 1993: 174–175); Casa di Paquius Proculus (amphora inscribed with "*Serapis dora*" [*RICIS* 504/0219b]; Tran Tam Tinh 1964: 177, no. 151; on such inscriptions, see Tran Tam Tinh 1964: 85; Malaise 1972: 265; de Vos 1983: 66); Casa del Fauno (sistrum, today unaccounted for, but noted in the excavation reports; Fiorelli 1860–64: 2:248). Wall paintings with small, isolated Egyptian motifs are not here taken as proof of participation in Isis cult (contra, e.g., Tran Tam Tinh 1964; see the critique of Petersen 2006: 42, 46, 246n80).

209. Versluys and Meyboom 2000; Versluys 2002; *pace* Schefold 1956; 1960: 90–91, 94–96; 1972: 176.

210. As also noted by Malaise 2005: 212–213.

211. Nilotic scenes from the Isis temple: De Caro 1992: 46, nos. 1.2, 1.24 (portico), 1.62, 1.63, 1.66–1.68 (*ekklesiasterion*); Versluys 2002: no. 61; and see chapter 5, this volume, "Introduction." Some other landscape vignettes from the portico do not appear to depict the Nile itself, but may include other allusions to Egypt: e.g., an obelisk (De Caro 1992: no. 1.27) and horned towers of a type common in, though not exclusive to, Nilotic scenes (De Caro 1992: no. 1.43; cf. Meyboom 1995: 29–31, 249n88). On the pilaster, see Blanc et al. 2000: 274; Kleibl 2009: 279. For other Nilotic scenes in cultic contexts outside Pompeii, see Malaise 2005: 212. On the possible relationship of the Pompeii "*purgatorium*" to Egyptian Nilometers, as well as other facilities for cultic water in Isiac sanctuaries, see most recently Saragoza 2017: 378–381, with n54. The terms "*ekklesiasterion*" and "*purgatorium*" have become attached to structures at the Iseum through modern convention, but need not correspond to ancient names or uses for these spaces.

212. Although some scholars have questioned whether first-century Roman audiences still perceived Isis as "Egyptian," textual and iconographic evidence suggests the continuing importance of her associations with Egypt: see notes 115–116, chapter 1, this volume, and see also chapter 6, this volume, "Isis Cult and Identity: Beyond 'Isiac' vs. 'Non-Isiac.'"

213. Three other houses—the Casa dei Fiori, Casa del Cinghiale I, and Villa di Diomede—also produced Nilotic paintings, but their room of origin is unknown (appendix A, this volume).

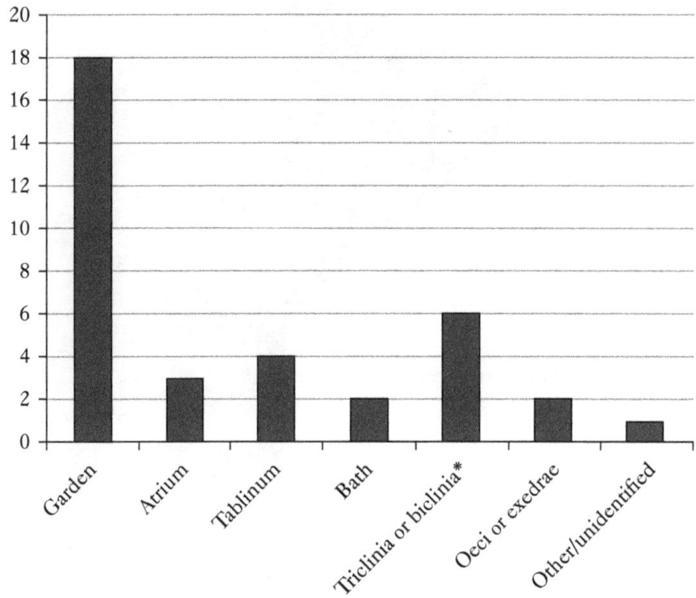

FIGURE 2.42 Graph showing contexts of Nilotic scenes from private houses at Pompeii (out of 36 domestic Nilotic scenes for which the room, as well as house, of origin is known). The *y*-axis shows numbers of Nilotic scenes from each room type.

Note: The figure for "triclinia or biclinia" here excludes freestanding triclinia/biclinia located within gardens.

Seven Nilotic scenes come from peristyle courts or partial peristyles, while 11 derive from other types of garden, usually near the back of the house. Of the other 18, three come from atria, four from tablina, two from private bath complexes, six from triclinia or biclinia, two from open-fronted oeci or exedrae located off gardens and probably used for dining (among other purposes), and one from a small room of uncertain architectural type.[214]

Garden contexts thus clearly predominate over other room types, and the emphasis on gardens becomes even more pronounced if we move beyond a

---

On the garden context of the Nilotic scene from the Casa dello Scultore, which does not appear in the Appendix A table, see the preface to Appendix A.

214. Vitruvius suggests that both oeci and exedrae can potentially be used for dining (6.3.8–10; see also Corlàita Scagliarini 1974–76: 23; Bek 1983: 84–85; Clarke 1991: 13, 175; 2007b: 161; Zaccaria Ruggiu 1995: 137–138, 140–141; Dickmann 1999: 32; Leach 2004: 42). *On dining in an exedra at the Casa del Fauno*, see chapter 5, this volume, "Encountering Alexander: The Nile Mosaic as Threshold." *On possible confusion between oeci and exedrae*, see Leach 1997: 60–62; 2004: 42–45, esp. 45. *On identifying triclinia based on room layout*, see Dunbabin 1991: 123–124; 1996: 67–68, 72–74; Roller 2006: 53–55. *On Latin terms for dining rooms, and ancient distinctions between triclinia and other dining spaces* (e.g., *cenationes*), see Leach 1997: 65–68.

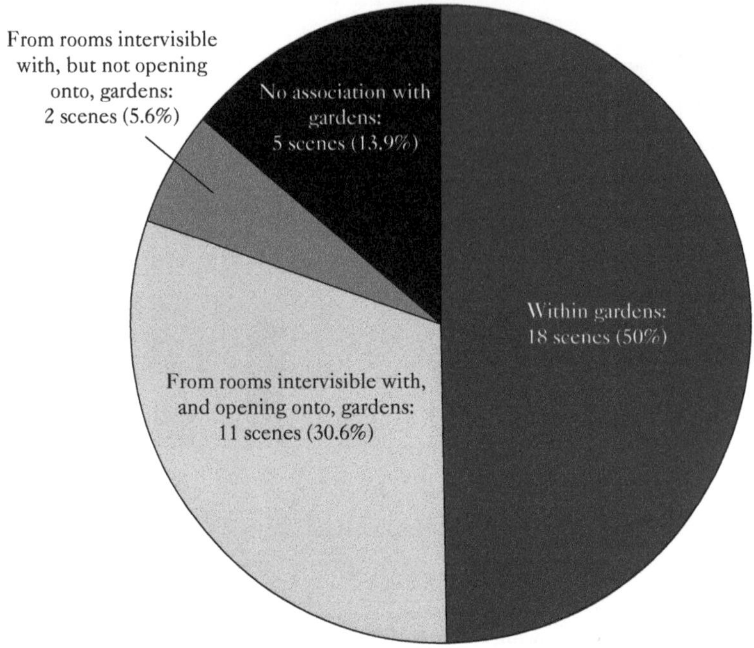

FIGURE 2.43 Pie chart showing association of Nilotic scenes with gardens in Pompeian domestic contexts: context, visibility, and accessibility. Of 36 Nilotic scenes from known contexts within a house, 32 (88.9%) come either from gardens or from rooms interaccessible or intervisible with gardens.

simple list of findspots to consider the larger visual and spatial setting of those rooms. Of the 18 scenes in non-garden rooms, a large majority came from rooms that looked and opened onto gardens. Thirteen (36.1%) came from rooms in which gardens were *visible* from the vantage point of someone standing directly in front of the Nilotic fresco or mosaic. Additionally, 11 (64.7%) of the 17 non-garden rooms with Nilotic scenes directly *opened onto* a garden, so that no more than one threshold separated the Nilotic imagery from garden space. For example, four triclinia or biclinia, one oecus, and one exedra decorated with Nilescapes opened directly onto gardens (see, e.g., figure 2.39), positioning diners to look out on the greenery. Of 36 domestic Nilotic scenes from secure contexts, then, 29 (80.6%) come from either gardens or rooms that immediately open on to gardens (figure 2.43).

In the light of recent research on contextual appropriateness, or *decor*, in Roman art,[215] we might then ask: What made Nilotic scenes appropriate to

215. See note 20, chapter 1.

gardens? As Versluys suggests, both garden settings and Nilotic imagery would have evoked general associations with *otium* and *tryphe*.[216] However, we can push this observation further: many additional parallels existed between these two-dimensional scenes and the three-dimensional spaces they help create. Despite their seeming exoticism, the landscapes and activities of Nilotic scenes also evoke many aspects of the settings and activities typical of Roman gardens. The painted landscapes and their real-world settings thus act as constant mirrors of, commentaries on, and responses to each other. The actions of the human inhabitants of these spaces would have further shaped those commentaries and responses, creating a dynamic relationship in which the meaning of both 2-D and 3-D landscapes was up for constant renegotiation.

Parallel Landscapes in the Garden

Both Nilotic scenes and Pompeian gardens confront the viewer with lush environments filled with water and vegetation. The iconographic emphasis on water in Nilotic scenes would have helped make them appropriate to watery settings.[217] Irrigation facilities were ubiquitous in Pompeian gardens.[218] The expansion of private gardens appears to have been made possible by the introduction of an Augustan aqueduct, and garden irrigation may indeed have accounted for the majority of domestic consumption of aqueduct water.[219] Almost all the gardens that contained Nilotic scenes also contained visible water features (figure 2.44). Over half (10 of 18, or 55.6%) had large, luxurious, display-oriented water features, such as fountains, pools, and canals. The additional consideration of less elaborate features, such as cistern mouths or gutters to collect rainwater, provides evidence for some type of water feature in 15 (83.3%) of the gardens with Nilotic scenes.[220] These less elaborate water features are not as visually impressive as

216. Versluys 2002: 257–258; followed by Malaise 2003: 314.

217. Also noted by Assmann 2000: 74; Versluys 2002: 258; Malaise 2003: 313; Clarke 2007b: 93–94; Balch 2008: 213–214.

218. On the presence of some type of water feature, even if only gutters or a cistern mouth, in almost all Pompeian gardens, see Jashemski 1979–93, vol. 2; Moormann 1996: 393–394; Allison 2004a: 84–87.

219. Jashemski 1979–93: 1:32–33, 54; 1981: 39, 48; 1994. On Pompeii's water supply and associated debates, see most recently Keenan-Jones 2015.

220. A large water channel at the Casa del Medico is here considered a display-oriented feature: see Chapter 4, this volume. Only three scenes—from the Casa del Gemmario (II 9.2), Casa dello Scultore (VIII 7.24–22), and Casa di Maius Castricius (VII 16 [*Ins. Occ.*].17)—come from gardens with no water features signaled in extant publications (see, e.g., Bragantini 1997: 887–889, 940–946; Sampaolo 1998b: 718–731), other than gutter

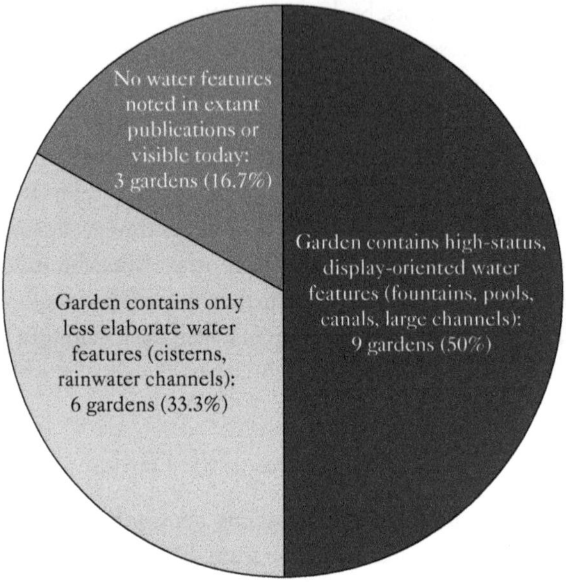

FIGURE 2.44 Pie chart showing distribution of water features in gardens with Nilotic scenes. Of 18 gardens containing Nilotic scenes, 9 gardens (50%) also contain high-status, display-oriented water features, and 15 (83.3%) contain some type of water feature.

fountains, but they would still have been economically significant. If the house owners could not afford to lease aqueduct water, rainwater collection units may have been their only water source aside from public wells and fountains. Even in rooms other than gardens, Nilotic scenes are often still associated with water. Eight (44.4%) of the domestic Nilotic scenes from non-garden settings come from rooms with prominent water features (figure 2.45): for example, baths (2 scenes), atria containing *impluvia* (3 scenes), or dining rooms associated with ornamental fountains (3 scenes).

In gardens, the splashing of real-life water on real-life plants would have formed a three-dimensional counterpart to the painted floods on painted fields. An example appears in the Casa del Granduca (VII 4.56), where the mosaic Nile on the bottom of a fountain was itself inundated by the real water flowing down the fountain (figure 2.46). Additionally, Nilotic imagery is sometimes associated with artificial

drains on eaves overhanging the Casa di Maius Castricius garden (Bragantini 1997: 945–946). However, the excavators of II 9.2 report evidence for "diverse piantine orticole a ridosso delle mura perimetrali" of the garden, within which they found 43 amphorae. Root cavities in the Casa di Maius Castricius garden suggest that it was planted as a vineyard (Jashemski 1979–93: 2:204), which would not require as much irrigation as some other garden types (cf. Jashemski 1973: 37; 1979–93: 1:214).

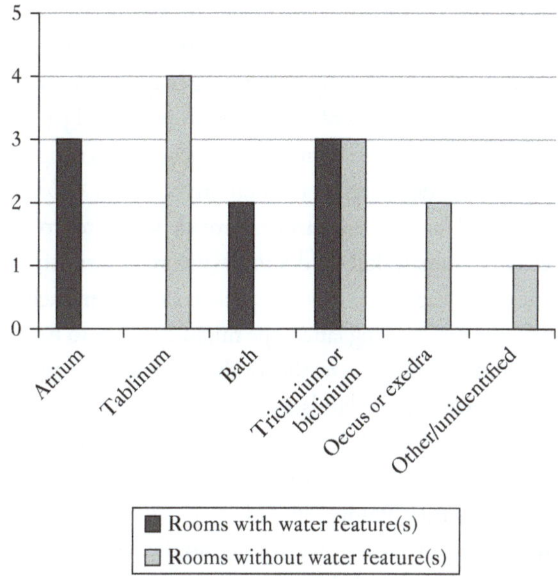

FIGURE 2.45 Graph showing the presence or absence of prominent water features in non-garden rooms with Nilotic scenes. Eight (44.4%) of the 18 scenes from non-garden rooms within domestic contexts were associated with prominent water features.

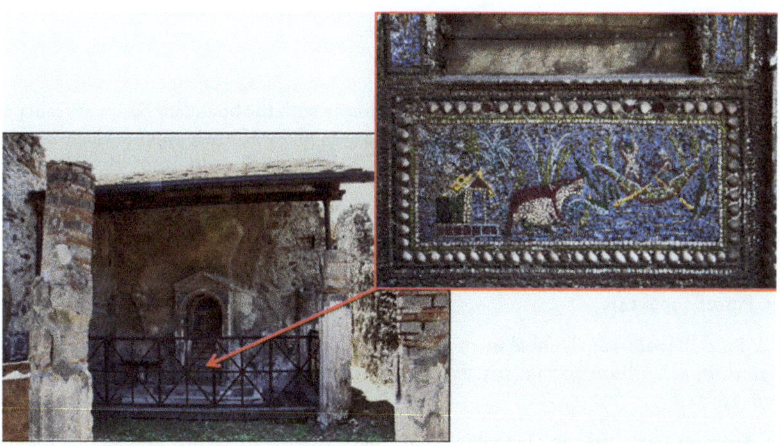

FIGURE 2.46 Reality and representation in the Casa del Granduca (VII 4.56): aedicular fountain with Nilotic mosaic in pseudo-peristyle courtyard. Inset shows closeup of Nilotic mosaic at base of fountain.

Photos: C.E. Barrett.

canals that provide three-dimensional complements to the two-dimensional Niles (see, e.g., chapters 3, 4). Such architectural contexts enlist Egyptian imagery into a very Roman artistic project: exploring the shifting borders between art and reality. Scholars often stress the paradoxes of naturalistic Roman wall painting, which frequently delights in tricking the eye while simultaneously asserting the fictive status of its own illusions.[221] As domestic representations of a *locus amoenus*—a "pleasant place" or idyllic natural landscape—gardens provide ideal sites to explore the relationship between *ars* (human skill) and *natura* (nature or intrinsic essence).[222] Part of the enjoyment guests took from Roman gardens came from their awareness that, in Purcell's words, the real-looking landscape might turn out to be in part "illusion, and actually fixed and artificial: the textiles or the grove itself might be a fresco."[223]

In addition to mirroring the landscapes of Pompeian gardens, Nilotic scenes provide parallels to—and possible commentary on—garden activities. At first glance, the typical behaviors of Nilotic pygmies may seem rather far removed from those of first-century Pompeians. However, if we set aside some of the more exotic indications of setting, we are essentially left with depictions of figures whose activities often include banqueting outdoors by the water, listening to music, carousing, and performing acts of rural cult.

Dining in or adjacent to gardens was an extremely widespread practice in good weather, well attested in both textual and archaeological records and practiced at a variety of social levels.[224] In 1950, Pietro Soprano published more than 40 dining installations (triclinia, biclinia, and one stibadium) in Pompeian gardens.[225] Subsequent scholarship has added many additional examples, bringing the total to 56.[226] Six gardens with Nilotic scenes contain

221. On painted decoration that plays self-consciously with the boundary between reality and artifice, see Bergmann 1991: 63; 1992: 34–36, 39–40; 1999: 85; Kuttner 1999: 8; Hales 2003: 68, 156–159; Leach 2004: 60, 89, 125; Elsner 2007; Platt 2009; Spencer 2010: 160–161; Newby 2012: 374, 378–380, 382.

222. Cf. Purcell 1987. On the garden as *locus amoenus*, see Kuttner 2003: 144; von Stackelberg 2009: 20–21.

223. Purcell 1996: 146.

224. Purcell 1996: 123; Dunbabin 1991: 124–125. *On finds of portable braziers for cooking in gardens*, see Allison 2004a: 103. *For literary references to outdoor dining*, see Dunbabin 1996: 66, 74.

225. Soprano 1950. I follow Dunbabin (1991; 1996: 70–72n22, 72n30) in reserving the term *stibadium* for sigma-shaped benches (*pace* Amedick 1993; 1994: 112–115).

226. Jashemski 1979–93 (esp. 2:346n1); Kockel 1986: 475–476, 481–483, 486–487, 489, 513; Richardson 1988b; Andersson 1990; Dunbabin 1991: 124–125, 139n23; 1996: 70–72; Amedick 1993, 1994: 112–113; Moormann 1996: 393–394; Allison 2004a: 84–87; Dessales 2013: 325–331.

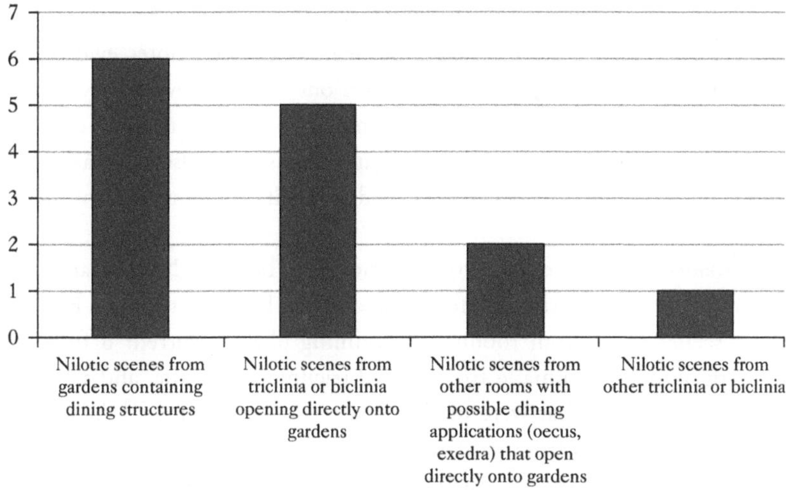

FIGURE 2.47 Graph showing the contextual association of Nilotic scenes with both gardens and dining. Fourteen Nilotic scenes (38.9% of the 36 Nilotic scenes from known rooms within domestic contexts) come from rooms or gardens that were likely used for dining.

dining structures (figure 2.47). Excavators at the Casa del Bracciale d'Oro garden (figure 2.39) even found faunal remains from one or more meals.[227] Seven additional scenes come from probable dining rooms opening directly on to gardens.[228] As discussed earlier, such placement of dining rooms around a peristyle or other garden space was very common, partly in order to ensure good views for the diners.[229] A total of 14 Nilotic scenes (38.9% of Nilotic scenes from known rooms within domestic contexts) come from rooms or gardens with surviving evidence for dining activities. Additionally, of the 29 Nilotic scenes that come either from gardens or from rooms that open onto gardens, 13 (44.8%) also share space with dining installations in the same room.

Although this latter number already represents almost half of Nilotic scenes associated with gardens, it is likely an underrepresentation of the number of such gardens where dining actually occurred. The garden triclinia preserved today

227. Jashemski 1979–93: 2:167; cf. Jashemski 1974: 403; 1977: 226 on meal debris in other gardens.

228. In addition to five triclinia or biclinia, one oecus and one exedra were likely also used for dining: see note 214, this chapter, and see further chapter 5, this volume, "Encountering Alexander: The Nile Mosaic as Threshold."

229. See note 200, this chapter.

are mostly built masonry structures, but many more gardens would undoubtedly have contained more perishable facilities. Andersson notes evidence for wooden triclinia in some gardens and points out that such evidence was probably overlooked in early excavations at Pompeii, so wooden triclinia may once have been much more common.[230] Dining in gardens could also have taken place without any permanent installation of garden furniture, as much furniture was portable.[231]

A broader look at the contextual relationships between Nilotic landscapes and dining strengthens the case for an association between the two. Fourteen Nilotic scenes come from rooms where dining likely occurred: 6 from gardens with outdoor triclinia, 6 from roofed triclinia or biclinia, and 2 from an oecus and an exedra that were probably also used for dining.[232] At least another 7 Nilotic scenes from gardens are intervisible with triclinia or biclinia, and one scene comes from an atrium intervisible with a probable garden triclinium (figure 2.48). Of the 35 Nilotic scenes for which we know not only the room of origin but also the spatial relationship of that room to other nearby rooms,[233] 22 (62.9%) thus come from locations that either served as dining space or commanded a view of dining space.

Turning from dining to dinner performances, we should recall that Roman dinner parties often involved mealtime entertainment, which might include music, dancing, dramatic performances, or literary readings.[234] Garden dining areas may have been particularly suited to such entertainment, since the outdoor setting provided more room.[235] Roofed triclinia that opened onto a garden provided similar opportunities, as the performers could use the garden space. Accordingly, when Pompeian dinner guests sat at a garden triclinium decorated with scenes of pygmy figures reclining, drinking, and listening to music in a watery, vegetation-rich environment, as at the Casa dell'Efebo, they would have gazed on scenes simultaneously

---

230. Andersson 1990: 226; cf. Morvillez 2018: 59. Soprano notes evidence for 7 wooden garden triclinia (1950: 307–308).

231. Bek 1983: 83; Farrar 1998: 37; Mols 1999: 139–141.

232. See note 214, this chapter, and see further chapter 5, this volume, "Encountering Alexander: The Nile Mosaic as Threshold."

233. Excluding the Casa dello Scultore, whose Nilotic paintings belong to an earlier phase of the peristyle and a stratum lower than that to which the rest of the house has been excavated (Maiuri 1956; Sampaolo 1998b).

234. See, e.g., Plin. *Ep.* 1.15; Ath. *Deipnosophistae*; Plut. *Quaest. conv.*; Jones 1991; Gowers 1993: 24–32; Dunbabin 1996; 2003: 43, 90, 130, 173–174, 198–199; Bradley 1998: 37–38.

235. Dunbabin 1996: 70–71.

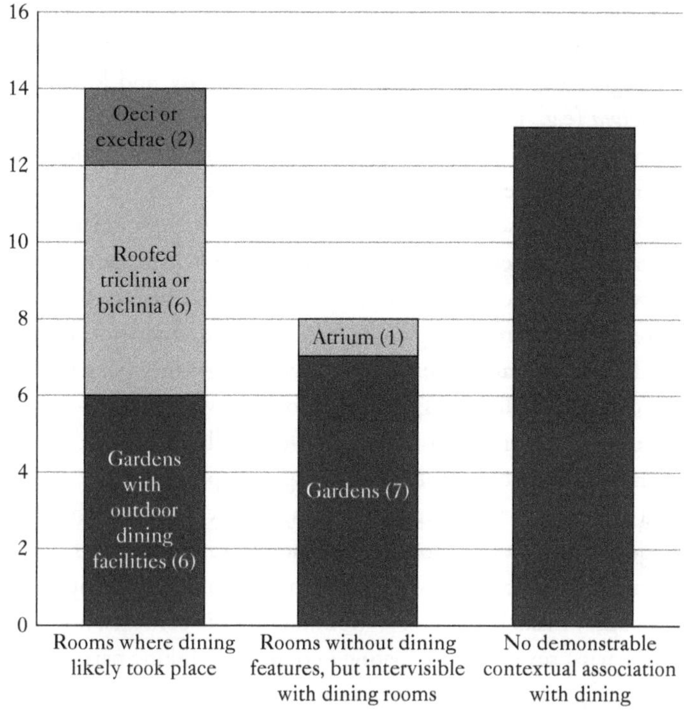

FIGURE 2.48 Graph showing the contextual association of Nilotic scenes with dining. Of 35 Nilotic scenes for which information is available on both the room of origin and that room's spatial relationship to the rest of the house, 22 (62.9%) come from rooms that either were likely used as dining space, or commanded a view of dining space.

exotic and familiar: distorted mirrors[236] showing figures simultaneously distant from, and yet parallel to, themselves.

236. While this book was in press, Andy Merrills also published an analysis of the Casa dell'Efebo Nilotica (2017: 106–149, esp. 107–115, 139–143). At one point in this discussion, Merrills independently employs similar language to describe the frescoes, saying they enabled diners to "[peer] back at themselves in the distorting fun-house mirror of these images" (2017: 140). Other aspects of Merrills's conclusions, however, differ from those proposed here. His reading of the frescoes emphasizes their celebration of Roman political dominance over Egypt. To this end—and despite his recognition of the parallels between dining pygmies and dining viewers—he places heaviest interpretive emphasis on scenes of violence and suffering. For Merrills, painted pygmies exist primarily to be "torn, penetrated and mocked for the entertainment of the viewer" (2017: 139). His interpretation of Nilotic scenes thus centers on "the representation of Egyptians as dwarfs, and the brutal depiction of their painful suffering" (2017: 146). Merrills's discussion of the Casa dell'Efebo also differs from mine in focusing exclusively on the Nilotic water-triclinium, rather than discussing the full assemblage of material and visual culture in this garden (see chapter 3, this volume).

As for the figures' more salacious activities, numerous literary and visual sources suggest Romans could perceive *convivia* as places of sexual tension and intrigue. Elegiac love poems refer to flirtation, jealousy, and lovers' secret signals at *convivia* (e.g., inter alia, Ovid, *Amores* 1.4), and a number of domestic wall paintings depict men and women flirting or kissing while dining.[237] However, any actual sexual encounters were expected to occur afterwards in private.[238] The public sexual activities of Nilotic scenes would thus be wildly inappropriate in a respectable Roman *cena*. The explicit images may thus have struck viewers as comically excessive, as Clarke and Roller argue.[239] However, they also represent figures acting on impulses that the banqueters themselves expected (at least potentially) to experience. Such images may poke fun not just at Egyptians or quasi-mythical pygmies but also at the viewers' own—presumed or aspirational—social identities as refined, self-controlled diners.

Certainly, erotic or convivial Nilotic imagery appears in contexts at Pompeii other than domestic gardens (appendix A, appendix B), and such imagery may play quite different roles in those other contexts. In public baths, for example, erotic pygmy scenes recall the well-documented humorous and apotropaic uses of erotic imagery in such settings.[240] Within domestic contexts, however, scenes of pygmies having sex and participating in outdoor drinking parties are strongly associated with gardens. Of 7 known rooms in Pompeian houses that produced erotic Nilotic scenes, 5 (71.4%) come from gardens.[241] Additionally, of 4 Nilotic scenes from Pompeian domestic contexts that depict outdoor drinking parties (Casa dell'Efebo, Casa dei Dioscuri, Casa del Medico, Casa dello Scultore), all

237. Gowers 1993: 37; Roller 2006: 142–145.

238. When Roman authors refer to dinnertime sexual intrigues going too far, they typically present this as inappropriate (e.g., Cato fr. 213m = Gell. 10.13.2; Suet. *Aug.* 69.1; Sen. *Ep.* 83.20; Roller 2006: 144).

239. Roller 2006: 146; Clarke 2007b: 103–104; 2007c: 163–164.

240. Apotropaic erotic imagery in baths: Clarke 1998: 212–240; 2007b: 194–215.

241. In the category of "erotic" scenes, I exclude those that simply show macrophallic or ithyphallic figures (appendix B, this volume). The Casa dei Fiori also produced a sexual Nilotic scene, but the room context is unclear (see note 5, appendix A). Another erotic Nilotic scene came from an unknown context in regio VI, insula 5 (Versluys 2002: no. 44). A now-lost Pompeian fresco drawn and described by Roux and Barré (1870: 5.46–47, pl. 22) and Reinach (1922: 161, fig. 1) also appears to have Nilotic and erotic features, but its context is unknown. Meyboom and Versluys (2007: 183) and Jacobelli (1995: 88n28) refer to sexual imagery in the Nilotic fresco from the garden of II 9.4, but I have been unable to verify the presence of such imagery here. Examination in situ reveals no surviving trace of sexual scenes, although the paintings are now faded. The original excavation report (Sodo 1988: 195–202) does not mention sexual imagery. However, if this fresco did feature a *symplegma*, its garden context would recall that of other erotic pygmy scenes.

three for whom the room of origin is known come from gardens. Furthermore, when Nilotic scenes with convivial or explicitly sexual imagery come from known contexts in Pompeian gardens, they are frequently located either on or adjacent to dining installations. Nilotic frescoes depicting such imagery appear in gardens at the Casa dell'Efebo, House II 9.2, the Casa del Medico, the Casa delle Quadrighe, and the Casa dello Scultore. At the Casa dell'Efebo and House II 9.2, the pygmy scenes were painted directly onto outdoor garden triclinia, while at the Casa del Medico, Casa delle Quadrighe, and Casa di Sallustio, a triclinium opened immediately on to, and was visible from, the garden. (The situation is harder to assess at the Casa dello Scultore, where the Nilotic frieze comes from Maiuri's excavations below the destruction layer of the peristyle, and the rest of the house was not excavated to the same level.[242]) When representations of pygmies banqueting or cavorting outdoors appear in close association with facilities for banqueting in or adjacent to gardens, they invite viewers to consider how the imagery may or may not relate to their own current situation.

Scholars usually interpret erotic pygmy scenes as "inter-group, or 'them versus us,' humor"[243]: mocking the "Other" or outgroup, not the "Self" or ingroup. Framed by their context, however, the pygmies of many Nilotic garden installations invite not only derision but also identification. When such scenes depict excess or overindulgence, Roman audiences may well have found them humorous—but viewers would potentially have found themselves laughing not only at imagined cultural or somatic "Others" but also at each other and at themselves.[244] In convivial settings, such humor might defuse social tensions, relieve anxieties, and create a relaxed, jovial atmosphere.

Further testifying to the appeal of such playful imagery in dining areas—and demonstrating that such humor need not only target social "Others"—are certain other paintings from dining contexts. Some Pompeian triclinia[245] contain images of Greco-Roman banqueters, of normal height and with idealized bodies, who engage in similar excesses. Once-elegant banqueters may appear hopelessly drunk, passed out, stumbling, cavorting amorously, even vomiting.[246] Triclinium

242. See note 233, this chapter.

243. Quoting Clarke 2014: 653; see also note 239, this chapter.

244. On "intra-group" Roman humor, see Clarke 2014: 652.

245. See Roller (2006: 45–84, 139–153, 195–196) on Pompeian paintings of banquets, including those from dining rooms. Figural imagery in triclinia covers a wide range of themes, especially mythological and/or erotic (Fredrick 2003: 311n4, 326–328); Dionysiac and/or banquet-related imagery appears in only a minority of triclinia (Ling 1995).

246. Dunbabin 1996: 52–63; Bergmann 2002a: 35–38; Clarke 2003: 227–233, 239–245; Roller 2006: 61–64, 70–77, 139–145. Clarke interprets the Casa dei Casti Amanti scenes as

paintings in the Casa di Marcus Lucretius show Cupids and Psyches drinking and dancing outdoors, watching musical performances, and engaging in amorous intrigues.[247] Frescoes from a possible tavern in Ostia reduce lofty figures to another sort of bodily exigency, depicting famous philosophers declaiming about defecation.[248] Playful tweaking of social conventions also appears in the Casa del Moralista triclinium, whose walls bore elegiac couplets forbidding guests from ogling each other's wives, fighting, or making messes.[249] As Clarke suggests, these warnings are probably "tongue-in-cheek humor, not actual injunctions to behave oneself."[250]

There is one notable difference between Nilotic scenes and other representations of exaggerated or excessive banquets: diners in non-Nilotic banquet scenes may flirt or kiss, and their drapery may slip suggestively off their shoulders,[251] but they typically do not engage in intercourse. Clarke and Roller attempt to explain this difference through pygmies' status as physical, as well as cultural, "Others": they can disobey social mores because their non-ideal bodies make them "designated vehicles for parody and satire."[252] However, Nilotic *symplegmata*, or scenes of sexual intercourse,[253] do not always feature protagonists with "non-ideal" bodies. For example, the Nilotic frieze from the peristyle of the Casa delle Quadrighe (VII 2.25) features a penetrative sex scene whose participants are all of normal height and proportions, with conventionally idealized bodies (figure 2.49).[254] The Nilotic paintings from the columbarium of the Villa Pamphili in Rome also include a depiction of outdoor, public intercourse between individuals of normal height

---

Greek *symposia* (2003: 227–233; cf. Dunbabin 1996: 52–56; Bragantini 2006: 163). However, Roller (2006: 65–68) argues persuasively that viewers would have seen the diners as "Roman," perceiving cultural similarity, not difference, between the figures and themselves. The Casa del Triclinio banqueters are generally understood as "Roman" (Clarke 2003: 239–245).

247. Bragantini 1999: 254–280; Dunbabin 2003: 59–60.

248. Clarke 2006; 2014: 653–654. Koloski-Ostrow and Moormann (2011) see the room as either a tavern or a latrine.

249. Bragantini 1991 (house plan, wall paintings, and mosaics); *CIL* IV 7698 (inscriptions).

250. Clarke 2007b: 59; cf. Clarke 2003: 233–239; Roller 2006: 146–148.

251. Roller 2006: 141–145.

252. Roller 2006: 146; cf. Jacobelli 1995: 88; Clarke 1998: 42–46.

253. This modern usage adapts Latin authors' use of *symplegma* to describe imagery of entwined or entangled, sometimes erotically entangled, figures (Plin. *HN* 36.24, 36.35; Mart. 12.43).

254. A different Nilotic sex scene from the same location (MANN 27702) depicts figures that more closely resemble dwarfs or pygmies (Versluys 2002: no. 50).

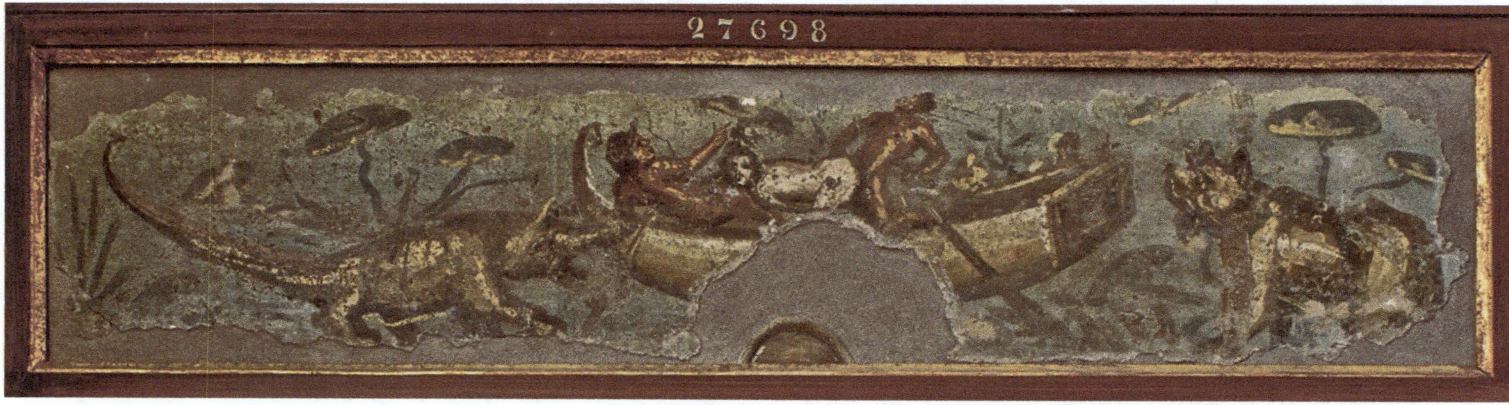

FIGURE 2.49 *Symplegma* scene aboard a boat, watched by crocodile and hippopotamus. One of two preserved Nilotic wall paintings from the pluteus wall of the peristyle at the Casa delle Quadrighe (VII 2.25), Pompeii. MANN 27698.

Photo: C.E. Barrett. Reproduced by permission of the Ministero dei Beni e delle Attività Culturali e del Turismo—Museo Archeologico Nazionale di Napoli.

and proportions.[255] Nor did Romans consider physical "otherness" necessary for sexual imagery to be humorous. With the exception of one figure with enlarged testicles, people with idealized physiques populate most of the Suburban Baths' outrageous sexual imagery, which Clarke has convincingly shown to be intentionally comical.[256] The sexual license allowed in Nilescapes may thus have less to do with somatic "Otherness" than with the imagined Egyptian context. As discussed earlier ("Egyptian Antecedents and Affordances"), textual and visual evidence from Greco-Roman Egypt suggests that during inundation festivals, exaggerated sexual behavior may sometimes have been temporarily appropriate. The characters in Nilotic scenes may thus carry on more explicitly than the Casa dei Casti Amanti banqueters not just because they are pygmies (if and when they are pygmies); the Nile festivals also provide them with an occasion to do so.

While the Casa delle Quadrighe and the Villa Pamphili columbarium depict Nilotic *symplegmata* between participants who are both (or, in the case of the Casa delle Quadrighe, all three) normally proportioned, the Nilotic sex scene at the Casa dell'Efebo (figure 1.8) presents a more complex scenario. Some of the figures in this vignette appear to be pygmies, but others (including, for one, the female protagonist of the *symplegma*) do not. Most of the figures in the Casa dell'Efebo paintings have proportions typical of Roman depictions of pygmies or dwarfs,[257] with relatively large heads and small limbs in proportion to their torsos (figures 1.3–1.8). In the *symplegma* scene, the man operating the water wheel (F) shares these proportions. However, others—the female partner in the sexual encounter (A), the aulos-player (B), the amphora-carrier (C), and possibly the seated spectator at right (E)—have more idealized proportions, with smaller heads and longer bodies. The male partner in the sexual encounter presents a somewhat ambiguous case; although the painting of his body is somewhat damaged, his limbs appear slightly shorter than his partner's. The standing woman (D) is also potentially ambiguous; the proportions of her upper body seem "normal," although her legs are rather short. Both the woman engaged in intercourse and the aulos-player are significantly taller than the figures around them. Comparing the woman in the sexual encounter to the woman standing to her right, the height difference is particularly dramatic; if the former were to stand up, she would be almost twice as tall as the latter. Although Nilotic scenes certainly do often feature sexual partners who both

255. Meyboom and Versluys 2007: 186–187; cf. Clarke 2003: 194–195.

256. Clarke 1998: 212–240; 2007b: 194–215.

257. On the difficulty of distinguishing pygmies and dwarfs in Roman art, see earlier this chapter, "Navigating 'Nilotic Scenes.'"

have "pygmy"-like features (e.g., at the Casa del Medico: figure 2.5), the female participant in the Casa dell'Efebo sex scene may not actually be a pygmy—or, if she is, the painter has gone to some effort to make her significantly more conventionally attractive than her compatriots.[258] It is hard to see much difference between this woman's proportions and those of the women in, for example, the sex scenes at the Suburban Baths.[259] While the Nilotic sex scene in the Casa del Medico (figure 2.5) may well have provoked amusement at the figures' "corpi sgraziati,"[260] those in the Casa dell'Efebo and Casa delle Quadrighe (figures 1.8, 2.49) might have inspired less ironic forms of aesthetic and erotic appreciation. This does not mean these scenes lacked a comic element; after all, the scenes in the Suburban Baths were also meant to be funny. However, the representation of (at least) the women in some Nilotic sex scenes as conventionally attractive may have prompted (male) viewers to another form of imaginative identification with those women's male companions, further complicating the complex play of identification and rejection that these images provoke.

The juxtaposition of Nilotic banqueting imagery with "real-life" banquets has a textual parallel in one of Propertius's elegies (4.8), where a garden scene features a Nilotically flavored combination of eroticism, alcohol, music, dance, and dwarfs. Jealous over the fickleness of his mistress Cynthia, the speaker tries to set up a tryst with two other women in a garden:[261]

> *There was one three-person couch in a garden set apart from view.*
> *You ask about the reclining arrangements? I was between the two girls.*
> *Lygdamus was cup-bearer, there was a summer service of glass*
> *and of Methymnian wine with a pleasant*[262] *flavor.*

---

258. The same is true for much of the Egyptian imagery with which Roman Nilescapes may be in dialogue. Note the size difference between the woman and her partners in a Ptolemaic *symplegma* statuette (figure 2.20), the less "grotesque" features of the female dancer in a terracotta pair-statuette (figure 2.24), and the more human features of "Beset" compared with Bes (figure 2.36). Bes's female companions only sometimes have dwarfism (Dasen 1993: 74). In the "Turin Erotic Papyrus," the females have much more idealized bodies and faces than their balding, paunchy male counterparts, two of whom have dwarfism (figure 2.37). Note also that pygmy women can appear attractive in Greco-Roman literature; Antoninus Liberalis recounts a myth about a beautiful female pygmy (*Met.* 16; cf. Ael. *NA* 15.29–30).

259. Jacobelli 1995.

260. Jacobelli 1995: 88 (on erotic pygmy scenes generally).

261. Prop. 4.8.35–40; my translation, from the edition of Heyworth 2007b.

262. Following Heyworth's reading of *grata saliua* (2007b: 177), rather than *Graeca saliua* (e.g., Butler 1967: 316). Nonetheless, as the wine is from Methymna, its flavor would indeed still have been "Greek," as well as "pleasant."

O Nile, the flute-player was yours; Orontes,[263] *yours the castanet-player –*
this girl ready to be strewn with roses, artlessly elegant[264] *–*
and Biggie himself, stunted[265] *in the limbs, was shortly*
throwing around his underdeveloped hands to the hollow boxwood flute.

Several elements in the passage recall Nilotic scenes. A small group of potential sexual partners reclines together, drinking and watching three performers. The flute-player is explicitly characterized as Egyptian ("Nile, tuus tibicen erat"), and the ironically named dwarf dancer Magnus ("Biggie") recalls his dancing cousins in Nilotic scenes.[266] The only thing missing is the Nile itself, and as if to make up for that omission, one line (4.8.39) directly addresses the Nile in vocative case. Yet a closer look reveals that the cultural referents of this passage are also much broader: set in Rome,[267] Propertius's garden party incorporates features from all over the empire. If we accept Heyworth's emendation of the text, the performers include a musician from the Orontes, as well as one from the Nile.[268] Other elements of the passage point elsewhere in the Greek world, such as the wine from Methymna.[269] As we shall see in chapter 3, Pompeian gardens often similarly evoke multiple geographical and cultural milieus.

Gardens, Water, and Dining

Nilotic scenes at Pompeii have strong contextual associations with gardens, water, and dining (figures 2.50, 2.51). Following the definition of "association" as comprising either (1) location within the same room or (2) visibility from the vantage point of the Nilotic fresco or mosaic, almost all Nilotica from known domestic contexts at Pompeii are associated with either gardens, dining, or water features, and about half have associations with all three.

---

263. Following Heyworth (2007a: 478–479; 2007b: 177).

264. *Munda sine arte*: as Richardson notes, "an oxymoron, the adjective *munda* implying *ars*" (2006: 467).

265. Heyworth reads *contractus* rather than *concretus* (2007b: 177; see discussion in 2007a: 479–480).

266. Quoting Prop. 4.8.39. On dwarfs as entertainment in Greek and Roman culture, see Dasen 1993: 230–233; 2013: 267–271; Garland 1995: 46–48, 116; Garmaise 1996; Stahl 2011: 722–723; Masséglia 2015: 266–302.

267. See, e.g., Prop. 4.8.1, 4.8.29–30.

268. See note 263, this chapter. Richardson's suggested reading of "Baetis" (2006: 467) is similarly internationalizing, suggesting a Spanish river and *crotalistria*.

269. Dee (1978: 47) notes the number of terms borrowed from Greek in this passage.

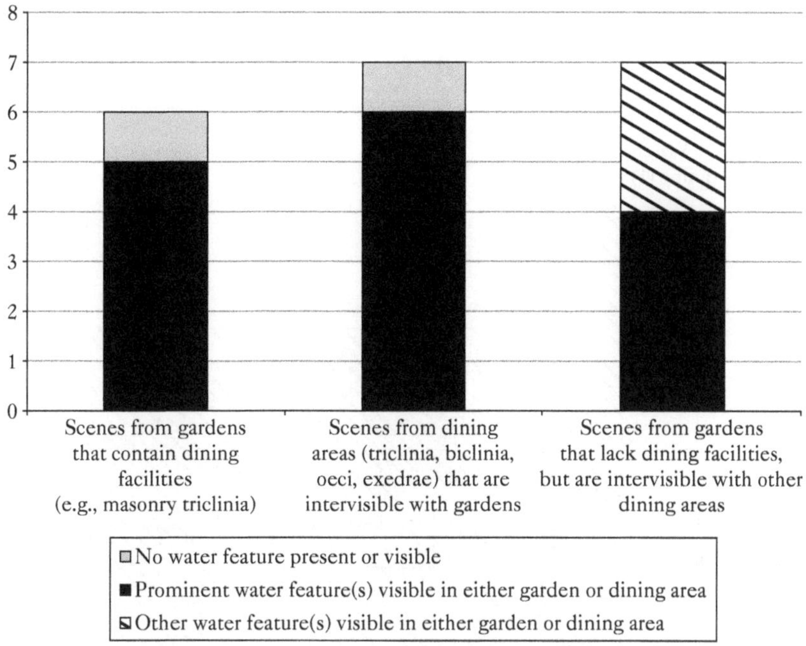

FIGURE 2.50 Graph showing the contextual associations of Nilotic scenes: a breakdown of the relationship between gardens, dining space, and water features for the 20 Nilotic scenes that come from gardens or dining rooms.

Of 35 scenes for which data exist on both the room of origin and the plan of the house more broadly, 34 (97.1%) share a room with, or are intervisible with, garden space, dining installations, and/or water features.[270] Twenty-one (60%) are associated with *both* gardens *and* dining (figure 2.51). Nineteen (54.3%) are associated with gardens, dining, and water features, and within that group, 15 (42.9%) are demonstrably associated with prominent water features: fountains, pools, unusually broad channels, or *nili/euripi*. For example, several Nilotic scenes come from garden "water-triclinia"—that is, dining couches, either inside a garden or opening on to one, that enclose a water feature, usually a fountain, basin, or canal (e.g., figures 1.1–1.2, 2.39).[271] Pompeian garden areas

270. The one exception—from the tablinum of I 2.24—actually could have been associated with dining and drinking during the last phase of its use-life, when the house was converted into a *caupona* (de Vos 1990a).

271. On "water-triclinia" (and "water-biclinia"), see Salza Prina Ricotti 1979; 1987; Richardson 1988b; Andersson 1990: 225–226; Dessales 2013: 326–327. Salza Prina Ricotti restricts the

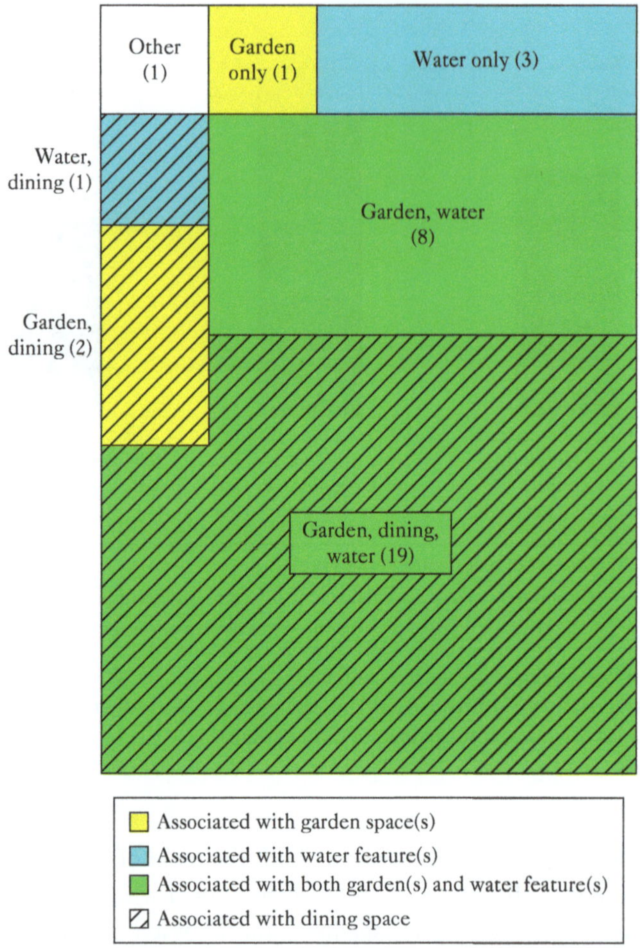

FIGURE 2.51 Graphic visualization of Nilotic scenes' contextual associations with gardens, dining space, and water features. The large rectangle represents the 35 Pompeian Nilotic scenes whose domestic contexts are most thoroughly documented (see notes 213 and 233, this chapter). This large rectangle is subdivided into smaller rectangles corresponding in size to the relative frequency of different contextual associations. "Association" is here defined as intervisibility from the vantage point of the Nilotic fresco or mosaic.

---

term only to installations in which the water between the benches reached a sufficient height that dishes could float on it (1987: 172). However, following more recent publications (e.g., Richardson 1988b; Andersson 1990: 225–226), I use the term for all biclinia/triclinia with water flowing between the benches.

have produced 13 such water-triclinia.[272] Of these, 5 (38.5%) are associated with Nilotic imagery.[273] Expensive to maintain, these combined garden-water-dining installations were effective displays of wealth.[274] Such structures are paralleled in elite textual descriptions and villas, including imperial villas.[275]

Clearly, Nilotic landscapes were not restricted to one single room type, and some contexts emphasize one or another possible contextual association to a greater or lesser degree. For example, in baths, an association with water would obviously be more relevant than an association with dining. Yet it is striking that an overwhelming majority (30, or 85.7%) of the Nilotic scenes from well-documented domestic contexts at Pompeii should be associated with gardens, and that a sizable majority (21, or 60%) should display strong contextual associations with *both* gardens *and* dining (figure 2.51).

While previous studies have suggested some links between Nilotica and domestic spaces associated with gardens, water, and/or dining, methodological considerations have caused the strength of these contextual links to be underestimated. Earlier research on the contexts of Nilotic scenes focused largely on assigning individual scenes to individual rooms—a process that generated useful initial data, but did not fully engage with the potential multifunctionality of those rooms or take account of their interrelationships with other forms of domestic space. Beyond simply attaching room labels to Nilotic imagery, a consideration of broader issues of access, visibility, and functionality can help produce more nuanced understandings of how Nilotic imagery interacted with larger domestic visual landscapes and activity areas.

Rooms in Pompeian houses comprised not only architecture (or even architecture plus paintings and mosaics) but also diverse assemblages that

---

272. To Andersson's 12 examples (1990: 225–226), add the Casa del Bracciale d'Oro; see Jashemski 1979–93: 2:166–167. As Andersson points out (1990: 236), indoor nymphaeum-triclinia are also closely functionally related to water-triclinia in gardens, as both permit dining in a watery environment. In house VIII 2.28, Nilotic motifs (Versluys 2002: 136–137) appeared within a subterranean triclinium containing a nymphaeum (Richardson 1988a: 231–233; Sampaolo 1998a: 227–228, 235–240).

273. *Watery dining installations inside garden*: Casa dell'Efebo, Casa di Octavius Quartio, Casa di Sallustio, Casa di Acceptus e Euhodia, Casa del Bracciale d'Oro. *Watery dining installations adjacent to garden*: Casa di Paquius Proculus, Praedia di Julia Felix. For two-dimensional Nilotic imagery from these structures, see Versluys 2002: nos. 34, 35, 38, 42, 48. The Nilotic imagery from the Casa di Octavius Quartio and Casa di Acceptus e Euhodia is sculptural; see chapter 6, this volume.

274. Water features as displays of wealth: Zarmakoupi 2014: 146–152; Bruun 2016; Morvillez 2018: 45–56.

275. See most recently Zarmakoupi 2014: 172.

incorporated statuary both large and small scale, furniture, so-called small finds, and all manner of humans and human activities. Within these dynamic settings, the relationships between viewer and images were constantly up for negotiation depending on the needs of the moment. The contexts of domestic Nilotic scenes frequently constructed parallels between the two-dimensional landscapes and their three-dimensional settings, challenging viewers to consider how they might themselves relate to the pygmy figures with whom they often shared these (real and imagined) spaces. However, the implications of these parallels could have varied not only from house to house but even from day to day and hour to hour within the same house. As different combinations of individuals came together in a single space at different times, or employed the space for different activities over the course of the day, the "meaning" of that space—and its associated assemblage of objects and images—would have been similarly fluid.

The following chapters thus take a closer look at how Nilotic images interacted with specific domestic settings, and how the "meanings" of any image might vary according to its framing. Chapters 3–6 provide contextual analyses of several specific case studies: the Casa dell'Efebo, Casa del Medico, Casa del Fauno, and Casa di Acceptus e Euhodia. These case studies build on the foregoing general discussion of the contexts of domestic Nilotica to explore the varied roles of Nilotic imagery within specific domestic ensembles.

3

From Egypt to Oikoumene

INTERACTIVE LANDSCAPES IN THE CASA DELL'EFEBO

Introduction

This chapter analyzes the frescoes from the garden triclinium of the Casa dell'Efebo (House I 7.10–12, Pompeii) as a case study in the recontextualization of Nilotica (figure 1.1).[1] These paintings provide some of the most extensive and detailed depictions of the Nile from Pompeii, but their setting has received relatively little discussion. Previous publications of the frescoes typically note their placement on a garden triclinium associated with a fountain.[2] However, this dining installation is just one part of a garden that also contained additional wall paintings, statuary, furniture, and a vegetable plot. This chapter recontextualizes the Nilescapes within this eclectic ensemble, examining how the contents of the garden work together to create a sense of place, construct real and imagined landscapes, and shape the experiences available to their human users.

The context of these frescoes—within a garden, on a triclinium, adjacent to flowing water—exemplifies many of the typical patterns that chapter 2 identified for the settings of Nilotica. The Casa dell'Efebo garden thus provides, among other things, a demonstration of the parallels that such contexts construct between representation and reality. Framing viewers' encounters with images, the setting challenges diners to consider their own relationships to these representations of carousal, commensality, and danger in a lush, "sacral-idyllic"

---

1. This chapter is adapted from a previous article by the author (Barrett 2017b) and is here republished by permission of the *American Journal of Archaeology*.

2. E.g., Della Corte 1930: 29; Tran tam Tinh 1964: 50; Michel 1980: 400; Appleton 1987: 164; Meyboom 1995: 62, 246n80, 254–255n112, 259n130, 262n139, 307n119, 320n154, 329n180; Versluys 2002: 101–105; Clarke 2007b: 103; Balch 2008: 230–232; Merrills 2017: 109–110.

landscape. Even as the Egyptian landscapes evoke the farthest fringes of the empire within the *domus*, other aspects of the garden assemblage familiarize and domesticate this seemingly exotic imagery, even prompting banqueters to consider possible parallels between themselves and the carousing pygmy-figures. Instead of a simple resolution of tensions between "Self" and "Other," the Nilotic installation thus enables viewers to explore those tensions for themselves and to respond, in potentially varying ways, through their own convivial behavior.

Beyond the relationships between viewers' activities and the activities depicted in the frescoes, however, the Casa dell'Efebo also provides an opportunity to explore the relationships between Nilotic imagery and other forms of domestic cosmopolitanism. In addition to Egyptian landscapes, this garden evokes a wide range of other cultural referents and imagined landscapes both distant and close at hand. Egyptian imagery is closely interwoven with allusions to Archaic, Classical, or Hellenistic imagery, as well as adaptations (and ongoing constructions) of specifically local traditions. This case study thus enables us to situate domestic representations of "Egypt" within the broader context of domestic representations of the empire. Simultaneously far away and familiar, the imagined landscapes of the garden transform domestic space into a microcosm of *oikoumene*. The resulting cosmopolitan domestic environment enables its occupants to construct their own understanding of, and negotiate their own position within, a wide range of imperial, local, social, and cultural identities.

The Garden of the Casa dell'Efebo

Excavated by Amedeo Maiuri, the Casa dell'Efebo (I 7.10–12) has a ground floor area of about 650 sq. m, within the largest 25 percent of Pompeian houses (figure 3.1).[3] Apparently composed of two originally separate dwellings, the house was seemingly under renovation in 79 CE to incorporate another house (I 7.19) to its

3. *Finds from the house:* Maiuri 1927; Della Corte 1930: 28–33; Tassinari 1993: 130–131; Allison 2004b, s.v. "Casa dell'Efebo." *Ground floor area:* Allison 2004b, s.v. "Casa dell'Efebo," following the quartile system of Wallace-Hadrill 1994: 79–82. Some scholars identify the owner with Publius Cornelius Tages, a wealthy freedman known from the Iucundus archives (*CIL* IV 3340.67, 3340.76, 3340.101, 3340.113, 3340.115; Andreau 1974: 88n1, 153, 191, 194, 175, 246, 251, 257–260, 263–264, 268–271, 306; see most recently Merrills 2017: 107–108, misprinting the *cognomen* of the man in question as "Tragus"). Inscriptions on two amphorae in the house refer to him, as do two political endorsements opposite the house entrance (Maiuri 1927: 81; Della Corte 1965: 316). However, Della Corte's methodology for identifying house owners cannot be supported, and the significance of the electoral inscriptions' placement is unclear (Mouritsen 1988: 13–27), making the identification uncertain.

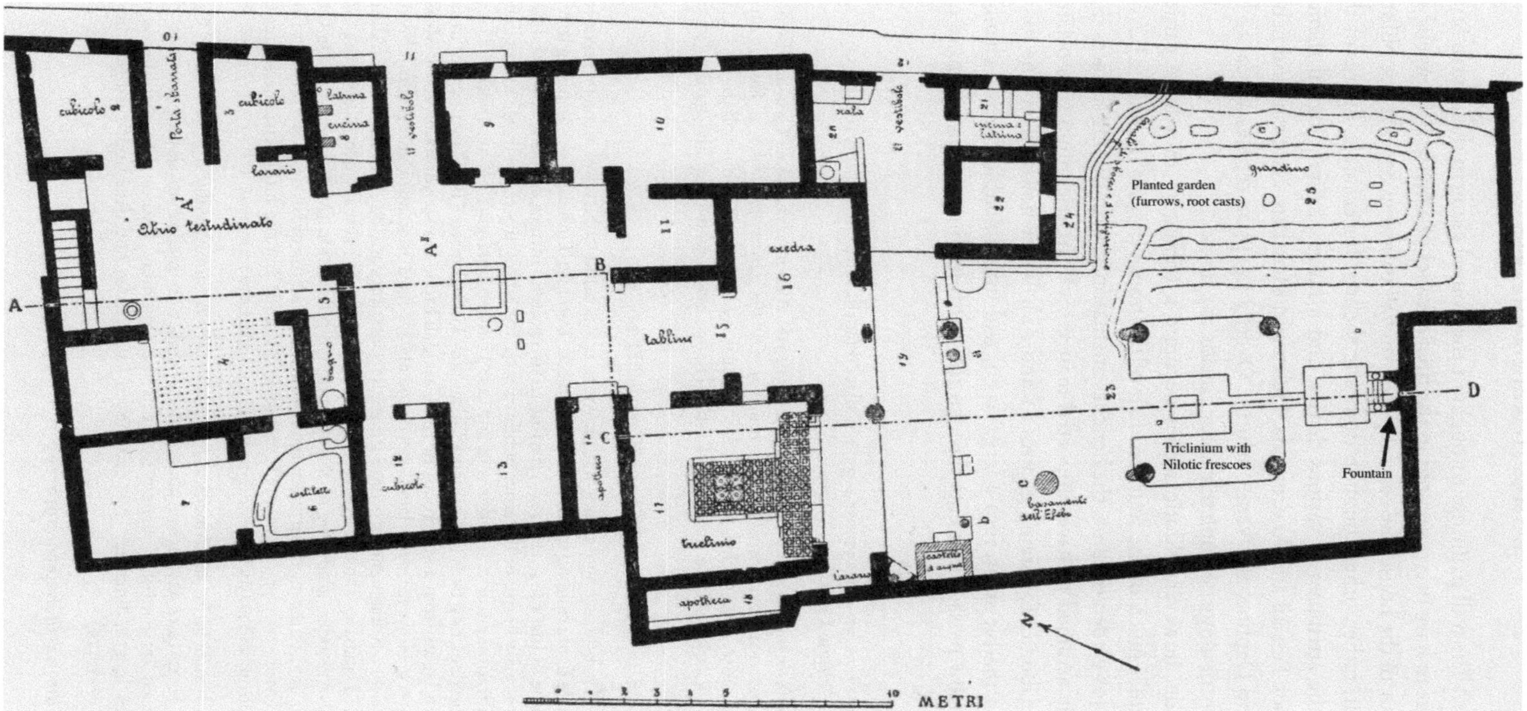

FIGURE 3.1 Plan of the Casa dell'Efebo (I 7.11).

Source: Adapted from Maiuri 1927: fig. 9.

southwest.[4] The wall paintings throughout the house belong to the Fourth Style and appear to have been executed in the final decades of activity at Pompeii.[5] In the interior of the house, the central panels of wall paintings contain not Nilotic motifs but, rather, other common domestic themes: for example, figures of the seasons, the attributes of various deities, and a range of mythological themes that include Perseus and Andromeda, Helen and Menelaus, Narcissus and Echo, Apollo and Daphne, and Venus fishing.[6] Overtly Egyptian themes appear to be completely or almost completely absent inside the house.[7]

Outside, however, Nilotic scenes decorate a triclinium in a large garden that occupies much of the southern half of the house (figures 1.2 and 3.1.). As is common for garden triclinia, the benches are masonry, with sloping surfaces that are usually assumed to have supported mattresses.[8] The triclinium is associated with an elaborate fountain[9] (figures 1.1 and 1.2). Water flowed from the fountain down marble steps, into a large basin, and into a marble-covered channel in the narrow passage that cuts through the south triclinium bench (figures 1.2,

4. Maiuri 1927: 32, 38–40; 1929: 354–356; Wallace-Hadrill 1994: 74; Allison 2004b, s.v. "Casa dell'Efebo," "House I 7,19."

5. *On the paintings generally:* Maiuri 1927: 32–83; 1938; Allison 2004b, s.v. "Casa dell'Efebo"; Versluys 2002: 101–105; Merrills 2017: 106–119. *On the garden paintings as Fourth Style:* Neuerburg 1965: 117; Michel 1980: 400; Versluys 2002: 101; Allison 2004b, s.v. "Casa dell'Efebo"; Clarke 2007b: 103; *pace* Barbet (2009: 205), who dates the garden to the Third Style.

6. Maiuri 1927, 1938; Allison 2004b, s.v. "Casa dell'Efebo."

7. One small subsidiary figure on a wall in room 2, a possible cubiculum, depicts a reddish-brown female figure holding up an indistinct object. Tran tam Tinh takes this painting to represent a statue of Isis (1964: 123–124, referring to two painted images, only one of which I have been able to verify in situ or in earlier publications). The ambiguity of the image makes definitive identification uncertain, and it is in any case a tiny and marginal element within a larger composition depicting fictive architecture. Tran tam Tinh also sees "Isiac" significance in two boat-shaped lamps from a storeroom of the Casa dell'Efebo (1964: 50; followed by McKay 1985: 129–131). However, such lamps need not always have Isiac associations (cf. Johnston 1985: 76, 79–81). A fragmentary bronze Isis statue sometimes attributed to this house (Allison 2004b, s.v. "Database: Table of Artifacts"; Mol 2015: 145, 265) actually comes from the Casa di Sacerdos Amandus, according to Maiuri's original publication (Maiuri 1927: 29; De Caro 2006b: no. 3.37). A recent dissertation refers to a marble statue of Isis or an Isis priestess from this house (Mol 2015: 135, 159, 170, 174, 175n466, 265, 478), but the author cites no source for this attribution, and no such statue appears in published site reports.

8. On the bench form, see Amedick 1993: 179–180. Although Allison (2004b, s.v. "Casa dell'Efebo") expresses skepticism about diners' ability to recline on sloping benches, the parallels cited by Amedick suggest that this was common. Rather than reclining along the benches' length, diners lay "mit den Füssen nach aussen und dem Kopf zur Mitte liegt" (Amedick 1993: 179).

9. Maiuri 1927: 52–61; Soprano 1950: 295–296.

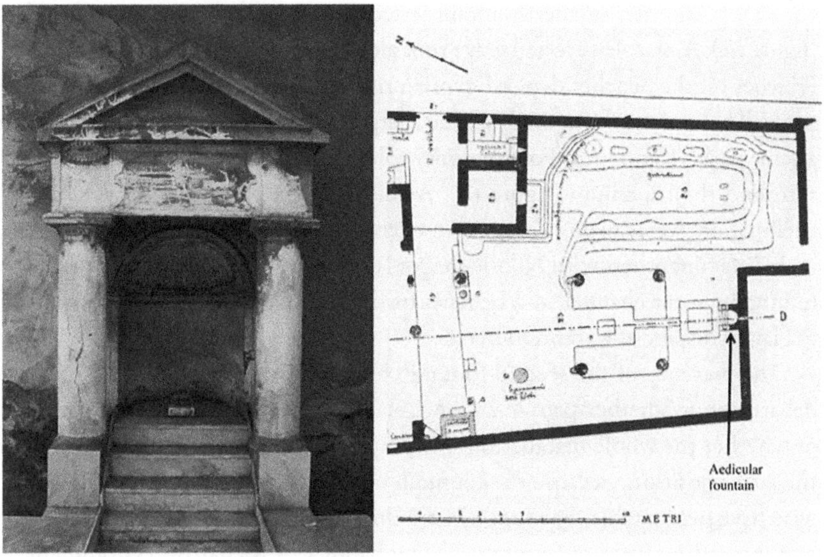

FIGURE 3.2 Aedicular fountain at the south side of the garden of the Casa dell'Efebo (after restoration). The fountain originally displayed a stucco frieze showing Diana between two deer. The garden plan to right (detail of figure 3.1) shows location.

Photo: C.E. Barrett.

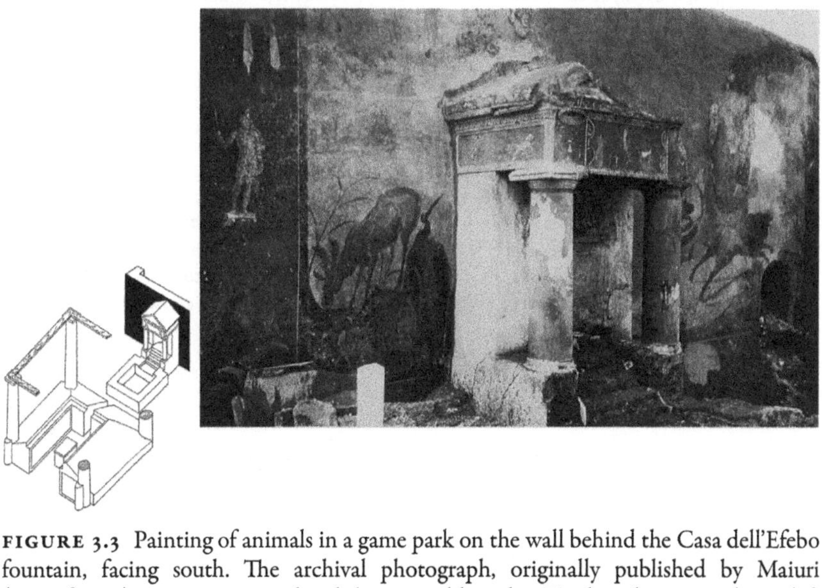

FIGURE 3.3 Painting of animals in a game park on the wall behind the Casa dell'Efebo fountain, facing south. The archival photograph, originally published by Maiuri (1927: fig. 29), preserves more detail than is visible today. The line drawing at lower left illustrates the image's spatial relationship to the fountain and triclinium.

Source: Photograph after A. de Vos 1990: fig. 158). Line drawing modified from Soprano 1950: fig. 31; location of painting shown in black.

3.1, 3.2).[10] Another, smaller fountain jetted from a table between the benches (figure 1.2). A marble-revetted ledge runs along the top of the triclinium couches. Frescoes on the benches depict Egyptian river landscapes (figures 1.3–1.8), while the wall behind the fountain displayed a larger-scale painting of exotic animals in a game park[11] (figure 3.3). Four masonry columns, once supporting a vine trellis,[12] surround the triclinium (figure 1.1). As discussed in chapter 2, similar "water-triclinia" or "water-biclinia" appear in twelve other Pompeian domestic gardens, including four others with Nilotic frescoes or mosaics.[13] The use of marble on the fountains, water channel, and benches further enhances the impressive effect of the lavish display of water resources.

The presence of the second fountain between the benches has led to some debate about whether people could actually use this triclinium for banquets, or whether the whole installation was simply ornamental.[14] Richardson argues the latter position, because the fountain occupied the place that might otherwise have held a table. For him, the triclinium complex was "essentially a new and interesting form of fountain."[15] Dunbabin sees Richardson's argument as "unconvincing," citing against him "the resourcefulness of the slaves in serving food and the Romans' evident propensity for dining damply."[16] Indeed, central tables do not appear to have been essential to Roman dining.[17] Pliny the Younger describes an outdoor triclinium where dishes float on a central pool of water in place of a table.[18] At the Casa dell'Efebo, the marble ledges built into the inward-facing sides of the masonry benches may have served for resting

---

10. For reconstructions of the garden's hydraulic system, see Jansen 2001: 36; 2005: 288–289; 2018: 411–413; Dessales 2013: 405–406.

11. Many scholars interpret depictions of large animals in open landscapes as game parks or *paradeisoi*, evoking the estates of Hellenistic rulers (e.g., Jashemski 1979–93: 1:73; Clarke 1991: 163; Zanker 1998: 178, 184, 186). Others see allusions to *venationes* in the arena (e.g., Leach 2004: 130–132; Merrills 2017: 120). In this case, the painted statues and *clipei* flanking the scene (Maiuri 1927: 60–61) may suggest a park. Such motifs appear elsewhere in paintings of gardens (Zanker 1998: 178).

12. Vines indicated by root casts at the column bases: Maiuri 1927: 52; Jashemski 1979–93: 2:39.

13. See notes 272 and 273, chapter 2.

14. On the functionality of the water features (even after the 62 CE earthquake), see note 44, this chapter.

15. Richardson 1988b: 310–311.

16. Dunbabin 1991: 139n25. For other "tables-fontaines" in Roman architecture, see Dessales 2013: 164.

17. Andersson 1990: 228.

18. Plin. *Ep.* 5.6.37–39; for discussion, see Salza Prina Ricotti 1979, 1987.

dishes, or people may alternatively have taken food from portable trays next to the benches. Such a set of trays, held by small bronze statuettes, was found inside the house.[19] If people perceived the spray from the fountain as a problem, they could simply turn off the fountain when they wanted to dine.[20] Indeed, the small fountain at the Casa dell'Efebo may have drawn on a different water source from the large fountain in the aedicula,[21] in which case diners might have been able to turn off the small fountain without affecting the large one. However, numerous examples of watery dining installations in Roman houses argue against assuming that people necessarily objected to spraying fountains.[22] Diners reclining outside in the heat of a southern Italian summer may well have welcomed some cooling mist. Since the Roman *cena* started in mid-afternoon, the day would still have been hot, and the frequent appearance of shade-arbors over outdoor triclinia[23] suggests a desire to seek cooling settings. Rather than positing a widespread practice of building elaborate, expensive outdoor dining installations that were inherently impossible to use, I consider this and other water-triclinia to have indeed been functional.

The current state of preservation complicates attempts to reconstruct some aspects of the hydraulic system. In particular, it is unclear how exactly water circulated within the triclinium. Excavators found a channel, no longer visible today, leading from the northern side of the eastern bench to the eastern part of the garden[24] (figures 1.2 and 3.1). Water must have traveled to this channel from the marble-covered channel that cuts the southern bench, perhaps powering the small central fountain en route. Maiuri states that the channel through the southern bench communicated with the central fountain, but he provides no description of the mechanism.[25] Based on similar excavated installations at Pompeii and a description in one of the younger Pliny's letters (*Ep.* 5.6.36–37), Richardson interprets the space between the benches as a flooded shallow basin.[26] However,

19. See later in this chapter, "Interactive Statuary: Social Performance and Service."

20. Cf. Amedick 1993: 181 (on a similar installation at Herculaneum).

21. Jansen 2001: 40.

22. See note 272, chapter 2, this volume.

23. For examples, see Soprano 1950; cf. Amedick 1994: 113. On the timing of the *cena*, see Mart. 4.6.8; cf. Dunbabin 2003: 137.

24. Maiuri 1927: 33, fig. 9; A. de Vos 1990: 708; Jashemski 1979–93: 2:41; Jansen 2005: 283, 288.

25. Maiuri 1927: 52, 56.

26. Richardson 1988b: 305–306; on other examples of this structure type, see note 272, chapter 2, this volume.

no traces of a waterproof surface have survived in this space.[27] Additionally, if the basin was flooded even to a very shallow depth, the water would have touched the frescoes, whose figured scenes begin just several centimeters above the ancient surface. While the placement of frescoes immediately adjacent to water does have parallels,[28] other means of water circulation may be more likely. A water channel or pipe might once have led from the southern bench to the central fountain, and thence to the channel north of the eastern bench.[29] Alternatively, Jansen suggests that the water flowed into a gutter around the triclinium, although it is not clear how such an arrangement would have powered the central fountain. She proposes that "the overflow was led into a gutter around an outdoor water *triclinium*, which had a fountain of its own in the middle of the *triclinium* couches."[30] Although Jansen does not state whether she envisions this gutter running around the outside or the inside of the benches, several considerations make the interior of the triclinium more plausible. The marble channel from the basin clearly leads inside the triclinium couches. Additionally, while surviving evidence does not provide convincing support for the flooded basin proposed by Richardson, he is right to point out that most of the archaeological and textual parallels for this assemblage involve water flowing between the benches. Accordingly, if Jansen is correct to envision a gutter around the base of the benches, it most likely occupied the triclinium interior. A sufficiently deep gutter might have also helped ensure that the water did not touch the paintings directly.[31]

While surviving evidence permits multiple possibilities for the circulation of water within or around the benches, the ultimate destination of that water appears more certain. Excavators found water channels leading from the triclinium to the eastern part of the garden, where a marble fence demarcated an area containing furrows and root cavities[32] (figures 1.2, 3.4, 3.5). Four painted marble herms topped the fence (figure 3.4). Jashemski identifies the furrows as "characteristic of

27. Traces of a white plasterlike substance may just possibly be visible between the benches in a photo from the excavations (Maiuri 1927: pl. V), but the image is too low-resolution for identification.

28. E.g., Versluys 2002: nos. 55, 56, 66; Clarke 2007c: 164; Dessales 2013: nos. 30, 84.

29. Dessales (2013: 405) describes "un canal traversant le *triclinium* d'été"; cf. Versluys 2002: 103.

30. Jansen 2005: 288; cf. Jansen 2018: 411.

31. For a gutter on the interior of another partially closed space decorated with Nilotic scenes, cf. the Casa del Medico (chapter 4, this volume, "Reinterpreting the Courtyard of the Casa del Medico").

32. *Water channels:* Maiuri 1927: 33, fig. 9; A. de Vos 1990: 708; Jashemski 1979–93: 2:41; Jansen 2005: 283, fig. 12, 288. *Root cavities:* Maiuri 1927: 61.

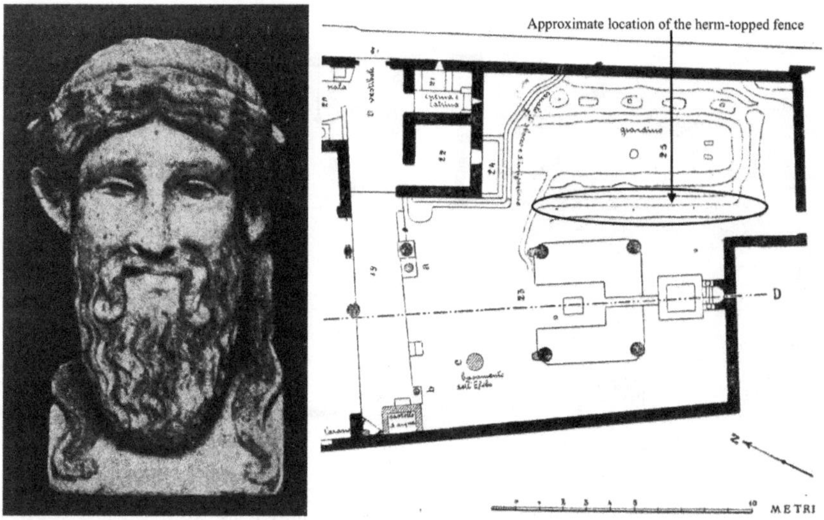

FIGURE 3.4 Head of a marble herm, 0.19 m high (Maiuri 1927: 74), once part of a fence dividing the eastern and western sides of the Casa dell'Efebo garden. Some paint preserved (yellow on hair, mustache, and beard; black on eyebrows and pupils; red on fillet). The herm is no. 4145 in the inventories from the excavations at Pompeii. For approximate location, see garden plan to right (detail of figure 3.1).
Source: After Maiuri 1927: fig. 37.

vegetable gardens."[33] The eastern garden area also contained a circular travertine table, a curved marble seat, and a small terracotta altar (figure 3.5).[34] Such tables and seats are paralleled in other gardens, and Pliny the Younger refers to marble seats scattered around the grounds of one of his villas.[35]

Another water basin at the northeastern edge of the garden, equipped with two taps, probably also served to irrigate the plants.[36] Stacked next to the basin

33. Jashemski 1979–93: 2:40; cf. Jansen 2005: 288.

34. Maiuri 1927: 61; Jashemski 1979–93: 2:40; Foss 1997: 214; Allison 2004b, s.v. "Database: table of artifacts"; Bassani 2008: 148.

35. Plin. *Ep.* 5.6.40; Farrar 1996: 15; 1998: 37–40.

36. Maiuri 1927: 61; Jashemski 1979–93: 2:41; Jansen 2005: 288. It remains unclear how this basin was supplied (Jansen 2001: 40n37). Allison (2004b, s.v. "Casa dell'Efebo") sees the presence of utilitarian activities as possible evidence for "a downgrading prior to the eruption." However, extensive evidence for the economic functions of Roman gardens (see note 38, this chapter) suggests such activities were not incompatible with the other purposes of the garden.

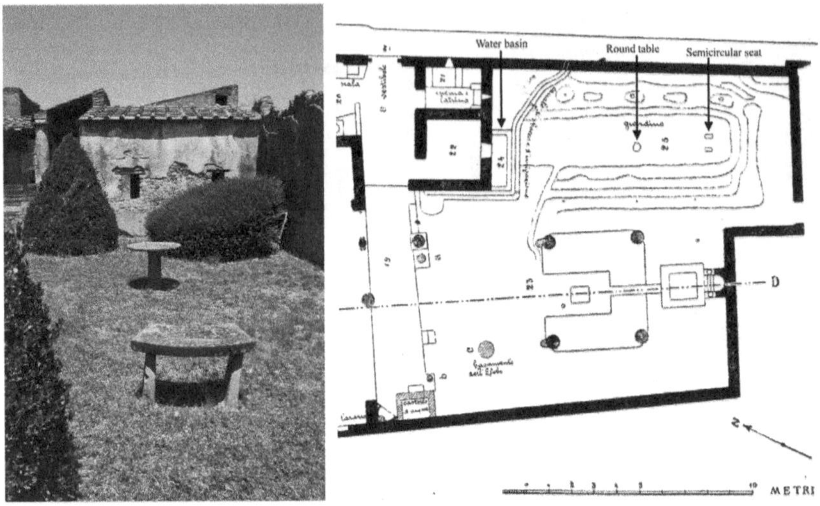

FIGURE 3.5 Eastern side of the garden of the Casa dell'Efebo, facing northwest toward the water basin (no. 24 on plan at right, detail of figure 3.1) and cubiculum (no. 22 on plan at right). A terracotta altar from this garden does not appear on Maiuri's plan, but Allison describes its findspot as "E side [of garden], centre, 1.53m from E wall" (2004b, s.v. "Database: table of artifacts").
Photo: C.E. Barrett.

were a group of storage amphorae and a ceramic pot.[37] These vessels, the basin, and the planted furrows may all testify to economic as well as leisure activities within the garden. Roman gardens served economic and luxury functions, and Roman intellectual discourse presented gardens not only as sites of Hellenizing *luxuria* and *otium* but also as agriculturally productive, traditionally Roman examples of successful human collaboration with nature.[38]

To the north of the water-triclinium, a portico separates the garden from the rest of the house. The western end of the portico leads to another dining area, the indoor Triclinium 17, which boasted an elaborate *opus sectile* floor mosaic of colored marbles and glass, as well as a view onto the garden (room 17 on Maiuri's

37. Allison 2004b, s.v. "Casa dell'Efebo."

38. *Economic functions of gardens:* Jashemski 1974; 1977; 1979–93: 1:167–181, 1:183–199, 1:233–243, 1:267–288; 1981; 1994: 53–57; 2018; von Stackelberg 2009: 10–16. *Gardens in Roman discourse:* Grimal 1969: 56–61; La Rocca 1986; Purcell 1987: 203; Beagon 1992: 79–91; Gowers 2000; von Stackelberg 2009: 10–16, 64–65; Spencer 2010: 139–142; Zarmakoupi 2010; 2014: 103–140; Myers 2018.

plan, figure 3.1).[39] Another, smaller room at the eastern end of the portico, sometimes identified as a cubiculum,[40] has a window looking out onto the garden. This combination of garden, triclinium, and (possible) cubiculum recalls the younger Pliny's description of a similar ensemble at one of his villas.[41] At the extreme western of the portico, a small aedicular shrine, with an empty base for one or more statuettes, is set into a wall niche[42] (figure 3.6). Similarities in form, construction, and stucco decoration between this shrine and the aedicular fountain suggest that they were constructed at the same time.[43] Immediately in front of that shrine, and partly obscuring it, stands a water tower whose reservoir fed the fountain and enabled the water-triclinium to operate[44] (figure 3.6a). A painting on the water tower depicts Mars and Venus (figure 3.6b).

Other pipes from the water tower led to the impluvium inside the house, as well as the garden of House I 7.19, which had seemingly just been added to the Casa dell'Efebo (figure 1.2). At the time of the eruption, renovations to incorporate I 7.19 appear to have been underway, as evidenced by this pipe rerouting. Construction materials and building debris are also present in several places in the garden.[45]

Either because of this ongoing construction or because of pre-eruption seismic disturbance, much of the garden statuary appears to have been temporarily stored in other rooms at the time of the eruption. Evidence for the former presence of garden statuary comes from five masonry bases for statues and statuettes: two on low columns adjacent to the northern columns of the pergola, two at the northern side of the garden near each end of the portico, and one large base, near the water-triclinium (figure 3.7). The large base fits the famous bronze ephebe statue that gives the Casa dell'Efebo its name

39. Maiuri 1927: 45–49. On the mosaic floor, see A. de Vos 1990: 682–685.

40. Amedick 1993: 183.

41. Plin. *Ep.* 5.6.36–40; Amedick 1993: 183.

42. Maiuri 1927: 51–52; Giacobello 2008: 254.

43. Dessales 2013: 263–264.

44. Maiuri 1927: 51; Jansen 2001: 36, 40n36; 2005; Dessales 2013: 263–265. Regardless of any damage the public water supply sustained in the 62 CE earthquake, this private tank probably still enabled the water-triclinium to function (Dessales 2006: 367–368; 2013: 264–265; Keenan-Jones 2015: 198, with n68). *On the painting:* Maiuri 1927: 49–51; A. de Vos 1990: 698–701. *Associated water pipes:* Jansen 2005: 288.

45. Piles of coarse plaster adjacent to the east and northwest walls of the garden; a pile of roof tiles and what Maiuri calls "blocchi di tufo" near the door to I 7.19 (Maiuri 1927: 65; Allison 2004b, s.v. "Casa dell'Efebo").

(a) (b)

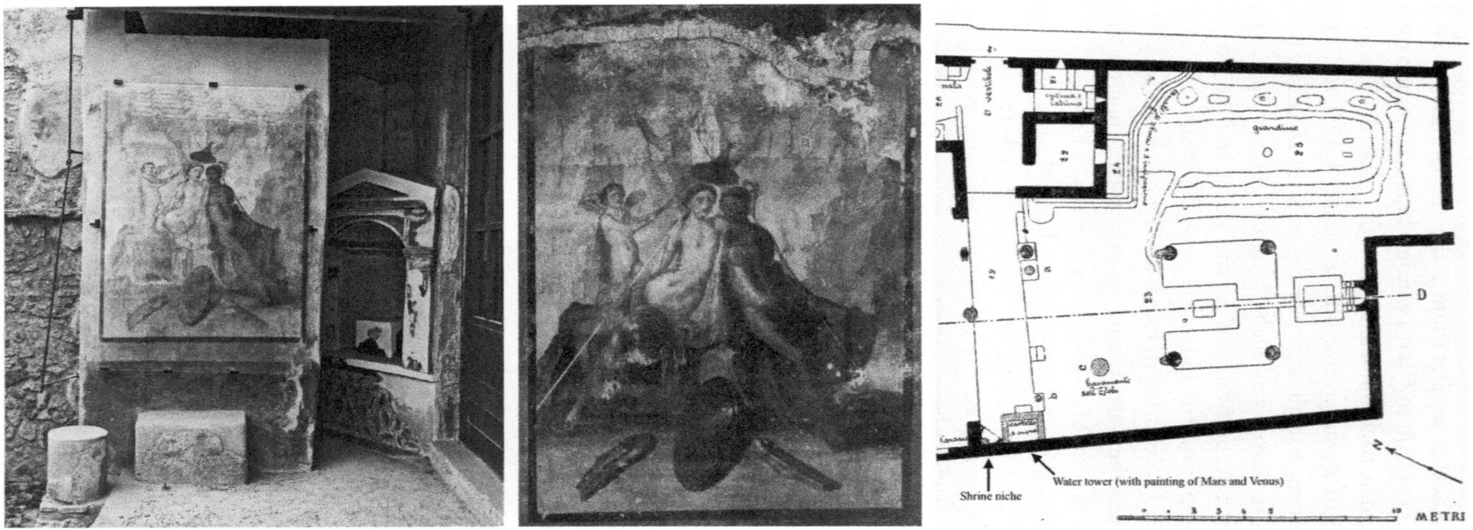

FIGURE 3.6 Water tower and aedicular shrine, in the northwest corner of the garden of the Casa dell'Efebo. (a) Photograph of both structures in 2016. (b) Photograph of the painting of Mars and Venus on the water tower shortly after excavation. The garden plan to right (detail of figure 3.1) shows locations.

Source: Photo (a) C.E. Barrett. (b) After Maiuri 1927: fig. 23.

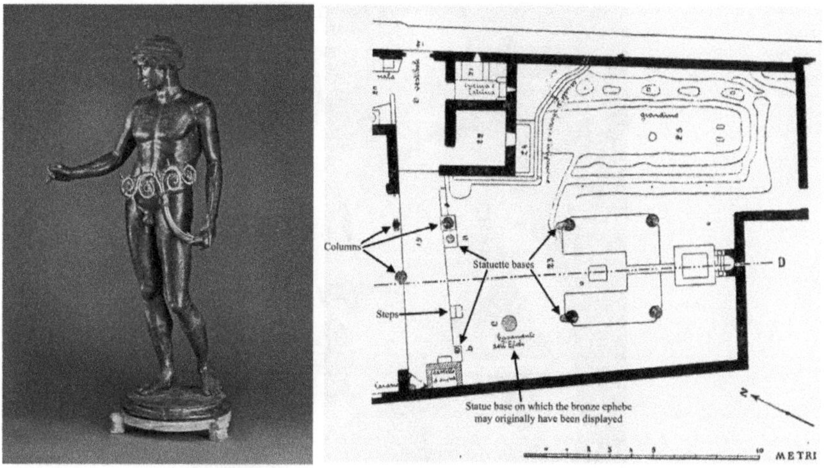

FIGURE 3.7 Bronze statue of an ephebe (MANN 143753), 1.49 m high (Maiuri 1926: 338). The right hand once also held a candelabrum. The plan to right (detail of figure 3.1) shows the location of statue and statuette bases in the garden of the Casa dell'Efebo.

Source: Adapted from Melillo 2013: fig. 5.1; photo reproduced by permission of the Ministero dei Beni e delle Attività Culturali e del Turismo—Museo Archeologico Nazionale di Napoli.

(figure 3.7).[46] This statue, whose hands originally each supported a candelabrum, was found broken inside the house under a large cloth together with disassembled furniture. It had seemingly been moved from its original display context, either because of garden renovations or for protection before or during the eruption.[47]

Additionally, four marble statuettes were found in the indoor triclinium (figure 3.1, room 17): a Pan (figure 3.8), a fragmentary figure with two caprid legs (undoubtedly another Pan), a reclining satyr, and a deer with a fawn—all

46. Maiuri 1926; 1927: 52–53, 63–66; Melillo 2013. While restoration work to the base makes it difficult to measure the diameter precisely today, its approximate diameter is 69 cm (measured in situ by author).

47. See Maiuri 1926; 1927: 52–53, 63–66; Jashemski 1979–93: 2:39; Melillo 2013. On the findspot in room A^{II} (figure 3.1), see Maiuri 1927: 63. On garden renovations as a possible motivation for moving the statue, see Maiuri 1927: 65. On protection from seismic disturbances (in the eruption or earlier) as possible motivation, see Allison 2004b, s.v. "Casa dell'Efebo"; Maiuri 1927: 65–66; Dwyer 2012: 306.

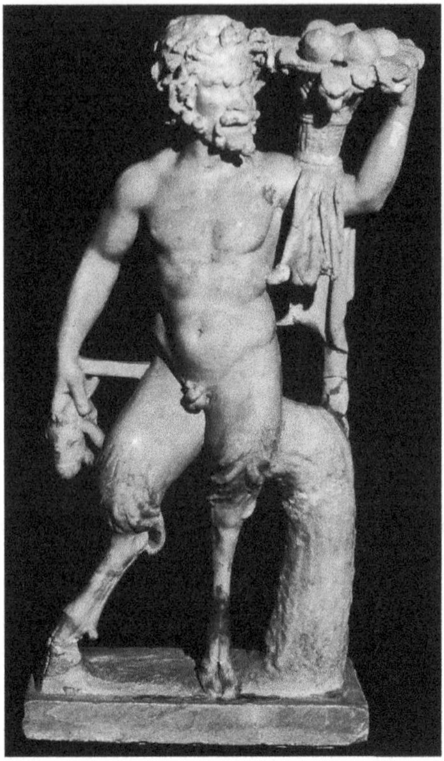

FIGURE 3.8 Pan statuette, marble, 0.56 m high (Maiuri 1927: 72). Found in triclinium 17 of the Casa dell'Efebo, but probably originally used in the garden. The statuette is no. 3682 in the inventories from the excavations at Pompeii.

Photo: Stanley Jashemski; courtesy of Papers of Wilhelmina Jashemski, Special Collections and University Archives, University of Maryland Libraries.

under a meter high.[48] Traces of gilding appear on the deer and fawn, satyr, and one Pan.[49] The original display context of these sculptures was probably the garden.[50] Statues depicting animals and members of the Bacchic *thiasos* are

48. *Pan*: height 0.56 m; base height 0.04 m. *Base of fragmentary Pan:* length 0.28 m, height 0.105 m. *Reclining satyr:* length 0.32 m, height 0.16 m, base height 0.27 m. *Deer and fawn:* length 0.35 m, height 0.22 m. (Maiuri 1927: 71–74).

49. Maiuri 1927: 72–74; Jashemski 1979–93: 2:40.

50. Jashemski 1979–93: 2:40; Allison 2004b, s.v. "Casa dell'Efebo." The house also produced several other statuettes: the arm of a bronze statue, perhaps a smaller *lychnouchos,* found in room 13; a bronze figurine of a nude soldier, near the entrance of room 2 and possibly fallen from an upper floor; five plaster statuettes (otherwise undescribed) from room 2; and a terracotta comic mask from room 14 (see figure 3.1 for room locations). Other finds from the

common in gardens, and the sculptures in the triclinium appear to have been already broken when placed there.[51] The match between these statuettes and the four empty, size-compatible[52] statuette bases in the garden reinforces the likelihood that the statuettes were originally garden sculptures. The indoor triclinium may have been temporarily repurposed for storing damaged statuary, perhaps following one of the seismic events preceding the eruption.[53] As the triclinium was next to the garden, it would have been a convenient place for such temporary storage.

Finally, another set of objects that may have seen use in the garden was found in room 13 (figure 3.1, room 13): an apparent storeroom filled with numerous vessels, lamps, and assorted domestic tools.[54] A carbonized wooden box in this room contained four bronze statuettes of emaciated, macrophallic male figures who clutch their throats while carrying silver serving trays[55] (figure 3.9). The figures were originally gilded, with silver eyes and bases.[56] Because their trays are functional, they were probably used to serve dishes at meals, likely including meals at the outdoor triclinium.[57]

Most of the architecture and Fourth Style paintings in the garden appear to be roughly contemporary. The stucco decoration on the aedicular shrine

---

house interior included (inter alia) numerous lamps, pieces of furniture, and vessels in bronze, glass, and ceramic (Maiuri 1927: 32–83; for a full list, see Allison 2004b, s.v. "Database: Table of Artifacts"). A marble relief of Cupid, found in room 17 with the Bacchic and animal statues (Allison 2004b, s.v. "Database: Table of Artifacts"), may also have been in temporary storage there, but its original location is unknown. It could also have come from the garden, but this is by no means certain, and (unlike the marble statuettes, whose likely bases were still in situ) there is no indication of precisely where in the garden it could have been located.

51. *Bacchic and animal statues in gardens:* Hill 1981: 86; Dwyer 1982: 123–124, 126–127. *Condition of the sculptures from this triclinium:* Allison 2004b, s.v. "Casa dell'Efebo."

52. While restoration work to the bases makes it difficult to measure their dimensions precisely today, the two bases attached to the triclinium columns are approximately 33 cm across at their widest point, while the two bases adjacent to the portico have diameters of approximately 36–38 cm (variation caused by slight irregularities in shape; measured in situ by author). For a plan of their locations, see figure 3.7.

53. Jashemski 1979–1993: 2:40; Allison 2004b, s.v. "Casa dell'Efebo." Two other rooms also seem to have been repurposed for storing damaged or salvaged artifacts (Allison 2004b, s.v. "Casa dell'Efebo").

54. Allison 2004b, s.v. "Casa dell'Efebo." Maiuri saw it as a tablinum repurposed for storage (1927: 43).

55. Maiuri 1927: 66–67; Dwyer 1982: 123; A. de Vos 1990: 620; Jashemski 1979–93: 2:40; Allison 2004b, s.v. "Casa dell'Efebo."

56. Maiuri 1927: 66.

57. On other uses of Roman statues as tray-holders, see also note 122, this chapter.

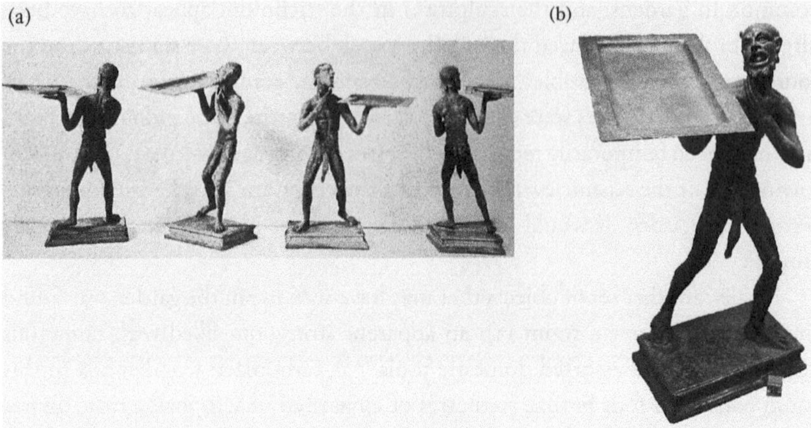

FIGURE 3.9 Bronze and silver statuettes of throat-clutching figures holding serving trays (MANN 143758–143761), found in room 13 of the Casa dell'Efebo. The height of each individual statuette is 255 mm, including the base (Maiuri 1925: 270). (a) Statuettes shortly after excavation. (b) Alternate view of one statuette.

Sources: (a) After Maiuri 1925: fig. 1; reproduced by permission of the Redazione del Bollettino d'Arte. (b) Photograph by C.E. Barrett; reproduced by permission of the Ministero dei Beni e delle Attività Culturali e del Turismo—Museo Archeologico Nazionale di Napoli.

and fountain, which appear to have been constructed simultaneously, has been stylistically dated to the Neronian period; and the interdependence of the fountain and triclinium suggests that they, too, were built at the same time.[58] The painting on the wall behind the fountain appears to have been designed around the fountain, in which case it would also be contemporary with, or later than, the water-triclinium. Many of the garden's most prominent features may thus have been added at the same time, or at least in relatively rapid succession. However, one slightly later addition must be the water tower, which must postdate the shrine in front of which it was built. Its construction may be a response to earthquake damage to the water system of Pompeii, either in 62 CE or (as some suggest) in the years just before the eruption.[59] The statues are also harder to date precisely, although the construction of masonry bases in the garden, including two attached to the triclinium, does suggest that the designers of the water-triclinium complex envisioned some form of statuary as an essential component of the ensemble. Regardless, the original date at

58. Dessales 2013: 263–265. Dessales notes, however, that one cannot rule out the possibility that the stucco came from a renovation rather than the initial construction.

59. Dessales 2013: 263–265.

which objects and images entered the garden need not have determined how, *once there*, they interacted as an assemblage. The very act of collecting originally disparate objects and images into one place would have generated new interrelationships and associations that would, in turn, prompt new viewer responses.

The Nilescapes of the Casa dell'Efebo Garden Triclinium

Painted on the triclinium benches are some of the most detailed and elaborate Nilotic scenes from Pompeii, including representations of pygmies, shrines and statues of Egyptian deities, sexual activity, drinking, musical performance, and conflicts with river beasts. The images on the interior faces of the benches form a single continuous frieze (figures 1.4–1.7), while the north-facing scenes on the eastern and western benches appear to be separate vignettes (figures 1.3, 1.8). These two scenes are most visible to someone entering the garden and approaching the triclinium from the north (producing the view shown in figure 1.1), while the continuous frieze would be most visible to reclining diners.

The northern face of the easternmost bench (figure 1.3) depicts a sanctuary containing a shrine with a statue of Isis-Fortuna (A) holding a cornucopia and either a *patera* or, more likely, a rudder.[60] The columns flanking her take the form of either Egyptian *djed*-pillars or sistra, with female figures at their bases. Outside the shrine is a statue of a bird (B), perhaps a falcon,[61] Horus's animal form. A worshiper lays a piece of cloth on altar (C). Slightly above and to the left of the figure at the altar, a bald male figure[62] (D), holding a staff, gazes at the shrine. His bald head may suggest priestly status. To the left of the sanctuary stands an obelisk (E), next to which two figures are walking. To the right of the sanctuary is a small aedicula (F) containing a statue. Flora and fauna include trees, among them two palms (G) and cypresses growing behind the shrine, and three ibises (H, I, J).

The images on the interior of the eastern bench depict a detailed river scene (figure 1.4). The river runs across the middle of the frieze, separating the far bank at the top and the near bank at the bottom. At the far left, a woman spins thread

---

60. On the statue as Isis-Fortuna, see Maiuri 1927: 53; Tran tam Tinh 1964: 124; Beaurin 2008: 289–290, no. 30. The object in her right hand resembles a *patera*, but close examination suggests it is more likely the handle of a rudder, the rest of whose form has been damaged by paint loss.

61. *Pace* Tran tam Tinh, who identified an ibis (1964: 124). A. de Vos's identification of a falcon (1990: 714) has better parallels (cf. Meyboom 1995: 307n119).

62. *Pace* Beaurin's identification of this figure as "une statuette féminine" (2008: 289).

beneath a canopy (A) while a dog (B) sits at her right (i.e., the viewer's left). Ducks swim amid lotuses and other water plants (C) and an ibis stands on the back of a hippopotamus (D). Farther right, a boat with an animal-headed prow (E) sails past. Numerous human figures traverse the banks, including one, just to right of center, who crosses a bridge over a crocodile (F). Above and slightly to the right of this figure, on the far bank, eight people recline at a stibadium (G) while a servant approaches from the left carrying two jugs to the table.[63] Farther right, two figures in the foreground may be fishing (H, I). Among the numerous structures are several cult installations and divine images. Two columned structures next to one of the possible fishermen (I) appear to be small shrines, including one pedimented structure (immediately behind the human figure) to which a *pinax* has been attached. Near the center, a tower (J) inside an enclosure may also be a shrine, as a statue outside its door represents a reclining animal deity (possibly feline, though the image is poorly preserved). A palm tree grows behind this building. A statue of another animal deity, the Apis bull (K), stands at the center right of the scene; the columned structures to the left and right of this statue may also be shrines. At the far right, a worshiper approaches an anthropomorphic statue in an aedicula (L). Below and to the right of the porticoed complex (M) in the upper right is an obelisk, visible in figure 1.5 (H). On the far bank of the river are two other architectural ensembles, each accompanied by two human figures: a temple within an enclosure (N) and a cluster of towers (O).

The southern bench is divided by the water channel. The eastern side (figure 1.5) depicts the continuation of the riverbank scene. Built into a long wall that divides the upper and lower halves of the frieze is a small aedicula containing a statue (A). Above, an obelisk (B) stands next to a poorly preserved building complex at the upper left (B). The scene also includes several walking human figures (C, D, E, F), the leftmost of whom (C) carries a staff or spear and wears a cap or helmet.[64] In the foreground, an amphora (G) rests on the lower roof of a two-story building and leans against the upper story. A herm outside this building, reported by Versluys,[65] is no longer visible. At the left of the photograph is the obelisk (H) at the rightmost edge of the eastern bench.

The fresco on the western side of the bench (figure 1.6a) is poorly preserved, with only traces of a male figure confronting a crocodile (A) who swims up from

---

63. Following Dunbabin (1991; 1996, 70–72n22, 72n30) in reserving the term "stibadium" for sigma-shaped benches (*pace* Amedick 1993; 1994, 112–115).

64. A. de Vos interprets this figure as a soldier (1990: 723).

65. Versluys 2002: 101.

the bottom right. Above the crocodile, the male figure crouches on the riverbank, holding a circular shield and spear (figure 1.6a, B, and figure 1.6b).[66]

The landscape continues on the inner face of the western bench (figure 1.7). To the left, a shrine containing a statue (A) stands inside a porticoed complex. Two women (B) make offerings at an altar, while another woman (C) feeds two ibises.[67] Beneath a canopy at right, five men recline on a stibadium (D) around a table that holds a *krater* and ladle, as a crocodile (E) approaches them. Nearby, a palm tree grows from a large *dolium* (F) embedded in the earth. Farther right, two figures walk on either side of a vase-topped column (G). Behind them is a line of monuments: an anthropomorphic statue[68] (H), a small aedicular shrine (I), a large horned altar (J), a tripod (K), and an aedicula (L). Two figures and a dog (M) approach what appears to be another aedicula (N) set into a wall in the background, while another figure with a dog walks briskly to the right of that wall. Below that figure is an obelisk (O) and, farther right, a traveler pausing before another possible shrine (P). At the far right, a statue in an aedicula (Q) is set into a wall patrolled by two helmeted figures (R, S) carrying staffs or spears, with a cluster of columned buildings (T) visible in the upper background. Assorted human figures traverse the landscape, and lotuses (U, V) bloom on the banks.

Finally, the northern face of the western bench (figure 1.8) shows a scene of sex, music, and conviviality. A couple (A) engages in sexual intercourse beneath an awning, a woman (B) plays the aulos, and a bald male (C) approaches with an amphora, while additional female (D) and male (E) figures look on to the right. To the left of this group, a man operates a water-screw (F), an irrigation device that moves water by rotating a screw-shaped device inside a pipe.[69] Meyboom and Versluys propose that the presence of this motif implies a relationship between the fertility of the couple and the fertilizing properties of the floodwater

66. *Pace* Maiuri (1927: 57), who identified the shield and spear as a walking stick and bag; I agree with A. de Vos 1990: 723; Versluys 2002: 103.

67. Her hand gesture, and the fact that one bird is pecking the ground, suggest feeding. Cf. Roman images of Egyptian priests feeding sacred animals (e.g., Versluys and Meyboom 2000: 116–121).

68. This object might alternatively be a *tropaion*, but the apparent presence of arms and legs suggests a statue.

69. Jacono 1927; Maiuri 1927: 60; Oleson 1984: 293, 296, 299–300; A. de Vos 1990: 727. The screw discharges water into a partly buried *pithos* behind the aulos-player; Oleson suggests the *pithos* is "probably a bottomless jar serving as a well-head for the discharge channel" (1984: 299). A Ptolemaic figurine from Naukratis depicts a similar water-screw (Thomas 2015c: fig. 54).

irrigating the fields.[70] Flanking the human figures, on either side of the scene, are a small prostyle temple shown in side view (G) and a tall horned altar (H) behind which palms are growing. The foreground depicts a dock, off whose banks three pairs of ducks swim amid flowering lotuses (I, J, K).

The Casa dell'Efebo frescoes thus provide a near-encyclopedic compilation of Nilotic motifs. The river is in flood, as indicated by blooming lotuses (figures 1.4, C, 1.7, U and V) and the irrigation of the fields (figure 1.8, F). Egyptian flora and fauna include crocodiles, a hippopotamus, ibises, palms, and lotuses. The riverine landscape is filled with religious activity, as indicated by sanctuaries, aediculae, divine statues, and people making offerings. Conflicts with river beasts include the spearman facing the crocodile on the southern bench (figure 1.6) and the possibly hostile encounter of crocodile and banqueters on the western bench (figure 1.7, D, E). Themes of sexual activity, musical performance, and alcohol consumption on the riverbanks are also prominent.

As already discussed in chapter 2, the proportions of most of the people in these frescoes characterize them as *pygmaei*, with large heads and short limbs relative to their torsos. A minority of figures have more "standard" proportions, appearing slimmer, longer limbed, and smaller headed (e.g., figures 1.4, A; 1.8, A, B, C, and probably E; and the statue of Isis in figure 1.3, A).[71] Additionally, even though the relative scale of people, buildings, and landscape is frequently inconsistent throughout the frescoes,[72] it is still worth noting that some of these more idealized figures are also depicted as taller. (Compare, for example, two figures in the foreground of figure 1.4: the seemingly non-pygmy woman, A, seated at the far left, and the much smaller male, L, standing at the aedicula to the far right.) As in some other Nilescapes, these scenes thus depict a population composed of both pygmies and non-pygmies.[73]

Dining on the "Nile"

The architecture, water installations, paintings, and statuary in the Casa dell'Efebo garden work together to evoke a *locus amoenus*. In so doing, they evoke a wide range of imagined landscapes, transforming the garden into a microcosm of the

70. Meyboom and Versluys 2007: 191. The combination of water-screw and erotic imagery also appeared in a now-lost Nilotic painting from the Casa di Sallustio (Versluys 2002: 114–115, no. 42; Meyboom and Versluys 2007: 191).

71. Pygmies and figures of average height co-occur in some other Nilotic frescoes as well: e.g., Versluys 2002: no. 60 (Casa dello Scultore).

72. See note 13, chapter 2.

73. See chapter 2, this volume, "Navigating 'Nilotic Scenes.'"

larger Roman world.[74] Some aspects of the setting provide three-dimensional echoes of the Nilotic banquet scenes. Other geographical and cultural allusions suggest additional imagined environments—cultivated to wild, local to faraway—and invite visitors to imagine themselves at once in the Egyptian countryside, the *paradeisos* of a Hellenistic dynast, the local environs of Pompeii itself, and the wild landscape roamed by Bacchus and his followers. Yet further aspects of the imagery, such as the miniature scale of the garden paintings and statuettes, simultaneously distance the viewer from the varied settings they evoke. Users of the garden would have found themselves transported to a place that was at the same time everywhere and nowhere: a space that was explicitly marked as taking place "somewhere else," but one that also provided an opportunity for engaging with some very immediate questions about culture, identity, and the relationship of the individual to the broader *oikoumene*.

As with many other garden Nilotica (chapter 2), the context of the frescoes creates parallels between the painted landscape and the real garden in which it appears. Real flowing water formed the context for banqueters' viewing of the frescoes' riverine imagery, bringing the painted Nile to life.[75] Although some details of the hydraulic system remain difficult to reconstruct, the dining installation is clearly constructed around displays of water: two fountains, a basin between the aedicular fountain and the triclinium, and channels carrying water first to the triclinium, then to the garden. Domestic water channels could be called *euripus* or *nilus*, names assimilating them to distant natural marvels: *nilus*, meaning the Nile River, and *euripus*, referring to a Greek strait (between Euboea and Boeotia) whose tides frequently reversed direction.[76] Here, the Nilotic frescoes reinforce such potential allusions to faraway natural *mirabilia*. Additionally, the sex scene and two outdoor banquet scenes place revelers beneath a trellis or pergola closely resembling the vine-shaded, pergola-covered structure on which they are painted. The setting of the triclinium thus shapes viewers' reception of the watery landscape it portrays, framing the frescoes as echoes of the real-life garden. The actual triclinium thus models the very scenes it depicts, framing the guests' own activities as a mirror of—and commentary on—the imagery around them. Indeed, given their placement on the structure supporting the diners' bodies, the Nilotic frescoes actually form a frame for their activities in a literal sense; everything that the reclining diners do or say takes place within a space circumscribed, defined, and supported by Nilotica.

74. On gardens as microcosms of empire, see Hales 2003: 156–157; Zarmakoupi 2014: 111–112.

75. For previous observations that diners might imagine themselves on a model "Nile," see Tran tam Tinh 1964: 50; Amedick 1993: 182; Versluys 2002: 103; Poole 2004: 237.

76. Cic. *QFr.* 3.9.7, *Leg.* 2.2; Dessales 2013: 59–62; Zarmakoupi 2014: 157–163. On *mirabilia* and faraway lands, see Naas 2011.

Such strategies of correlation and mirroring—in which painted structures form visual analogies with real structures in their vicinity, or motifs appear transformed in multiple media—are common in Roman domestic visual culture.[77] For example, paintings of gardens often appear within actual gardens. Similarly, Bettina Bergmann's work shows that at the elite villa at Oplontis, paintings of precious stones and metals would have echoed the real-life luxuries around them; frescoes of flower-entwined columns find three-dimensional parallels in real marble columns covered (according to archaeological evidence) in vines; and marble statues of centaurs in one garden reappear in fresco form in another garden.[78] The artistic recreation of natural marvels also recalls another theme well known to educated first-century audiences: the role of *ars* (human skill) in rivaling, shaping, and enhancing *natura* (nature or intrinsic essence).[79] The younger Pliny's letters describe outdoor water-triclinia as prompting pleasurable reflection on the marvels of technology, not just imitating nature but even outdoing nature.[80] The presence of artificial "grotto"-like decoration in some Pompeian water-triclinia further suggests that ancient viewers could perceive domestic water installations as emulating or rivaling natural water features.[81] Confronted with two complementary Nilotic images in different media, the viewer is encouraged to contemplate the relationship of humans to the landscape, reality to representation, and the familiar and domestic to the distant and marvelous. In the Casa dell'Efebo, the juxtaposition of complementary riverine or aquatic images—in different media and at different scales—encourages viewers to contemplate the relationship of humans to landscape, reality to representation, and the familiar and domestic to the distant and marvelous.

Painted directly onto the benches of the triclinium, the Nilotic fresco would have been most clearly visible to people approaching, reclining on, or standing

---

77. Cf. Bergmann 2002a: 22–24; 2002b: 103–120; Marzano 2014: 199.

78. Bergmann 2002a: 22; 2002b: 110–114; 2016: 99–100, 105; 2018: 298–299.

79. On the (sometimes complementary, sometimes rivalrous) relationship of *ars* and *natura*, see Newlands 1984: 164–174; 2002: 154–98; Leach 2004: 125 (garden paintings as both mimetic and "overtly fictive"); 2010; Bergmann 1991: 57–66; 1992; 2002b: 88–90; Isager 1991: 32–47; Beagon 1992: 55–91 (esp. 79–91 on gardens); Kuttner 2003: 106–107; Myers 2005: 107–108; Spencer 2010: 122–123; Tronchin 2011a; Bussels 2012: 26–31; Newby 2012: 356–363. On *nili/euripi* as (potentially problematic) artificial representations of natural features, see Cic. *Leg.* 2.2; Spencer 2010: 67–69; Zarmakoupi 2014: 162–163.

80. Plin. *Ep.* 4.30, 5.6.37. The latter letter goes on to praise the view from a nearby cubiculum as superior to its natural models: one can imagine oneself in a grove, *in nemore*, without having to worry about rainstorms (5.6.39). See further Spencer 2010: 122–123.

81. For examples, see Rakob 1964; Lavagne 1988: 634–636. On the construction of domestic water features to resemble artificial grottoes, see Prop. 3.2.14; Lavagne 1988; Amedick 1994: 113; Leach 2010: 66–67; Spencer 2010: 160; Dessales 2013: 120–126.

near the dining installation. Both the small size of the figures and the elaborate composition of the frescoes invite intimate, extended viewing by individuals spending prolonged time on or near the water-triclinium. Reclining diners could examine the scenes' details at leisure, and the slightly different views available to diners on different benches might prompt comparison and conversation. Meanwhile, the dining installation itself, with its own outdoor setting, artificial "river," and shady pergola, prompts its users to engage in many of the same activities that it depicts: outdoor drinking, dining, and probably also enjoying musical or other performances. Facilities for many similar activities, both convivial and cultic, appear in the surrounding garden.

Performing Pastoral Piety

The correlations between the Nilotic frescoes and their three-dimensional setting do not stop at the *convivium*. Although scholarly literature on the triclinium paintings in the Casa dell'Efebo mostly focuses on the depictions of banqueting and sex, the majority of the frescoes actually depict scenes of rural cult.[82] Indeed, despite frequent scholarly characterizations of pygmy scenes as representations of transgressive or comically uncouth behavior, only one small vignette—the open-air *symplegma*—depicts activity that might clearly strike Pompeian viewers as socially inappropriate. Much more of the painting consists of rural shrines and cult activity, set in what is essentially a "sacral-idyllic" landscape dominated by visual displays of *pietas*.[83]

The Casa dell'Efebo garden also contains visual evocations of piety, including two installations for cult performance: the shrine niche in the portico and the terracotta altar in the eastern garden (figures 3.5, 3.6).[84] The aedicular forms of the

82. Including, by one estimate, "approximately 20 country shrines with statues and altars" (Foss 1997: 214). For emphasis on dining and sex scenes, see Clarke 2007b: 103; Balch 2008: 230–232.

83. Although Clarke suggests some other Nilotic frescoes may parody "sacral-idyllic" landscapes (2007a: 220; 2007b: 102–103, 106–107; 2007c: 161), it is harder to see the parody in this case. The depiction of animal cults evokes a historically controversial aspect of Egyptian religion (Smelik and Hemelrijk 1984), but by Flavian times, even theriomorphic deities had some Greco-Roman apologists and worshipers; see, e.g., representations of theriomorphic divinities in the Iseum Campense (Lembke 1994: 221–230, 238–242) and the Pompeii Iseum (De Caro 1992: 57–62). However viewers might respond to animal cults, the fresco also depicts people worshiping Isis-Fortuna, whose cult was thoroughly integrated into, and uncontroversial within, Pompeian society (Beaurin 2008).

84. Foss 1997: 214; Allison 2004b; Bassani 2008: 148. No images of the (apparently now-lost) terracotta altar exist, but Allison describes its findspot as "E side [of garden], centre, 1.53 m from E wall" (2004b, s.v. "Database: table of artifacts").

portico shrine and triclinium fountain (figures 3.2; 3.6a) find visual echoes in the numerous aediculae of the frescoes (especially figure 1.3, A; 1.5, A; 1.7, I and L). Shrines are common in garden settings, and recent finds elsewhere in Pompeii attest to the performance of offering rituals in gardens.[85] However visitors may have responded to the painted depictions of Nilotic festivals, these visual displays of *cultus* and *pietas* emphasize the patron's own proper performance of ritual behavior.

The juxtaposition of such evocations of *pietas* with images of Nilotic festivals would then have been up to viewers to interpret. While some viewers might have interpreted the Nilotic frieze as an image of inappropriate ritual performance, others might have seen the nearby shrine and altar as real-life instantiations of the rural shrines, aediculae, altars, and offering scenes that fill the fresco. Like the garden triclinium itself, the shrine and altar echo the painted images in three dimensions, inviting viewers to respond through their own construction of connections—and, potentially, their own ritual performance.

Universal and Local

Other images in the garden universalize the scenes of rural piety in the Nilotic frescoes. Some of the paintings' religious resonances are geographically specific. Although Isis was at this point a highly international (and cultically "Hellenized") pan-Mediterranean deity, the paintings also include several allusions to animal cults, which more unambiguously evoke Egypt (e.g., statues of animal deities in figures 1.3, B; 1.4, J and K; possible feeding of sacred animals in figure 1.7, C).[86] The obelisks may also suggest a specifically Egyptian setting, even as the largely Greco-Roman forms of much of the rest of the architecture (e.g., prostyle temples and shrines, porticoed complexes) simultaneously render that landscape more familiar.[87] However, other garden paintings and sculptures construct complex

---

85. *Garden shrines:* Grimal 1969: 42–46; Jashemski 1979–93: 1:115–140; Foss 1997: 217; Giacobello 2008: 252–287. Soprano (1950: 290) remarks on the occasional proximity of such shrines to garden triclinia. Petersen argues (2012: 327–328) that shrines outside the kitchen area ("larari secondari": Giacobello 2008: 60) still saw actual cultic use. *Offering rituals in gardens:* Robinson 2002; Petersen 2012: 328–329, 332.

86. *Greco-Roman views of Egyptian animal cults*: Smelik and Hemelrijk 1984, and see also note 83, this chapter. *Isis as pan-Mediterranean figure*: see chapter 1, this volume, "Material Culture, Identity, and 'Authenticity,'" with notes 115–116.

87. The overall effect of the architecture thus strongly contrasts with the reed huts that characterize many other Nilotic scenes. The chiton-wearing figures of the Casa dell'Efebo friezes also contrast with the often naked pygmies of many other Nilotic scenes (see, e.g., chapter 4, this volume, on the Casa del Medico).

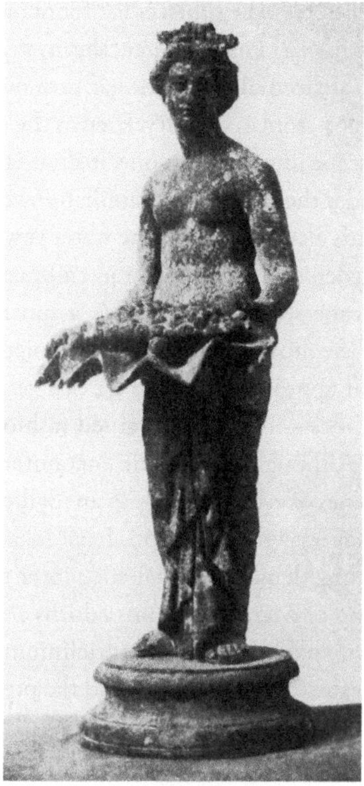

FIGURE 3.10 Bronze statuette used in the aedicular fountain of the Casa dell'Efebo, 0.382 m high (with base; see Maiuri 1927: 70). MANN 144276.
Source: After Maiuri 1927: fig. 32.

relationships between the water installation and the Greco-Roman pantheon, moving from locally specific to universalizing representations of divinity.[88]

The aedicular form of the fountain feeding the Nilotic water-triclinium visually echoes the shrine niche (figures 3.2, 3.6a).[89] Within that aedicular fountain, the source of this model "Nile" proved to be a Greco-Roman divinity, not a specifically Egyptian one. Water spouted from a fruit-filled shell held by a female figure, interpreted variously as a nymph, Venus, or Pomona[90] (figure 3.10). Regardless of specific identification, both Venus and Pomona were associated with gardens, and

88. Cf. Foss 1997: 214.

89. Dessales 2006: 263.

90. Maiuri 1927: 60, 70–71; Neuerburg 1965: 117; Jashemski 1979–93: 2:38; Dessales 2013: 137.

Venus and/or nymphs are frequent subjects for fountain statuary.[91] The Nilotic landscape thus proves to have a familiar source; the mysterious origins of the Nile, subject of such intense intellectual speculation,[92] turn out to be right here in the arms of a beneficent Greco-Roman divinity. Even as the Nilotic water-triclinium transforms a local landscape into a distant one, it simultaneously reveals that foreign landscape to be under the control of a familiar power.

The world of the gods also appears on the water tower next to the aedicular niche, where a painting depicts Mars and Venus embracing in front of a column in a rocky rural landscape while Cupid holds a sunshade over them (figure 3.6b).[93] Venus is one of the most common deities depicted in Roman gardens,[94] but beyond her general appropriateness to garden settings, the deities' amorous encounter also echoes—in more restrained fashion—the erotic scene on the Nilotic triclinium. Although the Nilotic encounter is much more explicit, both are, essentially, scenes of erotic activity in an idyllic landscape. Both scenes also associate erotic imagery with irrigation. Itself located next to real flowing water, the Nilotic painting depicts a sexual encounter next to active irrigation machinery. The painting of Mars and Venus adorns a water tower whose reservoir fed the fountain, enabling the water-triclinium to function.[95] In context, both scenes link sexuality to irrigation, and the pipe joining the two water installations created a (literal) connection between Greco-Roman deities and the Egyptian landscape.[96]

Within Pompeii, Venus has additional, specifically local associations. As the tutelary deity of Pompeii,[97] she reminds viewers that whatever other faraway

91. *Venus and Pomona in gardens:* Grimal 1969: 49–53; Jashemski 1979–93: 1:124–131; Dwyer 1982: 124; van Andringa 2009: 267; Bergmann 2018: 308–315. *Venus and nymphs in fountain statuary:* Kapossy 1969: 12; Dessales 2013: 137.

92. Beginning with Hdt. 2.27–2.34; for Roman discourse on the Nile, including its sources, see most recently Merrills 2017. See also note 143, this chapter.

93. On the painting, see Maiuri 1927: 49–51.

94. See note 91, this chapter.

95. See note 44, this chapter.

96. Although much of the piping is now lost, Jansen's work on the water system suggests that most of the pipes from the water tower were underground (2005: 283, fig. 12; note areas found by metal detector). However, even if the pipes themselves were not visible, the proximity of the water tower itself could still suggest visual connections to viewers. On water and sexuality, cf. the image of Venus above the Casa degli Amorini Dorati cisterns (Petersen 2012: 330) and the mythological couples on the Iseum "Nilometer" (Moormann 2007: 142–143, 149–150).

97. Mart. *Ep.* 4.44.5; Stat. *Silv.* 5.3.164; Schilling 1954: 288–289, 384–388.

vistas the garden might conjure, they are still here at home. Various Pompeian graffiti and dipinti invoke Venus Pompeiana,[98] demonstrating the significance of her cult to local identities. Yet the worship of Venus also connected Pompeii to the broader empire. Venus became the tutelary deity of the city only after Sulla's sack of Pompeii in 89 BCE made it a Roman colony.[99] Additionally, as Aeneas's mother, Venus was the ancestress of the Roman state and its first imperial family.[100] Venus can thus simultaneously evoke both the specific city of Pompeii and its incorporation into the larger Roman world.[101] Local and imperial identities coexist within her cult.[102]

The representations of deities in the Casa dell'Efebo garden are thus both universalizing and particularizing. From Venus to painted statues of animal gods, these images allude to religious geographies both local and distant while suggesting that the garden is under the protection of the same deities who guard the whole *oikoumene*. A parallel comes from the Casa degli Amorini Dorati (VI 16.7), whose garden contains two shrines: one with images of the Capitoline Triad, and one with images of Egyptian deities. As Petersen persuasively argues, this juxtaposition of Egyptian and imperial Roman deities—together with a nearby depiction of Venus, perhaps recalling Venus Pompeiana—not only shows the "reach of the Roman empire" but also creates a reassuring impression of "imperial, city, and domestic deities watching over the *domus*."[103] Pairings of Venus and Isis, who were sometimes identified with each other, also appear elsewhere at Pompeii, both in the Iseum and in private homes.[104]

---

98. E.g., *CIL* IV 26, 538, 2457, 4007; cf. Varone 2002: 23–31; Carroll 2010: 96. However, this painting lacks certain specific iconographic elements sometimes associated with Venus Pompeiana, such as more modest clothing. On Venus Pompeiana's specifically local traits and epithets, see van Andringa 2009: 35–36.

99. Recent archaeological work suggests the Temple of Venus in Pompeii was constructed post-Sulla (Carroll 2010).

100. Some Pompeian depictions of Venus may allude to Caesar's statue of Venus Genetrix (Carroll 2010: 96–97).

101. Some Pompeians might have understood Venus Pompeiana as the Romanized form of a local goddess (Carroll 2010: 96).

102. On the complexity of Italian identities in the republic and early empire, see Wallace-Hadrill 2008: 73–143.

103. Quoting Petersen 2012: 328, 330.

104. Petersen 2006: 36–37; Gasparini 2013: 191.

Safety and Danger in the Domestic Locus Amoenus

The familiarization of the foreign in the Casa dell'Efebo garden blurs the lines between "Self" and "Other," "here" and "there"; but this does not mean the landscape has been totally defanged. As the pygmy–crocodile conflicts of the Nilotic frescoes illustrate, the seemingly idyllic outdoors still contains its perils. Behind the fountain, the large-scale painting of animals in a game park (figure 3.3) provides another variation on the theme of wild, potentially dangerous beasts. The fresco is heavily damaged, but a deer, long-beaked bird, bull, bear, and large feline remain visible.[105] If correctly attributed to the garden, the marble sculpture of a deer and fawn would have provided a three-dimensional counterpart to the painted animals. Further echoing these animal depictions and reinforcing the implied hunting theme is the stucco frieze on the aedicular fountain. The central section of this frieze originally depicted a figure of Diana, carrying a bow and arrows, standing between two deer[106] (figure 3.2). The east side of the frieze depicted an anthropomorphic figure in a quadriga following a large quadruped, with a tree visible to the far left of the scene.[107] The imagery of the western side is mostly lost, although visible figures appear to have included a large, possibly feline quadruped and a smaller quadruped (perhaps a hunting dog or the other animal's young).

Diana, like Venus, is a frequent denizen of Roman gardens.[108] Yet while her presence near an artificial "Nile" may tell viewers that even Egyptian landscapes can be home to familiar deities, that realization may not be entirely reassuring; surprising Diana in the wilderness—as the viewer has apparently done, in suddenly spying her on the fountain—does not invariably lead to good results. Images of Actaeon's gruesome punishment for witnessing her bathing were popular in Pompeian houses, and Platt has analyzed such images as artistic explorations of the dangers of the voyeuristic gaze.[109]

---

105. The large feline is identified only as a "feline" in Maiuri 1927: 60 and Jashemski 1979–93: 2:316, while King calls it a tiger (2002: 441). Only the muzzle and paws of this figure survive today. On the bear, see King (2002: 441). The monkey reported by King (2002: 434) is no longer clearly visible.

106. Maiuri 1927: 60; A. de Vos 1990: 711.

107. A. de Vos (1990: 711) and Maiuri (1927: 60) identify the anthropomorphic figure as a Cupid. Attempted identifications of the quadruped vary widely (Maiuri 1927: 60; Schefold 1957: 34; A. de Vos 1990: 711). The current state of the frieze makes identification impossible.

108. Jashemski 1979–93: 1:131–133; Dwyer 1982: 125.

109. Platt 2002.

The garden landscape—at once Egyptian and Italian, haunted by gods both foreign and familiar—is thus also a place of simultaneous relaxation and danger. The installation seduces the viewer with its pleasant, *locus amoenus*–like qualities—bubbling water, greenery, shady arbors, and convivial opportunities, all present in both real and painted form—yet also reminds viewers not to relax completely: these scenes contain their perils, too. This seeming contradiction recalls Roman attitudes toward natural landscapes as, in Zahra Newby's words, both "sacred and potentially dangerous."[110] Addressing violent mythological imagery in gardens, Newby persuasively characterizes gardens as mediating spaces between domestic interior space and the world outside, sharing both pleasant and unsettling associations with the natural world.[111]

Additional objects suggest yet another sort of landscape, with associations similarly both appealing and potentially dangerous. Besides the deer and fawn, the garden probably also once contained three other marble statuettes: two Pans (figure 3.8) and a reclining satyr. Common in gardens, such images evoke what Spencer calls "Bacchic nature":[112] the wild settings haunted by satyrs and Pans, farther from human control than the pastoral world of "sacral-idyllic" landscape or the agricultural fields of the Nile. Not unlike Egyptian imagery, Bacchic imagery was thoroughly integrated into imperial Roman domestic visual culture, yet could still evoke—in a familiar, comprehensible way—concepts of the exotic, outré, and foreign.[113] Here, the Bacchic images echo the triclinium frescoes' representations of drinking and revelry, creating thematic correlations across media. Egyptian and Bacchic motifs frequently appear together in Roman domestic contexts (especially gardens), as well as in religious contexts.[114] Both have connotations simultaneously familiar and uncanny, standing both inside and outside the Roman world.

110. Newby 2012: 355; cf. Elsner 1999: 16–17. On dangerous aspects of the *locus amoenus* in Greek and Roman literature, see Hubbard 2010; Tronchin 2011a: 68–69.

111. Newby 2012.

112. Spencer 2010: 19–22. *On Bacchic imagery in gardens*, see Grimal 1969: 317–330; Jashemski 1979–93: 1:123–124; Dwyer 1982: 123–124, 126; Amedick 1994: 113; Hales 2003: 156–157; Dessales 2013: 139–140.

113. Wyler 2004; 2006: 224–228; 2011.

114. *On the co-occurrence of Egyptian and Bacchic motifs in gardens*, see Hales 2003: 156–157; Petersen 2006: 46. *In religious contexts*, see, e.g., the Dionysus statue at the Pompeii Iseum: Kleibl 2009: 281; Swetnam-Burland 2015: 113–114. An engraved silver object from Pompeii appears to pair Isis with Bacchus (Tran tam Tinh 1964: 171–172, no. 134).

Miniaturization and Distancing

Yet even as some garden images remind us of potentially threatening or uncontrollable landscapes, other features reassure us that we *are* in a safe space, where no real harm will befall us. These various models and representations of different types of landscape are ultimately just that: models and representations, created by human skill and controllable by humans. Even as various aspects of the installation seem designed to make the images seem "real" (e.g., the employment of a naturalistic representational style for the statuary and the painting of Venus and Mars, or the placement of the Nilotic scenes alongside actual water), other features undercut those mimetic illusions. The miniature size of many images—the waist-high statuettes, the approximately 0.4 m fountain figure, the half-life-size paintings of Mars and Venus, and the knee-high Nilotic landscape—emphasizes their artificiality and diminishes any unsettling associations, as the viewer towers reassuringly over them.[115]

The garden assemblage thus simultaneously entices viewers into marvelous landscapes and denies them access. Reclining and banqueting on masonry "riverbanks," diners effectively become participants in the scenes portrayed around them—yet the vast difference in scale separating them from those scenes simultaneously emphasizes the impossibility of complete participation.[116] The population of the landscape with quasi-mythical *pygmaei* (themselves human "miniatures"), rather than "real" Egyptians, further emphasizes the distant, fictive nature of the scene, even as other correspondences between the painted and "real" landscapes invite diners to imagine themselves actually within the Nilescape.

Other aspects of the garden assemblage reassure viewers that human *ars* and *cultus* are directing natural forces productively. Behind the fountain, the painted game park represents a landscape that may look wild but was in fact created and maintained for human benefit. Testifying to human power and agency are the statues and hanging *clipei* framing the scene, as well as the collection of a group of beasts that would not all coexist in nature.[117] The Nilotic fresco also shows human skill improving on nature. The flooding Nile irrigates its banks with human help,

115. *Dimensions of fountain figure and statuettes:* Maiuri 1927: 70–74. *Dimensions of Mars and Venus fresco:* 1.80 m × 1.425 m (Maiuri 1927: 49). *Height of triclinium benches:* approximately 0.5 m (Maiuri 1927: 53; Versluys 2002: 101). Many scholars emphasize the ability of miniatures to make their viewers feel relatively large and powerful: see note 133, chapter 1, this volume. The small scale of the Casa dell'Efebo Nilotic scenes is typical of Nilotic and "sacral-idyllic" landscapes (e.g., Hinterholler 2007: 33–35, 52, 54).

116. Cf. Bailey (2005: 42) on miniatures generally.

117. Cf. Hales (2003: 158) on human agency in other hunt scenes.

as the operator of the water-screw directs the waters. His work demonstrates the importance of human culture even to natural resources as powerful as the Nile, whose inundation Herodotus (2.14) famously claimed to eliminate much of the human labor of agriculture.

Fertility, Prosperity, and Stability

The result of this productive collaboration between *natura* and *cultus* is fertility and prosperity: in Egypt (where the floods produced famously abundant grain), in Campania (where the Egyptian grain trade made up a major part of the economy), in Rome (where Egyptian grain helped feed the population), and—in quite concrete fashion—in the Casa dell'Efebo garden. Since the water of the triclinium fountain ultimately flowed into a domestic *hortus*, the model Nile did not just create an illusion of fertilizing painted banks; it may also have actually fertilized a real-life crop. Recalling Pliny's (*HN* 18.167) characterization of the Nile as an ideal farmer irrigating the fields *coloni vice*, this domestic Nile generates prosperity and fertility within the home in a way that echoes the role of the "real" Nile within the broader empire.

The resulting association of Nilotic imagery with order, prosperity, and responsible human guidance of natural forces suggests a need to nuance Hales's characterization of Egyptianizing garden imagery as "transgressive display."[118] Hales argues that as liminal spaces between civilization and wilderness, gardens were suitable places for exotic imagery and the expression of otherwise socially inappropriate fantasies and sentiments. In the Casa dell'Efebo, however, the architectural and spatial setting actually appears to associate Egyptian imagery with the *maintenance* of social and natural order. Certainly, Roman painted landscapes often play with the line between order and disorder: garden paintings may juxtapose cultivated and wild growth, while *paradeisos* paintings put wild animals in controlled settings.[119] However, the use of Nilotic imagery to suggest domestic prosperity, order, and stability may also suggest changing attitudes toward the Nile and Egypt. In Flavian Pompeii, "exoticism" or "transgression" were by no means the only lenses through which people could be expected to view Egyptian imagery.

---

118. Hales 2003: 155; cf. von Stackelberg 2009: 122. On gardens as liminal space, see note 50, chapter 1, this volume.

119. For cultivated and wild plants in Roman paintings, see Kuttner 1999: 28. For *paradeisoi*, see note 11, this chapter.

Interactive Statuary: Social Performance and Service

Yet another set of cultural referents appears in the numerous retrospective allusions of the garden statuary to earlier Greek art. The pose of the famous bronze ephebe statue (figure 3.7) recalls Classical models, including Polykleitos's *Doryphoros*—here reworked to hold a candelabrum in each hand.[120] Such Classicizing allusions evoke the social prestige of connoisseurship and suggest further associations with Greek learning; Quintilian (*Inst.* 5.12.21) describes the Doryphoros as particularly appropriate to gymnasia, and Bergmann discusses the uses of similar statue types to convey "a sphere of cultivation" in Roman domestic spaces.[121] Although the employment of the statue as a lampholder might seem jarring to modern eyes, such use appears to have been common in Roman practice. At least ten other lamp-holding statues of beautiful archaizing or Classicizing youths exist (including three others from Pompeii), and Lucretius (2.24–26) refers to statues of lamp-holding youths as a standard way for elites to light nocturnal banquets.[122] For guests who knew their Homer, such *lychnouchoi* would have recalled the gilded torch-bearing *kouros* statues in the hall of Alcinous,[123] enabling their owners to display not only wealth but also *paideia* in equating their own homes to the palace of the Phaeacian king. At the Casa dell'Efebo, this nearly life-size (1.49 m) statue was likely the showpiece of the garden.[124] If it indeed stood on the base near the garden triclinium, it would have been one of the first things a visitor saw when entering the garden. It would also have been ideally positioned for visibility from the indoor Triclinium 17 (see figures 3.1, 3.7).

Nearby, the line of marble herms delimiting the planted, furrowed area (figure 3.4) provides another set of temporal and stylistic references, while also

---

120. Bergmann 1995: 81n5, 108, fig. 1. The head adapts a Classical female type, creating an eclectic and slightly androgynous whole (Lenski 2013: 138). For retrospective eclecticism and Polykleitan stylistic features in *lychnouchoi* and "sexy boy" statues, see Bartman 2002: 250, 260–261; Marvin 2008: 210–213; Lenski 2013: 140.

121. Bergmann 1995: 80.

122. On *lychnouchos* statues, see Bergmann 1995: 80–81; Lenski 2013: 136–141 (with findspots in table 1). On Lucr. 2.24–26 and *lychnouchoi*, see Lenski 2013: 139–140, 153n 57. Other statues of beautiful youths held not lamps, but trays (Clarke 2003: 245).

123. χρύσειοι δ᾽ ἄρα κοῦροι ἐυδμήτων ἐπὶ βωμῶν / ἕστασαν αἰθομένας δαΐδας μετὰ χερσὶν ἔχοντες (Hom. *Od.* 7.100–101).

124. Zanker 1998: 178. On such "collectible singletons," see Dwyer 2012.

recalling Greek uses of herms as boundary markers and cult objects.[125] Roman audiences admired the herm form for its perceived antiquity and associations with Greek gymnasia, philosophical education, and Athenian culture, and many Roman-period herms allude formally and stylistically to Archaic and Classical antecedents.[126] Accordingly, the herms in the Casa dell'Efebo garden may not only demarcate space but also mark the garden as suitable for learned discussion and philosophical contemplation.[127] The marble seat in the eastern garden may also construct a space for cultured conversation; Cicero refers to garden seats as convenient spots to rest and discuss philosophy.[128]

Drawing on models from yet another period of Greek art are the tray-holder statuettes, whose twisted, emaciated, and macrophallic forms take inspiration from Hellenistic "grotesque" figures (figure 3.9). Previous publications usually identify them as cake sellers or as caricatures of domestic slaves.[129] However, I suggest that they more likely recall Greco-Roman descriptions and depictions of personified Envy (Greek *phthonos*, Latin *invidia* or *livor*) as an emaciated, self-choking man.[130] The gaunt appearance and throat-clutching gesture of the statuettes find their closest parallels in Dunbabin and Dickie's catalog of such depictions, many of which are similarly macrophallic.[131] Furthermore, the

125. On Greek and Roman herms, see Wrede 1986; P. Stewart 2003: 62–64; Gaifman 2012: 181–242.

126. On the display, cultural associations, and valuation of Roman herms, see Dillon 2006: 30–33; cf. Cic. *Att.* 4.2, 5.2, 6.3, 9.3, 10.5. For herms in gymnasia and peristyles, see Wrede 1986: 34–36. In the Casa dell'Efebo, one bearded male head recalls Archaic herms (figure 3.4); the others, with female and juvenile heads, draw on later models (see Maiuri 1927: figs. 37–40).

127. On Roman gardens as places for philosophy and/or allusions to Greek gymnasia, see Grimal 1969: 69–70, 103, 359–362; Leach 2004: 35, 38–39; von Stackelberg 2009: 94–95; Zarmakoupi 2014: 103–111, 118–119.

128. Cic. *De or.* 1.28–29, alluding to Pl. *Phdr.* 229a–230c.

129. *Cake sellers:* Maiuri 1925; 1927: 66–67; Della Corte 1930: 31; Grant 1975: 120; Dwyer 1982: 123. *Domestic slaves:* Lenski 2013: 145.

130. For other depictions of *phthonos/invidia*, see Dunbabin and Dickie 1983; Mitchell 2013: 288–292. Iconographic renderings vary somewhat and sometimes include additional features, such as self-inflicted wounds. However, Dunbabin and Dickie argue that not all indications of *invidia* "need to be present simultaneously" in such images, and that "the iconographically crucial feature is the gesture of the hands, clutched to the throat as if the figure were choking" (1983: 19). My argument here was earlier presented in article form (Barrett 2017b). While that article was in press, it came to my attention that Sylvain Vanesse (2016) had simultaneously and independently developed a similar argument interpreting the Casa dell'Efebo figurines as apotropaic representations of Envy. However, Vanesse's article does not discuss the figurines' possible visual or functional relationships with other material culture from the Casa dell'Efebo.

131. For other macrophallic examples, see Dunbabin and Dickie 1983: 20–25, pls. 3b, 4a–b, 5a–d.

statuettes' protruding tongues suggest choking, not singing.[132] These statuettes may therefore serve, among other things, to ward off the evil eye—even as their presence, and the precious metals from which they were made, paradoxically proclaim to viewers that this household might well attract envy. If they were indeed used to serve food in the garden, their short size (0.233 m high) and oddly formed (though not dwarf-proportioned) bodies would have created a visual rhyme with the pygmies in the Nilotic frieze.[133] While such correlations might simply demonstrate a desire to pair artworks for formal and/or stylistic reasons,[134] the apotropaic value of the tray-holders could also have reinforced another level of meaning potentially present in the Nilotic pygmy scenes. As Clarke has suggested, pygmy imagery might also sometimes have functioned—among other things—as apotropaia.[135]

Like much of the rest of the garden, these statues and statuettes were essentially interactive, facilitating and encouraging various types of viewer behavior. The ephebe held candelabra that illuminated the garden, enabling people to use the triclinium complex after dark; the herms demarcated different spaces within the garden, shaping and patterning people's movement; and the tray-holder statuettes held food for diners to consume. These services suggest social, as well as functional, relationships between viewers and objects. Both the ephebe and the tray-holders implicitly act as servants, carrying lamps and food for visitors.[136]

These sculptures invite us to revisit the Nilotic pygmies' potential relationship to their viewers: Are the pygmies, too, here essentially to serve and entertain us? Dwarf dancers sometimes provided entertainment at Roman banquets, as the earlier-quoted passage from Propertius illustrates (4.8.35–40).[137] Such spectacles probably would have helped shape the expectations and opinions that Roman viewers brought to depictions of carousing dwarflike figures. Yet interpreting the pygmies solely as *objects* of entertainment and spectacle oversimplifies the composition in which they appear here. The convivial scenes in the Casa dell'Efebo frescoes depict pygmies not as banquet entertainment but, rather, as banqueters themselves: consumers, not only providers, of entertainment and service. This

132. *Pace* Lenski 2013: 145.

133. For measurements, see Allison 2004b, s.v. "Database: table of artifacts."

134. Cf. note 77, this chapter.

135. See notes 186 and 187, chapter 2, this volume.

136. On lampholders and tray-holders as slaves, see Bergmann 1995: 100–101; Bartman 2002: 268; Sabnis 2012; Lenski 2013: 136–146.

137. See chapter 2, this volume, "Parallel Landscapes in the Garden."

reversal of small persons' more usual role at Roman banquets sets up a structural parallel between the pygmies and their viewers that does not exist in the case of the ephebe or tray-holder statuettes, which are presented solely as objects (even tools) for the visitor's use and/or gaze.

This parallel playfully challenges viewers to consider their own social performance as banqueters. What type of social identity would one choose to project in response to the invitations, provocations, and challenges of one's setting? Additionally, the context invites viewers to put these images in dialogue not only with their own behavior but also with the behavior they expect from their companions. Would one's fellow-diners wind up resembling the denizens of an imagined Nile in their behavior—and if so, how would one respond: with amusement, disdain, arousal, disapproval, delight, piety, curiosity, or something else? How might one's companions choose to behave, and how would that influence one's own choices about social performance? The images thus prompt a self-reflexivity similar to that which Osborne has identified in Classical sympotic pottery, which similarly challenges its users to contemplate how they relate to images of festivity and (sometimes) overindulgence.[138]

Performing Cosmopolitanism

Finally, the visual culture in the garden may have facilitated another form of social performance: verbal performance of connoisseurship and cultural sophistication. The ability to talk intelligently and knowledgeably about art was a highly valued sign of social prestige, and dinner conversation could involve discussion of the artwork in one's immediate vicinity.[139] The possession of sculpture running a wide gamut of "retrospective" styles, including one particularly valuable showpiece, could help the house owner present himself as a knowledgeable sophisticate.

In addition to waxing eloquent about, for example, the bronze ephebe, such performance of rhetorical, ekphrastic, and/or interpretive skills might as readily address the Nilotic frieze. An ability to interpret its iconography, or expound on related Egyptian customs, could provide similar opportunities to display erudition. In order to hold one's own in discussions of retrospective art alluding to Greek traditions—to successfully perform membership, or at least aspirational

---

138. On social performance and symposium scenes on pottery, see Osborne 2007: 34–35, 50.

139. Roller 2006: 59–60; Bergmann 2007: 88; Squire 2009: 208, 217, 219–221, 224. Within the enormous bibliography on ekphrasis and image-oriented rhetorical performance, see within the past ten years Tanner 2006: 246–264; Bartsch and Elsner 2007; Bergmann 2007; Elsner 2007; Squire 2009: 176–293.

membership, in the Roman cultural elite—one needed, as Tanner puts it, "knowledge of the iconographic codes and the stories of Greek mythology."[140] If we are to take seriously the status of "Aegyptiaca" and "Nilotica" as part of *Roman* art, we must consider the likelihood of their affording similar opportunities for competitive performances of sophistication.

Greco-Roman literature had long presented Egypt as the land of esoteric knowledge par excellence.[141] Much imperial philosophical literature, from Plutarch to the Neoplatonists, makes a show of engagement with Egyptian (and pseudo-Egyptian) intellectual traditions of philosophy, theology, and esoterica.[142] Indeed, the invocation of Egyptian wisdom ultimately became enough of a trope to inspire parody. In Lucian's second century CE *Philopseudes* (31–37, 39), a pretentious philosopher boasts of his alleged knowledge of Egyptian magic and esoterica. The effectiveness of such satire suggests an audience used to hearing similar claims.

The social performance of knowledge about Egypt might address not only Egypt's reputation as a land of wisdom but also its unusual geography and natural history. Numerous works of Greco-Roman philosophy, natural history, geography, and literature testify to debates about the marvels of the Nile, especially its mysterious sources and yearly flood.[143] Such fascination even informed a major international event roughly contemporary with the final phase of the Casa dell'Efebo: Nero's expedition into Nubia, which tried but failed to find the source of the Nile.[144]

Nearly contemporary with the Casa dell'Efebo garden and similarly set on the Bay of Naples, Statius's (*Silvae* 3.2) poem on the departure of the young military officer Maecius Celer suggests that knowledge of the Nile and Egyptian customs could be appropriate and desirable for elite Romans. Bidding farewell to Celer, who was about to visit Egypt en route to a posting in Syria, Statius expresses the hope that Celer's travels will teach him about Egyptian geography, natural history, cultural practices, and the origins of the Nile (*Silv.* 3.2.107–120). Such learning appears to be primarily a matter of intellectual inquiry, rather than an expression of any particular religious identity. Although Statius calls on Isis to guide

140. Tanner 2006: 248.

141. E.g., Frankfurter 1998b: 217–221; Gruen 2011: 76–114; Tracy 2014: 3–6.

142. Fowden 1993; Hornung 2001; Steinhauer 2017: 67; Erler and Stadler 2017.

143. Hdt. 2.28–2.34; Luc. *Pharsalia* 10.171–331; Lucr. 6.712–737; Pompon. Mela 3.96–97; Sen. *QNat.* 4a, 6.8.1; Stat. *Silv.* 3.2.108; Tib. 1.7.23–24; Postl 1970; Huss 1990; Romm 1992: 149–156; Schrijvers 2007: 226–227; Manolaraki 2013.

144. Plin. *HN* 6.181; Sen. *QNat.* 6.8.3; cf. Romm 1992: 155–156.

the young man's inquiries (*Silv.* 3.2.101, 3.2.107), the goddess's role appears to be purely advisory, not "initiatory" in a ritual sense.[145] Knowledge about Egypt and Egyptian customs is thus desirable not because Celer is an Isiac initiate (which we have no reason to believe) but, rather, because such knowledge commands more general cultural value.[146]

Similarly, the Nilotic frieze could have provided an opportunity for diners to show off their own cosmopolitanism and sophistication. The extended digression on the Nile in Book 10 of Lucan's *Pharsalia* (10.172–192) provides an example of such dinnertime discussion: after a banquet, Caesar asks the Egyptian priest Acoreus to expound on the origins of the Nile. Acoreus's subsequent speech does far more than merely show off Lucan's knowledge of current theories of natural history and geography, serving also as a politically charged warning against imperial overreach and a reflection on Lucan's own poetics.[147] Yet it is also, among these other things, a literary example of a Nilotic ekphrasis[148] in a feasting context.

Among all the other reasons why house owners could have valued Nilotic scenes, then, we must also consider the opportunities they potentially provided for social performance. Prompted by representations of the flooding Nile, a host and guests might trade opinions on the sources of the river and the reasons for its flooding. Indeed, the pygmy population of the scene may signal that we are tantalizingly close to those elusive river sources; pygmies were said to live close to the Nile's origin, *inter paludes ex quibus Nilus oriretur* (Plin. *HN* 6.188). Since the quest for the source of the Nile was thought to typify imperial ambitions,[149] such images might also inspire discussions of politics and history. Representations of Egyptian cults and festival activities could additionally provide opportunities for diners to show off their ethnographic knowledge, discussing and debating foreign practices.

In addition to enabling house owners to *show* they possessed knowledge of natural history, famous marvels, or foreign customs, Nilotic landscapes like those of the Casa dell'Efebo actually *instantiate* that possession. If the pygmies'

145. Cf. Manolaraki 2013: 190.

146. Cf. the association between Egypt and culturally prestigious knowledge in the Villa della Farnesina, whose Egyptianizing judgment frieze is wreathed with *platanus* garlands evoking Aristotle's Lyceum (Platt 2009: 66–67).

147. *On Lucan's scientific knowledge and engagement with Seneca*, QNat. *4a*: see most recently Manolaraki 2011: 154, 167–171; 2013: 96–103; Tracy 2014: 153–169. *On the Nile digression as warning against imperial overreach*, see Barrenechea 2010; Manolaraki 2013: 81; Tracy 2014: 182–199. *On Acoreus and Lucan's poetics*, see Manolaraki 2011; 2013: 80–115.

148. Cf. Manolaraki 2011: 157–158; 2013: 85.

149. On the trope of mighty rulers trying to find the sources of the Nile, see note 150, this chapter.

presence suggests we are near the elusive origins of the river, then the owner of this garden literally possessed—at least in model form—something that had famously escaped ambitious rulers from Alexander to Nero:[150] the source of the Nile itself. The miniature, "collectible" size of the images further emphasizes their owner's authoritative possession of the knowledge, both cultural and geographic, that they imply.[151]

In this light, the very comprehensiveness of the paintings takes on new relevance. As noted earlier, the triclinium frescoes of the Casa dell'Efebo compile practically every major motif employed in Nilotic paintings and mosaics. This encyclopedic quality recalls not only the emergence of new practices of art collection in the early empire (including the collection and display of conquered peoples' works) but also the development of encyclopedias and handbooks.[152] Such works, both material and literary, not only compile knowledge but also assert imperial power.[153] In collecting material culture evoking the breadth of the empire, house owners effectively transformed the *domus* into a model of empire, appropriating imperial ideology for individual goals.

Yet even as such collecting practices express Roman imperial claims on the provinces, they also represent those provinces' claims on Rome. That which was once alien and potentially dangerous was now part of the *domus* itself, increasingly incorporated into—and constitutive of—"Romanness." The incorporation of once-foreign imagery into the household may well demonstrate pride in Roman achievements,[154] but it also contributed to changing conceptions of what "Rome" meant. Cosmopolitan domestic assemblages like the Casa dell'Efebo garden represent popular forerunners to the eclectic vision of empire embodied several decades later in Hadrian's villa at Tivoli, where Egypt—together with Greece—appears as a cultural ancestor of, contributor to, and source of legitimacy for the Roman world.[155]

---

150. Luc. *Pharsalia* 10.268–283; Romm 1992: 151–156; Barrenechea 2010: 267; Manolaraki 2013: 81.

151. Cf. note 133, chapter 1.

152. Chevallier 1991; Pearce and Bounia 2000: 97–162; Bounia 2004; Tronchin 2012b; Gahtan and Pegazzano 2015; Higbie 2017: 57–78. *On encyclopedism*, esp. Plin. *HN*, see Beagon 1992; Naas 2002; Carey 2003; Murphy 2004; Rutledge 2012: 202–204; König and Woolf 2013; Laehn 2013.

153. Bounia 2004: 204; Naas 2011; Laehn 2013.

154. Cf. Tronchin 2012a: 280.

155. See chapter 7, this volume, "Epilogue: From Pompeii to Tivoli."

Landscapes, Identities, and Viewer Responses

Encounters with the Casa dell'Efebo garden would have involved whole-body, multisensory, immersive participation. Visitors would have walked through and around the garden, smelled the growing vegetation (and possibly also some less pleasant smells[156]), touched and rested on the triclinium cushions, tasted and smelled food and drink during the *convivium*, felt cooling sprays of water from the fountains, listened to the gurgling water or mealtime musical performances, and engaged intellectually and emotionally in conversation with the friends around them. The effects of the garden would also have changed over the course of the day, as different people used the space for different activities (from garden maintenance, to children's games, to dinner parties) and as the light and temperature shifted. The garden in the full heat and brightness of the day—with fountain splashing, wall paintings clearly visible, and pergola providing inviting shade—must have provoked different perceptions and responses from the garden at evening, illuminated by the ephebe statue, whose lamps cast flickering plays of light and shadow across the paintings and his own metal surface.

This immersive environment evokes a wide range of imagined landscapes, including the agricultural riverscape of the Nilotic water-triclinium, the *paradeisoi* of Hellenistic potentates, the simultaneously local and universal landscape over which Venus presides, the more uncontrollable wildernesses prowled by Diana and the Bacchic *thiasos*, and the gymnasia frequented by Classical philosophers. The aspects of such settings range from reassuring to threatening, exotic to familiar. The material culture of the garden neither corresponds to one single iconographic program nor lends itself to one single, authoritative "reading." Instead, the interrelationships of images and objects create an open-ended space for individual viewers' reflections and social performances. Rather than solving the problems and paradoxes that these juxtapositions pose, the assemblage invites participants to grapple with them. As one component of this larger ensemble, the domestic Nile—alien-looking though its scenery and denizens may initially appear—ultimately proves to be fundamentally intertwined with, and fundamental to, local prosperity. Simultaneously distant and local, dangerous and reassuring, artificial and real, this Pompeian Nile provokes its viewers to reconsider binary constructions of "Self" and "Other."

The purposeful ambiguity of this foreign-yet-familiar Nile recalls similar treatments of the Egyptian river in post-Augustan literature. As Eleni Manolaraki

---

156. On waste water and sewage in Roman gardens, see Columella, *Rust.* 10.85; Pagán 2006: 20; Jansen 2018: 407. Cf. Jansen (2005: 288–289; 2018: 411) on waste in the Casa dell'Efebo garden.

has recently demonstrated, first and second century CE Latin authors portray Egypt and the Nile in multifaceted ways, transcending unambiguous "Othering" to make more complex statements about Roman identity: "As the dangerous Actian otherness of Egypt draws away, the Nile is re-imagined as an ambiguous space that is no longer as foreign nor is it yet domestic."[157] Such nuanced and shifting representations, whether iconographic, stylistic, or literary, evoke a "familiar alterity" similar to that which Wyler identifies in domestic Bacchic imagery.[158]

Certainly, some of the activities portrayed in the Casa dell'Efebo frescoes reinforce notions of Egyptian alterity. Also potentially distancing are the possible allusions to the Nile's faraway origins, evoking the truly alien lands beyond the increasingly familiar, reassuringly Roman-governed province of Aegyptus. The sources of the Nile belong to the farthest fringes of the world, an area inhabited by strange *monstra* not unlike the fantastic creatures of Third Style wall painting.[159] Yet at the same time, the larger context of this imagery also provides a means to incorporate and domesticate the formerly alien. Nero may have failed to find the sources of the Nile, but diners at the Casa dell'Efebo could trace their own household "Nile" to a fountain and water tank guarded by the protective divinity of their home city. Put to service watering a homely vegetable plot, this domesticized Nile becomes a useful and productive part of daily life. The garden itself becomes, among other things, a workshop where the "Other" is processed and transformed for inclusion within civilized, ordered, and familiar space.

Nilotic installations thus frame the experiences of their occupants in ways much more subtle than broadcasting a particular religious identity, as scholars once believed, or mocking the cultural failings of Egyptian "Others," as more recent arguments have proposed. In gardens like that of the Casa dell'Efebo, Pompeian viewers were not just gazing upon an objectified "Other" but also potentially *performing* the "Other." Within a space framed as simultaneously foreign and familiar, the scene transforms its setting and its viewers: an Italian garden becomes an Egyptian riverbank (and vice versa), and the host and his guests become the strange figures wandering those banks. The architectural and

157. Quoting Manolaraki 2013: 216. Indeed, even Augustan literature presents more multifaceted perspectives on Egypt than often assumed (Leemreize 2016).

158. On Dionysus as "familiar alterity," see Wyler 2006: 226; cf. chapter 6, this volume, "'Familiar Alterity': Egyptian and Bacchic Imagery."

159. On faraway *monstra*, see Plin. *HN* 6.195; Romm 1992. On Third Style *monstra*, see Yerkes 2000; Platt 2009. See also chapter 5, this volume, "Marginalia and *Monstra*."

social context of the Nilotic installation invites its visitors to imagine themselves among its pygmy protagonists and compare themselves to them. Yet even as these aspects of the garden suggest a virtual-reality-like immersion in its fictive world, other aspects—the miniature size of the imagery, the quasi-mythical status of its pygmy protagonists—emphasize the impossibility of viewers' fully entering the marvelous yet potentially dangerous land that it evokes. The Nilescape teases its viewers with simultaneous immersion and distancing.

The Nilotic garden triclinium thus provides no easy resolution of the tension between "Self" and "Other," or "here" and "there." At once highlighting and dissolving distinctions between "Roman" and "Other," the installation invites viewers to use their own social performances as a means to engage with the questions it raises. Are they to relate to the pygmies as jovial dinner companions, models of pastoral piety, embodiments of licentious *Kanobismos* (cf. Strab. 17.1.16), faraway curiosities, negative *exempla*[160] illustrating what not to do at dinner, or convenient prompts for displaying one's own intellectual and cultural sophistication? Visitors' subsequent choices, both in their behavior and in their conversation, would thus constitute their own creative collaboration with the domestic visual culture that framed those choices.

In inviting visitors to feast with the pygmies on the banks of a model "Nile," the Casa dell'Efebo garden prompts its users to engage through practice with an ever-evolving, society-wide discourse on the nature of cosmopolitan, imperial Roman identity. In the Flavian empire, even Egypt—now a useful and fertile component of the Roman world, rather than an external threat—could play a role in that discourse. Rather than reinforcing entrenched distinctions between "Roman" and "Egyptian," the Nilotic garden triclinium of the Casa dell'Efebo appears to problematize those barriers. At the same time that imperial propaganda and contemporary literature increasingly presented Egypt, the Nile, and its gods as productive contributors to Roman *imperium*,[161] the incorporation of culturally diverse motifs and themes into the material culture of the *domus* formed a popular, household-level contribution to this ongoing cultural dialogue. Simultaneously far away and familiar, the imagined landscapes of the garden enabled its occupants to engage in open-ended ways with changing constructions of imperial, local, and cultural identities.

---

160. On negative *exempla* in Roman art, see, e.g., Bergmann 1999: 85–89 (punishment for misbehavior); 2002a: 31–32 (bad wives and mothers); Platt 2002 (Actaeon as model of dangerous voyeurism); Squire 2012: 615 (Polyphemus as bad host); Elsner 2007: 132–176 (Narcissus as model of destructive gaze); Hodske 2007: 132, 137 (malefactors punished).

161. See chapter 1, this volume, "Flavian Pompeii as Case Study."

4

Self, "Other," and Beyond in the Casa del Medico

Introduction

The previous chapters have investigated several ways that Pompeian domestic contexts might frame Nilotic scenes. Irrigated garden landscapes and dining areas constructed parallels between two-dimensional Nilotica and three-dimensional space (chapter 2), prompting viewers to contemplate the relationships between faraway and familiar. While some iconographic features of Nilescapes appear to stress the alterity of the Egyptian landscape and its inhabitants, contextual considerations suggest more complex relationships between viewers and viewed. Garden assemblages like those of the Casa dell'Efebo (chapter 3) simultaneously assert and deconstruct conceptions of "Self" and "Other." Rather than imposing singular, invariant "meanings" on Nilotic motifs, this general framework provides wide scope for variation in the ways that individual gardens framed such imagery, as the case study presented in this chapter demonstrates.

Perhaps the most frequently discussed Nilotic paintings are those from the courtyard of the Casa del Medico (VIII 5.24) (figures 2.4, 2.5, 4.1–4.4).[1] These frescoes display some of the most seemingly over-the-top Nilotic imagery at Pompeii, including an explicit sex scene and blood-soaked animal combats. An adjacent pygmy scene is often interpreted as a parody of the biblical Judgment of Solomon (figure 4.1). This combination of lurid imagery and

1. On the prominent representation of these scenes in general discussions of Nilotic landscapes, see, e.g., the inclusion of pygmy scenes from the Casa del Medico as representatives of "Nilotica" in major recent museum exhibitions on Greek and Roman representations of landscape (Sena Chiesa and Pontrandolfo 2015: no. 99, one of three examples of "Nilotica") and Greco-Roman interactions with Egypt (Poole 2016: no. 129, as the sole representation of "Nilotica"; Spier et al. 2018b: no. 153, as one of two examples; see further Barrett 2018b; 2018d).

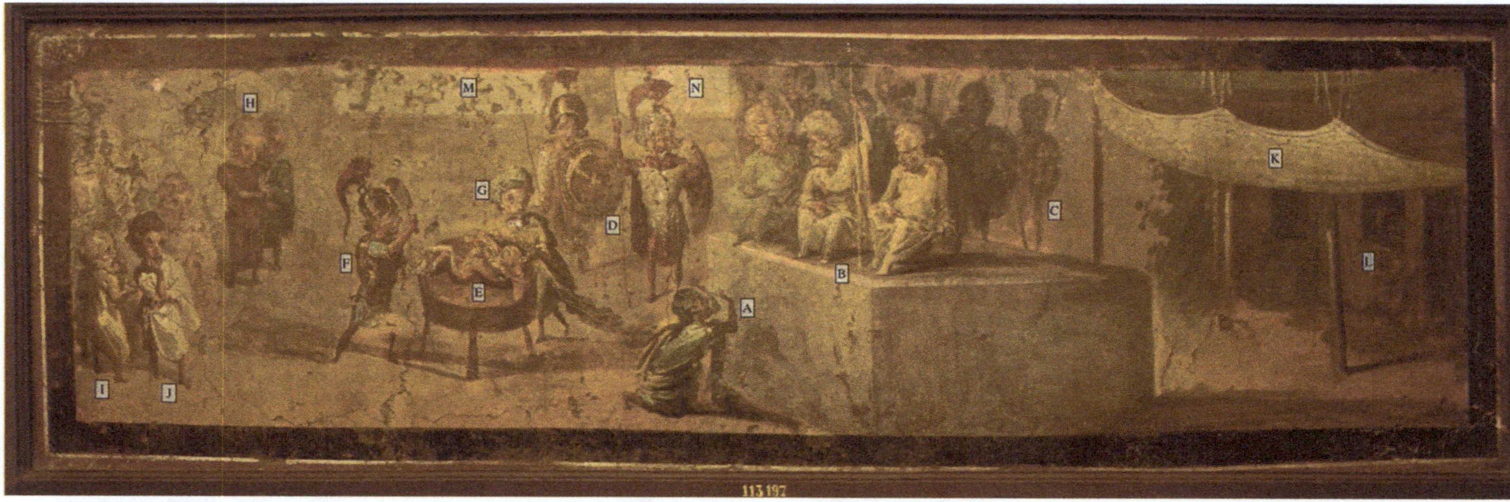

FIGURE 4.1 Wall painting with a pygmy scene often interpreted as a reference to the "Judgment of Solomon." Pompeii, Casa del Medico (VIII 5.24), pseudo-peristyle. MANN 113197. Letters have been added to show the location of particular features discussed in the main text.

Photo: C.E. Barrett. Reproduced by permission of the Ministero dei Beni e delle Attività Culturali e del Turismo—Museo Archeologico Nazionale di Napoli.

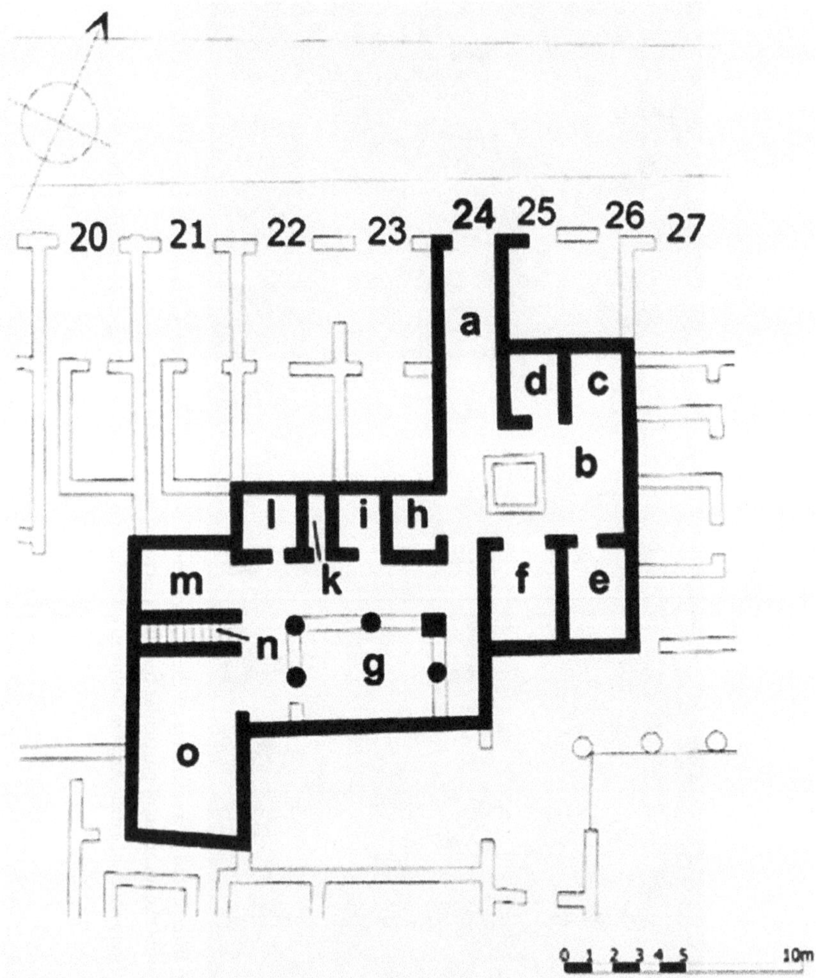

FIGURE 4.2 Plan of the Casa del Medico.
Source: After Clarke 2007b: fig. 44; reproduced by permission of John Clarke.

apparent biblical references has inspired scholarly debate since the nineteenth century.

Practically every aspect of the Casa del Medico assemblage has been subject to dispute, from the interpretation of the iconography to the function of the courtyard space itself. However, research on this house has followed two largely separate trajectories. While art historians have examined the paintings, a quite different scholarly debate has addressed the artifactual assemblage. However, the latter debate has focused on only one subset of the finds: the supposed medical

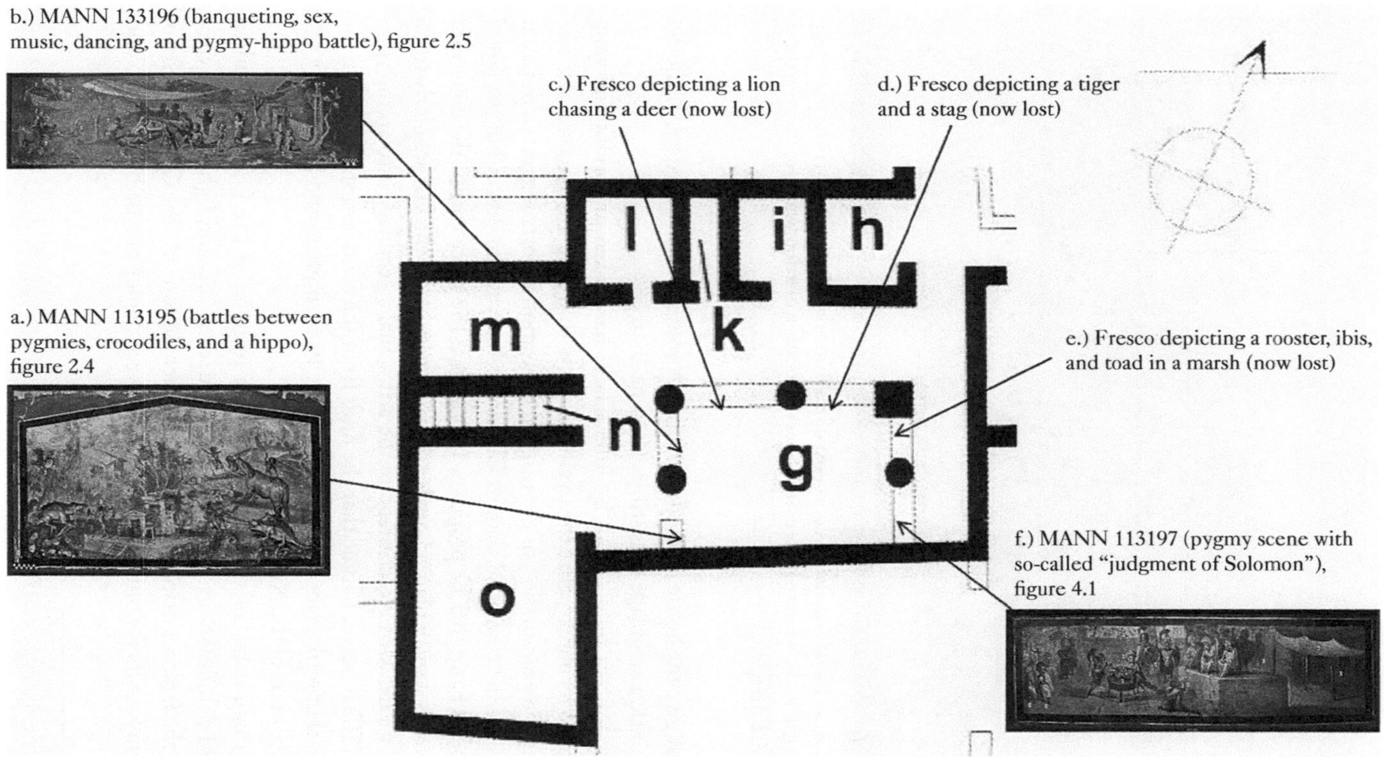

FIGURE 4.3 Detail of the plan of the Casa del Medico (figure 4.2), showing the partially colonnaded court and surrounding rooms. Insets illustrate the position of the frescoes, both those now in the Naples Museum and those now lost.

Source: Plan detail adapted from Clarke 2007b: fig. 44; reproduced by permission of John Clarke. Inset photos: C.E. Barrett.

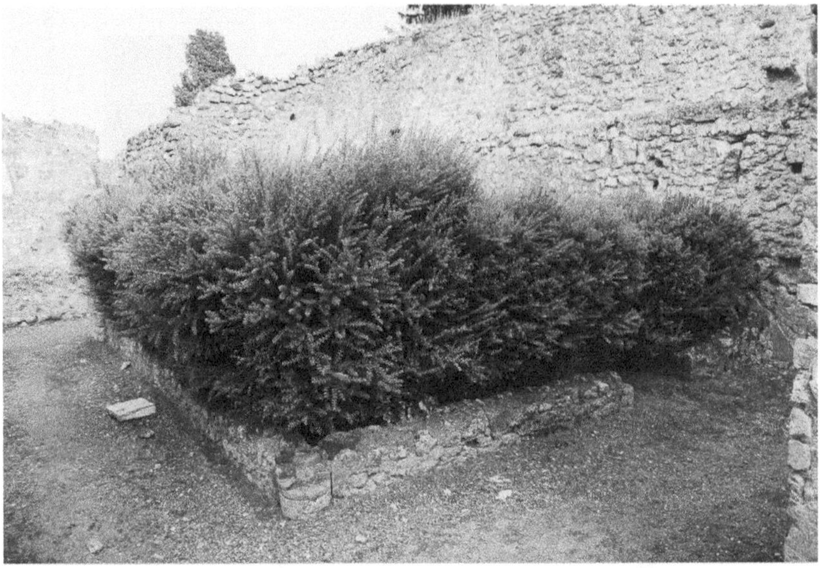

FIGURE 4.4 Pseudo-peristyle courtyard of the Casa del Medico in 2012 (facing southeast). The gap in the western pluteus indicates the location of a doorway into the courtyard. *Photo:* C.E. Barrett.

tools that give the house its modern name.[2] A comprehensive account of the entire domestic assemblage thus still remains necessary.

Several recent publications by John Clarke have taken important and valuable steps to recontextualize the pygmy scenes within their architectural setting.[3] However, a reexamination of the original excavation reports suggests a different function for the courtyard than in Clarke's reconstruction. Instead of the paved and flooded basin that he proposes, I argue that the space within the courtyard was probably a small planted garden. This chapter is also the first scholarly study of the Casa del Medico to integrate the artifactual assemblage into an interpretation of the frescoes. I argue that the finds from the courtyard shed light on the activities that took place in the house during its final phase and help clarify the viewing context of the Nilescapes.

Although many of the activities of the pygmies at the Casa del Medico and Casa dell'Efebo are similar (e.g., banqueting, sex, and fighting with river beasts), the frescoes display marked differences in tone. At the Casa dell'Efebo, the pygmy–animal conflicts include no overt depictions of bloodshed, and

2. See esp. Breton 1870: 456; Eschebach 1984: 45–47; Bliquez 1994: 81–82; Geissler 2011: 18–20.

3. Clarke 2007a; 2007b: 98–105; 2007c.

none of the pygmies appear defeated or injured. The Casa del Medico frescoes place far more emphasis on violence and death, representing multiple pygmies as dead or dismembered. The pygmy combatants also appear more overtly ridiculous, gesticulating wildly or assuming sexually provocative postures while hippopotami devour them. Furthermore, the Casa del Medico sex scene appears more comical than erotic. In contrast to the conventionally attractive female participant at the Casa dell'Efebo,[4] both protagonists in the Casa del Medico sex scene display bodies that Romans would have perceived as non-ideal. Finally, depictions of cultic activity—arguably the most prominent feature of the Casa dell'Efebo frescoes—are much less visible at the Casa del Medico.

Ever since 1932, the Casa del Medico scenes have served as Exhibit A in the argument that Nilotic imagery functioned as "caricature" or "Othering" of Egyptians.[5] The prominence of these frescoes in general discussions of Nilotica derives partly from their dramatic imagery, which appeared well suited for adaptation to the interpretive paradigms dominating scholarship on Nilotica in the 1990s and early 2000s. Additionally, because these paintings were removed to the Naples Museum at an early date, they are now among the best preserved Nilotic frescoes.

However, the Casa del Medico frescoes are by no means generic representatives of all Nilescapes. Far from being typical, their emphasis on violent and gory imagery is actually fairly unusual among Pompeian Nilotic scenes. Additionally, a comprehensive analysis of these frescoes should treat them not in isolation but, rather, as one component of a larger domestic assemblage. Rather than assuming that the contents and framing of these paintings simply represent the "default option" for Nilotic scenes, we need to reexamine the ways they interacted with their settings to generate location-specific viewer responses.

Another major difference between the Casa del Medico and the Casa dell'Efebo concerns socioeconomic status, as the Casa del Medico is much smaller and less lavishly furnished than the Casa dell'Efebo. Both houses feature entertainment complexes uniting garden and dining areas, but the scale and contents of these complexes suggest a significant disparity in resources. These houses thus provide an opportunity to compare the uses of Nilotic imagery at different points along the ancient socioeconomic spectrum.

In both houses, Nilotic pygmy scenes are associated with garden space, water use, and dining. However, other aspects of both the context and the content

---

4. See chapter 2, this volume, "Parallel Landscapes in the Garden."

5. McDaniel 1932: 270–271; Walker 2003: 192–193; Versluys 2002: 138–140; Clarke 2007a; 2007b: 98–105; 2007c: 162–169.

of the Casa del Medico frescoes differ substantially from the Casa dell'Efebo frescoes and provoke different viewer responses. Chapter 3 argued that the assemblage from the Casa dell'Efebo garden constructs a cosmopolitan, elite identity for the householders, inviting (or challenging) guests to perform their own *paideia*. While such associations may also be relevant at the Casa del Medico, this assemblage places greater emphasis on human–environment relations, the dangers endemic to natural landscapes, and the role of human civilization in controlling and minimizing those dangers.

The Casa del Medico assemblage also differs from that of the Casa dell'Efebo in separating the *depiction* of dining from the *act* of dining. At the Casa dell'Efebo, the location of the Nilotic banquet scenes on the banks of a triclinium stages a direct confrontation between image and performance. However, the Casa del Medico creates more distance between viewing and experiencing. Visitors would have first encountered the Nilotic frescoes not *within* a dining space, but immediately *before* entering a dining space. Standing within view of the painted banquet scene, guests could also obtain a partial view of the triclinium where actual banqueting was about to occur—but they could not yet see the entirety of that dining space, nor had they themselves yet joined the banquet. Viewers' own relationships to representations of banqueting were thus still notional and anticipatory. Instead of comparing the images to their own current social performances, they were prompted to imagine what forms their performances might soon take. Their initial encounters with the paintings thus created a confrontation, not between representation and reality but, rather, between representation and anticipation.

The Casa del Medico

The Casa del Medico is a modest house of irregular plan (figure 4.2), excavated in 1841 and 1882.[6] The 1841 campaign explored the general region around the cul-de-sac now known as the via della Maschera, uncovering portions of the Casa del Medico as well as other houses.[7] Antonio Sogliano subsequently excavated the house in a more systematic fashion in 1882.[8]

6. For an overview of the house, see Bragantini 1998b.

7. Fiorelli 1860–64: 2:395–399, 3:166–171. These volumes of Fiorelli's *Pompeianarum Antiquitatum Historia* compile and publish the notes from excavations at Pompeii before 1860. Museum inventories for these finds are unfortunately nonexistent (see Allison 2004a: 32).

8. Finds and excavation daybooks from this season: Fiorelli 1882; Sogliano 1882: 322–324, 435–441; Mau 1883b: 228–330.

Relatively little information survives from the 1841 excavations. Excavation journals recorded finds from this campaign without precise proveniences, leaving unclear whether those finds came from the Casa del Medico or one of its neighbors.[9] Further complicating matters is a chronological inconsistency in the numbering of insulae and houses. Since the excavators expected the via della Maschera to be a through street rather than a cul-de-sac, they initially assumed that the houses on either side would belong to separate insulae. As a result, early reports label the Casa del Medico as VIII 6.6 rather than its current designation, VIII 5.24.[10]

However, the 1882 excavation produced much more detailed records. Mau's 1883 report records the finds according to house and room.[11] Fortunately for present purposes, the courtyard and adjacent rooms were excavated in 1882, so these finds were relatively well recorded.

The courtyard was a partial peristyle or "pseudo-peristyle,"[12] columned on three sides. Clarke suggests that the partial peristyle and reception rooms at the southwest end of the house were later additions to the original structure.[13] Two of the rooms adjacent to the courtyard, (m) and (o), have been architecturally identified as triclinia.[14] Three smaller rooms around the courtyard (h, i, l) could be either cubicula or storage rooms, although they lacked finds that could confirm these functions. Room (l) is sometimes identified as a kitchen,[15] although preserved remains provide little decisive evidence. Two staircases (k, n) originally led

---

9. See note 7, this chapter. For example, Eschebach proposes that a marble fountain figure of a satyr or Silenus might have come from the impluvium of the atrium of the Casa del Medico (1984: 46). While excavation records do state that the statue came from an impluvium, they do not identify a specific house in the insula (Fiorelli 1862: 397). Eschebach also suggests that a marble group depicting a satyr and faun could have come from the Casa del Medico courtyard (1984: 47). However, the only provenience that Fiorelli gives for this statue group is the vague designation "a destra la strada dei Mercatanti" (1860–64: 3:169). Furthermore, the Casa del Medico courtyard appears to have been excavated in 1882 rather than 1841 (see note 11, this chapter).

10. See discussion in Mau 1883a: 170–171; Eschebach 1984: 45. As late as the 1880s, Sogliano refers to the house as "Fiorelli isola 6, n. 6" (1882: 322), while Mau refers to it as no. 24 in what he calls "Isola VIII 5 e 6" (1883b: 228; on the insula designation, see 1883a: 170–171). Today, this insula is known as 5, with VIII 6 now referring to a separate insula.

11. Mau 1883b: 228–230.

12. On the terminology for partially colonnaded open spaces, see note 49, chapter 1, this volume.

13. Clarke 2007a: 219.

14. Mau 1883b: 229; Bragantini 1998b. Clarke refers to (o) as a "reception space" (2007a: 219) or "dining room" (2007a: 221; 2007b: 102; 2007c: 164, 167).

15. On room (l) as a possible kitchen, see Eschebach 1984: 47, though using the room label (h); Kastenmeier 2007: 158.

to an upper story. Third Style paintings were still visible at the time of excavation in both triclinia.[16] According to Mau, the paintings in triclinium (o) depicted fantastic architecture and a "sacral-idyllic" scene of worshipers at a rural shrine, while those in (m) were mostly destroyed.[17] Mau also notes Fourth Style paintings on white ground in the tablinum (f) and a painting of a bird at the entrance to the atrium (b).[18] The walls of the atrium itself bore only rough plaster, while (d), (e), and (h) were white-plastered, (c) had rough plaster, and (i) had traces of red paint or plaster.[19] The atrium contained a marble-revetted impluvium.[20]

The Frescoes of the Casa del Medico Courtyard

Fourth Style paintings in the courtyard included animal scenes and pygmy scenes.[21] Because the three pygmy scenes were removed to the Naples Museum, they have survived better than the other paintings. Previous interpretations have thus focused almost exclusively on the pygmy scenes (figures 2.4, 2.5, 4.1), ignoring or paying less attention to the now-lost animal scenes. However, a reexamination of the courtyard frescoes reveals numerous echoes and relationships between the images, suggesting that their intended effects depended at least in part on the interactions between scenes.

South Side of West Wall (MANN 113195)

As visitors approached the courtyard from the house entrance, the first visible fresco occupied the west wall of the pluteus (figure 4.3a).[22] The south side of this

16. Mau 1883b: 229; Bragantini 1998b: 607–610; Ehrhardt 2012: 69. Eschebach suggests that the 62 CE earthquake may have required repair to the walls (1984: 45–46). However, the surviving Third Style paintings suggest he may go too far to suggest that "Fast alle inneren Trennmauern mußten erneuert werden" (1984: 46).

17. Mau 1883b: 229.

18. Mau 1883b: 228.

19. Mau 1883b: 228; Ehrhardt 2012: 70. Schefold asserts that the atrium had a painting of Venus and Adonis, as well as floating nymph motifs (1957: 227; cited, with the images described as "verloren," by Eschebach 1984: 46). However, the only source that Schefold cites is Mau, who clearly states that "le pareti dell'atrio pare che non fossero rivestite che d'intonaco grezzo" (1883b: 228).

20. Eschebach 1984: 46.

21. On the date, see Clarke 2007a: 221; 2007b: 100. Clarke (2007b: 101) notes the animal scenes, but his analysis focuses mainly on the pygmy scenes.

22. On the placement of the courtyard frescoes, see Sogliano 1882: 322–324; see also note 88, this chapter.

wall presented a jarring juxtaposition of religious imagery and violent conflict (figure 2.4; MANN 113195).

At the center of the foreground is a cluster of shrinelike structures (figure 2.4, A). To the left is a small aedicula flanked by vases on elevated podia. Behind and to the right of the aedicula is a taller rectangular building with pointed acroteria at its corners. Two sculptures on raised podia guard the right side of this building. Both statues depict winged, recumbent animals. Traces of downward-pointing beaks or snouts are visible on both heads, suggesting that they are not sphinxes but, rather, some other composite animal. Behind the complex of buildings grows a tall tree.

Immediately behind and on both sides of the small shrine complex (A), blue paint indicates water. To the upper left is a building (B) with a cylindrical tower, thatched roof, and small enclosure fence. A long-necked black bird (C), probably a black stork, approaches from the left. Many Nilotic "Campana reliefs" show storks perching on thatched houses and towers that resemble the structure (B) in this painting.[23]

At the top center right, a large porticoed complex (D) stands on the far riverbank. As at the Casa dell'Efebo, the depiction of a *porticus* complex evokes Roman rather than indigenous Egyptian architectural traditions.

Despite the presence of some Egyptianizing features (e.g., the composite animals), the architectural features in the fresco are fairly typical of Roman landscape painting, with numerous parallels in "sacral-idyllic" and villa scenes.[24] The placement of a shrine complex at the center of the composition suggests rural piety and prepares viewers for a tranquil pastoral vista like those of "sacral-idyllic" landscapes. However, the frenzied, violent activity surrounding the sanctuary immediately subverts those expectations. On both sides, pygmies fight to the death against enraged river monsters.

At left, a naked figure (E) lunges at a crocodile (F) emerging from a clump of water plants and lotuses (G). The oversized head and torso, spindly limbs, and slight swayback of the human combatant suggest the achondroplastic dwarfism attributed to Nilotic *pygmaei*. He wields two disc-like objects, possibly cymbals.[25] In other Nilotic scenes, pygmies sometimes frighten animals by beating sticks or clappers to make noise.[26]

---

23. Nilotic scenes in "Campana reliefs": see Rauch 1999; Barrett 2013; 2018c.

24. On these categories of Roman painting (and critique thereof), see note 168, chapter 2.

25. On cymbals and castanets in Nilotica, cf. Garmaise 1996: 58–61; Boissel 2007: 350, 385; Vucetic 2014: 148; Trentin 2015: 48–49. Rattles are also possible (cf. Garmaise 1996: 65–66).

26. Boissel 2007: 273–279. The crossed sticks often held by Nilotic pygmies are musical instruments; see Garmaise 1996: 61–64.

To the right of the sanctuary, two other groups of naked pygmies fight against two other hostile river animals. A crocodile (H) approaches three pygmies (I) who pull ropes held by a fourth pygmy on the crocodile's back. All four work together to subdue the crocodile.

Above this group, a hippopotamus (J) battles several pygmies. One stands on the hippo's back, drawing blood with a spear. The hippo meanwhile capsizes a boat (K), swallowing one of its two occupants. The victim stands bent over, disappearing down the hippo's bloodstained mouth. The other pygmy stands up, arms flailing. In other Nilescapes, when two figures in a boat adopt these postures—one standing, one bent over—they are typically engaged in sex.[27] Ancient viewers might thus understand this vignette as a *symplegma* gone gruesomely wrong.

In front of the hippo floats another pygmy (L), probably dead; red streaks on his face and limbs appear to be blood. Another pygmy (M) swims away. Farther in the background is a larger boat (N) manned by at least four, possibly five, figures.

Some features of this fresco may reflect Greek and Roman accounts of crocodile hunts at the Upper Egyptian city of Tentyra (Dendera).[28] The image of pygmies using ropes to trap or lead a crocodile (figure 2.4, H and I) recalls ancient claims that the Tentyri caught crocodiles with ropes.[29] The pygmy on that crocodile's back also resembles Pliny's description of Tentyri riding on the backs of crocodiles (*HN* 8.93). Also recalling Pliny is the depiction of a pygmy seemingly frightening a crocodile by creating a loud noise (figure 2.4, E).[30] Ancient references to human casualties among the Tentyri parallel some of the pygmies' bloody ends.[31]

This implied conflation of "real-life" Tentyri and fantastical *pygmaei* contrasts with other Roman portrayals of *pygmaei* as a completely different population

---

27. At Pompeii, see figure 2.49; Versluys 2002: nos. 20, 42, 50, 63; probably also MANN 8566 (published, though without comment on the sexual imagery, as Versluys 2002: no. 30) and probably also a now-lost fresco published in drawing form (Roux and Barré 1870: 5.46–47, pl. 22; Reinach 1922: 161, fig. 1; Trentin 2017: 241–242). More generally, see Meyboom and Versluys 2007: 183–185; Vucetic 2014, esp. 146–150, 154.

28. For sources, and on the possible Egyptian background for such claims, see note 45, chapter 2, this volume. Merrills has recently proposed that textual accounts of the Tentyri/Tentyritae actually derive *from* Nilotic scenes, rather than the other way around (2017: 144–146). However, this is unlikely. Strabo's description of the Tentyritae (17.1.44) predates the large-scale popularity of Nilotic pygmy scenes (see note 26, chapter 2, this volume), and Greek and Roman accounts of Egyptians battling crocodiles at Tentyra (Dendera) more likely derive from the actual practice of ritual crocodile hunts at Dendera and Edfu (Wilson 1997). Roman popular interest in the Tentyri likely also derived from public exhibitions; according to Strabo (17.1.44), at least one group of "Tentyritae" was brought to Rome to subdue crocodiles in the arena.

29. *Lassos:* Sen. *QNat.* 4a.2.15. *Nets:* Ael. *NA* 10.21; Strab. 17.1.44.

30. Pliny says the Tentyri used their voices (*HN* 8.93).

31. Plin. *HN* 8.93; Sen. *QNat* 4a.2.14.

from "typical" Egyptians.[32] Greek and Roman accounts of the Tentyri do not characterize them as somatically abnormal.[33] Strabo's description of Tentyrite performers in Rome (17.1.44) says nothing of their physical appearance, even though any anomaly would surely have added to the spectacle. One probable visual representation of a Tentyrite employs idealized proportions. Depicting an African youth performing a handstand atop a crocodile, a marble statue (first century BCE—first century CE) may reflect the type of performance that Roman audiences expected from shows like the one Strabo describes.[34]

As a site in Upper Egypt, Dendera was relatively close to the southern regions where Romans imagined the Nile to originate and *pygmaei* to dwell.[35] Some Romans may thus have imagined the (actual) Tentyri as another quasi-mythical population on the edges of the known world and conflated stories about one group with another. However, the parallels between the Casa del Medico scene and the textual descriptions of crocodile-fighting Tentyri may alternatively indicate that, because *pygmaei* were imagined to live in or near Egypt, Romans assumed that their crocodile-fighting techniques would be similar to those used by "actual" Egyptians. Rather than identifying Nilotic pygmies *as* Tentyri, the fresco may thus show them fighting crocodiles *in the same fashion* as Tentyri.

North Side of West Wall (MANN 113196)

North of a gateway on the west wall is a second Nilotic pygmy scene (figures 2.5, 4.3b; MANN 113196). Much of this fresco depicts a riotous scene of banqueting, sex, music, and dancing, but violence also erupts again in another gory pygmy–hippo battle.

On a riverbank stands a canopy-shaded stibadium (A). Four male pygmies recline, while a fifth sits on the ground at right. His himation has mostly fallen off, revealing his torso. In front of these figures are a mixing bowl and two ladles on a table, and a garlanded amphora in a stand. The banqueters watch a sexual encounter between a prone naked pygmy (B) and a squatting pygmy (C) wearing a floral wreath. Although the gender of the squatting figure has recently been

32. See chapter 2, this volume, "Navigating 'Nilotic Scenes'".

33. Pliny describes them as small in size (*mensura eorum parva: HN* 8.92), but in the context of the passage, he likely means that they (like other humans) are small compared to crocodiles.

34. British Museum 1805.0703.6 (Roman, findspot unknown); see Barrett 2018a.

35. For example, Pliny locates pygmies near the origin of the Nile (*HN* 6.188).

questioned,[36] I interpret this individual as female. A swelling breast is visible beneath the right armpit, and the curving hips and buttocks suggest female proportions. This figure's light pink skin is also paler and cooler in tone than that of the other figures around the stibadium, who have either dark brown or light orange-brown skin.[37] Finally, while Nilotic *symplegmata* are not always heterosexual, couplings between a prone male and a squatting partner who faces away from him are invariably heterosexual in other Nilotic scenes.[38] To the left of the couple, a musician (D) plays a double flute; to their right, another pygmy (E) dances with long sticks in his hands. His energetic, twisting posture has parallels not only in Nilotic frescoes and mosaics but also in numerous Hellenistic and later figurines (from Egypt and elsewhere) depicting dwarfs or pygmies.[39] Behind him stands an ibis (F).

Several onlookers react to these sexual and musical performances. The man at the leftmost end of the couch (A) raises his right hand in surprise or enthusiasm, while the indecorous posture and dishabille of the man seated right of the couch may suggest inebriation, arousal, or both. To the right of the banqueters, two standing figures carrying staffs (G, H) extend their right arms outward in apparent surprise.

The bottom left of the fresco depicts a violent hippo–pygmy combat (I, J, K), echoing themes from the southern half of the wall. A hippo (I) devours a pygmy whose lower body disappears down its throat, spilling gouts of blood. Another pygmy (J), wearing a white conical cap, attempts to pull the victim out. Such caps are common in Nilotic scenes and are often (though not always) worn by dancers.[40] Atop the hippo's back, another pygmy (K) raises a round object, possibly a cymbal or castanet. Both his stance atop the hippo and his possible weaponization of an instrument have parallels in the southern painting (figure 2.4, E and J). Since one pygmy wears a cap associated with dancers and another may hold an instrument, viewers might conclude they were originally participating in the musical performance (D). The gory end of their performance thus parallels the hippo attack in

36. Kimel, forthcoming.

37. The contrast with the squatting figure's skin color is less pronounced for the figures at the left and right sides of the fresco (J, K, M, N, O) but is clearly visible with all of the figures around the stibadium (A, B, D, E, H).

38. On the use and attestations of this position in Nilotic sex scenes, see Meyboom and Versluys 2007: 182–202.

39. See chapter 2, this volume, "Egyptian Antecedents and Affordances."

40. On conical caps as an attribute of dancers in Nilotic scenes, see Meyboom and Versluys 2007: 186–187; cf. Dunbabin on "pointy-hatted stick dancers" (2016: 127–133). See also note 121, chapter 2, this volume.

the southern painting, which seemingly brings a similarly abrupt end to another form of outdoor enjoyment: a riverine *symplegma* (figure 2.4, J and K).

Another such *symplegma* is occurring in the upper left corner of the present fresco. A boat with a donkey-headed prow[41] (L) sails away from the riverbank, laden with amphorae. In the stern are two pygmies (M, N). The leftmost figure (M), a male with a receding hairline, stands naked, leaning back and gazing upward. His right hand holds a brown shape, probably a wineskin or amphora. The other figure (N) bends over with exposed buttocks; a blue cloth conceals the upper body. The figures' postures suggest a sexual encounter.

At the right of the fresco, two pygmies (O) in himations converse near a low wall attached to a roofed structure (P). A tree-entwined pillar supports statuettes of standing figures (Q).

North Wall

Although the frescoes on the north wall are now lost, this wall is the longest section of the pluteus (figures 4.3c, 4.3d). Left in situ when the pygmy scenes went to the Naples Museum, the remaining paintings faded away. However, first-century viewers would have encountered the pygmy paintings as part of an ensemble that also included the other frescoes.

Sogliano's report lists two different scenes for the north wall. For the first—and hence the more westerly, since his description moves from west to east—he states: "Sul fondo rosso un leone che insegue un daino o capriuolo." For the second scene, he notes: "Sul fondo rosso tigre e cervo accovacciati; ambedue però mancanti del capo."[42]

Even though Sogliano's description of these scenes is less detailed than his lengthy discussion of the pygmy scenes, his report provides us with enough information to detect some thematic parallels with the surviving frescoes. Both scenes on the north wall portray predator–prey interactions: to the west, a lion chasing a deer; to the east, a tiger confronting a stag. These images of tense, potentially violent confrontations would have echoed the pygmies' battles with river animals. The red ground of the animal scenes continued the red ground of the pygmy banquet scene (figure 2.5), encouraging viewers to form connections between the images.

---

41. Compare the animal-headed boat in one of the frescoes from the Casa dell'Efebo triclinium (figure 1.4, E). On animal-headed boats in Egypt, see Meyboom 1995: 254–255n112.

42. Sogliano 1882: 323; see also 322 for a discussion of the progression of his description from west to east. Clarke's description of the north wall (2007b: 101) mentions only the first of these two scenes.

North Side of East Wall

The fresco on the north side of the east wall is also lost (figure 4.3e). However, Sogliano again summarizes the imagery: "Sul fondo rosso una pianta palustre: a dritta un gallo e a sinistra un ibis, che becca un rospo stante sul suolo."[43] The red ground visually integrated this painting with the adjacent wall, while the iconography combined themes from both west and north walls. Like the north wall, this panel depicted an animal scene: a rooster, ibis, and toad in a marsh. The ibis's pecking at the toad reinforced the theme of predator–prey interactions. Another ibis appears on the northern panel of the west wall (figure 2.5, F), directly across from this panel. The repetition of the Egyptian bird visually linked the two frescoes, inviting viewers to situate the marsh scene in Egypt.[44] The scene thus synthesized themes from both the west and north walls, but in miniature. Rather than encompassing an entire riverscape, this panel presented a single wetland vignette; and rather than showcasing large, powerful megafauna, this panel dramatized the smaller-scale encounter of two birds and a toad.

South Side of East Wall (MANN 113197)

Finally, the south side of the east wall (figure 4.3f) bore the fresco that first directed art historians' interest to this house: the "Judgment of Solomon" scene (figure 4.1). However, the identification of this scene is less clear-cut than often assumed. While it does show a judgment scene involving the life or death of a child, several aspects differ substantially from the biblical account. Some evidence also suggests that similar stories may have circulated outside Jewish or Christian traditions. Ancient viewers of this fresco may not necessarily have known or cared whether the central figure sitting in judgment was Solomon or some other legendary judge, such as the Egyptian Bocchoris.

As on the west wall, most of the figures display disproportionately large heads and torsos and small, spindly limbs. Roman viewers probably would have understood these figures as *pygmaei*. However, this scene does not include any riverine or otherwise Egyptianizing features. Additionally, while one of the figures (figure 4.1, A) has the dark skin and black hair that characterize some "Nilotic" pygmies, the other characters in this scene all have (what Roman viewers imagined to be) less geographically specific beige skin and brown or

43. Sogliano 1882: 323.

44. On Romans' association of the ibis with Egypt, and on paintings and sculptures of ibises at Pompeii, see Watson 2002: 397.

white hair. While this is a pygmy scene, it is not a "Nilotic" or "Egyptianizing" scene per se. Some viewers might conceivably have taken the other frescoes' Nilotic imagery as an indication of setting for this scene too, but the painting itself does not compel this conclusion.

The center of the fresco depicts a petition scene. A figure (A) with dark brown skin and short curly black hair, wearing chiton and himation, kneels with hands upraised below a tall dais or podium. This figure is probably male. While close-cropped hair is not always gender-specific, the himation is draped in a fashion more characteristic of males.[45] Atop the dais sit three white-haired, aged-looking males (B) in himations. The staff of the bearded man in the center may suggest particular authority.

Behind these three seated figures stand six soldiers (C) with shields, spears, and plumed helmets. To the left, two other standing soldiers (D) gaze at the three men on the dais with slightly open mouths, possibly connoting surprise. The soldiers' costume resembles that of the first century CE Roman army.[46]

To the left, a prone form (E), usually identified as a naked baby, lies on a table. Visible are two spindly legs, a large round head, and a thin arm. The torso displays unusual proportions, with a huge rib cage, narrow waist, and pronounced sway-back. This shape has inspired one commentator to question the humanity of the figure.[47] However, the figure's proportions are probably simply an attempt to render achondroplastic features on an infant.

Next to the table, another soldier (F) lifts a large cleaver, seemingly about to chop the outstretched infant. He gazes upward at the bearded man at the center of the dais. A woman (G) leans over the table with her left hand on the baby's head and her right hand on the baby's shoulder. Her legs are shown mid-stride, with the left leg advanced, and her himation billows out behind her. These indications of rapid movement suggest she has rushed to the table. She twists her head backward to gaze at the petitioner.

---

45. *Pace* the typical interpretation of this figure as female (see note 56, this chapter). For women in postures of supplication, see Brilliant 1963: 100, 116, 150, 158, 190–191, 189–191, figs. 2.115, 3.29, 3.112, 3.134, 3.135, 4.68–4.72, 4.85b.

46. Fuentes 1987: 52–53. For the helmets, compare the types in Bishop and Coulston 1993: 92–94; Robinson identifies the form as "Montefortino Type F" (1975: 16). The Roman army of the early empire employed multiple shield types, including oval and round shields like those in this fresco (Bishop and Coulston 1993: 81–84). Some commentators have assumed that the soldiers must be Ptolemaic, based on the assumption that the fresco must be a copy of a lost Alexandrian original (e.g., Sumner 2002: 20). However, there is no evidence that the fresco is a copy, or that any scenes similar to this fresco ever existed at Alexandria.

47. Kimel, forthcoming.

To the left stands a crowd of spectators. Only the figures in the front row are depicted in detail (H, I, J). One figure (H) points toward the baby. At the bottom left corner stands a bald, white-bearded man (I), wearing a himation over a bare torso, who makes a gesture of surprise. To his right stands a man (J) in a chiton and himation who rests his chin on his right hand.

The kneeling man's act of petition appears central to the implied narrative of the fresco. The other figures all gaze either at him or at the seated man on the dais. The composition implies that the seated man's response to the kneeling man's petition will determine the child's fate.

A canopy (K) suggests that the scene takes place outdoors. Behind the canopy is a built structure (L) with two square windows. The sharp corners and vertical lines of the windows and walls suggest stone, brick, or masonry. Also visible in the background are indications of a monumental structure. The human figures stand on a yellow-brown surface, on which they cast dark brown shadows. At the top of the scene, however, two pale beige rectangles (M, N) break into this yellow-brown background. These spaces appear to be windows or open spaces between pillars. The open-air petition scene must thus be taking place either within a monumentally large courtyard, or immediately outside a monumental built structure.

A Pygmy "Judgment of Solomon"?

Ever since its discovery, most scholars have interpreted this fresco as a reference to the biblical "Judgment of Solomon" (1 Kings 3:16–28). Sogliano claimed to see an "esatta corrispondenza col racconto biblico" in all respects except for the use of Roman costume.[48] Some scholars have seen the scene as evidence for anti-Semitism, and possibly the presence of a Jewish community, at Pompeii.[49] Feder, in contrast, takes the painting as evidence for Roman appreciation of Jewish wisdom.[50] Clarke accepts the identification as the Judgment of Solomon, but sees the fresco as making fun "not of Jews but rather of the high-minded seriousness of judgment scenes."[51]

48. Sogliano 1882: 323, followed shortly thereafter by Mau 1883: 230; de Rossi 1883; Martha 1885: 260.

49. For recent overviews of the evidence for Jews in Pompeii, and attempts to connect that evidence with the Casa del Medico fresco, see Giordano and Kahn 2001: 41–61; Small 2007: 195; Gunderson 2013 (going back ultimately to the early work of Mau 1908: 16–17).

50. Feder 2008 (not addressing the figures' representation as pygmies).

51. Clarke 2007b: 105; see more generally Clarke 2007a: 221; 2007b: 101, 104–106; 2007c: 164.

Ever since Sogliano's publication, however, dissenting voices have questioned the identification of the fresco with the biblical episode. Fiorelli appended a note at the end of Sogliano's article expressing doubts about the Solomon hypothesis.[52] Several scholars have suggested that instead of Solomon, the fresco might depict the legendary Egyptian king Bocchoris, who was similarly famous as a judge and lawmaker.[53] A frieze at the Villa della Farnesina is sometimes interpreted as depicting Bocchoris as a judge, and one scene from this frieze may depict another variant of the Solomon story.[54] The pygmy fresco in the Casa del Medico might represent a playful reimagination of such solemn depictions of judgment scenes.[55]

Certain features of the iconography cast doubt on the supposed "esatta corrispondenza" with the biblical story. Although previous commentators almost universally identify the kneeling figure (A) as one of the child's competing "mothers,"[56] the figure actually appears to be male, not female. Additionally, the woman (G) holding the baby does not clearly fit the biblical description of the unconcerned false mother. Both Mau and Clarke view this woman through the lens of the biblical text; Mau sees her as gesturing toward the real mother to take her half, and Clarke believes that she "indicates the middle of the baby to the soldier-butcher with her right hand—so she'll get her fair half—and turns toward the real mother with a quizzical expression on her face."[57] However, a close examination of the woman's posture and gestures suggests that she is more likely trying to protect the baby. Her striding legs and the himation billowing out behind her—the only billowing garment in the painting—indicate rapid movement. Additionally, rather than pointing to the baby's midsection, her right hand actually grips the child's right shoulder while her torso leans

52. Sogliano 1882: 324n1 (though without further explanation of the objection).

53. Engelmann 1904: 146; McDaniel 1932: 260, 267, 271; Gesztelyi 1989; Hansen 2002: 232; Gunderson 2013: 67.

54. On the "Solomon" scene: Gesztelyi 1989; Bragantini and de Vos 1982: 234, 236; van Aerde 2015: 156. For recent questioning of attempts to identify the scenes and figures in the Villa della Farnesina frieze, see van Aerde 2015: 156–157. On the frieze itself, see Bragantini and de Vos 1982: 22, 234–283; Bragantini and Pirelli 2006–2007.

55. Ling 1991: 166; Clarke 2007a: 221; 2007b: 105.

56. Descriptions of the kneeling figure as female: Sogliano 1882: 323; Mau 1883: 230; Giordano and Kahn 2001: 59; Clarke 2007a: 221; 2007b: 105; Feder 2008: 32–33. Nineteenth-century drawings of the scene (Mau 1908: 16, fig. 6; Martha 1885: 260) even render the figure's hair as long and flowing, rather than (as it is in reality) short and curly. While there is some dark pigment on the figure's upper back, in-person examination suggests these dark lines represent folds in the drapery, not hair. Kimel (forthcoming) correctly identifies the figure as male.

57. Mau 1883: 230; Clarke 2007b: 105 (quoted).

over the table. In other words, this woman has not been standing calmly; she has come running and is now practically hurling herself onto the table, leaning forward to grab the baby while her legs are still in motion. Additionally, while her slightly open mouth, raised eyebrows, and gesture toward the kneeling petitioner (A) might suggest confusion at his distress, these features might just as readily express an equally intense distress of her own. If the biblical text had not framed modern perceptions of this scene, it is unlikely that scholars would have interpreted figures (A) and (G) as two women competing for the baby. Based solely on the painting, we might instead suppose this agitated man and woman to be the baby's parents, both trying to intercede to save their child.

If the scene is not the Judgment of Solomon (or at least not the canonical version of that story), then a reference to the Egyptian lawgiver Bocchoris does present a possible alternative. The juxtaposition with Nilotic landscapes, and the representation of the characters as pygmies, might have prompted viewers to locate the scene in Egypt. The similar reputations of Solomon and Bocchoris might have encouraged some popular conflation of the two. Some Romans might also have assimilated both Egyptians and Jews to generalizing, proto-Orientalist stereotypes of Eastern "Others," confusing Egyptian and Jewish traditions.[58]

However, we should be wary of trying to "match" the Casa del Medico fresco to any single traditional story, whether biblical, Egyptian, or otherwise. Multiple stories about wise kings may have circulated in the first century CE, drawing on traditions from multiple cultural and geographical origins within the empire.[59] Some evidence also suggests the existence of multiple stories in which a judge resolves a dispute over a child by threatening to cut the child in two. A Roman-period Greek papyrus fragment refers to an unnamed judge threatening to cut a baby in half to resolve a dispute about parentage, and in Petronius's *Satyricon*, one of the young Giton's two lovers proposes to resolve their rivalry by cutting the boy in half.[60] A fresco from the columbarium of the Villa Pamphili depicted a naked baby on a table, a man raising a knife, and a woman pleading on her knees.[61] Another now-lost fresco from a tomb in Rome appears to have originally depicted a seated judge, a baby on an altar, and two women.[62] Similarly, an engraved gem

58. On Roman conflations of "Eastern" foreigners and cultures, see Alvar 2008; Versluys 2013b.

59. As also similarly observed by Mitchell (2008, citing a personal communication from Katherine Dunbabin) and already anticipated by Radin (1917: 173–175).

60. Petron. *Sat.* 79–80; *POxy.* 2944; discussed in Hansen 2002: 228–232.

61. Engelmann 1904: 147–148; Bendinelli 1941: 12–13; Gesztelyi 1989: Abb. 2; Ling 1993: 131.

62. Engelmann 1904; Gesztelyi 1989: Abb. 5.

of imperial date depicts a seated man holding a baby, two pleading figures, and a soldier lifting a sword.[63] Although these images resemble the Solomon story, it is not certain that their ancient viewers would have interpreted the imagery as specifically "Jewish" (or, if there were a similar legend about Bocchoris, as specifically "Egyptian"). Recent scholarship has similarly questioned the identification of specific narratives about either Bocchoris or Solomon in the Villa della Farnesina frieze, which may present a more open-ended depiction of good governance and judgment.[64] Attempts to explain potentially multivalent images exclusively through literary sources risk reducing visual culture to illustration. Artists, like writers, could alter older models or recombine existing motifs to new ends. Finally, viewer responses could further alter the import of the image. Depending on their own backgrounds, different viewers might have perceived the judgment fresco as referring to Solomon, Bocchoris, or neither.

Violence and "Otherness" in the Casa del Medico Nilotica

While both the Casa del Medico and Casa dell'Efebo Nilotic frescoes depict beast combats and sexual encounters, they depict these scenes in very different ways. In the Casa dell'Efebo paintings, no blood is shed and no pygmies show signs of injury. Not only are pygmy–animal battles more prominent at the Casa del Medico (especially figure 2.4), but those combats are also much more violent. Two pygmies are devoured alive, blood oozing from their mangled forms (figures 2.4, K; 2.5, I). Another bloodied pygmy seems already dead (figure 2.4, L). The pygmies also deal out violence in their turn, spearing and clubbing hippopotami; one hippopotamus (figure 2.4, J) bleeds as profusely as his victim. Indeed, the pygmies do not always get the worst of it. One crocodile–pygmy battle (figure 2.4, E and F) is a standoff without obvious victor, and another group of pygmies has successfully trapped a crocodile (figure 2.4, H and I). When they are defeated, however, the pygmies' inelegant postures imply a cruel slapstick humor. They die in undignified fashion, succumbing to extreme physical mutilation, one even bending over in a sexually provocative position (figure 2.4, K; 2.5, I).

The Casa del Medico sex scene is also more obviously "grotesque" in its depiction of both protagonists (figure 2.5, B and C) as somatically non-ideal. At the Casa dell'Efebo, the female sexual partner (figure 1.8, A) noticeably lacks achondroplastic features. Her body instead conforms to Roman conventional ideals, potentially capable of generating desire not only in painted pygmies but also

63. Gesztelyi 1989: Abb. 3.

64. Van Aerde 2015: 156–157.

in real-life viewers.[65] At the Casa del Medico, however, the woman's body type resembles the other pygmies around her, with a large head and somewhat stocky trunk. Like her partner, she is coded less as "desirable" than "comical."

Although scholarly discussions of the Casa del Medico paintings often treat their violence and grotesquerie as typical for Nilotic scenes,[66] the reality is more nuanced. Conflicts between pygmies and river animals are extremely common, appearing in nineteen (40%) of Pompeian Nilotic frescoes or mosaics.[67] However, these representations usually lack graphic violence, with neither pygmies nor animals displaying visible injuries. I here use "graphic violence" to mean any representation of either the act of wounding a living being or visible blood or injury to a living being.[68] Most often, Nilotic frescoes depict pygmies either fleeing an approaching crocodile or hippopotamus, or arming themselves against such an approach.[69] On other occasions, the pygmies ride on top of subdued hippopotami or crocodiles (e.g., figure 2.4, H).[70] Images of actual battle or its aftermath, however, are rare. Outside the Casa del Medico, graphic pygmy–animal violence appears in only three other frescoes at Pompeii: the Temple of Apollo, the Casa di Maius Castricius, and the Casa dello Scultore. Nilotic frescoes at the Temple

65. See chapter 2, this volume, "Parallel Landscapes in the Garden."

66. E.g., Walker 2003: 192–193.

67. See appendix B, this volume. The figure of 40% derives from all 47 Nilotic scenes from known contexts at Pompeii, including those from non-household contexts. I here define "conflicts between pygmies and river animals" as depictions of pygmies and dangerous macrofauna taking defensive or aggressive action toward each other. Examples include fleeing, taking up arms, holding rocks over one's head, or riding a captured animal. Not included are scenes that simply depict pygmies and macrofauna in the same general vicinity, fishing scenes, or scenes in which two animals approach each other with seemingly hostile intent. Two scenes depicting pygmy–animal fights also include depictions of armed pygmies approaching each other with apparently antagonistic purposes (Versluys 2002: nos. 33, 63).

68. Omitted from this category are depictions of pygmies and animals approaching or fleeing each other, or pygmies brandishing weapons; such scenes might foreshadow violence, but they do not actually depict the violence or its consequences.

69. *Fleeing:* see, e.g., the Casa dei Pigmei, where one pygmy climbs a tree to get away from an approaching crocodile, while another pygmy walks away from the crocodile while looking back over his shoulder (Versluys 2002: no. 62, confirmed through personal observation). *Arming themselves:* see, e.g., the pygmy with spear and shield awaiting an approaching crocodile at the Casa dell'Efebo (figure 1.6), the pygmy holding up a shield in front of a crane in II 9, 2 (Versluys 2002: no. 39, confirmed through personal observation), or the pygmies raising stones above their heads to menace approaching river animals at the Casa del Granduca (figure 2.46) and the Casa di Maius Castricius (Versluys 2002: no. 54, confirmed through personal observation).

70. E.g, at the Casa della Caccia Antica tablinum (published as Versluys 2002: no. 51; today badly faded and no longer discernible in situ).

of Apollo and the Casa di Maius Castricius both depict a crocodile devouring a pygmy.[71] At the Casa dello Scultore, one pygmy spears a panther, while another pygmy spears the backside of a crane, who simultaneously pecks his buttocks.[72]

Out of forty-seven Nilotic scenes from known contexts at Pompeii, then, only four (9%) depict graphic pygmy–animal violence, and only three (6%) show pygmies receiving fatal injuries.[73] Within domestic contexts, graphic violence appears in only three (8%) of thirty-nine houses containing Nilotic scenes, and only two (5%)—the Casa del Medico and the Casa di Maius Castricius—depict pygmies dying violently. The vast majority (92%) of Nilotic scenes from known domestic contexts at Pompeii omit any direct depictions of pygmy–animal violence. While scenes of pygmy–animal *encounters* are extremely common, these usually consist simply of images of pygmies reacting fearfully or defensively to the animals.

The sexual imagery in Nilotic scenes also spans a broad tonal spectrum, with some frescoes depicting relatively idealized participants (as at the Casa dell'Efebo and Casa delle Quadrighe: figures 1.8, 2.49) and others emphasizing somatic "Otherness" (as at the Casa del Medico: figure 2.5). Since some erotic Nilescapes are now lost and known only from decorous published references to, for example, "scenette erotiche," it is impossible to provide precise percentages for these types.[74] However, of the ten preserved or well-recorded Nilescapes from Pompeii that depict explicit sexual encounters, five portray either the female or both partners with idealized bodies, while five portray both partners with achondroplastic features (appendix B).[75] Both options thus appear to have been equally available

71. Versluys 2002: no. 53, now lost (Temple of Apollo), and no. 54, confirmed through personal observation (Casa di Maius Castricius). Given Apollo's connections with Augustus and Actium, viewers at the Temple of Apollo might have viewed Nilotic scenes through a more "Actian" lens than elsewhere at Pompeii (cf. chapter 1, this volume, "Flavian Pompeii as Case Study"). A painting at the Terme del Sarno may also have once depicted a pygmy in the mouth of a crocodile or hippopotamus, but the painting is poorly preserved and the animal (if ever present) is no longer visible (Versluys 2002: no. 55, with discussion on p. 135). Written accounts of the now-lost pygmy scenes from IX 5.14-16 also leave unclear whether or not those pygmy-animal conflicts portrayed graphic violence (e.g., Schefold 1957: 261; for further references, see Versluys 2002: 150).

72. Sampaolo 1998b: figs. 15, 16; Versluys 2002: no. 60. The pygmy–crane battle draws on earlier Greco-Roman geranomachy traditions; see chapter 2, this volume, "Antecedents and Affordances in Greek and Roman Visual Culture."

73. Here omitting the dubious case from the Terme del Sarno (note 71, this chapter), as it is not certain whether the pygmy is being attacked by an animal, or—if so—how severe the pygmy's injuries might be.

74. On "scenette erotiche" in II 9, 2, see Sodo 1988: 200; see also Versluys 2002: no. 39. On "pigmei in atteggiamenti osceni" in the Casa con Ninfeo, see Sogliano 1888: 512; see also Versluys 2002: no. 56.

75. *Either female or both partners physically idealized:* Casa del Efebo, Casa di Sallustio, Casa dei Fiori, Casa delle Quadrighe, Terme Stabiane (see appendix B). *Both partners shown as*

to artists and patrons. Other types of sexual imagery that Roman audiences probably understood as comical rather than arousing (for example, depictions of grotesquely, disproportionally macrophallic males) appear in five additional scenes (appendix B).[76] In total, images of sex, ithyphallism, and/or macrophallism appear in thirteen (27.8%) of forty-seven Nilotic scenes from known contexts at Pompeii. Many of these images depict their protagonists as somatically atypical, with either achrondroplastic features, exaggerated genitals, or both. However, Nilotic sex scenes appear approximately equally likely to portray somatically "normal" or idealized figures. Furthermore, the majority (72.3%) of Nilotic scenes from known contexts at Pompeii omit any sexual imagery whatsoever.

The prominent violence and "grotesque" sexuality at the Casa del Medico is thus not simply the "default option" for Nilotic scenes; rather, these features demonstrate the purposeful selection of some available Nilotic motifs rather than others. Many Nilotic pygmy scenes are devoid of either sex or violence, depicting innocuous scenes of rural religious practice or daily life. Indeed, a substantial number of Nilescapes omit pygmies entirely. Sixteen (34%) of the Nilotic scenes from known contexts at Pompeii depict only the Nile itself with its characteristic flora and fauna (appendix B). The Casa del Medico frescoes also differ from those of the Casa dell'Efebo in what they omit. Scenes of rural cult occupy the majority of the Casa dell'Efebo frescoes. Besides sanctuaries and aediculae, the Casa dell'Efebo scenes portray ongoing ritual activity; pygmies approach shrines (figures 1.4, L; 1.7, P), make offerings at altars (figures 1.3, C; 1.7, B), and feed sacred animals (figure 1.7, C). The Casa del Medico paintings include one rural

---

achondroplastic: Casa del Medico, Casa delle Quadrighe, Casa con Ninfeo, IX 5.14–16, and a now-lost fresco from an unknown house (see appendix B). The Casa delle Quadrighe peristyle paintings fall into both categories, containing scenes of intercourse between both normally proportioned and dwarflike couples (see appendix B; see further chapter 2, this volume, "Parallel Landscapes in the Garden," with figure 2.49 and note 254). For the Casa con Ninfeo, I take Sogliano's reference to "pigmei [plural] in attegiamenti osceni" to suggest that all participants were pygmies (1888: 512), but since the paintings are lost, this supposition can no longer be verified. The Terme Stabiane scene is now lost, but the excavator's description does not portray the participants as pygmies (Minervini 1856: 34). "Explicit sexual encounters" are here defined either as images of sexual penetration (of any orifice), or images in which one partner bends over and another prepares to penetrate.

76. Additionally, the Nilotic scene from the Casa dei Ceii shares a room with a "sacral-idyllic" scene that includes an ithyphallic Priapus (Versluys 2002: no. 33). On the humorous and apotropaic potential of macrophallic figures, see Clarke 1998: 119–142; 2007b: 91–96; 2007c. *Pace* Versluys (2002: 159), the pygmies from the tomb of Vestorius Priscus are not ithyphallic or macrophallic; in-person examination indicates that the limbs in question are their legs. Four other Pompeian scenes depict another form of "grotesque" bodily display: defecation or urination (appendix B, this volume, and see further note 156, chapter 6, this volume).

sanctuary (figure 2.4, A), but no ritual activity is currently taking place there.[77] The architecture of cult is visible, but not its practice.

Furthermore, if Richardson is correct to attribute the Casa del Medico pygmy frescoes to the same painter who created the Nilotic pygmy scenes in the Casa dei Ceii, Praedia di Julia Felix, and Casa dei Pigmei[78], then the exaggerated violence and sexuality of the Casa del Medico pygmy scenes may not even be typical of this artist's work. At the other three houses, no sex scenes and no visible blood or injury appear in any of the surviving paintings[79]. While partial preservation conditions make it impossible to be certain that *no* such elements were ever present in those houses, the substantial surviving portions of the paintings suggest that at the very least, they lacked the Casa del Medico's overwhelming emphasis on violent and sexual motifs. This apparent thematic divergence across houses may represent meaningful selections from the available repertoire of Nilotic motifs in order to create unified compositions with adjacent associated imagery; flexibility in response to patron preferences; or random variation between commissions.

From the wide array of available Nilotic tropes (chapter 2), the Casa del Medico frescoes emphasize some themes while minimizing others. While rural cults are downplayed, pygmy-animal combats and outdoor sex are prominent and comically exaggerated. Rather than setting the standard for all Nilotica, the Casa del Medico scenes demonstrate just one of many available contemporary options for Nilotic iconography. Roman humor frequently incorporated themes widely considered unacceptable for comedy today: for example, extreme violence or mockery of physical or cultural "Others."[80] Some Nilotic pygmy scenes, including the Casa del Medico paintings, demonstrate the ability of Nilescapes to serve as vehicles for such themes. However, some of the most prominent motifs in the Casa del Medico paintings are actually relatively rare in Pompeian Nilotica. These frescoes thus appear *unusual*, not typical, in placing such emphasis on violence and degradation. Their association with the so-called "Judgment of Solomon" scene is even more extraordinary—unparalleled in other Nilescapes at Pompeii or elsewhere.

77. Though note that in some parts of Egypt, crocodile hunting was itself ritualized (Wilson 1997).

78. Richardson 2000: 147–153, based on similarity in "the proportions of figures and the treatment of landscapes" (2000: 150) and in the type and proportions of animals depicted. For a more skeptical take on the possibility of identifying distinct "hands" in Pompeian frescoes, see Beard (2008: 125–126).

79. One vignette from the Casa dei Ceii depicts a pygmy confronting a hippopotamus, but no injuries are apparent (Versluys 2002: no. 33, confirmed through personal examination, though the painting is today much faded). At the Casa dei Pigmei, two pygmies flee from a crocodile, but again neither has been harmed (Versluys 2002: no. 62, confirmed through personal examination).

80. Clarke 2007b, 2014; Beard 2014.

The Casa del Medico Frescoes as Ensemble

Although the Casa del Medico frescoes are often taken to exemplify "standard" Nilotic imagery, they are actually relatively atypical in both their contents and their association with the pygmy judgment scene. To understand how ancient viewers might have reacted to these images, we need to consider them in context. We will first examine the courtyard paintings as an ensemble and then relate that ensemble to its architectural and artifactual setting. As an ensemble, the frescoes from this garden present a range of perspectives on outdoor landscapes, human society, and the relationship between civilization and the (potentially dangerous) natural world.

Similar pairings of pygmy scenes with paintings of animal-filled landscapes are well-paralleled at Pompeii and elsewhere.[81] Here, the north wall and the north side of the east wall depicted natural landscapes devoid of humans and dominated by predator–prey conflict. On the west wall, humans do appear (in pygmy form), but their desperate, violent existence resembles that of the animals in the other frescoes. Often naked or semi-naked, the Nilotic pygmies fight wild animals or comport themselves in hedonistic fashion.

Juxtaposed with images of animal predator–prey interactions, the Nilotic pygmies might strike ancient viewers as only a few steps removed from the dangerous and primitive "state of nature" imagined by philosophers such as Lucretius.[82] The pygmies resemble Lucretius's characterization of primitive humans as unclothed, indiscriminately engaging in outdoor sex, and constantly struggling for survival with wild beasts. Indeed, Lucretius's description of early humans' bloody deaths might just as easily describe the Casa del Medico Nilescapes: "Each one was then more like to be caught and devoured alive by wild beasts, torn by their teeth . . . as he saw his own living flesh buried in a living tomb."[83]

Certainly, the paintings of the west wall provide ample evidence of what modern anthropologists call "complex society." The landscape is filled with monumental architecture, much of it resembling contemporary Greco-Roman buildings (figure 2.4, A and D; figure 2.5, P and Q). However, the Nilotic pygmies themselves appear to occupy a marginal role within this landscape. In contrast, the Casa dell'Efebo frescoes depict pygmies not only revelling and fighting,

81. At Pompeii, e.g., the Casa dei Ceii (Versluys 2002: no. 33). Outside Pompeii, see, e.g., Clarke (2003: pl. 16) on tomb 22 at Ostia.

82. Lucr. 5.925–1010. For other "progressivist" Roman narratives of human cultural evolution, cf. Varro *Rust.* 16, and see Moatti 2015: 69; Campbell 2016: 504–508.

83. *unus enim tum quisque magis deprensus eorum / pabula viva feris praebebat, dentibus haustus, / . . . viva videns vivo sepeliri viscera busto* (Lucr. 5.990–993, trans. Rouse 1924: 411).

but also—in even greater numbers—pursuing activities central to Roman conceptions of ordered, settled society: irrigating fields (figure 1.8, F), spinning thread for cloth (figure 1.4, A), acting as soldiers or guards (figure 1.7, R and S), or performing religious rituals (figures 1.3, 1.4, 1.7). No such activities appear in the Casa del Medico Nilotica. Although these pygmies clearly live within (or at least adjacent to) a society similar to that depicted at the Casa dell'Efebo, their own activities consist almost exclusively of carousing and battling wild animals.[84] Their frequent complete or partial nakedness (especially in figure 2.4, but see also figure 2.5, A, B, C, K, M, and N) further separates them from both the Casa dell'Efebo pygmies and their Pompeian viewers.

In contrast, the south side of the east wall also presents a scene of conflict and potential violence, but the figures presented here are full participants in "civilized" urban society. The judgment fresco (figure 4.1) includes almost no "natural" landscape features, aside from one tree to the left of structure (K). Instead, we see the workings of a centralized government, with judges (B) handing out decisions and soldiers (C, D, F) enforcing state power against a backdrop of monumental architecture (L, M, N). The imagery does not impose a geographic location on the scene, which might take place in Judea, Egypt, or elsewhere, depending on viewers' interpretation of the story. However, several features of the painting evoke specifically Roman institutions. The soldiers' armor appears Roman, and the supplicating figure has close parallels in representations of individuals supplicating Roman state representatives (e.g., the emperor, generals, or magistrates).[85] Furthermore, rather than going naked like many of their counterparts on the west wall, the pygmies in this scene all wear either Greek clothing (chiton and himation) or Roman military costumes. The himation and bare shoulder of one bearded onlooker (I) even resemble Roman depictions of Greek orators.[86]

While both the west and east walls depict pygmy scenes, these pygmies appear to occupy very different social roles. The framing of these scenes reinforces

84. Possible exceptions are the figures manning the ship (figure 2.4, N); Lucretius (5.1006) characterizes navigation as an art unknown to early humans. The boat in the other Nilotic scene (figure 2.5, L) is a smaller, short-distance craft. Of course, it is unlikely that the painter was literally trying to illustrate Lucretius's account; he more likely drew on more widely shared tropes about "primitive" existence.

85. *On the armor:* see note 46, this chapter. *On supplication images*: Brilliant 1963; Naiden 2003. *Generals and magistrates receiving supplication:* Brilliant 1963: 158; Naiden 2003, esp. 50. Robinson actually interprets this fresco as a "caricature of a Roman magistrate's court" (1975: 16).

86. Although Feder (2008) considers this figure and his companion (J) to be portraits of Socrates and Aristotle, I consider the figures insufficiently distinctive for such specific identification.

the sense of intentional juxtaposition. Most of the frescoes on the pluteus were painted onto a red background. The only two exceptions are the judgment scene (figure 4.1) and the scene directly across from it on the south side of the west wall, depicting pygmies fighting river beasts (figure 2.4). Both these scenes are painted onto light backgrounds, with black frames distinguishing them from the surrounding red ground. The special framing singles out these two scenes as particular visual foci within the courtyard. This parallel treatment, as well as the scenes' placement opposite each other, directs viewers to consider the relationship between these two images.

As in the pygmy–animal battles, the action in the judgment scene revolves around the threat of violence—but that violence is now checked by the institutions of civil society. The soldier's upraised knife threatens the baby with the same gory death that hippos inflict on pygmies in the opposing painting: cleaving in two (figures 2.4, K; 2.5, I). This time, however, we see no landed blow. Rather than striking indiscriminately like the wild beasts, the soldier, a representative of the state, waits for the decision of judicial authority. In the biblical story of Solomon, and probably also in other variations of the story,[87] the result is a happy ending: the child lives on unharmed, in the custody of the most appropriate parent, and the judge's wisdom wins widespread admiration. The spectators' reactions may in fact show their amazement at this wise pronouncement; two soldiers (figure 4.1, D) gaze open-mouthed at the judges, while other onlookers gesture with surprise (figure 4.1, H, I, and J).

The Nilotic scenes depict a dangerous landscape in which the threat of violence and dismemberment is vividly realized. In the ordered, civilized world of the judgment scene, however, state institutions successfully protect against such threats. Danger comes no longer from the unpredictable aggression of wild animals, whose own violent "state of nature" occupies the north wall and the north half of the east wall. Instead, danger comes primarily from conflicts between people; civic institutions, including the judicial system and the military, ensure that such conflicts find satisfactory resolution.

All the humans in the courtyard frescoes, including the judgment scene, appear as pygmies. This pygmy population may cast the paintings as humorous or playful, prompting ancient viewers to respond with amused appreciation rather than solemn reverence. However, humorous images are no less freighted than others with cultural assumptions, ideologies, and values. Even as the Casa del Medico frescoes provoked amusement, they also prompted their viewers to think

87. Certainly in the *Satyricon* episode (Petron. *Sat.* 79–80), Giton survives unscathed.

about the relationships between human societies and the various environments they sought to inhabit, create, or control.

Reinterpreting the Courtyard of the Casa del Medico

All these paintings occupied the interior of a low wall (pluteus) spanning the intercolumniations of the partial peristyle[88] (figure 4.3). Most commentators initially assumed that the space within these walls was a planted garden, as is usual for Roman peristyles.[89] More recently, Clarke has reinterpreted this space as a paved basin for collecting water. However, nineteenth-century documentation suggests that the paving did not actually cover the entire courtyard but, rather, was limited to a water channel at the base of the pluteus. I thus interpret the rest of the courtyard as probable garden space.

Clarke takes Mau's 1883 report as evidence that "the open area was not a planted garden ('viridario') but a 'doccia' or basin, paved in *opus signinum*, and that it had two drainpipes, one leading to a cistern and the other out to the street."[90] Clarke notes that a shallow layer of water would have enabled the Nilotic frescoes to "seem to float on the Nile," and he proposes that guests might sit on the pluteus to splash their feet in the water.[91]

However, Mau's original text actually supports a different conclusion. In full, Mau's reference to paving reads as follows: "Il portico del peristilio è sorretto da quattro colonne ed un pilastro, congiunti da un podio; gira intorno una larga doccia rivestita di *opus Signinum*."[92] The word *doccia* does not typically mean "basin" but, instead, "shower" or—more relevant here—"gutter." Furthermore, Mau describes the "larga doccia" not as filling the entire courtyard but as "gira intorno"—"going around" it. Instead of characterizing the entire courtyard as paved, Mau is describing a *paved gutter* that surrounds the courtyard. Furthermore, far from demonstrating that the courtyard was not a "viridario," Mau himself uses that term for the open space within the columns. When describing the finds from

88. Sogliano 1882: 322; Clarke 2007a; 2007b: 100–101; 2007c: 162–169; cf. also Mau 1883b: 229–230.

89. E.g., Eschebach 1984: 47; Jashemski 1979–93: 2:217; Walker 2003: 192.

90. Clarke 2007a: 219; cf. Clarke 2007b: 100–101; 2007c: 164.

91. Quoting Clarke 2007a: 219; see also Clarke 2007b: 102; 2007c: 164.

92. Mau 1883b: 228. For a parallel case in which Mau clearly uses the term *doccia* to refer to an elongated water channel running along the side of a garden, compare his description of the Casa di Acceptus e Euhodia (1884: 128).

the courtyard, he describes this space as a "viridario": "Furono trovati nel portico 5 caldai, una patera e una lagena di bronzo, nel viridario un focone e due grandi anfore prive del collo, ripiene di calce e lava frantumata per fare mortaio"[93] (see further discussion of these finds below).

The interpretation of the courtyard as a planted space surrounded by a water channel or gutter has much stronger archaeological parallels. While a fully paved courtyard would be highly unusual for a Pompeian house, gutters are ubiquitous features of garden perimeters.[94] This interpretation also matches several twentieth-century scholars' observations of the Casa del Medico courtyard. Today, as Clarke notes, modern plantings inside the courtyard have destroyed any ancient surfaces originally present[95] (figure 4.4). However, in the later twentieth century, Wilhelmina Jashemski was still able to observe a water channel "on the east, north, and west" of the basin interior.[96] Eschebach also seems to have observed water channels around the basin, reporting in 1984 that "Das Wasser vom Dach der Porticus wurde in Rinnen aufgefangen und in die Zisterne geleitet."[97] Both accounts appear consistent with Mau's observations.

Close observation of the courtyard further argues against a flooded basin. First, any standing water within the frescoed walls would have risked destroying the paintings. While paintings were sometimes located in close proximity to water,[98] habitual direct contact with water appears unlikely. As at the Casa dell'Efebo, however, a sufficiently deep gutter beneath the pluteus would have kept water from direct contact with the frescoes above.

Second, a break in the western pluteus wall—visible on old plans and still in situ today—indicates a doorway into the courtyard[99] (figure 4.4). The dimensions of the paintings confirm the antiquity of this feature; the fresco south of the doorway (figure 2.4) is significantly narrower than that to the north, in order to

93. Mau 1883b: 228.

94. Von Stackelberg 2009: 39; Dessales 2013: 188–189; and see also note 218, chapter 2, this volume.

95. Clarke 2007a: 219; 2007b: 101.

96. Jashemski 1979–93: 2:217.

97. Eschebach 1984: 47. Although neither Eschebach nor Jashemski refers to *opus signinum* in the water channel, it is perhaps relevant that the floor surfaces in this house appear to have suffered substantial loss after excavation; according to Eschebach (1984: 47), the mosaic floors originally reported by Mau were "nicht mehr sichtbar" by 1984.

98. See note 28, chapter 3.

99. As also observed by Clarke 2007b: 101.

fit between the doorway and the exterior house wall.[100] This break would have prevented the pluteus walls from holding water. Clarke's digital reconstructions of the site insert an elevated threshold at this doorway.[101] If such a threshold were present, the basin could indeed have retained a shallow layer of water. However, no such feature is visible in situ (figure 4.4) or referenced in earlier reports. By contrast, a channel dug below the floor level would eliminate the problem of water leakage, collecting small quantities of water without spilling out onto the portico.

This gutter would also have served a practical purpose, collecting rainwater from the portico roof and steering it toward one of two pipes. Mau describes two drainpipes, one leading to a cistern and one leading to the street outside.[102] Similar arrangements are well-paralleled at Pompeii.[103] Mau also records a lead plate, pierced with holes, which would originally have accompanied one of the pipes. Such plates were commonly used for water filtration.[104] The inhabitants could thus open or close the pipes to the cistern or street, depending on their needs.[105] They could send clean rainwater to the cistern, but if the water in the gutter were soiled, they could divert it to the street.

Finally, artifacts inside the courtyard further confirm that this space was a site of human activity rather than an ornamental pool or basin. According to Mau, the "viridario" contained a brazier (*focone*) and two amphorae.[106] This space was thus not a flooded basin to be admired from a distance but, rather, an activity area, accessible through the doorway in the west wall.

Parallels suggest this space would have been planted. The nineteenth-century excavators did not record, and would not have been looking for, botanical remains or root cavities. However, colonnaded courtyards in Roman homes were

100. The fresco to the south of the doorway (MANN 113195) is 1.23 m wide, while the fresco to the north of the doorway (MANN 113196) is almost twice that size at 2.17 m wide. For measurements, see Versluys 2002: 138 (though note that Versluys's attribution of MANN 113195 to the north wall—followed recently by Mol 2015: 270—contradicts the nineteenth-century excavation reports; for the location of both paintings, see Sogliano 1882: 322–323).

101. Clarke 2007a: fig. 4; 2007b: pl. 5; 2007c: fig. 6.

102. Mau 1883b: 228; cf. Eschebach 1984: 47; Clarke 2007a: 219; 2007b: 100–101.

103. Jansen 2007: 259; 2018: 407; Dessales 2013: 77–79, 188.

104. *On the Casa del Medico lead plate:* Mau 1883b: 228 (asserting uncertainty about which pipe the plate closed). *On the use of such plates for water filtration:* Dessales 2013: 77–78, 188–189.

105. On the use of stoppers to close drainpipes, see Dessales 2013: 79–80.

106. Mau 1883: 228.

typically planted with vegetation.[107] The frescoes of the Casa del Medico thus almost certainly faced not a flooded basin but, instead, a small planted garden surrounded by a water channel.

That said, the channel around this garden may indeed have been unusually broad. Although none of the scholars who observed this feature in situ provides dimensions, Mau describes it as "larga."[108] His further characterization of the channel as "rivestita di *opus Signinum*" may also suggest unusual size. Small gutters around Pompeian peristyles are usually stone, masonry, lead, or terracotta.[109] In contrast, and as Clarke's interpretation acknowledges, *opus signinum* is more common in water basins.[110] The channel around this garden may thus have boasted relatively large dimensions. Accordingly, even if the entire courtyard could not be flooded, this water feature probably still served display purposes, as well as practical functions. If the house owners closed both drains and filled the channel with water, this feature could even have served as a less expensive version of the *euripi* that ornamented many larger gardens.

The Artifactual Assemblage from the Casa del Medico

Despite substantial art-historical interest in the peristyle paintings, discussions of the accompanying artifactual assemblage have been largely confined to studies of ancient medicine and have dealt only with a small subset of the total finds from the house. The find of a vaginal speculum, as well as several other possible medical items, has inspired several scholars to identify the Casa del Medico as a "doctor's house."[111] However, the evidence for this identification is slim, and many of the supposed medical items may not actually have come from the Casa del Medico. We may thus make more progress by (1) limiting analysis to those objects that have a secure provenience within the house, and (2) considering the domestic assemblage as a whole, rather than focusing solely on the possible medical finds.

107. See note 49, chapter 1, this volume.

108. Mau 1883b: 228. On Roman "gutter basins" that "circulated water around the perimeter of the peristyle," see more generally von Stackelberg 2009: 39.

109. On lead channels, and for some representative dimensions, see Dessales 2013: 188–189. On terracotta and masonry channels, see, e.g., Jashemski 1979–93: 2:222, 2:229. On stone channels, see Jansen 2007: 259.

110. Dessales 2013: 64–75.

111. Breton 1870: 456; Eschebach 1984: 45–47; Bliquez 1994: 81–82.

The first of these criteria limits us to finds from 1882, as the artifacts recorded in 1841 could potentially come from many different houses in Insula 5. The excavation reports from 1882 likely do not provide us with a complete list of finds from the house. However, they do provide a record of major finds from the courtyard and surrounding rooms (albeit with the caveat that we only know about those finds that the excavators considered worth recording[112]).

Was the "Casa del Medico" Really the House of a Doctor?

Eschebach and Bliquez identify the house as a doctor's residence or clinic on the basis of several finds: a bronze speculum, two strigils, three probes, two tweezers, two cases for tools, part of a scale, two mortars, a pestle, a round box that supposedly contained pills, and two iron hooks that Bliquez compares to birthing hooks.[113] However, Geissler has recently shown this evidence to be problematic. First, many of these finds derive from the 1841 excavations and cannot actually be tied to the Casa del Medico.[114] Second, many of the finds may have served nonmedical purposes as readily as medical ones.[115]

The only unambiguously medical artifact reliably provenienced to the house is the speculum, which came from triclinium (o).[116] While such a tool might well

---

112. On the nineteenth-century excavators' lack of interest in objects without perceived artistic value or metal content, such as "fragmentary glass and pottery," see Allison 2004a: 31–32.

113. Eschebach 1984: 47 (though Eschebach thinks the speculum may belong instead to the Casa di Acceptus e Euhodia; see further note 116, this chapter); Bliquez 1994: 81–82. The finds are also summarized by Geissler (2011: 19n83). Although Eschebach mentions only one strigil, Fiorelli records two (1860–64: 3:170, entries for April 26 and 29).

114. For the original report, see Fiorelli 1860–64: 2:395–398, 3:169–171. Writing some years later, Breton claims that the Casa del Medico contained "quelques drogues et un assez grand nombre d'instruments de chirurgie renfermés en partie dans le petit nécessaire de bronze que nous avons vu au musée de la rue du Forum," but he provides no further details on objects or precise provenience (1870: 456, with a description of the bronze box on p. 361). It is unclear whether Breton understood the "instruments de chirurgie" to have all been *found* in the bronze box in which they were displayed at the museum, or whether this display brought together objects from originally disparate contexts.

115. Geissler 2011: 18–20.

116. Mau 1883b: 229. Eschebach questions the provenience of the speculum on the basis of the excavation journals, which attribute this artifact to "casa n. 6 dell' isola 5, reg. VIII" (Fiorelli 1882: 420; Eschebach 1984: 50). However, in the same issue of *Notizie degli Scavi*, Sogliano refers to the speculum and the three amphorae as coming from "una spaziosa stanza all'angolo sud-ovest del viridario, in cui tornò a luce il *giudizio di Salomone*" (1882: 439). As Bliquez points out (1994: 81–82), this reference to the "giudizio di Salomone" can only mean the scene from the Casa del Medico, and the reference to a large room at the southwest corner of a "viridario" corresponds to the position of triclinium (o) in that house. While the house number cited

appear in a doctor's house, a single piece of medical equipment might alternatively signify the residence of a patient who was receiving medical treatment.[117] Accordingly, there is no proof that the Casa del Medico was a doctor's house or clinic, and the occupations of the house's inhabitants remain unknown.

The Artifactual Assemblage from the Courtyard

More productive information comes from the finds within the courtyard itself. In the portico, excavators found five cauldrons (*caldai*), a shallow bowl (*patera*), and a jug or flask (*lagena*), all of bronze.[118] Within the courtyard itself, Mau records a brazier and two large amphorae whose necks had been removed. The amphorae contained a mixture that he describes as "calce e lava frantumata per fare mortaio." These finds provide evidence for both commensality-related activities and utilitarian tasks.

Mau's report does not provide inventory numbers for any of the finds, and their whereabouts and identification in the Naples Museum are uncertain. However, Allison's studies of Pompeian vessels provide a starting point for interpreting Mau's descriptions. Mau's references to *caldai*, a *patera*, and a *lagena* become clearer in the light of Allison's analysis of the usage of these terms in other nineteenth- and twentieth-century publications on Pompeii. In such publications, the term *caldai* (sing. *caldaio*) typically describes either "wide, rounded bronze vessels with constricted necks and handles for suspension" or "round-based buckets."[119] The other two vessel terms derive from ancient sources, but their usage in nineteenth- and twentieth-century Pompeianist literature differs somewhat from ancient Roman usage. While the Latin *lagena* appears to describe function rather than form (referring to a wine-pouring vessel), archaeologists at Pompeii have typically applied this term to a specific form: a handled, spouted jug.[120] People may sometimes have used such jugs to pour wine, but they need not always have done so. Finally, ancient texts often

---

in the excavation journals remains problematic, a similar inconsistency of numbering appears elsewhere in the journals from this period and probably testifies to changes in the numbering system over the course of excavation.

117. Geissler 2011: 20. Geissler also points out that the house lacks the *taberna* that literary sources and archaeological parallels associate with "Arzthäuser."

118. For all finds described in this paragraph, see Mau 1883b: 228.

119. Allison 2006: 25.

120. On ancient uses of the term *lagena*, see White 1975: 161–164. On the modern use of the term for spouted, handled jugs at Pompeii, see Allison 2006: 20.

associate *paterae* with religious libations, but archaeologists have applied the term to a variety of shallow bowls, including some that likely served utilitarian purposes. Nuber argues that some of the vessels labeled today as *paterae* were actually used for hand washing.[121]

Commensality and Courtyard Space

The cauldrons, jug, bowl, and brazier may have served several possible purposes. However, most possibilities connect them in some way to dining, probably in the adjacent triclinia (m) and (o). These objects thus link the courtyard to commensal practices, asserting a functional as well as visual connection between the courtyard and adjacent dining rooms.

The brazier and metal vessels may have been used for food preparation within the courtyard. Cooking and serving equipment is well attested in Pompeian courtyards. After indoor kitchens, gardens are the most frequent findspots for food-preparation equipment.[122] Portable braziers, like the one found here, are common in gardens.[123] The Casa del Medico portico also opened onto the possible kitchen, (l), providing ready interaccessibility. Since Pompeii was destroyed in late summer or fall, the warm weather would have made outdoor food preparation appealing. The jug (*lagena*) and bowl (*patera*) could have been used for consumption.

Alternatively, the vessels in the portico might have been used for water ablutions before, during, or after meals. Roman dining etiquette required hand and foot washing.[124] This explanation is particularly plausible for the *lagena* and *patera*. Handled jugs and *paterae* are often found together at Pompeii, leading Nuber and Allison to suggest that they formed "sets . . . used for washing hands" at mealtimes.[125] Numerous paintings, mosaics, and reliefs depict such sets in use.[126] This jug-*patera* pair may represent such a set.

---

121. Nuber 1972: 117–121; followed by Tassinari 1993: I.232; Allison 2004a: 58; 2006: 21; Wallace-Hadril 208: 404–405.

122. Allison 2004a: 126.

123. Allison 2004a: 103, 126; Nevett 2010: 106–107. Some evidence suggests Pompeii was destroyed in October rather than August (Cooley and Cooley 2014: 43), but October temperatures in the Bay of Naples are frequently still quite warm, facilitating outdoor cooking and dining.

124. *Hand washing:* Ath. 9.408–411; Nuber 1972: 117–121; Dunbabin 2003: 86, 121, 150, 156, 165, 169, 178, 192, 197, 239n45. *Foot washing:* Petron. *Sat.* 31; *CIL* IV 7698a.

125. Quoting Allison 2006: 21; see also Nuber 1972.

126. Dunbabin 2003: 86, 121, 150, 156, 165, 169, 178, 197, 239n45.

Additionally, excavation records at Pompeii sometimes use *caldaio* to describe a round-bottomed vessel type that may have been used for carrying water[127]. Since this courtyard lacks running water, mealtime ablutions would have required bringing water in vessels. The cauldrons, jug, and shallow bowl might thus all represent ablution equipment. While five cauldrons of water might sound excessive, Mau does not provide vessel dimensions, and some vessels labeled in Pompeian records as *caldai* are fairly small.[128] Since Roman diners might wash their hands during, as well as before and after, meals (e.g. Sid. Apoll. *Epist.* 1.11.14), a medium-sized dinner party could require a substantial amount of water.

The vessels' location in the portico affords proximity not only to the dining rooms but also to the paved water channel and drain inside the courtyard. These drainage facilities make the courtyard the obvious location for performing ablutions. With the cistern opening closed and the drain open, the channel would expel waste water onto the street.

Building Equipment, or Garden Fertilizer?

The truncated amphorae filled with lime and crushed lava testify to other activities, either construction-related or horticultural. Lime and powdered lava or volcanic ash could be used to make waterproof mortar,[129] so the location of these amphorae inside the courtyard might suggest, for example, repairs to the water channel. However, Mau's account provides no other indications of ongoing renovation in the garden or elsewhere. The continued presence of dining-related equipment in the portico further suggests that any repairs must have been only minimally disruptive.

Alternatively, the amphorae may have served horticultural purposes. Lime is an effective fertilizer, and lime-filled amphorae are attested in other Pompeian gardens, where Allison proposes that they might represent garden fertilizer.[130] Volcanic ash and rock dust are also rich in nutrients for plants, and gardening

127. Allison 2006: 25. On bronze cauldrons and buckets at Pompeii, see also Tassinari 1993: 1.102–105, 1.231, 2.274–294 ("caldaie") and I.108–113, I.231, 2.300–343 ("secchie e vasi globulari e ovoidali").

128. E.g., one *caldaio* from the Casa del Menandro is 230 mm rim diameter, 324 mm height to rim (Allison 2006: 67).

129. Vitr. 2.6, 5.12.2–3. Truncated amphorae could be used for a variety of purposes, including food preparation, carrying water, and containing building material (Allison 2004a: 101, 114, 125; 2006: 17, 45, 86).

130. Allison 2004b, s.v. "Casa del Sacerdos Amandus" and "House I 7, 19"). On lime-filled amphorae in gardens, see also Allison 2004a: 86.

stores today still sell volcanic rock dust as a soil additive. If the amphorae in the courtyard did contain fertilizer, their presence would further support interpretations of the courtyard as a planted garden.

Whether used for home repairs or garden maintenance, the lime- and lava-filled amphorae testify to utilitarian, not just display-related, activities in the courtyard. The brazier in the garden and bronze vessels in the portico suggest close connections between the courtyard and dining areas: these interaccessible, intervisible rooms all functioned as part of a broader entertainment space for guests. However, the amphorae of mortar or fertilizer demonstrate that the courtyard also saw more practical activities. Either repair or garden work would have served to maintain the display value of the space when guests were present. However, the performance of utilitarian tasks in the courtyard would also have constituted another reception context: different viewing circumstances, and potentially also different viewers, for the frescoes.

Representation, Anticipation, and Reality in the Garden

The context of the Casa del Medico frescoes includes many of the same features commonly associated with other Pompeian Nilotica: garden space, proximity to dining, and proximity to water. Although the total assemblage is much simpler and less luxurious than at the Casa dell'Efebo, it produces similar effects on a smaller scale.

Most immediately apparent at the Casa del Medico is the connection between garden and dining space. The dining rooms were interaccessible with the courtyard, and viewers of the Nilotic imagery could also see into the triclinia from some angles. For example, a visitor standing at the northeast entrance to the courtyard could view a Nilotic pygmy scene depicting an outdoor banquet and *symplegma*; approximately 30 percent of the space within triclinium (o); and the majority of the space within triclinium (m).[131]

Further suggestive of a functional connection between garden and dining space is the location of the only direct entrance into the garden. The doorway in the pluteus is directly opposite the entrance to triclinium (o), the largest dining room (figure 4.2). Visitors leaving the triclinium thus confronted the entrance to the garden, and visitors leaving the garden confronted the threshold of the

131. Cf. Clarke's digital reconstruction of the space: Clarke 2007a: figs. 3, 4; 2007b: pls. 4–5; 2007c: figs. 3, 6.

triclinium. This linkage would have encouraged people to experience the garden and triclinium as integrated components of a larger entertainment space.

Unlike the Casa dell'Efebo, the Casa del Medico garden lacks facilities for running water. However, water still likely played significant roles in both utilitarian and display-oriented activities in this space. The water channel inside the courtyard performed the important economic function of funneling rainwater into the cistern. However, this channel—which, as discussed earlier, may have been unusually large—could also serve display purposes when its drains were closed. Flooded with rainwater or water carried from public fountains, the channel could become a temporary *euripus* or *nilus*: a cheaper, ready-to-hand version of the elaborate water installations in elite gardens. The flooded channel would form a real-life complement to the aquatic imagery (Nilescapes and a marsh scene) on the adjacent pluteus walls. The artifacts from the garden and portico may also suggest a role for water in preparations for meals. If the jug, *patera*, and round vessels are equipment for dinnertime ablutions, then guests probably began and ended their banqueting by cleaning their hands in the garden.

Wealth, Resources, and Goals at the Casa del Medico and Casa dell'Efebo

The Casa del Medico and Casa dell'Efebo clearly occupied different points on the socioeconomic spectrum of first-century Pompeii. The owners of the Casa dell'Efebo invested extensive resources into an elaborate outdoor space with a water-triclinium, *euripus*, two fountains, and a private water tower; a sizable statuary collection, including one spectacular bronze showpiece; a separate planted area; and two aedicular structures. The garden in the Casa del Medico, by contrast, consists solely of a small, simple courtyard with a water channel. The Casa del Medico house plan is small and irregular, and outside the "entertainment complex" of courtyard and triclinia, many walls had minimal or no adornment beyond simple plaster. Mau characterized the now-lost triclinium paintings as poor quality, although his criteria for this assessment are unclear.[132] The excavators reported no statuary, and the bronze vessels and speculum are the highest-value objects recorded.

However, within their seemingly more limited resources, the owners of the Casa del Medico created a domestic entertainment area structurally similar in many ways to the Casa dell'Efebo garden. Both installations forge strong functional and visual links between garden space, dining, and water use, and both

132. Mau 1883b: 229 ("non è molto bella").

facilitate similar forms of social performance. These parallels suggest that householders of different economic means might nonetheless share similar concepts of the "desirable" or "ideal" house, and, to the best of their ability, might strive for similar goals.

Both garden complexes served as display sites for art, plantings, and (to the degree possible) water resources. The visual culture of both gardens also stresses themes of wilderness and danger: chaotic forces, whether animal, human, or divine. As we have seen, Roman gardens could be constructed as domesticated, miniaturized model "wildernesses," and their visual culture often includes images of gods, animals, or activities associated with the untamed spaces beyond cities and fields.[133] Nilotic scenes are appropriate to such spaces not only because they often feature dangerous animals and outré behaviors but also because they depict one of the frontiers of empire and borderlands of the known world.

Order and Chaos in the Garden

The appearance of such themes in domestic space would not just have offered guests a *frisson* of excitement but also made a statement about the house owner's own ability to impose order onto chaos. As at the Casa dell'Efebo, the Casa del Medico garden depicts *images* of potential danger, but they appear as harmless miniatures within a domestic space that is itself to be understood as well ordered and securely governed by the paterfamilias. The inclusion of potentially unsettling imagery within domestic space thus asserts an ability to control chaotic and mysterious forces: to transform wilderness into garden and foreign into familiar.

At the Casa del Medico, the peristyle frescoes perform a series of variations on the theme of tension between (dangerous) wilderness and (ordered) human civilization. Three of the six original frescoes depicted landscapes devoid of humans, with wild animals fighting each other and struggling for survival. In the two Nilotic scenes, humans enter the fray as pygmies who battle hippopotami and crocodiles in a rural riverine landscape on the edge of *oikoumene*. However, the Nilotic pygmies' unusual behaviors place them outside Roman norms, and their bloody battles with river beasts demonstrate that they are still very much prey to environmental dangers. By contrast, the pygmies in the judgment fresco inhabit a more urbanized and socially regulated space, overseen by civic magistrates and soldiers in Roman military costume. While violence again threatens, that threat now appears to be checked by human law. No wild

133. See chapter 3, this volume, "Safety and Danger in the Domestic *Locus Amoenus*."

animals are present, and the institutions of civilized society visibly constrain and mediate conflicts between citizens.

One of the "messages" potentially available to Pompeian viewers, then, would have been reassuring: as Roman citizens and inhabitants of a well-governed city, they need not fear dangers like those afflicting the unfortunate Nilotic pygmies. This illustration of the benefits of order and authority might implicitly reflect positively on the paterfamilias, the primary representative of hierarchical authority within the home. The house owners might also hope for guests' lived experience to itself demonstrate the difference between the painted scenes and reality. While the painted landscapes brim with danger and violence, the real domestic garden was manifestly a place of safety and relaxation. Additionally, while the Nilotic pygmies' banquet is a wild cacophony of public sex, carousing, and rampaging beasts, the actual banquets in the Casa del Medico triclinia were (presumably) rather more sedate, socially acceptable affairs. From the house owners' perspective, these discrepancies between image and reality could have offered visible proof of the orderly, well-run nature of their own household.

Interacting with Images, Interacting with People

However, guests and hosts need not necessarily have perceived domestic visual culture in the same ways. For visitors to the Casa del Medico, images of socially transgressive activities might raise complex questions about social performance. When guests arrived to dine in either triclinium (o) or (m), they passed first through the courtyard (figure 4.2). Before they entered the dining room, their gazes would first fall on the no-holds-barred, uninhibited banquet of the pygmies on the pluteus wall. This association would be further reinforced if guests stopped to wash their hands inside the courtyard before dinner, as proposed earlier. Looking from this vantage point toward either triclinium (o) or (m), they would have also seen the Nilotic frescoes, which present banqueting as inextricable from sex, violence, and excess.

What relationship between "real life" and art do these scenes construct? As at the Casa dell'Efebo, contextual considerations blur the boundaries between reality and representation, enticing the viewer to imagine that the painted scenes are "real." If the water channel in the courtyard were filled, it would have enlivened the painted riverbanks with actual water. Yet at the same time, the images' miniature scale demonstrates their own artificiality, reminding viewers that they cannot fully participate in the wild festivities on display.

Issues of visibility toy further with viewers' perceptions of image and reality. Within the courtyard (figure 4.3), the Nilotic banquet scenes are visible only from viewing positions that also permit partial, not full, views into triclinia

(o) and (m). The paintings thus tease guests approaching the triclinia: Can they be sure that when they enter, they will *not* see a real-life equivalent to the pygmies' carousing? If they did encounter such a scene, how would they react? While most Pompeian dinner guests would (presumably) be fairly confident that they were not, in fact, about to walk in on some Nilotic free-for-all, the framing of the painting nonetheless prompts viewers to consider the line between permissible and impermissible behavior. The threat (or temptation) of excessive, undecorous, and unacceptable conduct rises before their eyes as they approach the *convivium*, even as images of riverine violence and danger simultaneously remind them of the risks associated with such lifestyles.

Accordingly, while visitors to the Casa del Medico probably understood the Nilotic frescoes as humorous, that humor targeted not only an outgroup of imagined foreigners but also the ingroup of invited guests. The real-life behavior of host and guests would have been more immediately socially relevant than the putative behavior of faraway and possibly fictional pygmies. These images thus provide opportunities not only for expressing views about distant lands but also for reflecting playfully and humorously on one's own social circle.

The frescoes likely interacted with real-life social performance not only before dinner, but also afterwards. As Clarke has observed, the judgment scene would be most visible to diners exiting triclinium (o) after dinner (figure 4.3). He interprets this image as parodying the "well-established high-art theme" of judgment scenes like that in the Villa della Farnesina.[134] Expanding on Clarke's suggestion, I propose that this scene—and indeed the entire fresco sequence—represents a playful twist on the high-minded visual culture often associated with the convivial practice of ekphrastic rhetorical performance. The placement of the scenes in a partially colonnaded space further supports this interpretation, as porticoes and peristyles were considered appropriate settings for "Hellenizing" philosophical and literary conversation.[135]

Clarke sees the judgment fresco as parodying the "elite culture" of more serious judgment scenes.[136] However, I am not convinced that the house owners sought to mock social, economic, or cultural *elites* as such. As discussed here, the architecture and decoration of the courtyard and triclinia suggest a sustained attempt, carried out as best as possible with limited resources, to aspire toward such elite status. Additionally, the employment of a humorous rather than solemn tone

134. Clarke 2007b: 104–106, quoting 104; cf. Clarke 2007a: 221.

135. O'Sullivan 2007, and see also chapter 3, this volume, "Interactive Statuary: Social Performance and Service" and "Performing Cosmopolitanism."

136. Quoting Clarke 2007a: 221.

need not necessarily mock the *act* of speech in response to visual culture. Even humorous frescoes might inspire guests to propose interpretations of the imagery, much as modern scholars have been doing since 1882. The use of humor could, however, help relieve social anxieties and put guests at ease. If the paintings provide a comical twist on accepted subjects for convivial conversation, the ultimate target of their humor is the social milieu of the dinner-party guests themselves. Entering and exiting the triclinium, guests encountered images that teased their own pretensions and defused tensions about living up to social expectations.

As at the Casa dell'Efebo, the evocation of distant lands and peoples invites viewers to compare themselves to the figures portrayed. However, the Casa dell'Efebo leaves those comparisons more open-ended; different viewers might come to different conclusions about their relationship to the images. The Casa dell'Efebo frescoes place great emphasis on rural piety, omit any depictions of pain or suffering, and showcase a *symplegma* scene whose female participant and aulos-player seem eminently capable of attracting viewers' own erotic interests. The Casa del Medico installation appears less ambiguous in presenting the Nilotic pygmies as negative examples. Their battles with river animals frequently end in gory defeat, and their sexual encounters appear more comical than erotic. However, the judgment fresco presents its pygmy protagonists in a more positive light; they live in a well-ordered society, engage in careful judicial deliberations, and employ a military that looks very much like the Roman army. As an ensemble, the frescoes present civilization and civic institutions as welcome protection against the savage state of nature.

Ultimately, both the Casa dell'Efebo and Casa del Medico installations employ images of distant or fantastic landscapes to prompt viewers to reflect on their own experiences. By framing and mediating social performances, domestic visual culture constructs its users' experience not only of faraway "Others" but also of the here and now.

5

Making Meaning on the Margins

NILOTIC MARGINALIA IN THE CASA DEL FAUNO

Introduction

In the houses considered so far, Nilotic imagery is notable partly for its *prominence* within garden ensembles. Nilotic frescoes in both gardens acted as framing devices for guests' own experiences—both as prompts for humorous reflection on real-life social interactions, and as literal frames for those interactions. Visitors standing within the small garden of the Casa del Medico were surrounded by the pluteus frieze, and if they gazed toward the triclinia (figure 4.3), the Nilotic pygmy scenes formed the lower "border" for their view of those rooms. At the Casa dell'Efebo, Nilotic landscapes formed not only a physical support for reclining diners' own bodies but also a visible pedestal literally underlining their views of their fellow diners.

At many other houses, however, Nilescapes occupy less visible positions. When Nilotic scenes appear outside gardens, they are often smaller and less prominent, occupying only a small portion of a wall otherwise devoted to other imagery.[1] An example appears in the Villa dei Misteri atrium, where one of the earliest Pompeian Nilescapes—a first century BCE Second Style frieze—occupies the upper zone of walls otherwise painted with imitation marble[2] (figure 5.1). Positioned far above eye level, the miniature Nilotic landscapes encircle the room but do not constitute its primary visual focus.

1. Tybout 2003: 511; cf. the typically subsidiary functions of pygmy scenes (Clarke 2003: 192–193; 2007b: 94–95), non-Nilotic Egyptian motifs (Wyler 2006: 227; Platt 2009: 68–69), and images of *adynata* and *monstra* (Platt 2009).

2. Versluys 2002: no. 67. On the date, see Esposito 2007, esp. 449–450. On the faux marble, see Eristov 1979: 739.

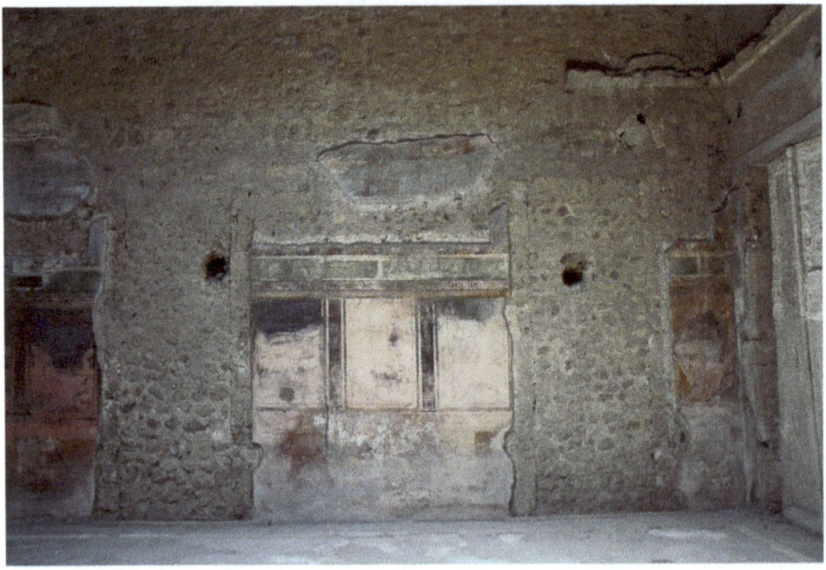

FIGURE 5.1 Atrium of the Villa dei Misteri, north wall. Remains of painting on the wall include Second Style painted imitations of marble, as well as a Nilotic frieze in the upper zone. Nilotic imagery is preserved only on the north and south walls but probably originally occupied the upper zone of all four walls (see Versluys 2002: no. 67).
Photo: C.E. Barrett.

Nilotic pygmy scenes from nondomestic contexts are often similarly marginal. For example, in the portico of the Temple of Isis, Nilotic landscapes (including one pygmy scene) appeared alongside other landscape scenes as small *pinakes* inserted into a framework of fantastic architecture[3] (figure 5.2). Together, the architecture and landscape panels framed a series of central vignettes depicting Isiac priests.[4]

Even in gardens, some Nilotica function not as centerpieces but as marginalia: smaller motifs that adorn, or perhaps trouble, the borders of larger scenes. For example, Nilotic frescoes appear in the Casa delle Quadrighe not only in the peristyle (where they occupied the pluteus wall, as at the Casa del Medico; figure 2.49) but also in an adjacent garden room, the so-called viridarium.[5] There,

---

3. On Nilotic scenes from the Isis temple, see further note 211, chapter 2, this volume.

4. On the portico paintings, see Sampaolo 1992; Moormann 2007: 143, 150–151.

5. Sampaolo 1996b: figs. 7, 8, 25; Versluys 2002: no. 50. The space in question adjoins the peristyle. On the term "viridarium" (or "viridiarium") as problematic, see note 47, chapter 1, and note 196, chapter 2, this volume.

(a)

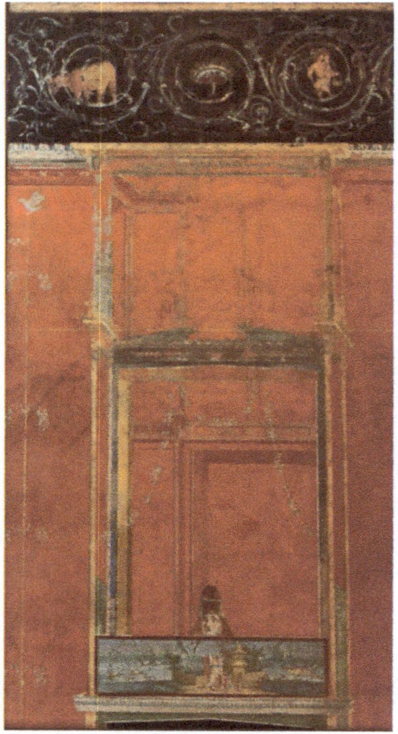

(b)

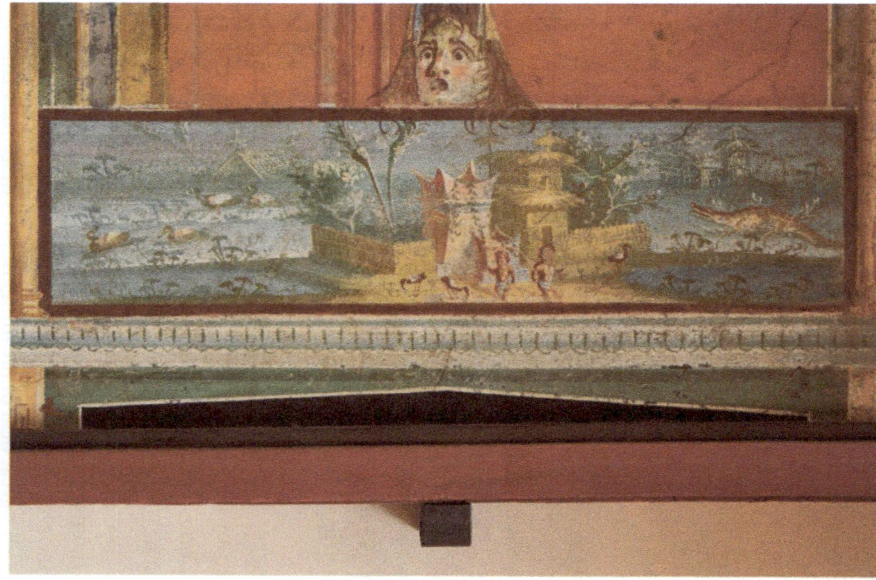

FIGURE 5.2 One of two Nilotic scenes from the portico of the Isis temple at Pompeii (MANN 8607). (a) In the central zone of the wall, a Nilotic landscape appears on a pinax set into an assemblage of fantastic architecture; the architectural forms are decorated with garlands and a theatrical mask. Vignettes in the upper zone include a hippopotamus and a pygmy, both surrounded by vegetal forms. (b) Detail of the previous image: closeup view of a Nilotic landscape whose features include two pygmies, a crocodile, reed and masonry structures, ibises, ducks, lotuses, and palm trees. A graffito, "VENI HOC" ("I came here"), is scratched above the head of the pygmy to the right.

Photos: C.E. Barrett. Reproduced by permission of the Ministero dei Beni e delle Attività Culturali e del Turismo—Museo Archeologico Nazionale di Napoli.

Nilotic landscapes occupied the socle of a wall whose central zone depicted a painted garden. A Nilotic mosaic in the Casa del Granduca (figure 2.46) similarly occupies a subsidiary role within a larger composition, decorating the bottom of an aedicular fountain.[6]

Another large subset of Nilotic "marginalia," attested both in gardens and elsewhere, focuses exclusively on the river and its flora and fauna (appendix A, appendix B). Sixteen (34%) of the previously identified Nilotic frescoes and mosaics from known contexts at Pompeii omit pygmies or any indications of human activity (appendix B). Indeed, some of these images are so simplified as to problematize the very concept of "Nilotica." For example, at the Casa delle Nozze d'Argento, miniature paintings of ducks, swans, and lotuses adorn the upper walls of the peristyle ambulatory[7] (figure 5.3). Such images of birds and lotuses are often interpreted as "Nilotic." However, lotuses were also cultivated in first-century Italy, and ducks and swans are common to both Italy and Egypt[8]. Evidence suggests that Roman audiences associated lotuses with Egypt,[9] so ancient viewers might conceivably interpret such vignettes as references to "the Nile." Associated statuary at the Casa delle Nozze d'Argento may have also primed viewers to interpret the river scenes as Egyptian; the garden contained four faience or imitation faience figurines of crocodiles and frogs, seemingly Egyptian imports.[10] On their own, however, the Casa delle Nozze d'Argento frescoes do not demand a Nilotic interpretation as forcefully as do images with more specifically Egyptian fauna.

Even when Nilotic scenes play a peripheral role to larger images, we should not take peripherality as synonymous with insignificance. Recent work on Roman art suggests that viewers' interactions with marginal, "ornamental" images may often have produced a great deal of meaning. Besides describing embellishment, equipment, or ornamental decoration, the Latin word *ornamentum*

---

6. See further chapter 2, this volume, "Parallel Landscapes in the Garden," and appendix A.

7. Versluys 2002: no. 41; Barrett 2017a: fig. 8.

8. *Lotuses in Italy:* Jashemski et al. 2002: 131–132. *Swans in Egypt:* Ikram 1995: 27. Mol (2015: 277–280) and Pearson (2015: 4n15) also doubt that ducks and lotuses alone suffice to mark a scene as "Nilotic." Versluys notes "a large number of doubtful cases" of Nilotica (2002: 41).

9. Jashemski et al. 2002: 131–132. Lotus crowns often characterize Egyptian deities in Roman art (see chapter 6, this volume, "Visual Culture, Cult, and Identity at the Casa di Acceptus e Euhodia").

10. See chapter 1, this volume, "Material Culture and the 'Material Turn,'" and chapter 6, "Model Landscapes: Water Features, Statuary, and 'Egyptian' Installations."

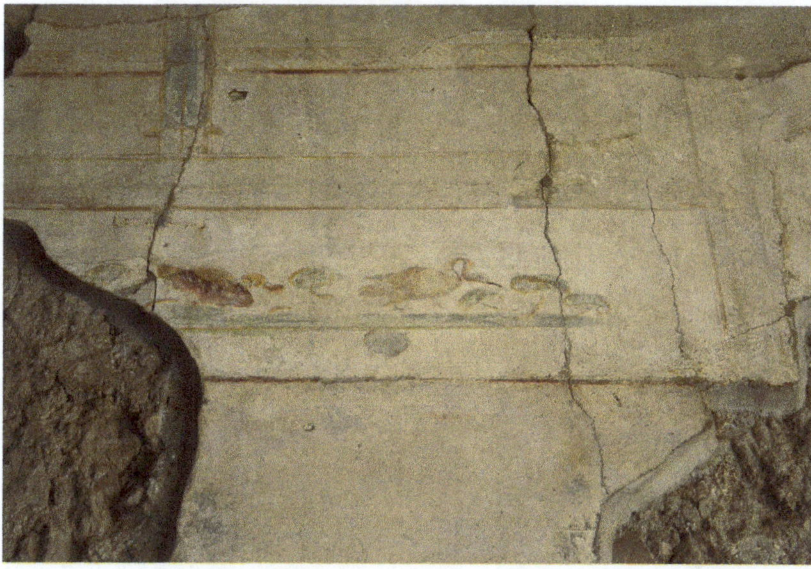

FIGURE 5.3 Fresco from the upper zone of the ambulatory of the peristyle, Casa delle Nozze d'Argento (V 2.1). Detail of east wall, depicting a duck, a swan, and lotuses. *Photo*: C.E. Barrett.

can also carry ideological weight, conveying both aesthetic appeal and what Hölscher calls "connotations of social and moral values."[11] From the fabulous *monstra* of Third Style wall painting, to vegetal motifs on monumental architecture, the iconography of domestic lamps, and framing devices of many types, "ornamental" imagery could communicate numerous ideologically loaded concepts concerning, for example, order and chaos, cultural identity, fertility, renewal, transgression, political ideals, fantasies, and the nature of representation and reality.[12]

Accordingly, we should treat the peripherality of much domestic Nilotic imagery not as a reason to dismiss such images but as a phenomenon worth investigating in its own right. This chapter examines a mosaic (figure 5.4) from the Casa del Fauno as a case study in the uses of Nilescapes as marginalia. This mosaic is famous as one of the earliest Nilotic scenes in the Mediterranean, and

11. Hölscher 2009: 62. On *ornamentum,* see also Swift 2009: 16; Platt and Squire 2017a; Dietrich and Squire 2018.

12. *On* monstra: Platt 2009. *Vegetal motifs:* e.g., on the Ara Pacis, Castriota 1995; cf. Sauron 2000 (with comments by Platt 2003). *Lamps:* Bielfeldt 2014. *Framing devices:* Platt and Squire 2017b.

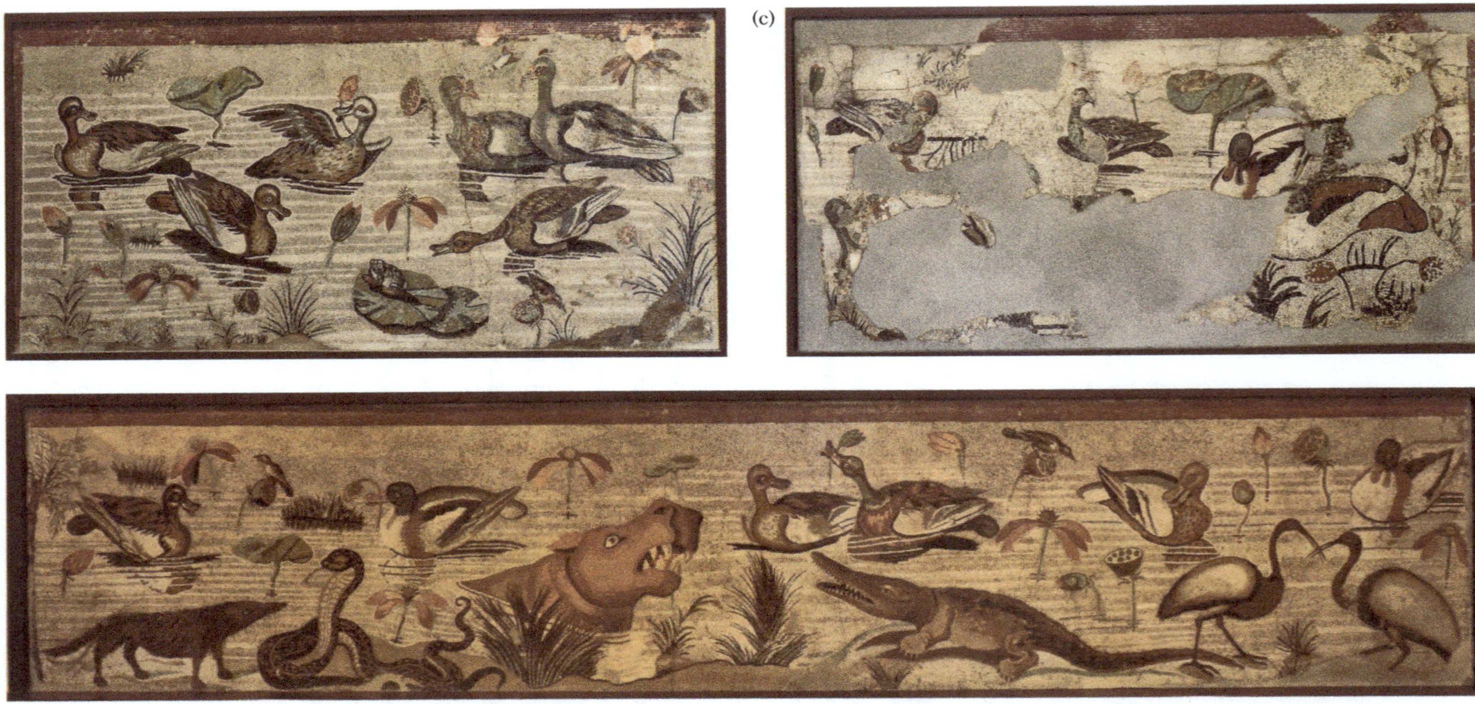

(a) (c) (b)

FIGURE 5.4 Three segments of the Nilotic mosaic from the threshold of Exedra 37 in the Casa del Fauno (VI 12) (MANN 9990; images not to same scale). (a) Leftmost segment (from the perspective of a viewer facing the exedra), 1.323 m long. (b) Central segment (from the perspective of a viewer facing the exedra), 3.33 m long. (c) Rightmost segment (from the perspective of a viewer facing the exedra), 1.31 m long. Apparent color differences between segments reflect variable lighting conditions in the museum.

Photos: C.E. Barrett. Reproduced by permission of the Ministero dei Beni e delle Attività Culturali e del Turismo—Museo Archeologico Nazionale di Napoli. Measurements: Pernice 1938: 167.

possibly the earliest at Pompeii.[13] As a border for the larger and more elaborate "Alexander mosaic" (figure 5.5), the Nile mosaic is a subsidiary element within a larger composition. However, this very marginality enables the Nilescape to shape visitors' experience of that composition, acting as both a literal and a conceptual frame for the central scene.

The Casa del Fauno

The Casa del Fauno (VI 12.2) is the largest dwelling at Pompeii: over 3000 sq.m in area, occupying an entire insula, and equipped with two atria and two peristyle courtyards (figure 5.6).[14] This was a truly palatial house, larger than the royal palace of Hellenistic Pergamon.[15] The unusual number of fine *opus vermiculatum* and *opus sectile* mosaics exceeds any other Pompeian dwelling. The initial excavations in 1831–1832 were only minimally published, but twentieth-century excavations by the Deutsches Archäologisches Institut have provided more information on the phasing, layout, and contents of the house.[16]

Constructed around 180 BCE, the house was renovated in the late second or early first century BCE and decorated with First Style paintings and mosaics that still remained largely in situ in 79 CE.[17] Some evidence, including an Oscan inscription on an honorific statue base within the house, may suggest that the inhabitants responsible for the First Style decoration belonged to the Satrii gens, an influential Samnite family.[18] Pesando interprets the famous statuette of

---

13. See, e.g., Versluys 2002: no. 47. On the date, see this chapter, "The Nile Mosaic," with note 24. If the Nile mosaic dates to the early first century BCE, it may be roughly contemporary with the Nilotic frescoes in the Villa dei Misteri atrium (on which see Versluys 2002: no. 67), which date to approximately 80 BCE.

14. *On the house:* Hoffmann and de Vos 1994; Faber and Hoffmann 2009. *Mosaics:* Zevi 1998a, 1998b; Christensen 2006. *Nile mosaic:* Versluys 2002: 121–123; Andreae 2003: 111–125. *Alexander mosaic:* Andreae 1977, 2003: 63–77; A. Stewart 1993: 130–150; 2003: 42–44, 56; Cohen 1997; 2010: 162–170; Pfrommer 1998; Moreno 2001; 2009: 422–425. *House area:* Clarke 1991: 81.

15. Castrén 1975: 40; Leach 2004: 34.

16. On DAI excavations, see Hoffmann and Faber 2009; on their proposed phasing, note the cautions of Flohr 2010. On finds from the early excavations, see Bechi 1832; Niccolini and Niccolini 1854, s.v. "Casa del Fauno"; Fiorelli 1860–64: 2:240–255; 1875: 154–159.

17. Following Faber and Hoffmann 2009; see notes 16 and 24, this chapter.

18. On this and two other Oscan inscriptions from the house, see abundant bibliography in Zevi 1998a: 40n53. The use of Oscan suggests a pre-Sullan date. The statue base names "V. Sadiriis" as an aedile (Pesando 1996: 218–219). On the activities of this gens outside Pompeii, see Pesando 1996: 221; Wallace-Hadrill 2008: 136–137. On their continued political prominence post-Sulla, see Mouritsen 1988: 151.

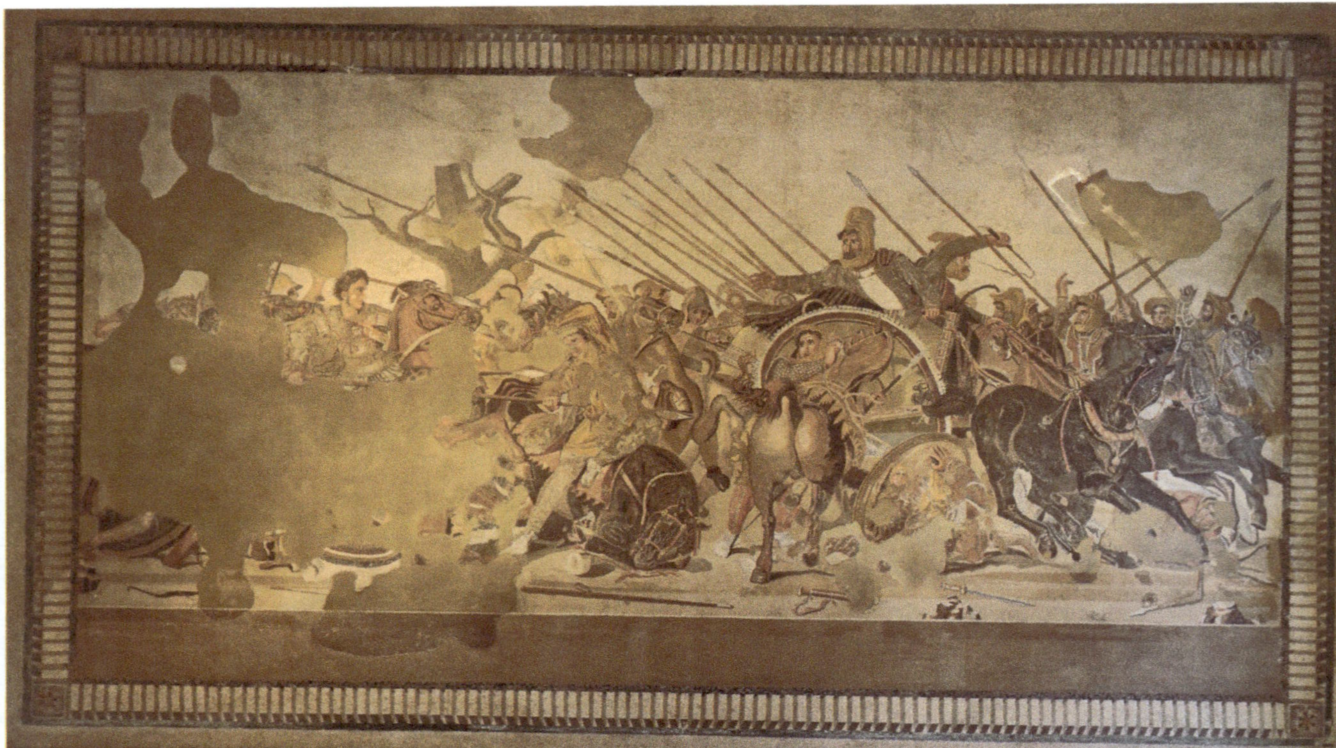

FIGURE 5.5 The "Alexander mosaic" from Exedra 37 in the Casa del Fauno (VI 12) (MANN 10020). The dimensions of the mosaic are almost 6 m × 3 m (Zevi 1998a: 41).

Photo: C.E. Barrett. Reproduced by permission of the Ministero dei Beni e delle Attività Culturali e del Turismo—Museo Archeologico Nazionale di Napoli.

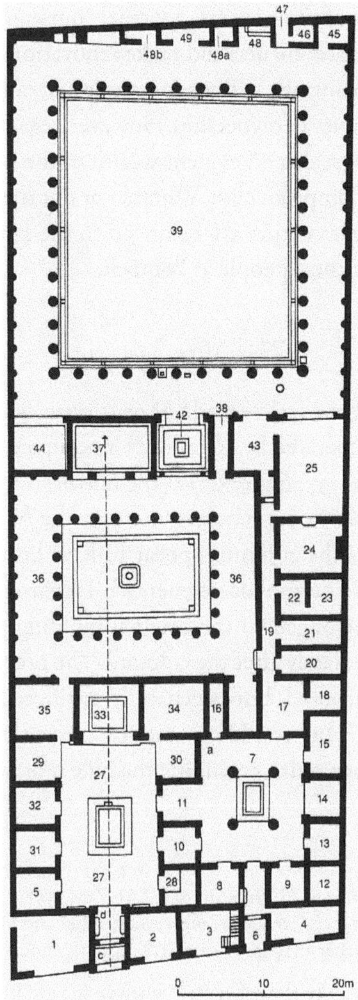

FIGURE 5.6 Plan of the Casa del Fauno (VI 12), with visual axis from the doorway to the Alexander exedra (room 37) marked.
Source: After Clarke 1991: fig. 22; reproduced by permission of John Clarke.

a bronze satyr, displayed in the Tuscan atrium in 79 CE, as a visual pun on the owners' name.[19] Scholars often assume that the same family retained the house until 79, since the later inhabitants conserved the earlier decoration, including the Oscan-inscribed statue.[20]

19. Pesando 1996: 217–223; see recently Wallace-Hadrill 2008: 136.

20. See, e.g., Zevi 1996: 132–134; 1998a: 63; Leach 2004: 68–69.

In fact, descriptions of the house as a virtual "museum" to the First Style are not quite accurate; the later owners did make renovations, as well as repairs and alterations to the wall paintings.[21] The conservation of older decoration may also not always imply continuity of ownership. However, regardless of the inhabitants' precise identity, the vast scale and evident wealth of the house suggest they were prominent among the Pompeian elite. Whether or not the inhabitants in the first century BCE and the first century CE belonged to the same family, they were at both times among the richest people at Pompeii.

The Nile Mosaic

The Nilotic mosaic (figure 5.4) is made of *opus vermiculatum* and divided into three parts, originally separated by columns.[22] It occupied the threshold to an exedra (figure 5.6, room no. 37; figure 5.7) at the northern end of a peristyle garden (figure 5.6, room no. 36). Some scholars interpret the Nile mosaic as a later addition to the exedra, as the columns appear to have been cut down slightly to enable the insertion of the mosaic segments.[23] However, it is equally possible that the Nile mosaic belonged to the original building program of the exedra and was installed immediately after the columns. The precise date of the peristyle renovations remains disputed, but is generally attributed to the late second or early first century BCE. Faber and Hoffmann have recently proposed an early first century BCE date for the exedra, including the Nile mosaic.[24]

21. *Comparisons to a museum:* e.g., Hoffmann 1986: 495; Zevi 1998a: 22, 62–63; Leach 2004: 68. *Architectural renovations after the early first century BCE*: Faber and Hoffmann 2009. *Alterations to the surface decoration:* Ehrhardt 2012: 115, 117–118, 144–147, 192–193, 206.

22. Length of segments is 1.323 m (left of center) when facing exedra, 3.33 m (center piece), 1.31 m (right of center when facing exedra) (Pernice 1938: 167).

23. Pernice 1938: 167–168; Meyboom 1995: 337n9; Zevi 1998a: 44. Pernice and Meyboom also base their dating on subjective assessments of the mosaic as less artistically impressive than the adjacent Alexander mosaic. Donderer (1990: 27–28, 31) further contends that neither the Nile nor the Alexander mosaic was made in situ.

24. Faber and Hoffmann 2009: 107–108; cf. Laidlaw 1973: 45. The extraction of the mosaics predated stratigraphic excavations at the site (Faber and Hoffmann 2009: 107–108; Hoffmann 2009: 39). The date of the Nile mosaic also has potential consequences for the dating of the Nile mosaic from Palestrina, given stylistic similarities between the rendering of the animals and plants on both mosaics (Meyboom 1995: 17–19, 81, 92; cf. Zevi 1998b: 37–40; Andreae 2003: 112). Daszewski further remarks on the similar materials used in the Palestrina mosaic and at the Casa del Fauno (1985: 20). These parallels are sometimes invoked to assign a Hellenistic date to the Palestrina mosaic. However, I am not convinced (*pace* Meyboom 1995: 17–18) that the similarities are sufficiently distinctive to warrant attribution to the same workshop. On the date of the Palestrina mosaic, see further note 23, chapter 2, this volume.

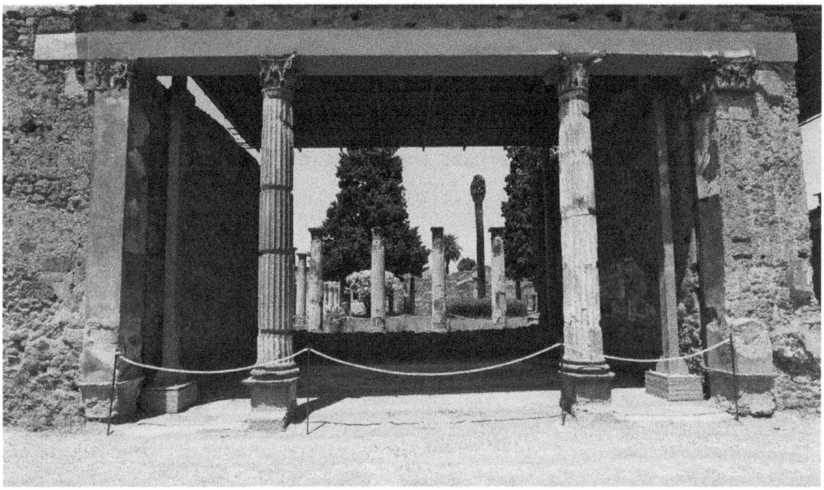

FIGURE 5.7 Exedra 37 in the Casa del Fauno (VI 12). A replica on the floor of the exedra illustrates the original location of the Alexander mosaic. The three segments of the Nile mosaic were positioned between and on the sides of the Corinthian columns.
Photo: C.E. Barrett.

The three segments of the mosaic depict a riverscape filled with animals and plants. Many are native to Egypt, marking the river as the Nile. The materials of some of the tesserae further enhance these Nilotic associations, as many of the blue and green tiles appear to be made of faience imported from Egypt.[25] Animals visible in the foreground of the central mosaic segment (figure 5.4b) include (from left to right) the ichneumon (Egyptian mongoose), cobra, hippopotamus, crocodile, and two ibises. The six animals constitute three pairs, each braced for combat: ichneumon versus cobra, hippopotamus versus crocodile, and ibis versus ibis. The cobra rears up at the advancing ichneumon, the hippopotamus and crocodile approach each other with gaping mouths and bared teeth, and the ibises lock beaks in battle. Behind these animals stretches the river, filled with lotuses (in multiple stages of flowering) and reeds. Six ducks, divided into three male–female pairs, swim in the shallows. The ducks probably represent mated couples, whose peaceful pairings contrast with the hostile encounters of the beasts at the riverbank. Two smaller birds perch on water plants.

25. Boschetti 2004: 9–10. Faience was a relatively common material in Campanian mosaics of this period, and craftsmen may have imported it in plaque form (Boschetti 2011: 66–68). Accordingly, its presence does not necessarily indicate that a mosaic itself was imported or that the mosaicists were Egyptian.

To the left and right of the columns that originally separated the three portions of the Nile mosaic, additional mosaic segments (figure 5.4a and c) depict more water birds, riverine plants, and a frog. The burgeoning abundance of vegetation and animal life evokes the fecundity of Egypt's riverbanks, inundated every year by the Nile. For Greeks and Romans, this fertilizing power was one of the Nile's most impressive features; Herodotus famously described Egypt as the "gift of the river" and claimed that the flood removed much of the labor of agriculture (Hdt. 2.5, 2.14).

Encountering Alexander: The Nile Mosaic as Threshold

The Nile mosaic occupied the threshold between the smaller peristyle (figure 5.6, room no. 36) and the exedra that held the famous "Alexander mosaic" (figure 5.6, room no. 37; figure 5.7). Often thought to emulate an earlier Hellenistic painting, this large floor mosaic depicts Alexander confronting the Persian king Darius[26] (figure 5.5). Interpretations of this exedra as a dining room have recently been confirmed, at least for the final use-phase, by the publication of faunal remains (chicken and fish bones) in the drains leading out of this room.[27]

The mosaic is thus located immediately between a planted garden and a dining space, and it is intervisible with several prominent water features in the adjacent peristyle: a fountain, pool, and basins.[28] Not all these associations were necessarily present when the mosaic was first installed: for example, the faunal remains testify to food consumption in the first century CE but cannot prove that the exedra was also used for dining one hundred years previously. However, by the time of the eruption if not before, the Casa del Fauno was typical of first century CE Pompeian houses in linking Nilotic imagery both visually and spatially to gardens, dining, and water.

Yet while guests might admire the Nile mosaic while walking across it, the larger decorative program of the exedra sought to draw their attention elsewhere. The main attraction of this space was the extraordinary Alexander mosaic. Multiple features of this mosaic, including the four-color scheme of Classical painting, indicate the emulation of an earlier painting. Similar images elsewhere, in other media, may suggest that the "original" was widely renowned.[29] One

26. On the Alexander mosaic, see bibliography in note 14, this chapter.

27. Faber 2009: 98; Faber and Hoffmann 2009: 107; Hoffmann 2009: 41.

28. On these water features, see Jashemski 1979–93: 2:145.

29. A. Stewart 2003: 42.

frequently cited candidate is a lost work by Philoxenos (Plin. *HN* 35.110). Other scholars have proposed works by Apelles, Helena, or an unknown painter associated with the Ptolemaic court.[30] Regardless of specific identification, however, viewers of the Alexander mosaic probably interpreted it as referring to a famous "Greek" artwork.

Positioned on the edge of this larger and more elaborate composition, the Nilotic mosaic functioned not as main attraction but as border, delineating the threshold of the exedra. Within this defined space, visitors were invited to participate in the cultured viewing of a Hellenistic "masterpiece." From patterns of damage to the tesserae and ancient repairs to the mosaic, Martin Beckmann concludes that people habitually walked on some parts of the Alexander mosaic more than others.[31] He argues for traffic patterns that centered around, but avoided stepping on, the figure of Alexander and several other figures. These patterns of use-wear suggest that visitors actively engaged with the mosaic, walking around the space purposefully and focusing their viewing on certain areas. Building on Beckmann's theory, I propose that such a use of space suggests self-conscious viewing of images as something very like "art."[32] As we have seen, Hellenistic and imperial Roman elites valued such viewing as an opportunity to project the socially desirable persona of an educated sophisticate, and conversation about nearby visual culture was considered particularly appropriate when dining.[33]

The visual framing of the Alexander mosaic further indicates a sustained effort to direct attention toward this showpiece. The positioning of both the Alexander and Nile mosaics implies an intended viewer standing in the peristyle just *outside* the exedra, from which point the images on the exedra floor would be wholly visible and "right side up." Within the peristyle, a series of nested frames steers visitors' gazes toward the mosaic. Indeed, since the house plan features a direct sightline from the entrance to the exedra, almost the entire Casa del Fauno might

30. *Apelles:* Moreno 2001. *Philoxenos and Helena:* Pfrommer 1998: 13–19. *Unknown Ptolemaic artist:* Pfrommer (1998) argues for a Ptolemaic painter because the mosaic depicts some artifacts datable to the Hellenistic period, and symbols of the Seleukids (frequent enemies of the Ptolemaic dynasty) appear on the saddlecloths of some Persian soldiers.

31. Beckmann 2010 (arguing against attributions of the damage to the 62 CE earthquake).

32. For debate over the existence of an ancient concept of "art," see Platt and Squire 2010. On the role of aesthetic considerations in the production and consumption of ancient visual culture, see also Tanner 2006; Porter 2012.

33. See chapter 3, this volume, "Performing Cosmopolitanism," with note 139.

be understood as a frame for Alexander.[34] The exedra occupies the center of a visual axis stretching from the fauces of the house, through the Tuscan atrium, through the large window at the back of the tablinum, and into the peristyle, at whose northern end the Alexander exedra looks out onto an even larger peristyle (figure 5.6). Guests at every stage of entry from the doorway to the first peristyle were visually directed toward the Alexander exedra.[35]

At the exedra, additional visual and spatial frames mediated guests' encounters with the Alexander mosaic. These included the tripartite entrance to the exedra, divided by red-painted Corinthian columns;[36] the Nile mosaic at the exedra threshold; the white mosaic pavement, which likely held dining couches, surrounding the Alexander mosaic; and the dentellated border outlining the battle scene. Even the wall paintings flagged the contents of the exedra as special. Underneath painted imitation-alabaster panels, a ring of illusionistic painted drapery—a slightly later Second Style addition—suggests a curtain surrounding the mosaic.[37] Within the mosaic itself, a complex play of gazes, bodies, and spears forms yet another set of frames for the central encounter between Alexander and Darius. Finally, when the exedra was used for dining, the guests' own reclining bodies would have formed yet another (living) frame around the mosaic.

Several of these framing devices surrounding the mosaic (stone walls, curtains) are themselves self-consciously *images* of framing devices. Since the palatial scale and luxurious decoration of this house demonstrate that the owners could well afford real stone and curtains, the choice of illusionistic painting (featuring even painted "fossils" embedded in the "stone"!) must evoke the high cultural value placed on trompe l'oeil art.[38] Furthermore, the faux alabaster panels included not only painted "fossils" but also hidden figural images emerging from the swirls of "stone" veining: in one panel, a bird, and in another, a depiction of centaurs at

34. Cf. Zevi's assertion of a decorative program centered around the Alexander mosaic (1998a: 45; 1998b: 99), although I do not follow Zevi in seeing a single "program" in the decoration of the house.

35. On this visual axis and the views afforded along its length, see Clarke 1991: 82–83; Zevi 1998a: 32–35; Dickmann 1999: 144–146; Christensen 2006: 31–33.

36. Zevi connects the color to the royal associations of porphyry (1998b: 98), while Andreae suggests it imitates red marble (2003: 111).

37. On the date of the painted drapery, see Ehrhardt 2012: 117, 192. On the design of the wall paintings to emphasize the floor mosaic, see Clarke 1991: 84–85; Leach 2004: 62.

38. Compare Pliny's story that Parrhasius won a painting competition by painting a curtain so real that his rival Zeuxis tried to look behind it (*HN* 35.65–66). On painted "fossils," see Barry 2011: 98.

a symposium.[39] Such images may recall ancient traditions of "spontaneous figuration," through which natural forces might generate objects that looked like artworks.[40] The wall paintings are thus art imitating nature imitating art.

As if this play of *ars* and *natura* were not complex enough, the exedra also presents different *artes* emulating each other. Pliny characterizes mosaics as "painting in stone" (*HN* 35.3), and this mosaic may replicate a specific famous painting. Yet even as the Alexander mosaic is a "painting in stone," the walls around it render stone in paint—and that "stone" itself produces further paintings.[41] These plays on the relationship between reality and representation invite visitors to engage with the mosaic as *image*. The exedra encourages interpretations of the Alexander mosaic not only as, for example, an interpretation of Hellenistic history,[42] a display of surplus wealth, or a private claim to princely glory, but also as a triumph of visual sophistication.

Despite the fineness of its craftsmanship, the Nile mosaic was not itself the focus of guests' attention but, rather, just another means of framing Alexander. Use-wear and ancient repairs to the Nile mosaic may further indicate that visitors did not linger at this threshold but instead passed through en route to the Alexander mosaic.[43] Yet this peripheral role itself invites further examination. If the Nile mosaic served primarily to mediate viewers' encounters with the Alexander mosaic, then how did it actually affect those encounters?

"Egypt" as Place vs. "Egypt" as Culture

In contrast to the Casa dell'Efebo and Casa del Medico frescoes, the Casa del Fauno Nile mosaic depicts no humans (pygmies or otherwise)[44] and no signs of human activity. We see only the river with its flora and fauna. Commentators sometimes explain the absence of pygmies in chronological terms, asserting that pygmy imagery did not become popular in Italy until after the conquest of

39. Laidlaw 1985: 33–34, 194; Barry 2011: 98–99, figs. 2.23–2.25.

40. Barry 2011: 94–96, 99; Platt 2018.

41. Platt 2018. On Plin. *HN* 35.3 and mosaics as "painting in" or "painting on" stone," see Barry 2011: 96–98, 129–130, 533–534.

42. The mosaic is unusual in depicting a relatively recent historical event (Dunbabin 1999: 299).

43. Both damage and ancient restorations are heaviest at the right side of the Nile mosaic, suggesting visitors usually entered the exedra from this direction (Beckmann 2010, and personal communication from Martin Beckmann).

44. As also noted by Zevi 1998a: 45–46; 1998b: 54–55, 58–59.

Egypt.[45] As we have seen, however, the connection between Actium and pygmy scenes is more tenuous than often assumed.[46] Nilotic scenes with human figures (both pygmies and non-pygmies) are already attested in the first century BCE, and paintings of achondroplastic figures, both Nilotic and otherwise, appear in other houses from Samnite Pompeii.[47]

Accordingly, the depiction of a human-free Nile was not a "default" option but, rather, a choice. The mosaic omits not only humans but also all signs of human presence, with none of the temples, huts, boats, fields, or stibadia that characterize so many other Egyptianizing landscapes. In place of these indications of human activity, the mosaic presents only geography and ecology: the river and the animals, birds, and plants that inhabit its ecosystem.

Furthermore, all these animals, birds, and plants are wild rather than domesticated. In reality, the banks of the Nile were perhaps the most famous farmlands in the world. The economic importance of Egyptian grain, and its trade at nearby Puteoli,[48] would arguably have made agriculture the most salient feature of the Nile for many Pompeians. No grain or other domesticates appear in the Nile mosaic, however. The emphasis on wild species underlines humans' absence from this vision of Egypt: not even human-grown plants or animals are present.

The Nile mosaic thus evokes not Egypt as *culture*, but Egypt as *place*. While visitors to the house might have espoused all sorts of opinions about Egyptian society, this mosaic does not primarily seek to engage with such opinions. Visitors beheld not the people of the Nile, but the river itself, whose unknown origins and mysterious flooding were perennial objects of wonder. The exotic appearance of the animals locates the river on the edge of the known world, while those animals' hostile interactions recall the risks associated with wild animals and landscapes.

For Italians in the Flavian period, the concept of "the Nile" might evoke various associations ranging from reassuring (Nile as breadbasket and source of empire-wide prosperity) to thrilling (Nile as scientifically unexplained

45. E.g., Versluys 2002: 288.

46. See Chapter 1, this volume, "Flavian Pompeii as Case Study."

47. On Nilotic scenes from first century BCE Pompeii and Delos, see note 62, chapter 1, this volume; notes 23 and 167, chapter 2, this volume; and note 13, this chapter. Pygmies appear in the Second Style paintings of the Casa dello Scultore and Casa di Maius Castricius, on whose dates see note 167, chapter 2, this volume. Non-achondroplastic figures and buildings appear in the Nilotic frieze from the Villa dei Misteri atrium, perhaps contemporary with the Casa del Fauno mosaic (see note 13, this chapter). On non-Nilotic pygmy imagery at Pompeii, see chapter 2, this volume, "Antecedents and Affordances in Greek and Roman Visual Culture," with note 183.

48. Not only in the Imperial period, but also in the late Republican period (Rossi 2014).

phenomenon, embodiment of the faraway and mysterious, and home of dangerous creatures). A similar range of associations is plausible for the first century BCE, when the mosaic was originally installed. At that point Egypt was subject to a Greco-Macedonian dynasty rather than Roman *imperium*, but the Nile was already a major grain source for Italy. However, the Casa del Fauno mosaic emphasizes some of the Nile's available cultural connotations while playing down others. The depiction of an exotic landscape filled with battling animals activates associations with distance, mystery, and danger, even as the very proliferation of animals and plants also suggests fertility and fecundity.

This emphasis on Egypt as *place,* and especially as a faraway and potentially unsettling place, assumes further significance through proximity to the Alexander mosaic. In context, the Nile mosaic functions partly as a geographic marker recalling Alexander's Eastern conquests.[49] The aggressive interactions of the Egyptian animals echo the antagonistic encounter of Alexander and Darius in the larger mosaic, enabling the two mosaics to explore violent conflict in both human and animal spheres. The juxtaposition of images of Alexander and Egypt also suggests further associations with Alexander's career. It was in Egypt that the oracle of Zeus-Ammon declared Alexander's divine parentage, and in Egypt that Alexander founded the city, Alexandria, that came to embody Roman notions of Hellenistic luxury and sophistication. The Nile mosaic thus evokes not Egypt as *contemporary reality* (whether Ptolemaic kingdom or Roman province, depending on the viewers' era), but Egypt as *attribute of Alexander*.

The embodied viewing of both mosaics may even construct a relationship between the viewer and Alexander himself. Since the mosaics were located on the floor, viewers had to walk over them in order to interact with them. To enter the room and examine the Alexander mosaic more closely, visitors first had to step across the Nile mosaic: in effect, to cross "the Nile." In miniature, they left the Mediterranean (manifest in the "real" Pompeian garden outside the exedra), traveled through Egypt, and arrived at a battleground deep in the Persian Empire.[50] Each viewer thus retraced the journey—and implicitly stepped into the role—of Alexander himself.

49. Cf. Zevi 1998b: 55, 59.

50. Cf. Kuttner: "Visitors stepped over this running Egyptian 'river' to see the Alexander Mosaic, which illustrates a battle in Asia to which Alexander had come from Egypt" (2003: 110). Much debate surrounds the identification of the battle as Issos or Gaugamela (see sources cited in note 14, this chapter, for various perspectives), but whether we are in Anatolia or Mesopotamia, stepping into the exedra implies a mental journey.

From Hellenistic Koine *to Roman* Imperium

Other aspects of the assemblage identify the house owner with Alexander. The sightline from the fauces to the Alexander exedra would also have crossed the paterfamilias seated in the tablinum,[51] creating a visual rhyme between Macedonian king and Pompeian patron. The "palatial" scale and luxurious appointment of the Casa del Fauno allude to kingly splendor, asserting the owner's claim to what Dillon characterizes as "private royalty."[52] The presentation of the Alexander mosaic may further identify the house with the royal palaces in which famous paintings were sometimes displayed.[53]

Accordingly, even though the Nile mosaic does not depict Egyptian culture per se, it still carries profound cultural implications. The exedra asserts its owner's Alexander-like claim to possession, at least in miniature, of the Hellenistic East. In the first century BCE, this possession would have suggested not only identification with Alexander himself but also embrace of the Hellenistic *koine* that his conquests initiated. The exedra mosaics thus proclaim the house owners' membership in an international category of cosmopolitan elites.

In the Hellenistic period, increased connectivity, mobility, and the presence of Greco-Macedonian dynasties throughout the eastern Mediterranean all encouraged what Versluys describes as "a common repertoire of types and styles of material culture available all around the Mediterranean and Near East."[54] This repertoire of internationalizing material and visual culture is sometimes characterized as a cultural *koine*.[55] As the term "*koine*" implies, this repertoire of shared forms and styles provided a common visual "language" for its users, albeit one with many local "dialects." Certainly, the internationalization of some forms of material culture by no means indicates universal "Hellenization" or cultural homogenization. The employment of certain material forms throughout the eastern Mediterranean coexisted with the continued use and development of traditional forms, as well as new "hybridized" forms. Nonetheless, a shared repertoire of styles, forms, and images enabled people across the Mediterranean

51. Zevi 1998a: 35.

52. Dillon 2006: 48. On Casa del Fauno as private "palace," see Hoffmann 1996; Zanker 1998: 33–41; Christensen 2006.

53. Pesando 1996: 224.

54. Versluys 2015b: 154.

55. Trimble 2011: 146, 236; Papantoniou 2012: 28–54, esp. 41–43, 52–54; Versluys 2013a: 429–431; 2015; 2017b; van Aerde 2015: 49, 52–53, 140, 162; van Eck et al. 2015: 12.

to participate in transregional communication networks and claim transregional cultural identities.

The material culture of late Republican Italy was profoundly entangled with this international *koine*.[56] Such entanglements are clearly visible in the Casa del Fauno, where "Hellenistic" allusions enabled the owners to display both their wealth and their appreciation of international "Greek" culture. In Samnite Pompeii, the architectural form of the peristyle would have alluded to multiple Greek precedents: palaestras, sanctuary stoai, and the parks and peristyles of Hellenistic palaces and some elite houses.[57] The exedra may also have struck contemporary audiences as a "Hellenizing" architectural feature.[58] The Alexander mosaic evokes the cultural prestige of Greek art, and the prominence of Dionysiac imagery throughout the house recalls the culture of *tryphe* promoted by Hellenistic rulers.[59]

Some scholars argue that the mosaic assemblage of the Casa del Fauno points not just to "Greek" culture generally but also to Ptolemaic Egypt specifically. The Nile mosaic obviously references Egypt, and the Alexander mosaic has been (controversially) argued to reproduce a Ptolemaic painting.[60] Egyptian parallels have also been proposed for several other mosaics from the house. A mosaic *emblema* from an *ala* of the Tuscan atrium depicts a cat stealing a chicken from a pantry shelf. A lower shelf holds additional equipment for a feast: ducks, songbirds, fish, shellfish, and lotuses. The combination of ducks and lotuses is reminiscent of Nilotic scenes (although the Nile itself is not depicted), and Egyptian parallels have been proposed for the cat as well.[61] In a cubiculum off the same atrium, an *emblema* depicting sex between a satyr and a nymph or maenad resembles a similar mosaic from Thmuis in Egypt.[62] Alexandrian influence has also been proposed for a fish-themed mosaic *emblema* in Triclinium 35 (figure 5.6, room

56. Wallace-Hadrill 2008, 2011; Versluys 2013a: 429–431; 2015.

57. Dickmann 1997; Zarmakoupi 2010; 2014: 103–140; Christensen 2006.

58. Dickmann 1999: 145–146.

59. Zanker 1998: 37.

60. See note 30, this chapter.

61. On the ducks and lotuses as potentially Nilotic, see Pesando 1996: 207; Zevi 1998a: 36; Meyboom and Brouwer 2005: 235–238. On the cat as Egyptian or an adaptation of an Egyptian subject, see Hoffmann and de Vos 1994: 104–105; Meyboom 1995: 92; Tammisto 1997: 78; Pfrommer 1998: 210–211; Zevi 1998a: 36; Meyboom and Brouwer 2005: 236.

62. Daszewski 1985: 160–163; Meyboom 1995: 92–93; Pesando 1996: 205–207; Pfrommer 1998: 210; Zevi 1998a: 36.

no. 35), as Meyboom argues that all such fish mosaics at Pompeii derive from a single workshop with Alexandrian connections.[63] In another dining room, Triclinium 34 (figure 5.6, room no. 34), scholars have suggested Ptolemaic precedents for a mosaic depicting an Eros-like figure riding a tiger and carrying Dionysiac attributes.[64] Finally, some authors argue that the Alexander mosaic was made by Alexandrian artisans. Donderer proposes that the mosaic was imported from somewhere in the Hellenistic East, while other scholars attribute it and other mosaics from the house to a workshop located in Italy but staffed by expatriate Alexandrian mosaicists.[65]

Although the exedra mosaics do recall Alexander's relationship with Egypt, I am more skeptical of some other claims for "Alexandrian" influence at the Casa del Fauno. Alexandria was one of the major artistic centers of the Hellenistic world, but recent scholarship has subjected older claims of a recognizable "Alexandrian style" to much critical scrutiny.[66] Small mosaic *emblemata* were portable, and good evidence demonstrates that *emblemata* in Pompeian houses were often transported intact from offsite workshops.[67] However, it remains unclear whether the Alexander mosaic could have been transported in this way. Many scholars have argued that the mosaic must have been made in situ, as it is too large to transport safely.[68] Others contend that the mosaic might have been divided into smaller pieces for transport.[69] The origins of the mosaic tesserae also remain controversial; one early study suggested that the tesserae came from the Vesuvian region, but later research has disputed this claim.[70] The present study will therefore remain agnostic about the status of the Alexander mosaic as import versus local product. However, even if the mosaic were originally constructed

63. Meyboom 1977; 1995: 17; Meyboom and Brouwer 2005: 235; cf. Pfrommer 1998: 211; Zevi 1998a: 36.

64. Meyboom 1995: 92; Pesando 1996: 213; Pfrommer 1998: 211–214; Zevi 1998a: 36–39.

65. On the Alexander mosaic as import, see Donderer 1990. On attributions to Alexandrian artists in Italy, see Meyboom 1995: 91–95; Pesando 1996: 206–207, 226–227; Tammisto 1997: 162–168; Zevi 1998a: 43.

66. See note 181, chapter 2, this volume.

67. Clarke 1991: 41; Dunbabin 1999: 39.

68. Clarke 1991: 84; Zevi 1998a: 42–44; 1998b: 86–87; Dunbabin 1999: 43.

69. Donderer 1990. For an earlier argument that the mosaic was assembled from separate pieces, see Ippel 1930: 80–98.

70. Fuhrmann 1931: 110–114, 277–278; challenged by Donderer 1990: 19–20. For a more recent analysis of the materials used in Pompeian mosaics (including some materials, including faience, that were likely Egyptian imports), see Boschetti 2003, 2011.

elsewhere, we lack good evidence for where exactly that original location might have been.

Beyond speculative assumptions that only Alexandrians would create images with Egyptian themes, we know very little about the backgrounds of the craftsmen who worked on the Casa del Fauno mosaics. Mosaics depicting a particular theme (e.g., seafood) need not always share a single prototype, and the diners in Triclinium 35 likely appreciated the fish mosaic less as a reference to Alexandria than as a foretaste of the lavish meal they expected to enjoy. Additionally, some Casa del Fauno mosaics have parallels not only in Ptolemaic Egypt but also at other Hellenistic sites, such as Delos.[71]

It is thus probably impossible to determine the mosaicists' origins—but I contend that this question about the *producers* is not terribly important for the *consumption* or *reception* of the mosaics. Whether viewers understood the mosaic assemblage as specifically "Alexandrian" or more generally "Greek"-looking, the images projected cosmopolitanism and international sophistication. The Nilotic mosaic participated in an assemblage that asserted not only its owners' material resources but also their participation in a broader, transregional elite *koine*.

Rather than displacing local identities, however, such participation coexisted with their continued assertion. Oscan inscriptions within the house proclaimed the owners' rootedness within their local setting.[72] Indeed, some traditions cast the Samnites as descended from Greeks and Persians, so a mosaic depicting Greeks and Persians might refer not only to faraway lands but also to the imagined realities of Samnite Pompeii.[73]

Furthermore, this family asserted its ties not only to the Greek East but also to Rome. Outside the grand doorway was a mosaic with the word *HAVE*, greeting potential clients in Latin rather than Oscan.[74] The axial design of the house around the view from that doorway recalls Roman social conventions of patronage and public access to great men's houses.[75] In late Hellenistic Italy, Alexander's conquests could be understood as a precedent for the Roman Empire.[76] Wallace-Hadrill has recently proposed that "the most important motor

---

71. For example, parallels exist on Delos for the winged, tiger-riding figure (Dunbabin 1999: 43–44; Christensen 2006: 92–93) and the Nilotic mosaic (Versluys 2002: no. 121).

72. See note 18, this chapter.

73. Serv. 8.638; Salmon 1967: 29–30; Pfrommer 1998: 209.

74. See discussions (and arguments for the pre-Sullan dating of this mosaic) in Pesando 1996: 199–200; Zevi 1998a: 24–26.

75. Zevi 1998a: 32–33.

76. Cohen 1997.

behind competitive luxury" in late Hellenistic Italy was "the desire of these Italian elites to assert their belonging to the Roman project of conquest."[77] While competitive luxury likely had motors aplenty, the Casa del Fauno does make a powerful claim for the compatibility (even inextricability) of local, pan-Italian, and pan-Mediterranean visual vocabularies of wealth and prestige.

In the final phase of this house, political changes imbued its visual culture with altered implications. The establishment of the principate entailed significant limitation on private claims to heroic or kingly status.[78] However, internationalizing images could now take on new significance as allusions to the expanse of the Roman *imperium*. Visitors may now have responded to the visual culture of the Casa del Fauno less as an attempt to equate the paterfamilias to a Hellenistic *basileus*, and more as a domestic model of Roman *imperium*. For Flavian viewers, the Alexander exedra and its associated peristyle would appear to be (among other things) a grander version of the Casa dell'Efebo garden: a model of *oikoumene* and demonstration of cosmopolitanism.

The Flavian-period inhabitants of this house may also have developed a more specific interest in Egypt, or at least Egyptian-derived cults. Among assorted artifacts from the tetrastyle atrium, Fiorelli records a bronze sistrum.[79] Other (non-Egyptianizing) finds in this area included two altars with Oscan inscriptions; bronze statuettes of a female figure, a Silenus, and another female figure holding a *patera*; a candelabrum; two couch legs; and a female skeleton.[80]

Since sistra functioned as ritual instruments in Isis cults,[81] it is possible that one or more inhabitants of the Casa del Fauno had a personal interest in Isis or other Egyptian deities. If some members of this *familia* were involved in Egyptian-derived cults, this involvement might well have influenced their perceptions of the Nile mosaic. However, the finds in the house testify only to activities in its final phase. Evidence for Isiac activity in 79 CE has little bearing on viewers' perceptions of the mosaic circa 80 BCE. Indeed, such evidence should also not necessarily determine our understanding of viewers' perceptions in 79

77. Wallace-Hadrill 2008: 439; cf. Zevi 1998b: 89–90.

78. Zanker 1988.

79. Fiorelli 1860–64: 2:248; noted also by de Vos 1983: 69. Published catalogues of Pompeian sistra do not attribute any to the Casa del Fauno (Tran Tam Tinh 1964: 181–185; De Caro 2006b: 168–172), but this object may be one of several unprovenienced sistra at the Naples Museum.

80. Fiorelli 1860–64: 2:247–248.

81. On sistra in Hellenistic and Roman Isis worship, see references in Barrett 2011: 129n408.

CE—but that is a subject for another chapter. Chapter 6 will explore the complex relationship between Aegyptiaca and Isiac practice.

Margins, Frames, and Borderlands

The impact of the Nile mosaic on viewers was inextricable from its own marginality: not only its location on the periphery of the Alexander mosaic, but also its role as threshold between the peristyle garden and the more intimate space of the exedra. As an actual *limen*, the Nile mosaic occupies "liminal space" in the most literal possible sense. Paradoxically, however, this very liminality becomes a source of power, enabling the mosaic to generate new meanings with the visual culture it adjoins. As border to the Alexander mosaic, the Nile mosaic reframes viewers' experience of the larger image. It not only activates additional associations with Alexander's career but even encourages viewers to identify with the conqueror by retracing his journey.

The Nile mosaic thus collaborates with adjacent visual and material culture to construct social, economic, and cultural identities for its owners. A member of the local Samnite elite becomes, at least aspirationally, a Hellenistic *basileus*; a Flavian garden becomes a demonstration of cosmopolitan tastes and a model of *oikoumene*. From the first century BCE to the first century CE, the inhabitants of the Casa del Fauno used domestic material culture to assert their membership in a pan-Mediterranean cultural *koine*. During this period, the expansion of Roman power dramatically altered the geopolitical context of that *koine*. Egypt, still an independent kingdom at the time of the mosaic's installation, became part of the empire, and Pompeii itself became a Roman *colonia*. Throughout these changing political circumstances, however, the Casa del Fauno attests to its owners' continued efforts to assert their participation in a shared, transregional elite culture. In both Hellenistic and Flavian Pompeii, domestic cosmopolitanism enabled people not only to display their wealth but also to assert their sophistication and "Hellenizing"—or, perhaps better, internationalizing—*paideia*.

As a border for the Alexander mosaic, the Nile mosaic played multiple roles: marking the limit of the larger image, calling attention to the image, and helping to shape viewers' understanding of the image. Recent art-historical research explores the functions of the "frame," where "framing" is defined broadly as any device that "defined, demarcated and contained the field of vision."[82] As Platt and Squire assert, frames "do not just articulate boundaries" but also

82. Quoting Platt and Squire 2017a: 3; see generally Platt and Squire 2017b.

"actively negotiate them."[83] Like the other visual and spatial devices that drew attention to the Alexander mosaic, the Nile mosaic fits this definition of visual frames. Collectively, these frames called attention to the Alexander mosaic as *an image* and prompted viewers to respond to it as such. The framing of the mosaic encouraged a mode of viewing similar to modern approaches to "art"—even as such viewing remained compatible with simultaneous valuation of the mosaic as, for example, a prompt for reflection on historical events or the character of "great men," a reminder of the cult of Alexander, a commentary on viewers' own relationship to "Greek" culture, or a self-aggrandizing gesture on the part of the house owner.

Marginalia and *Monstra*

As one of the earliest Nilotic landscapes at Pompeii, the Casa del Fauno mosaic was among the first of many such scenes to create meaning with their own marginality. Nilotic frescoes and mosaics often act as peripheral ornamentation for larger scenes. Indeed, even the more visually prominent Nilotic frescoes in the Casa del Medico and Casa dell'Efebo can be interpreted this way. Although they do not frame larger *painted* scenes, these frescoes act as frames for the real-life tableaux constituted by human activity in these gardens.

Positioned at the periphery of larger images, Nilescapes recall marginalia of other types. In their frequent emphasis on fertility, abundance, and excess, they resemble medieval marginalia, whose playful or outrageous imagery similarly generates wild overabundance on the (literal) margins of ordered space.[84] Nilotic marginalia play a similar structural role to the *monstra*, marvels, and *adynata* that destabilize and enliven the edges of late Second and Third Style wall paintings.[85] As Platt discusses, Augustan painting typically treats "marvellous or paradoxical elements" as peripheral framing elements or subsidiary motifs.[86]

Nilotic scenes resemble such imagery not only in their frequent peripheral relationship to larger compositions but also in their iconographic emphasis on the bizarre and the marvelous. The Nile itself was one of the most renowned natural *mirabilia* of antiquity, famous for its floods and mysterious sources, as well as its astonishing fertility. Other common Nilotic motifs suggest particular interest in

83. Platt and Squire 2017a: 47. On the "concept of boundary" within Roman gardens, see Bergmann 2014.

84. On medieval and Renaissance marginalia, see Camille 1992; Zorach 2005: 140–158.

85. On *monstra* and *adynata*, see notes 1 and 87, this chapter.

86. Platt 2009: 70.

those features of (real or imagined) Egyptian landscapes that would have struck Italian audiences as most unusual. These "curiosities" may be natural (exotic animals), somatic (achondroplastic *pygmaei*), or cultural (inundation-related festival practices). However, they all share the quality of strangeness or novelty.

The earliest Nilotic frescoes and mosaics, including the Casa del Fauno mosaic, predate the most extreme of the *monstra* excoriated by Vitruvius.[87] However, the popularity of such imagery in the early principate may have helped lay the groundwork for the increased frequency of Nilotica in Fourth Style domestic paintings. While Nilotic scenes do not typically include the fantastic architecture or impossible hybrids of Third Style marginalia,[88] Nilotica and other forms of fantastic border decoration do sometimes appear in close proximity. For example, at the portico of the Temple of Isis[89] (figure 5.2), the buildings *within* the Nilotic landscapes all appear fairly architecturally plausible. However, those landscapes are themselves incorporated (as hanging *pinakes*) into a larger framework of architecturally impossible structures.

Like some other forms of border decoration in Roman art, then, Nilotic scenes confront viewers with fantastic and novel imagery. Paradoxically, however, the very commonness of such scenes in domestic and other contexts also familiarized viewers with this imagery. Additionally, Nilescapes make different types of truth claims than do other painted marginalia. For Augustan observers, the hybrid creatures and impossible architecture of contemporary painting were manifestly contrafactual; they depicted things that, as Vitruvius put it, "neither exist, nor could exist, nor ever existed."[90] However, Nilotic scenes—whether pre-Augustan, like the Casa del Fauno mosaic, or later, like most other Pompeian Nilotica—assert a different relationship to their own reality. Nilescapes present their *mirabilia* not as contrafactual but, rather, as *far away*: existing not in Italy, but in the marvelous land of Egypt.[91] It was possible to debate the reality of certain wonders attributed to Egypt: for example, ancient authors disagreed on

87. Vitr. 7.5.3–4. However, predecessors of these late Second and Third Style motifs appear already in earlier Second Style painting and emulate earlier Classical and Hellenistic forms (Yerkes 2000; Platt 2009: 62).

88. Though hybrid animals can appear in Nilotic scenes as *statues* of composite deities: e.g., figure 1.4, J; 2.4, A.

89. On these paintings, see note 211, chapter 2, this volume.

90. *nec sunt nec fieri possunt nec fuerunt* (Vitr. 7.5.4). Cf. Horace's comparison of painted composite creatures to a sick man's deceptive dreams (Hor. *Ars P.* 1).

91. On Egypt as land of fantasies and marvels, see Assmann 2000.

whether or not pygmies actually existed.[92] However, the Nile itself was clearly not only real but also essential to Italy's own prosperity, and public displays of crocodiles in Italy proved that Nilotic *monstra* were no fantasy.[93] Even the seemingly extraordinary behavior on display in Nilotic scenes often referred to demonstrable ethnographic realities.[94] Where Augustan marginalia envisioned impossibilities, Nilescapes tantalized Roman viewers with images of a land where impossibility became reality.

The Garden as Margin

As representations of the borders of the Roman world, Nilotic scenes are appropriate to other border zones: the edges of walls, or of household space itself. Away from gardens, Nilotic landscapes typically serve as subsidiary decoration to larger installations or painted panels. In or adjacent to gardens, however, Nilotic landscapes may cover entire dining installations (as in the Casa dell'Efebo), walls (as at the Casa del Medico), or rooms (as at, e.g., the Casa dei Pigmei or Praedia di Julia Felix).[95] Certainly, Egyptian landscapes can also play peripheral roles in garden ensembles, as at the Casa del Fauno. However, in garden settings, considerations of *decor* no longer appear to dictate explicit marking of Nilotic imagery as "marginal" relative to particular walls or frescoes. While such subordination of Egyptian themes may sometimes still occur, it is no longer necessary. The garden setting itself suffices to frame these images as liminal.

The frequent association of Egyptian imagery with gardens thus recalls the conceptualization of gardens as transitional spaces between the *domus* and the wider world. As an artificial "landscape" within a domestic environment, the garden "mediates between outside and inside" and "between the concepts of wild and cultivated."[96] In such spaces, Egyptian scenes often join other images suggestive of faraway, marvelous, and potentially dangerous settings, such as Bacchic motifs and representations of wild animals. Even as these eclectic domestic landscapes evoke spaces far beyond the house or city, their incorporation

---

92. See sources in note 4, chapter 2.

93. Strab. 17.1.44, and see also chapter 4, "South Side of West Wall (MANN 113195)", this volume.

94. See chapter 2, this volume, "Egyptian Antecedents and Affordances."

95. Compare the observations of Clarke 2003: 192; 2007b: 94. For further examples, see entries in appendix A, this volume. On Casa dei Pigmei and Praedia di Julia Felix, see Versluys 2002: nos. 38, 62.

96. Quoting Von Stackelberg 2009: 53; see also note 50, chapter 1, this volume.

into constructed domestic space marks them as civilized and safe.[97] Gardens thus create an ideal environment for grappling with cultural contradictions, exploring—and renegotiating—the boundaries of the domestic and the civilized.

Marginal spaces, whether at the borders of larger images or on the (conceptual) borders of the *domus* itself, thus become simultaneously places of distancing but also of incorporation and transformation. This process had social and cultural implications. The domestic display of internationalizing imagery gave Campanian elites of the first century BCE a way to perform participation in Hellenistic *koine* culture. Later, by the first century CE, the domestic display of foreign landscapes took on new significance in a Pompeii now identified with the empire that ruled most of the known world. Even as people in Rome's other provinces grappled with the implications of adopting (or attempting to adopt, or refusing to adopt, or only periodically adopting) "Roman" identities, so did people in Italy grapple with the question of what "Rome" meant in an age of global empire. The increasing use of international imagery in domestic contexts attests to a phenomenon inextricable from the much-debated "Romanization" of the provinces: the concomitant internationalization, or globalization, of daily life in Italy.

97. Hales 2003: 153.

6

Model Worlds

THREE-DIMENSIONAL NILOTICA IN THE CASA DI ACCEPTUS E EUHODIA

Introduction

While previous chapters have emphasized the three-dimensionality of the *settings* of Nilotic imagery, they have focused on Nilescapes that are themselves largely two-dimensional: frescoes and mosaics. Indeed, "Nilotica" are usually studied as a form of surface decoration. Versluys's catalogue of Nilotic imagery focuses primarily on frescoes and mosaics, along with some stuccoes, terracotta reliefs, and decorated vessels.[1] Certainly, not even surface decoration should be understood as wholly "flat"; the previous chapters demonstrate the importance of situating frescoes and mosaics within three-dimensional space. However, recognizably "Nilotic" imagery also appears outside of surface decoration: for example, in statues and figurines of Nilotic animals or pygmies, or domestic canals sometimes known as *nili* ("Niles").

Such statuettes and model "Niles" have attracted much study under the general heading of "Aegyptiaca." However, they are usually categorized separately from Nilotic frescoes and mosaics. While scholars have noted iconographic parallels between such objects and two-dimensional Nilotica,[2] these parallels still await systematic exploration. Less remarked, and also requiring investigation, are contextual parallels. As with two-dimensional Nilescapes, these artifacts and features are common in domestic contexts, especially domestic gardens.

1. Nilotic motifs on lamps, sarcophagi, and statuettes are collected in an appendix of "doubtful cases and parallels" (Versluys 2002: 451–456, quoting 41).

2. E.g., Assmann 2000: 72–73; Amedick 1993: 183, 185; Mol 2015: 307 (on the Casa di Octavius Quartio garden as "a three dimensional Nilotic scene").

This chapter thus examines the use of domestic water features, statues, and statuettes to create three-dimensional Nilotic landscapes within Pompeian homes. An initial example comes from the most famous such assemblage, at the garden of the Casa di Octavius Quartio. However, our primary case study addresses a smaller, seldom-studied house: the Casa di Acceptus e Euhodia. This dwelling provides new perspectives on sculptural phenomena more often studied in wealthier contexts. The Casa di Acceptus e Euhodia also points to new directions in longstanding debates concerning the relationship of "Egyptianizing" garden decoration to "Isiac" religious identity. Although a shrine inside the Casa di Acceptus e Euhodia suggests the inhabitants of this house incorporated a form of Isis into domestic cult, I argue against allowing this evidence to dominate interpretations of the garden assemblage. Participants in Isis worship constituted an extremely heterogeneous group, and such participation need not have mandated one universal response to Nilotic imagery.

In place of artificial dichotomies between "religious" and "decorative" material culture or between "Isiacs" and "non-Isiacs," the Casa di Acceptus e Euhodia assemblage points to more nuanced relationships between material culture and religious identity. Contrary to prevailing assumptions that "Isiacs" and "non-Isiacs" must have used Aegyptiaca in fundamentally different ways, this assemblage suggests that participants in Isis cult might employ domestic Nilotic imagery in ways very similar to other Pompeians. These finds further pose an opportunity to question received constructions of "Isiac" religious identity. Rather than a binary distinction between "Isiacs" and "others," Pompeian society more likely furnished a wide spectrum of engagements with Egyptian-derived deities.

Imagining "Egypt" in Three Dimensions

Three-dimensional Aegyptiaca in garden contexts take numerous forms, including "Nile"-like canals, statuettes in various media, and shrines dedicated to Egyptian deities. However, this book focuses on representations of Egyptian landscapes, not Aegyptiaca as a comprehensive category. Accordingly, we have to begin with a question: What kinds of artifacts or assemblages would have evoked Egyptian landscapes for Pompeian viewers?

For example, shrines to Egyptian-derived deities (most commonly Isis, Sarapis, Harpocrates, and Anubis) are common in Pompeian houses, both in gardens and in other domestic areas.[3] However, shrines or images of gods need not have suggested Egyptian *landscapes* to Pompeian viewers. By the first century CE,

3. See appendix D, this volume, for garden shrines.

Isis and her circle were pan-imperial deities, whose cults and iconography derived from Hellenistic and Roman transformations, as well as Pharaonic traditions.[4] Isis temples were equally at home in Rome as in Alexandria.[5] Although the Roman Isis retained strong iconographic, ritual, and theological ties to (Roman constructions of) Egypt,[6] her cult was no longer exclusively *geographically* situated in Egypt.

Accordingly, this chapter will *not* address as "Egyptian landscapes" assemblages like that of the Casa degli Amorini Dorati garden, which contained a shrine to Egyptian gods.[7] Also not surveyed here are the numerous domestic depictions of "Isis-Fortuna" in multiple media.[8] Such images and assemblages provide rich evidence for Roman receptions of Greco-Egyptian cults, but they did not necessarily prompt viewers to consider Egypt as *landscape*.

Of the numerous Aegyptiaca found in Pompeian domestic gardens, some types have particularly close parallels to two-dimensional Nilotica. These include statuettes of Nilotic river animals, statuettes of dwarfs or pygmies, and certain water features. In isolation, such artifacts and features need not always evoke "Egypt." Occasional descriptions of some domestic canals as *nili* do not imply that all domestic canals were model Niles, and similarly, not all dwarflike figurines are Nilotic *pygmaei*. When found in association, however, these objects and features take on new implications. As an *assemblage*, a collection of water features, dwarflike figures, and Egyptian river animals creates a three-dimensional parallel to the Nilotic landscapes of mosaic and painting. Assemblages, not individual artifacts, are thus the chief focus of this chapter.

Model Landscapes: Water Features, Statuary, and "Egyptian" Installations

We have already considered two cases in which proximity to "Egyptianizing" imagery framed domestic water features as "Nilotic." At both the Casa dell'Efebo and the Casa del Medico, a water feature—whether an elaborate *euripus*, as at the Casa dell'Efebo, or a simpler channel, as at the Casa del Medico—ran immediately beneath painted Nilescapes. Without the frescoes, neither water feature

4. See chapter 1, this volume, "Material Culture, Identity, and 'Authenticity,'" with note 115.

5. On Isis temples in the Roman world, see Kleibl 2009.

6. See note 116, chapter 1, this volume.

7. Seiler 1992: 37–49; Petersen 2012; Mol 2015: 332–391.

8. Beaurin 2008; Mol 2015: 169–176.

would have appeared inherently "Egyptianizing." However, the frescoes cast the real-life water as a three-dimensional extension of the painted Nile. Association with three-dimensional Aegyptiaca could produce similar effects. For example, the garden of the Casa delle Nozze d'Argento originally contained a fountain decorated with four green faience or imitation faience figurines that served as waterspouts: two crocodiles (figure 1.9) and two frogs.[9] The crocodiles allude iconographically to Egypt, and the faience material provides a further visual suggestion of Egyptian production.[10] Adorned with material culture suggestive of Egypt and Egyptian river creatures, the domestic water feature takes on Nilotic associations.

Besides Egyptian animals in faience or stone, a number of Pompeian gardens also contained statuettes of dwarflike beings.[11] Most are faience or imitation faience, although at least one marble example is known.[12] Some statuettes depict the Egyptian god Bes, who is recognizable by his squatting posture, semi-leonine face, protruding tongue, achondroplastic proportions, and frequently plumed headdress (e.g., figures 6.1, 6.2).[13] Others depict a different figure with achondroplastic proportions and fully anthropomorphic features (e.g., figures 6.3, 6.4).

Seven or possibly eight faience or imitation faience statuettes of Bes, and at least five faience statuettes of fully anthropomorphic dwarfs, have been recorded from Pompeii.[14] Where their original findspot is known, these statuettes

---

9. See chapter 1, this volume, "Material Culture and the 'Material Turn.' "

10. On chemical evidence of Egyptian production for at least one of the figurines, see note 132, chapter 1, this volume. A minority of other faience or imitation faience objects at Pompeii may have been produced in Italy (Mangone et al. 2011: 2866). This assemblage is also discussed in Barrett 2017a. A minority of other faience or imitation faience objects at Pompeii may have been produced in Italy (Mangone et al. 2011: 2866).

11. See appendix C, this volume.

12. A marble statuette from an interior room in the Casa di Oppius Gratus depicts a pygmy or dwarf (Carrella et al. 2008: 178, no. D 22). Jashemski suggests that this and an associated marble statuette of Aphrodite may have been originally garden decorations (1979–93:2:238).

13. On the Bes image, see Tran tam Tinh 1988; Romano 1989; Dasen 1993: 55–83; Malaise 2004: 267–268.

14. *Bes statuettes:* Di Gioia (2006: 115–119) records six extant faience or pseudo-faience examples, but one of these (no. 5.5) is not actually a Bes but a baboon (as correctly noted by Mol 2016b). To those examples, add a now-lost Bes figurine from the Casa di Octavius Quartio (Tronchin 2011b: 35–37) and a now-lost possible (though not certain) Bes figurine from the Iseum (see note 15, this chapter). Four known examples come from domestic contexts, of which three were domestic gardens. Besides the two examples from the Casa di Acceptus e Euhodia garden (see note 65, this chapter), one comes from the garden of the Casa di Octavius Quartio (see this chapter, "Garden Luxury: The Casa di Octavius Quartio") and one from the right *ala* in the Casa di Marcus Lucretius (Kekulé 1880: 61; Dwyer 1982: 31; de Vos 1983: 62; Di Gioia 2006: no. 5.6; Mol 2015: 158, 479; for the context, see Dwyer 1982: 31). *Statuettes of anthropomorphic dwarfs (so-called Pataikoi):* Di Gioia (2006: 110–114) records five examples. A faience

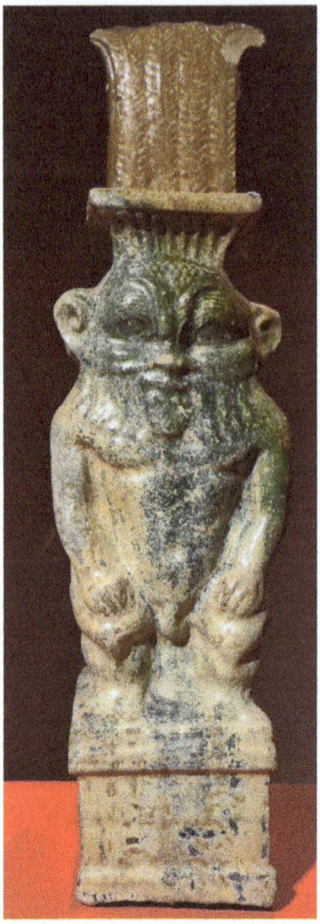

FIGURE 6.1 Pseudo-faience (glazed terracotta) figurine of Bes, from the garden of the Casa di Acceptus e Euhodia (VIII 5.39). The figurine is bifrontal, with the same imagery on front and back. Height: 40 cm (Di Gioia 2006: 116). MANN 116665.

Photo: C.E. Barrett. Reproduced by permission of the Ministero dei Beni e delle Attività Culturali e del Turismo – Museo Archeologico Nazionale di Napoli.

frequently derive from garden contexts.[15] This association with gardens resembles the distribution pattern of two-dimensional Nilotica.

---

or pseudo-faience "Ptah" mentioned in excavation reports from the garden of the Villa delle Colonne del Mosaico (Schulz 1838: 192; Kekulé 1880: 61; Jashemski 1979–93: 2:278) may represent a sixth figurine or may be identical with one of the unprovenienced examples in Di Gioia's publication. See further appendix C, this volume.

15. Di Gioia 2006: 110–114; Mol 2015: 211; and add also the additional examples in note 14, this chapter. The three examples from non-garden contexts are a so-called Pataikos from VI 1.2, a small *caupona* (figure 6.4; Di Gioia 2006: no. 4.1; Pagano and Prisciandaro 2006: 68; Poole 2016: no. 116); a Bes from a room adjacent to the atrium in the Casa di Marcus Lucretius (see notes 14 and 119, this chapter); and a possible Bes or "Pataikos" figurine found in the 1766 excavations of the Iseum (Fiorelli 1860–1864: 1.1:192; Tran tam Tinh 1964: 165, no. 115). Although the latter figurine no longer survives, Fiorelli's description of it as "di gesso, o di qualche altra mistura bianca dipinto di verde" suggests faience. Mol doubts that this object would have depicted Bes (2015: 189), but Fiorelli's description of the figure as "molto informe

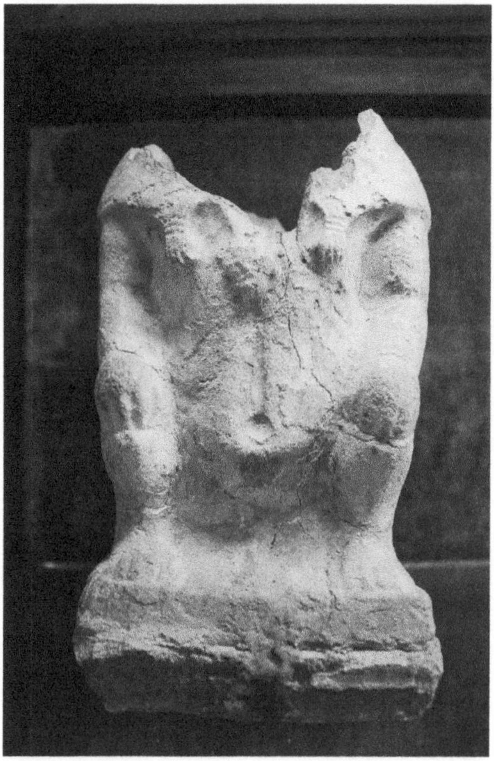

FIGURE 6.2 Faience figurine of Bes, from the garden of the Casa di Acceptus e Euhodia (VIII 5.39). The figurine is bifrontal, with the same image on front and back. Height: 21.5 cm (Di Gioia 2006: 117). MANN 117178.

Photo: C.E. Barrett. Reproduced by permission of the Ministero dei Beni e delle Attività Culturali e del Turismo—Museo Archeologico Nazionale di Napoli.

e ridicola" (1860–1864: 1:182) suggests a figurine type that eighteenth- and nineteenth-century observers would have found "grotesque." Among types attested in the faience repertoire at Pompeii, this would almost certainly be either a Bes or a "Pataikos." In a separate publication, Breton describes a figurine found in the Iseum as "une petite figure accroupie de porcelaine verdâtre, les mains et les bras appuyés sur les genoux, et le corps couvert d'une légère draperie" (1870: 52). Kekulé suggests that Breton and Fiorelli were describing the same figurine and interprets it as a Bes (1880: 61). Another Iseum figurine, described by Tran tam Tinh (1964: 165, no. 115; followed by Malaise 2004: 283n137) as a Bes of "porcelaine verdâtre," is in fact not Bes but a seated, bearded deity (De Caro 1992: 79, no. 6.2). It is conceivable that Breton was actually describing this object rather than Fiorelli's "informe e ridicola" figurine. Either way, however, Fiorelli's description does seem to point to a figurine of "grotesque" type. Mol attributes another Bes statuette to a *caupona* (I 14.8), citing "*Notizie degli Scavi* 6 October 1770" (Mol 2015: n558): an impossibility, as *Notizie degli Scavi di Antichità* was first published in 1876 (see Allison 2004a: 30). However, Bourbon-era excavation notes do record the find of a "Pataikos" in a different *caupona* (VI 1.2) on October 6, 1770 (Pagano and Prisciandaro 2006: 68; see also discussion earlier in this footnote).

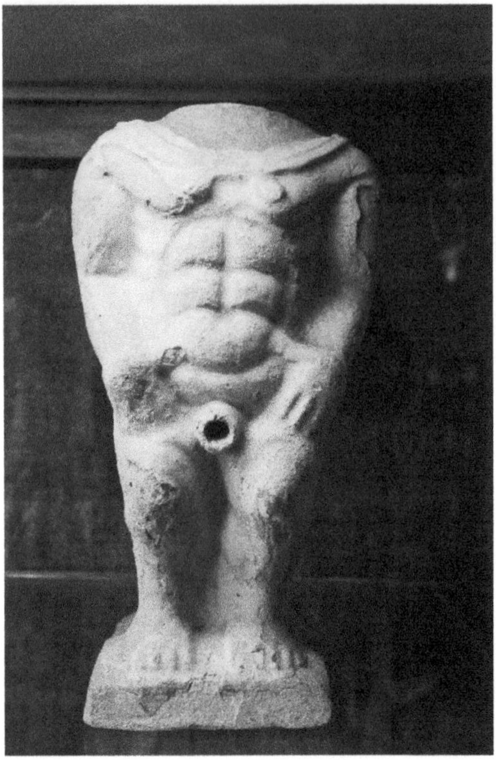

FIGURE 6.3 Faience figurine of a pygmy-like figure, from the garden of the Casa di Acceptus e Euhodia (VIII 5.39). Height: 27.3 cm (Di Gioia 2006: 112). MANN 116666. *Photo*: C.E. Barrett. Reproduced by permission of the Ministero dei Beni e delle Attività Culturali e del Turismo—Museo Archeologico Nazionale di Napoli.

Although scholars typically identify fully anthropomorphic faience dwarfs at Pompeii as "Pataikos" or "Ptah-Pataikos,"[16] this identification may be misleading. In Egyptology, "Pataikos" is a conventional scholarly designation for representations of the Memphite god Ptah as a dwarf.[17] However, the Pompeian figurines lack divine attributes. Scholarly identifications of these faiences as "Pataikoi" appear to rely on two features: their dwarflike proportions and their

16. On the identification of the latter type of figurine at Pompeii as "Pataikos," "Ptah-Pataikos," or "Ptah," see, inter alia, De Caro 2006b: 209; Di Gioia 2006: 110–114; Mol 2013: 127; 2015: 182–219.

17. See chapter 2, this volume, "Egyptian Antecedents and Affordances."

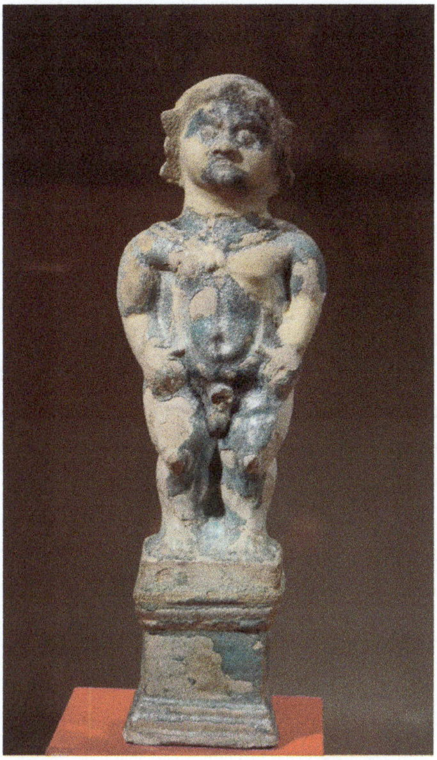

FIGURE 6.4 Faience figurine of a pygmy-like figure wearing a crown of ivy or vine leaves, found in a *caupona* (VI 1.2). MANN 22607. Height: 47.8 cm (Di Gioia 2006: 110).
Photo: C.E. Barrett. Reproduced by permission of the Ministero dei Beni e delle Attività Culturali e del Turismo—Museo Archeologico Nazionale di Napoli.

probable production in Egypt.[18] As discussed in chapter 2, however, Egyptian images of dwarfs take many forms and represent many different beings. Some Egyptian statuettes of dwarfs feature unambiguous divine attributes, such as special crowns.[19] However, many others lack indications of divinity and may more probably represent humans. Such figurines often wear wreaths and other attributes suggestive of festivals (e.g., figures 2.22, 2.26, 2.34, 2.35) and may be cult attendants or festivalgoers rather than divinities. It is these figurines that the

18. On the Egyptian origins of faience at Pompeii, see Mangone et al. 2011. All the known faience "Pataikoi" from Pompeii have a siliceous fabric characteristic of "true" faience (Di Gioia 2006: nos. 4.1–4.5).

19. See chapter 2, this volume, "Egyptian Antecedents and Affordances," with notes 114, 115, and 117.

so-called Pataikoi from Pompeii most closely resemble. The Pompeian statuettes' achondroplastic proportions, naked or nearly naked bodies, and floral or ivy wreaths all closely parallel Egyptian figurines of seemingly human dwarfs.[20]

The achondroplastic proportions and naked or mostly naked bodies of many of these figurines, both those depicting Bes and those depicting anthropomorphic dwarfs, recall many Roman depictions of Nilotic *pygmaei*. Their (usually) faience material would have reinforced this association with Egypt. Most of the green-glazed statuettes from Pompeii are made of siliceous, quartz-based faience, not terracotta.[21] Of the extant glazed figurines of Bes or anthropomorphic dwarfs from Pompeii, all have a siliceous fabric except for two Bes figurines in clay-based fabrics.[22] Chemical testing suggests that while the glazed terracottas may be locally made, most of the "true" faience artifacts at Pompeii are Egyptian imports.[23] Their brilliant blue-green surfaces visually signaled these Egyptian origins. Especially when found in association with water features or other Aegyptiaca, faience dwarf figurines would have appeared to be three-dimensional equivalents of the protagonists of Nilotic frescoes and mosaics: *pygmaei* who had stepped out of their pictorial frame and entered the garden itself.

Garden Luxury: The Casa di Octavius Quartio

The best-known example of Egyptian or "Egyptianizing" garden statuary at Pompeii comes from the so-called Casa di Octavius Quartio (II 2.2) (figure 6.5).[24]

---

20. Two faience "Pataikoi" from Pompeii are naked except for a lion skin around the shoulders (figures 6.3 and 6.4; cf. Di Gioia 2006: nos. 4.1, 4.2; one of these is MANN 116666, discussed in this chapter). Of three with preserved heads, two wear wreaths of ivy leaves (Di Gioia 2006: nos. 4.1, 4.4) and one wears a floral wreath (Di Gioia 2006: no. 4.5).

21. Di Gioia 2006: 133–138; Mangone et al. 2011.

22. The two exceptions: MANN 116665 (described by Di Gioia 2006: 116 as "pasta silicea," but identified as clay-based through chemical testing: Di Gioia 2006: 138); and MANN 13586 ("argilla": Di Gioia 2006: 117).

23. Mangone et al. 2011.

24. Della Corte 1930: 60–81; 1931–32; Spinazzola 1953; de Vos 1991; Tronchin 2006, 2011b. On the garden ensemble, see also (within the past 15 years) Platt 2002; von Stackelberg 2009: 101–125; Mol 2015: 392–446; 2016a; Poole 2016: nos. 104–115. The house takes its modern name from an inscription on a bronze seal found in one of the rooms; however, identifications of the house's owner as either Octavius Quartio or Loreius Tiburtinus are unreliable (see discussion in Platt 2002: 108n7). Von Stackelberg (2009: 102–104) argues persuasively against Salza Prina Ricotti's (1987: 170) interpretation of the house as a *caupona* or *hospitium* during its final phase.

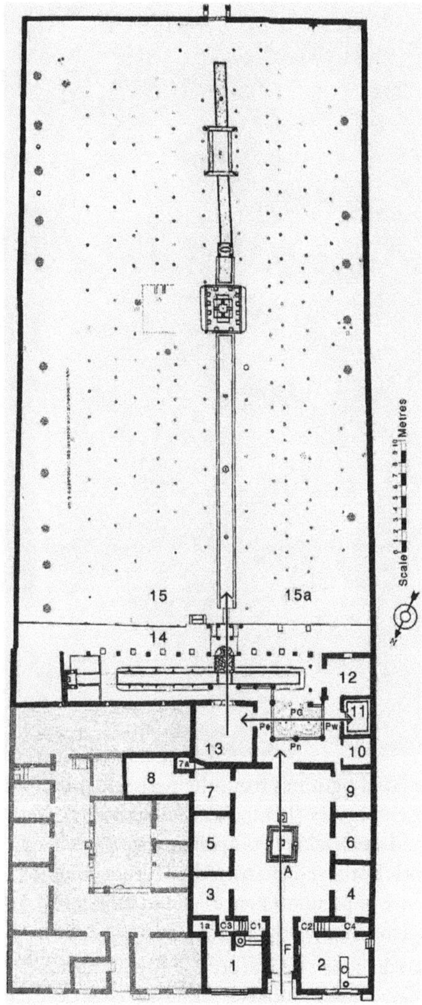

FIGURE 6.5 Plan of the Casa di Octavius Quartio (II 2.2), with sightlines marked. *Key*: A = atrium, F = fauces, Pn = peristyle north, Pe = peristyle east, Pw = peristyle west, PG = peristyle garden. Space (14) represents a portico garden on an upper terrace, while spaces (15) and (15a) represent the lower garden area.

Plan: K. von Stackelberg and A. Perri (von Stackelberg 2009: fig. 4.1); reproduced by permission of Katharine von Stackelberg.

The garden of this house is one of the largest and most elaborate at Pompeii. Taking up two-thirds of an insula, the garden comprises an upper and a lower terrace, two artificial canals, two waterside dining installations, an extensive array of statuary, and an ensemble of frescoes depicting animal landscapes and

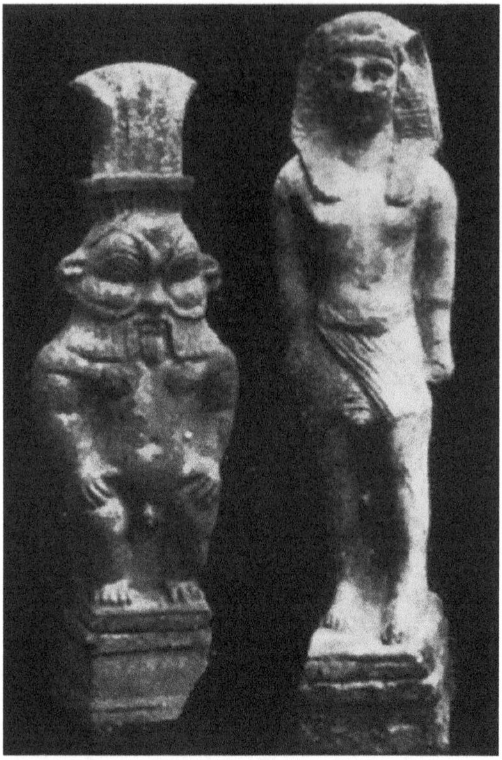

FIGURE 6.6 Green-glazed figurines from the peristyle garden of the Casa di Octavius Quartio (II 2.2), representing Bes (Pompeii inv. no. 2897; height: 51 cm) and a standing figure in *nemes* crown and *shendyt* kilt (Pompeii inv. no. 2898; height: 54 cm), both restored from multiple fragments. Both were destroyed in Allied bombing in 1943. The statuettes' material was either faience or pseudo-faience (glazed terracotta). For dimensions, description, and findspot, see Tronchin (2006: 45–52).

Photo © Società Tiburtina di Storia e d'Arte (previously published as Della Corte 1931–32: pl. 30, fig. 14; Tronchin 2011b: fig. 2.3); reproduced by permission of Società Tiburtina di Storia e d'Arte.

mythological scenes set in landscape settings.[25] The garden statuary includes multiple examples of Aegyptiaca, including a marble statuette of a *nemes*-wearing androsphinx (a male sphinx of Egyptian type) and at least seven heavily damaged faience or pseudo-faience statuettes. Two of these statuettes (lost in World War II) represented a striding pharaoh wearing a *nemes* crown and *shendyt* kilt and a standing figure of Bes (figure 6.6). Possibly also from this house is a faience

25. For discussions of the location of specific objects, see Tronchin 2006; 2011b.

statuette of a seated baboon, sometimes erroneously identified as Bes.[26] A small wall painting in a room overlooking the garden (figure 6.5, room no. 11) depicts a bald Isis priest holding a sistrum.

Because of these Egyptian allusions, the Casa di Octavius Quartio features prominently in most discussions of Pompeian Aegyptiaca.[27] However, the eclectic visual assemblage from this garden also presents viewers with many motifs and associations unconnected with Egypt. The statuary assemblage also includes Bacchic images, depictions of animal combats, and references to Greek *paideia* (e.g., two statuettes of Muses). The mythological frescoes on the upper terrace depict scenes from Greco-Roman, not Egyptian, narratives: Narcissus, Pyramus and Thisbe, and Diana and Actaeon.

Two debates dominate discussions of this assemblage. One addresses the relationship between Aegyptiaca and religious identity: Did this house belong to an "Isis worshiper"? Della Corte interpreted the paterfamilias as an Isis priest who used the *euripus* for rituals celebrating the Nile flood.[28] However, as we have seen, individuals might possess Egyptian or "Egyptianizing" material culture for many reasons other than religious identity. Additionally, the small painted image of the Isis priest in the room adjoining the garden should not be overinterpreted.[29] This vignette is just one tiny part of a much larger composition that also depicts nymphs, fantastic animals, and other common motifs of Fourth Style art. Rather than indicating the house owner's religious affiliation, the painting appears

---

26. See note 14, this chapter. On the association of the baboon with this house, see Di Gioia 2006: 117; Mol 2015: 119; 2016b (though Tronchin omits this figurine from her studies of the assemblage: 2006; 2011). Another marble statuette, often described as an ibis, is probably a heron (Tronchin 2006: 157–162; 2011: 37–38), though iconographic confusion with an ibis remains possible (Mol 2015: 404–405). Numerous amphorae were buried in the garden; Della Corte relates these to Nilotic water rituals (1930: 76; 1931–1932: 197–198), but they may in fact have been flower pots or irrigation equipment (von Stackelberg 2009: 121; Mol 2015: 414–415).

27. Beginning with Della Corte 1930: 60–81; 1931–32; Tran tam Tinh 1964: 43–46. For subsequent scholarship, see note 24, this chapter.

28. Della Corte 1930: 74–76; 1931–1932; still followed recently by Capriotti Vittozzi 2002a: 27.

29. Compare the similar skepticism of Salza Prina Ricotti 1987: 172; Mol 2015: 397–399. Clarke (1991: 193–201) still connects the image to religious identity, specifically that of the *domina* of the house. Tronchin (2006: 295–299) suggests that the priest may represent not the paterfamilias but one of his grown children. However, the image does not contain obvious portrait features. A nearby graffito writes a name of dubious reading, often taken as that of the priest (see the review in Tronchin 2006: 297–298). However, the relationship between image and graffito is not wholly clear.

typical of the widespread use of Egyptian motifs as subsidiary decorations within larger painted compositions.[30]

Another debate addresses Zanker's influential interpretation of the Casa di Octavius Quartio as a "miniature villa" whose lavish displays exemplified the social striving of nouveau riche Pompeians.[31] Like all Roman dwellings, this house undoubtedly participated in practices of social display and competition. However, more recent scholarship has compellingly critiqued subjective modern characterizations of ancient domestic decoration as "kitsch," "bad taste," or typical of "freedman style."[32]

Recent studies by Tronchin, von Stackelberg, Mol, and Platt[33] have thus sought to free the Casa di Octavius Quartio from interpretive paradigms emphasizing either "Isiac" cult affiliation or the "kitschy" taste of ambitious arrivistes. Instead, this garden assemblage is more productively considered within the context of mainstream elite domestic display. The eclecticism of the statuary assemblage is typical of Roman houses of this period,[34] and as chapters 3–5 have shown, Pompeians of multiple socioeconomic backgrounds treated Egyptian themes as appropriate (*decorus*) to domestic gardens. Accordingly, neither nouveau riche status nor Isiac religious affiliation is necessary to explain the appearance of Aegyptiaca at this house.

Rather than a site of Isiac ritual or a monument to tasteless social striving, I interpret the Casa di Octavius Quartio garden as a particularly luxurious example of the type of domestic installation explored in chapters 3–5: an outdoor entertainment complex combining a planted garden, dining facilities, and water features.[35] The appearance of three-dimensional Egyptian imagery in this context is thus consistent with the frequent appearance of Nilotic imagery in other such complexes, including the Casa dell'Efebo, Casa del Medico, and Casa del Fauno.

The Casa di Octavius Quartio is one of the most famous and elaborately decorated houses at Pompeii, and its garden assemblage is one of the most thoroughly studied. However, the sustained attention focused on this house risks obscuring the existence of similar assemblages of "Nilotic" statuary in smaller, less affluent

---

30. For Egyptian motifs as marginalia to larger paintings, see note 1, chapter 5, this volume.

31. Zanker 1998: 145–156; followed by Clarke 1991: 23–25, 193, 207.

32. Petersen 2006; 2012: 320–321; Tronchin 2011a; 2012a: 275–278.

33. See note 24, this chapter.

34. E.g., Bartman 1991: 73–74; Trentin 2014: 23, 63. For a similarly lavish statuary collection, compare the garden assemblage of the Casa di Marcus Lucretius: Trentin 2014: 318–339.

35. Cf. Salza Prina Ricotti: "this part of the house [i.e., the garden] was a water triclinium of the well-known Plinian type and not a shrine for a foreign cult" (1987: 172).

houses.[36] Such assemblages provide a fuller perspective on the use and valuation of Nilotica across social and economic boundaries. The remainder of this chapter thus considers the garden assemblage from the much smaller Casa di Acceptus e Euhodia. This assemblage also provides new evidence on one of the longest-standing controversies at the Casa di Octavius Quartio: the relationship between domestic Aegyptiaca and religious identity.

The Casa di Acceptus e Euhodia

Excavation and Description

The Casa di Acceptus e Euhodia (VIII 5.39) was excavated in 1881–82, followed by publications by Sogliano and Mau.[37] The house is small (150 sq. m) and irregular in plan, lacking an atrium (figure 6.7).[38] Most of the ground floor rooms open onto the columned entrance hall (figure 6.7, a) adjoining the garden (figure 6.7, b).

The wall paintings throughout the house, now mostly faded, were Fourth Style, and the statuary (discussed later) is stylistically datable to the first century CE.[39] Modern scholarship often characterizes the wall paintings as having been simple and unimpressive, but Mau's 1884 publication includes detailed descriptions of now mostly lost scenes, some of which he characterizes as painted "non senza diligenza."[40] Most of their contents were fairly typical of the Fourth Style repertoire: for example, images of animals, fantastic architecture, flying Amores, female busts, young boys, and gorgoneia[41]. Outside of the garden, the only recorded Egyptian themes appeared in the kitchen (e), where a lararium contained a painting of Isis-Fortuna.[42]

---

36. See appendix C, this volume.

37. Sogliano 1882: 241, 278, 280–281, 324, 359–360, 439; Mau 1884: 126–132; 1908: 360–362. For more recent descriptions of the house and finds, see Eschebach 1984: 48–50; Jashemski 1979–93: 2:218; Bragantini 1998c; Kastenmeier 2007: 159–160. The room that Carrella et al. describe as an "atrio" (2008: 151) is elsewhere identified more plausibly as a triclinium.

38. On the house area, see Kastenmeier 2007: 159.

39. On the date of the paintings, see Mau 1884: 127 (followed by subsequent scholarship; see most recently Kastenmeier 2007: 159). Carrella et al. assign all the marble statuary from the house to the first century CE, with two herms, the satyr mask, and a *labrum* specifically datable to the Flavian period (2008: 148–152).

40. Quoting Mau 1884: 129. On descriptions of the decoration as "molto semplice," see Bragantini 1998c: 648; Carrella et al. 2008: 149.

41. On descriptions of the paintings, see Mau 1884: 128–132; Schefold 1957: 229–230.

42. Boyce 1937: 372; Fröhlich 1991: 292; Beaurin 2008: 284.

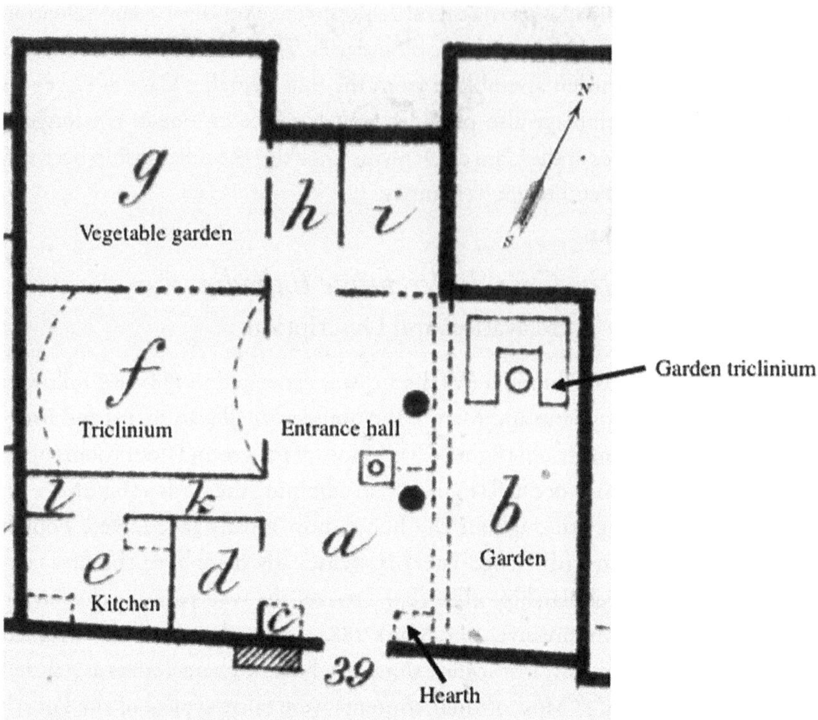

FIGURE 6.7 Plan of the Casa di Acceptus e Euhodia (VIII 5.39).
Source: Adapted from Mau 1884: foldout.

Several human remains were associated with this house. In the corridor (k), excavators found three skeletons with a hoard of valuables: three pairs of gold earrings, a lapis lazuli ring set in gold, a coral phallus amulet, an emerald, twenty-seven silver coins, five bronze coins, a bronze weight, and a bronze cooking pot.[43] No further research appears to have been done on these skeletons, whose current whereabouts are uncertain.[44] Immediately outside the house, two additional skeletons—a woman and child—lay four meters above the street level.[45] As Eschebach points out, this is approximately the same height as the upper story

43. Mau 1884: 129; Eschebach 1984: 48.

44. On the historically poor storage conditions of skeletal remains from Pompeii, and the difficulty in assigning often disarticulated remains to specific individuals today, see Lazer 2009: 99–116.

45. Sogliano 1882: 281; Mau 1884: 127; Dwyer 2010: 100.

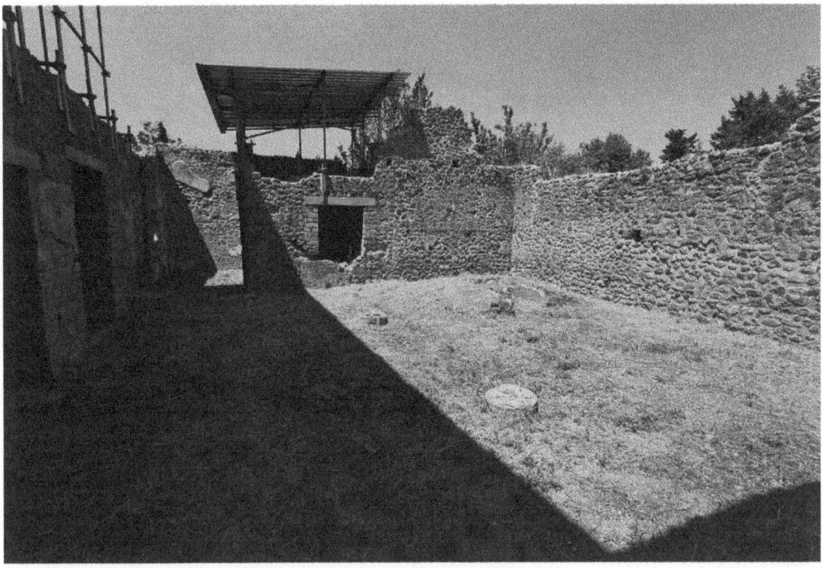

FIGURE 6.8 Garden (b) of the Casa di Acceptus e Euhodia, facing magnetic north. The column bases of the porticoed hall (a) remain visible, although the pluteus wall, water channel, marble pavements, and most of the upper story are no longer preserved today. For the position of spaces (a) and (b) within a plan of the house, see Figure 6.7.
Photo: C.E. Barrett.

of the Casa di Acceptus e Euhodia.[46] The woman was holding the child's legs, perhaps trying to pass him out of the building to safety.[47] She wore two gold bracelets, an emerald ring, and an amethyst ring.[48]

Portico and Garden

The columned entrance hall (figure 6.7, a; 6.8) originally supported a terrace off which opened upper-story rooms.[49] Four terracotta antefixes with female protomes once adorned the edge of the terrace.[50] Separating the portico and garden were two columns and two pilasters, originally connected by a low pluteus

46. Eschebach 1984: 48.

47. Dwyer 2010: 100.

48. For a description of the jewelry and finds, see Mau 1884: 127; Dwyer 2010: 100–103.

49. Mau 1884: 127, 132.

50. Sogliano 1882: 278; Mau 1884: 128.

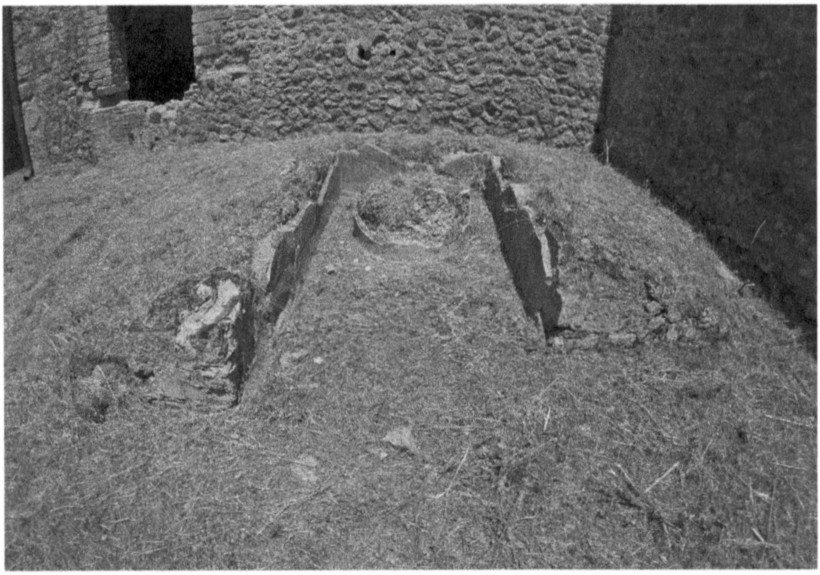

FIGURE 6.9 Masonry triclinium at the north end of garden (b) in the Casa di Acceptus e Euhodia.
Photo: Kaja Tally-Schumacher, reproduced with permission.

wall.[51] A water channel ran along the garden side of this wall. People entered the garden through an opening in the center of the pluteus, where a small bridge of multicolored marble slabs spanned the water channel.[52]

A masonry triclinium stands in the north of the garden (figure 6.9).[53] The upper surface of the benches was originally covered in yellow-painted plaster, while the walls were of red-painted plaster.[54] In 1950, Soprano noted that the red walls were further painted with "pitture oggi quasi totalmente svanite."[55] These paintings, which are not described elsewhere, are now lost. The space between the benches was paved with *cocciopesto*.[56]

51. The pluteus no longer stands today; see Mau 1884: 128; Eschebach 1984: 48.

52. Mau 1884: 128.

53. Dimensions (Jashemski 1979–93: 2:218): *lectus medius* 2.63 m; *lectus imus* 3.45 m; *lectus summus* 3.10 m.

54. Mau 1884: 128; Soprano 1950: 301.

55. Soprano 1950: 301.

56. Bragantini 1998c: 649.

A circular masonry table topped with a colored marble mosaic once stood inside the benches.[57] By 1998, this table had lost its marble covering, and by 2016 the remaining portion was substantially damaged and overgrown with weeds.[58] Traces of yellow-painted plaster were still visible around the base of the table in 2018. Soprano, who observed the table when it was still intact, records that it was equipped as a fountain: "Il foro al centro di esso serviva al passaggio di una fistula per getto d'acqua."[59]

Soprano's assertion has not gone unchallenged. The original reports of Mau and Sogliano do not mention a hole or water pipe in the tabletop, and Richardson is dubious that one existed: "That [proposed pipe] is possible, but it is reported by no one else. The absence of any basin around the foot of the mensa makes it seem unlikely."[60] Jashemski refers to a water pipe in the tabletop, but her use of past tense ("had a hole in the center for a fountain jet") leaves uncertain whether she verified this feature in situ or followed Soprano's account.[61] However, despite Richardson's objection about the absence of a basin, some other table fountains lack attached water basins. A very similar installation appears at the Casa delle Nozze d'Argento, where a lead pipe emerged from a hole in a marble-covered table.[62] The table fountain at the Casa dell'Efebo similarly lacked an associated basin (see chapter 3, "The Garden of the Casa dell'Efebo").

Accordingly, I follow Soprano's reconstruction of a fountain. Soprano provides a careful description of the table as it appeared when still intact. His recording of other details not present in the original excavation reports, such as the painting traces on the benches, attests to close firsthand observation. The triclinium in this garden should thus be understood as a water-triclinium, albeit one less elaborate than that of, for example, the Casa dell'Efebo.

The garden also contains another, much simpler water feature: the gutter along the base of the pluteus wall. As at the Casa del Medico, people could direct the water in this channel either into the street or into a cistern.[63]

---

57. Diameter: 0.59 m (Jashemski 1979–93: 2:218).

58. For the condition of the table in 1998, see Bragantini 1998c: 649, fig. 1.

59. Soprano 1950: 301.

60. Richardson 1988b: 308–309.

61. Jashemski 1979–93: 2:218.

62. As Richardson himself notes, this installation provided "no provision for collecting the water thrown ... onto the mensa; it was apparently simply allowed to run off into the garden" (1988b: 308); see also Allison 2004b, s.v. "Casa delle Nozze d'Argento."

63. Mau 1884: 128. Jashemski (1979–93: 2:218) also mentions the gutter.

The Artifactual Assemblage from the Garden and Portico

Despite the small size of the house, the garden contained a surprisingly large array of statuary and furniture in faience, pseudo-faience (glazed terracotta), marble, and alabaster (figures 6.1–6.3, 6.10–6.13). Within this assemblage, Egyptian allusions appear in multiple media. Both the imagery and the materials of the faience and pseudo-faience statuettes evoke Egyptian origins, and two marble herms wear crowns associated with Egypt. Surviving documentation on this assemblage suggests multiple parallels with other Nilotic installations at Pompeii.

Three figurines depict dwarflike beings, all covered in a greenish glaze resembling Egyptian faience.[64] Two bifrontal figurines are identifiable as the dwarf-god Bes.[65] One depicts a Bes figure with protruding tongue, naked, wearing a platform crown and four yellow-glazed plumes, and squatting with hands on hips atop a statuette base (MANN 116665; figure 6.1). Most of the head of the second figure is lost, but this statuette also depicted a Bes figure, with protruding tongue, naked aside from a leopard skin on his shoulders, squatting with hands on knees atop a low base (MANN 117178; figure 6.2).[66]

A third, headless faience figure with achondroplastic proportions stands with hands on hips atop a low plinth, naked aside from a feline (lion?) skin around his shoulders[67] (MANN 116666; figure 6.3). His phallus is erect and hollow. The body type closely parallels several other faience statuettes from Pompeii (e.g., figure 6.4). When the heads of such statuettes survive, they resemble the example in figure 6.4: round-headed and plump-cheeked, wearing wreaths of

64. Mau refers to five faience figurines (1884: 129; 1908: 361; seemingly followed by Beaurin 2008: 284), but only three are attested in Sogliano's excavation reports or today accounted for. The discrepancy likely arises from the bifrontality of the two Bes figurines, so that each is effectively two figurines in one. Sogliano's report (1882: 360) clearly describes *three* statuettes "con patina verdastra," of which two were "bifronte."

65. Di Gioia (2006: 116) and Mol (2015: 199) assign only one of these (MANN 117178) to the Casa di Acceptus e Euhodia. However, MANN 116665 appears to derive from this house as well (de Vos 1983: 62; Bragantini 2006: 167n1; 2012: 23n13; De Caro 2006b: no. III.127; Poole 2016: no. 117).

66. For the reverse side, see the image in Di Gioia (2006: fig. 244). Except for small traces, the protruding tongue in figure 6.2 is mostly lost to damage, but the tongue is better preserved in Di Gioia's photograph. The skin is not a lion, *pace* Di Gioia (2006: 116); the head is visible between the dangling paws (both on figure 6.2 here and, albeit more indistinct, on Di Gioia's fig. 244). On the leopard skin as an attribute of Bes, see Romano 1989: 1.218.

67. Di Gioia (2006: 112) does not attribute this object to the Casa di Acceptus e Euhodia, but other documentation connects this figurine with this house (de Vos 1983: 62; Bragantini 2006: 167n1; 2012: 23n13; De Caro 2006b: no. III.126; Mol 2015: 184).

flowers or leaves.[68] Scholars often identify MANN 116666 as "Pataikos" or "Ptah-Pataikos,"[69] but as with other figurines of this type, this identification may be misleading. No features explicitly identify him as a deity according to either Egyptian or Roman iconographic conventions, and if a similar figure were to appear in a Nilotic painting, few would hesitate to identify him as a typical *pygmaeus*.

Chemical analysis of two of these figurines suggests that although their glazed surfaces would have looked similar, their materials and origins are different. The pygmylike figure (MANN 166666; figure 6.3) has the siliceous fabric of true faience, but one of the two Bes figurines (MANN 166665; figure 6.1) is glazed terracotta.[70] The second Bes figurine (MANN 117178; figure 6.2) has not seen chemical analysis, but in-person macroscopic examination suggests a faience fabric.[71] Recent research on "true" faience at Pompeii indicates that such objects were Egyptian, perhaps specifically Memphite, imports.[72] The glazed terracotta Bes, in contrast, is "pseudo-faience": a fired clay object whose glaze imitates the gleaming blue-green surface of faience. Chemical analysis of such "pseudo-faiences" at Pompeii suggests local production.[73] Accordingly, two of the glazed figurines are probably Egyptian imports, while the third appears to be a locally produced emulation of such artifacts. When intact, however, all three would have presented a similar shining blue or green surface to viewers.[74]

The largely Egyptian origins and frequently Egyptian iconography of glazed figurines at Pompeii likely encouraged viewers to associate all three objects with "Egypt,"[75] whether or not they recognized a distinction between imports and local production. Ancient consumers might well have valued Egyptian faiences for their distant origins, but they might alternatively have valued them for, among other things, their brilliant coloration and sheen—in which case local and imported versions might be equally desirable. Either way, it was probably

68. Compare Di Gioia 2006: figs. 230, 236, 237.

69. E.g., de Vos 1983: 62; De Caro 2006: no. III.126; Di Gioia 2006: 112; Bragantini 2012: 23n13; Mol 2015: 184.

70. Di Gioia 2006: 137–138.

71. Cf. Di Gioia's description of the fabric as "pasta silicea" (2006: 116).

72. Di Gioia 2006: 133–138; Mangone et al. 2011.

73. Mangone et al. 2011: 2866. (Note that this 2011 study does not include MANN 116665, which appears in the earlier work of Di Gioia 2006: 138.)

74. Some minor color variation is also present. MANN 116665 is largely green but has yellow plumes, and part of the glaze on MANN 116666 has a yellowish appearance, which may either be original or the result of post-depositional processes (Di Gioia 2006: 138).

75. On the likelihood that Pompeians associated blue-green glaze with Egypt, see Mol 2015: 197–210.

(a) (b) (c)

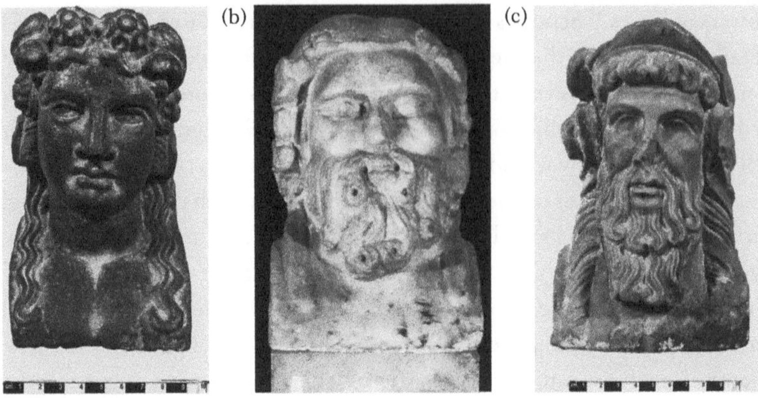

FIGURE 6.10 Three marble herms with proposed "Bacchic" features from the garden of the Casa di Acceptus e Euhodia (note: photos not all to same scale). (a) Herm depicting a youthful Dionysus, crowned with ivy leaves and berries. MANN 120038. (b) Herm with features resembling Silenus. Height: 206 cm (Carrella et al. 2008: 149). MANN 120036. (c) Bearded herm wearing *taenia* and possible *mitra*. MANN 120040.

Sources: (a) Photo by Giorgio Albano; reproduced by permission of the Ministero dei Beni e delle Attività Culturali e del Turismo—Museo Archeologico Nazionale di Napoli. (b) Measurement and photo after Carrella et al. 2008: 149, fig. C 42; image courtesy of "L'Erma" di Bretschneider. (c) Photo by Giorgio Albano; reproduced by permission of the Ministero dei Beni e delle Attività Culturali e del Turismo— Museo Archeologico Nazionale di Napoli.

difficult to tell the difference by sight alone. When their glaze was intact, both the terracotta and faience Bes figurines presented viewers with similar imagery and shining blue-green surfaces. To ancient onlookers, both the imported and the local product may have appeared equally "Egyptian."

Marble garden statuary included five herms (MANN 120036–120040; figures 6.10, 6.11). Although Mau describes all five as "Bacchic,"[76] the reality appears more complex. Three of the herms do appear to have plausible "Bacchic" characteristics, but the other two appear less "Bacchic" than "Egyptianizing."

Sogliano identified three of the herms as "Bacco indiano": one of *rosso antico* marble, wearing a crown of ivy, and two of *marmo grechetto*, one of which Sogliano described as "diademato."[77] Although he does not provide inventory numbers or images of the objects, three herms from this house do have features resembling either Bacchus or members of his entourage. An ivy-crowned herm of red marble (MANN 120038; figure 6.10a) depicts a youthful Dionysus

76. Mau 1884: 129, 1908: 361; followed by Jashemski, who provides the museum inventory numbers for the herms (1979–93: 2.218).

77. Sogliano 1882: 359–360; Carrella et al. 2008: 150, no. C 44.

type.[78] Another herm, representing a broad-faced older man crowned with ivy, flowers, and a *taenia*, should probably be understood as Silenus (MANN 120036; figure 6.10b).[79] A third, bearded herm wears a raised cap or crown that may perhaps recall the Dionysiac *mitra* (MANN 120040; figure 6.10c), although definite identification here remains more tentative.[80] This herm is undoubtedly the one Sogliano describes as "diademato, mancante della punta del naso."[81] On either side of the face, the dangling ends of a *taenia* entwine two long locks of hair that fall over the shoulders.[82]

However, two other herms from the garden evoke associations more Egyptian than Bacchic, wearing crowns associated with pharaohs and Egyptian or Egyptianizing gods. One wears what appears to be a short *modius* crown similar to that worn by the bearded god Sarapis in Greco-Roman art[83] (MANN 120039; figure 6.11a). Finally, the last herm depicts a bearded man wearing a Pharaonic *nemes* crown with a uraeus serpent (MANN 120037; figure 6.11b).[84] Despite these Egyptian allusions, stylistic considerations suggest production outside of Egypt. The mustache and beard of the *nemes*-wearing herm are unusual in Egyptian depictions

78. MANN 120038 might also be taken to resemble a Maenad, but the rendering of the chest suggests male pectoral muscles rather than female breasts. For the type, see, e.g., Berlin Antikensammlung nos. Sk 126, 127, 128, 132.

79. Carrella et al. also recognize the herm as a Silenus type (2008: 149, no. C 42). For parallels, see, e.g., Berlin Antikensammlung no. Sk 112, no. 2872 x.

80. Viewed frontally, the headgear resembles a *stephane*, but parallels come from some vase paintings of Dionysus wearing the *mitra* (e.g., Gasparri 1986a: no. 399). A number of other herms from the first and second centuries CE wear similar headgear (e.g., Berlin Antikensammlung nos. Sk 106, 109, 111; I thank Annetta Alexandridis for these references); compare also a Hellenistic example (Getty Villa no. 79.AB.138). Some Classical sculptures and plastic vases identified as Dionysus wear other, superficially visually similar forms of projecting headgear, such as the *stephane* or *polos* (Gasparri 1986a: nos. 14, 17, 18, 20, 56, 180, 181; see also no. 136 for a Classicizing Roman-era statuette wearing a *polos*).

81. Sogliano 1882: 359.

82. Compare the *taenia* worn by a stylistically archaizing Dionysus/Bacchus herm of the first century CE (Gasparri 1986b: no. 61).

83. Shaped like a grain measure, the *modius* crown is also sometimes worn by other Egyptian deities, including Isis (e.g., Arslan 1997: no. V.68; De Caro 2006b: nos. III.28, III.30, III.31, III.34, III.41, III.43, III.100) and Bes (e.g., De Caro 2006b: III.49). However, the depiction of a *modius* crown on a bearded male deity finds its most frequent parallels in images of Sarapis (see discussion in Barrett 2011: 414, with n1706).

84. Sogliano 1882: 360 (misidentifying the *nemes*-crowned figure as "un pastoforo barbato, col capo coperto da berretto rigato, e col fiore di loto sulla fronte"). Carrella et al. (2008: 149–150) correctly identify the *nemes* crown.

(a) (b)

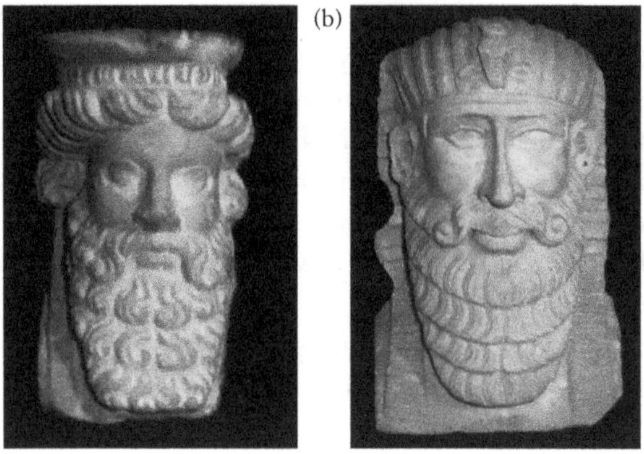

FIGURE 6.11 Two marble herms with "Egyptianizing" features from the garden of the Casa di Acceptus e Euhodia (note: photos not to same scale). (a) Herm wearing *modius* crown. Height: 16.5 cm. MANN 120039. (b) Herm in "Pharaonizing" style, wearing *nemes* crown. Height: 20.6 cm. MANN 120037.

Sources: Measurements and photos after Carrella et al. 2008: 149–150, figs. C 43, C 44; images courtesy of "L'Erma" di Bretschneider.

of *nemes*-wearing figures and more closely resemble Greek archaizing or classicizing herms.[85]

Other statuary depicts water animals. The hollow bodies and mouth holes of two marble statuettes—a frog (MANN 120042; figure 6.12) and a tortoise (MANN 120043; figure 6.13)—identify them as waterspouts for a fountain.[86] While frogs and tortoises are not geographically distinctive, the juxtaposition of river animals with Aegyptiaca (the faiences, pseudo-faience, and "Egyptianizing" herms) recalls Nilotic landscapes.

Because the excavators did not record findspots within the garden, it is unclear where these waterspout sculptures were placed. We might perhaps associate them with the fountain near the triclinium. They might alternatively have belonged to another, now-lost water installation, but extant reports provide no other evidence for such a construction. It is also possible that the frog and tortoise no longer functioned as waterspouts in 79 CE and had been repurposed simply as statuary. However, both do appear to have been functional at some point, as marks around the holes at the lower backs of both sculptures suggest attachment to pipes.

85. Carrella et al. (2008: 149) suggest that this herm could have come from a neo-Attic workshop. They also note stylistic parallels for the *modius*-wearing herm at Pompeii (2008: 150).

86. Sogliano 1882: 360; Appleton 1987: nos. 74, 188 (identifying the tortoise as a turtle); Mau 1884: 129; 1908: 361; Jashemski 1979–93: 2:218; Carrella et al. 2008: nos. C 46, C 47. On Egyptian parallels for frog fountains, see note 132, chapter 1.

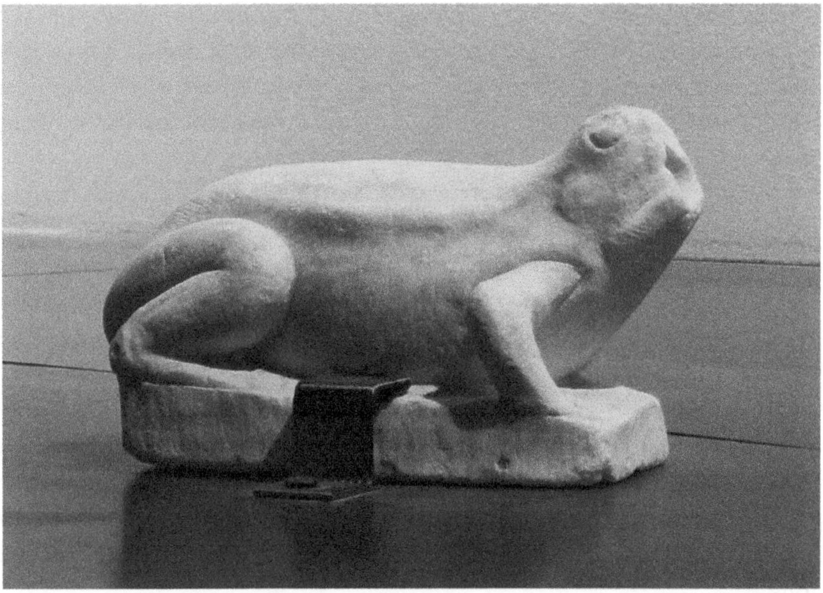

FIGURE 6.12 Marble frog, used as a waterspout, from the garden of the Casa di Acceptus e Euhodia. Height: 14 cm (Carrella et al. 2008: 150). MANN 120042.

Photo: C.E. Barrett. Reproduced by permission of the Ministero dei Beni e delle Attività Culturali e del Turismo—Museo Archeologico Nazionale di Napoli.

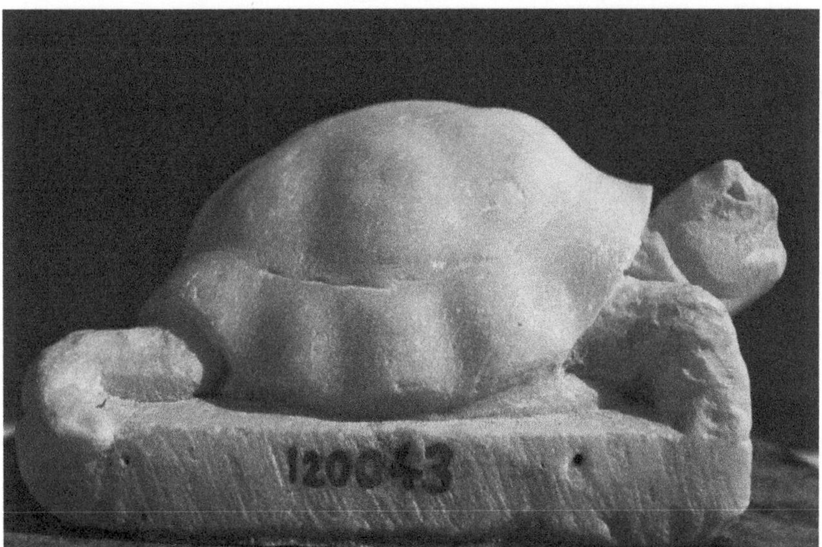

FIGURE 6.13 Marble tortoise, used as a waterspout, from the garden of the Casa di Acceptus e Euhodia. Length: 12 cm (Carrella et al. 2008: 151). MANN 120043.

Photo: C.E. Barrett. Reproduced by permission of the Ministero dei Beni e delle Attività Culturali e del Turismo—Museo Archeologico Nazionale di Napoli.

The final example of marble sculpture from the garden is an *oscillum* in the form of a mask (MANN 120041; figure 6.14). The mask depicts a beardless, agitated-looking, angular-eared male whom Carrella et al. persuasively identify as a young satyr.[87] His presence reinforces the Bacchic themes of the marble herms in figure 6.10.

In addition to these marble objects, two small alabaster bases (both 11.5 mm high)[88] once supported statuettes that are no longer present. Sogliano found a fragment from an alabaster statuette of a female figure, 12.0 cm high and "tutto corroso," in the portico (a).[89] Mau speculates that this statuette might once have stood in the garden,[90] in which case it might have belonged to one of these bases.

The garden also contained two tables. The excavators found two marble monopodia (one-legged table supports): one was fluted and inscribed with CΑΡΑΠΙΩΝΟΣ, "of Sarapion," while the other was shaped like a tree trunk.[91] Also found were two small marble tabletops, presumably mates to the monopodia. Both were rectangular, decorated with leaves, and 18 cm long.[92]

Several finds in the portico may also pertain to garden activities. Adjacent to the pluteus was a small hearth area[93] (figure 6.7). Mau plausibly suggests that people could have heated food here for consumption in the garden.[94] Parallels for such facilities exist in or near many other Pompeian gardens.[95] Across from

87. Carrella et al. 2008: no. C 45. Earlier publications identify the object as a tragic mask: Mau 1884: 129; 1908: 361; Dwyer 1981: 281 no. 106; Jashemski 1979–93: 2:218. Sogliano also describes the mask (1882: 360).

88. Sogliano 1882: 360; Mau 1884: 129; 1908: 361.

89. Sogliano 1882: 360.

90. Mau 1884: 129; 1908: 361.

91. Sogliano 1882: 360; Mau 1884: 129; 1908: 362. Some later publications imply that both monopodia had the same inscription (Appleton 1987: 160; Jashemski 1979–93: 2:218; Carrella et al. 2008: 149), but Sogliano and Mau clearly attribute the inscription to only one of the two.

92. Sogliano 1882: 360.

93. Described as "un piccolo focolare a due fornelli" (Mau 1884: 128). In Pompeian scholarship, references to "fornelli" typically describe open-fronted masonry hearths (Allison 2006: 26).

94. Mau 1884: 128.

95. Allison 2004a: 103, 126.

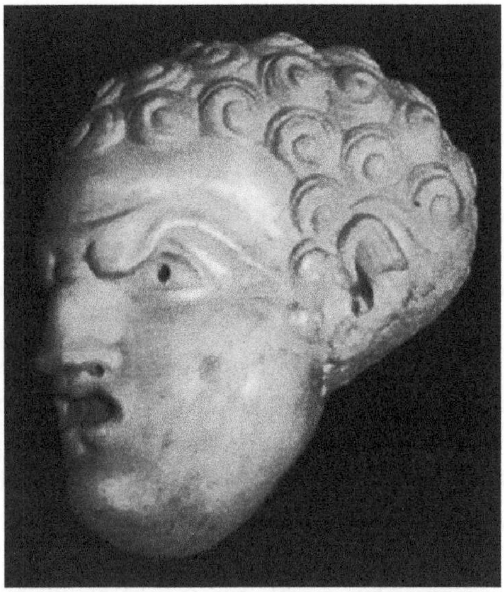

FIGURE 6.14 Marble *oscillum*, probably depicting a young satyr, from the garden of the Casa di Acceptus e Euhodia. Height: 15 cm (Carrella et al. 2008: 150). MANN 120041. *Photo* after Carrella et al. 2008: fig. C 45. Image courtesy of "L'Erma" di Bretschneider.

the stove, a cabinet (figure 6.7, c) contained numerous bronze, ceramic, and glass vessels for food preparation and consumption.[96] This cabinet was roughly equidistant from the kitchen (figure 6.7, e), the indoor triclinium (figure 6.7, f), and the outdoor garden triclinium (Figure 6.7, b), and the vessels may have been used in all three.

Display and Performance in the Garden

The placement, layout, and relatively lavish furnishings of the garden suggest that in this house, garden space took on particular importance for social presentation and performance. The house is quite small and lacks certain standard features for the reception of guests, most notably an atrium. In lieu of owning impressive interior architecture, the inhabitants appear to have concentrated most of their display-oriented efforts in the garden.

96. As well as several other small household objects: Sogliano 1882: 241; Mau 1884: 128.

Underlining those efforts is the location of the garden at the front of the house (figure 6.7, b), directly accessible from the first room guests entered (Figure 6.7, a). Pompeian gardens are often positioned toward the back of a house, a location associated with more restricted access.[97] At the Casa di Acceptus e Euhodia, however, direct access to the garden was not just possible but indeed scarcely avoidable for anyone entering the house. The house interior is also largely oriented toward the garden, with most of the rooms opening onto the portico (figure 6.7, a) that adjoins it. People passing in the street could also view much of the garden through the doorway.

Additionally, the concentration of statuary and marble objects in the garden suggests that this was by far the most lavishly decorated part of the house. The only stone objects found elsewhere in the house were the fragmentary female statuette from the portico and a marble basin in the indoor triclinium.[98] In the garden, however, visitors encountered a profusion of statuary in multiple eye-catching materials: marble, faience and pseudo-faience, and alabaster. Clustered into this small space were five marble herms, two marble tables, two marble waterspouts, a table topped with a multicolored marble mosaic, a bridge of multicolored marble, a marble *oscillum*, and three faience or pseudo-faience statuettes. Additionally, the materials of the two faiences and the pseudo-faience visibly asserted connections with Egypt. By using materials associated with Egyptian production, the house owners could broadcast not only their cosmopolitan tastes but also their ability to obtain imported (or at least, in the case of the glazed terracotta figurine, imported-looking) luxuries.[99]

Although the garden was walled on three sides, the portico on the west side still alluded—in abbreviated form—to the luxury and cultivation associated with colonnaded gardens. The pluteus wall between the portico and the garden would have been low enough to render the portico and garden largely intervisible.

Finally, this house may also assert a physical and functional separation between two types of garden space: one dedicated to social display, and one dedicated to more utilitarian functions. In addition to the garden at the east of the house, there was a second garden in the northwest corner (figure 6.7, g). However, garden (g) lacks most of the luxurious or entertainment-related elements of garden (b). The excavation reports assign no sculpture, dining installations, marble objects, or furniture to garden (g). Garden (g) was also

97. Von Stackelberg 2009: 67–71.

98. On the basin, see Mau 1884: 131; Carrella et al. 2008: no. C 48.

99. On the desirability of Egyptian imports as collectible luxury goods, see Pearson 2015.

invisible from the street and invisible from most positions within the "entertainment areas" of the house, portico (a), and garden (b).[100] Rather than a social center, garden (g) may have been a site of horticultural production. Mau reports that the ground level was upraised all around the sides of this walled garden for a space about 0.55 m wide, leaving only a strip of land in the center of the room at the same floor level as the rest of the house.[101] Unusually for Roman gardens, this central strip of land was paved. Mau and Jashemski interpret the upraised soil as plant beds, and Kastenmeier identifies this space as a vegetable garden.[102] In many Roman gardens, vegetable plots often coexist with entertainment facilities.[103] Here, however, garden (g) lacks the dining furniture and elaborate visual culture of garden (b). If garden (g) were reserved for more utilitarian activities, then this spatial partitioning of the *otium*-related and economically productive functions of Roman gardens would have further reinforced the importance of garden (b) as a site of relaxation, display, and social performance.

Positioning the Casa di Acceptus e Euhodia on the Socioeconomic Spectrum

The mixed evidence for the socioeconomic status of the inhabitants of this house points to the limitations of binary categories like "elite" and "non-elite."[104] Some features of the house suggest a relative lack of resources. The ground plan is quite small, and the layout is seemingly unsuitable for a family of any prominence, as

---

100. This location at the back of the house corresponds to what von Stackelberg (2009: 21) considers the traditional location of gardens in Roman houses before the introduction of the peristyle.

101. Mau 1884: 132.

102. Mau 1884: 132; Jashemski 1979–93: 2:218; Kastenmeier 2007: 160. On raised planting beds in Roman gardens, see further von Stackelberg 2009: 38.

103. See, e.g., chapters 2 and 3, this volume, on the Casa dell'Efebo garden.

104. It should be noted that, *pace* Eschebach (1984: 48–50), I interpret this house solely as a residence and not a medical clinic. The major medical artifact that has been attributed to this house, a vaginal speculum, almost certainly comes from the Casa del Medico (see chapter 4, this volume, "Was the 'Casa del Medico' Really the House of a Doctor?"). Other artifacts that Eschebach interprets as medical, such as glass flasks, a razor, a tweezer, and two strigils, could easily have functioned within daily life (e.g., as personal grooming items). Eschebach also asserts that a relief on a marble basin from this house depicted a young boy holding a vial (1984: 50). The medical implications of such an image would be ambiguous at best, and in any case, the iconography does not support this interpretation. As Carrella et al. state (2008: 151–152), the boy holds not a vial but the hem of his garment. See further discussion of the marble basin later in this chapter, "Visual Culture, Cult, and Identity at the Casa di Acceptus e Euhodia."

it lacks an atrium or tablinum. If the paterfamilias did not require such features, then we might suspect he was not wealthy or influential enough to serve as a patron (cf. Vitr. 5.5.1).

However, the finds appear to nuance this picture of relatively straitened circumstances. In terms of sheer numbers, this house contained more statuary than did the much larger and otherwise more generously appointed Casa dell'Efebo. Certainly, the Casa dell'Efebo also contained one extremely valuable object—the bronze ephebe—whose large scale, precious metal, and fine workmanship are unmatched by any sculpture from this house. Nonetheless, many objects at the Casa di Acceptus e Euhodia also commanded substantial value. The garden assemblage includes multiple objects made of valuable stones, marble and alabaster, as well as two faience objects that are probably long-distance imports. The owners even employed marble in multiple colors for the table fountain, marble bridge, and statuary.[105] Such colored marbles were often imported and desired partly for their association with distant lands.[106] Additionally, the presence of a fountain suggests potentially substantial expenditure on running water.[107] The two marble fountain sculptures further testify to the conspicuous display of water.

Certainly, by the first century CE, marble was no longer an exclusive possession of the wealthiest householders. Marble—even colored marble—is fairly common in the Vesuvian cities. Indeed, the ubiquity of marble-clad bar counters in Pompeii and Herculaneum testifies to the accessibility of this stone even to "relatively humble shopkeepers."[108] Fant et al. suggest that individuals of modest means could have obtained discounted marble from salvage dealers, who recycled older building materials from renovated or demolished houses. At Pompeii, the 62 CE earthquake would have produced a glut of such materials just seventeen years before the eruption.[109] However, even if marble had become more widely accessible than before, many people at Pompeii still seem to have treated the stone as a valuable commodity. As Fant et al. point out, when shopkeepers made use of

105. The statuary included one herm of red marble (figure 6.10a) in addition to stones identified by Sogliano as "marmo grechetto" and "marmo bianco" (1882: 359–360).

106. Barry 2011: 61–71; Fant and Barker 2016; McAlpine 2016: 111–112; Müskens 2017: 47–56.

107. On the probable costs associated with access to the public aqueduct, see Richardson 1988a: 55, 62; von Stackelberg 2009: 40. Even if the aqueduct was nonfunctional in 79 CE (a continuing subject of controversy: see recently Keenan-Jones 2015), the provision for a fountain suggests the house owners expected to be able to pay for such access whenever it did become available.

108. Quoting Packer 1975: 142; see more recently Fant et al. 2013. On the increased availability and affordability of imported marble in the first century CE, see also McAlpine 2016: 112, 116.

109. Fant et al. 2013: 202–205.

marble, they did so sparingly and with an eye toward maximum visibility: "the most lavish decorative schemes and colourful materials were placed on the vertical faces of counters facing directly on to important streets," suggesting that the owners "understood the value of these materials and took care to display them conspicuously."[110] The availability of reused stone may have introduced marble to a wider range of consumers, but within that new group of sub-elite or middling consumers, marble still remained a precious material. It is perhaps within this sphere that we should understand the display strategies of the Casa di Acceptus e Euhodia garden.

Helping to round out this impression of the family's socioeconomic status are the artifacts found with the individuals who died in this house. We do not know whether these people were the house owners, household slaves, friends or neighbors, or some combination thereof. However, the objects they carried provide some indication of the resources circulating within (or at least, if they were friends or neighbors, adjacent to) this *familia*. The three skeletons inside this house, as well as the remains of the woman possibly trying to escape the upper story, wore fine jewelry of precious metals and gemstones. According to a recent study, jewelry appears on only 10.5 percent of skeletons from the Vesuvian area.[111] Although gold jewelry may have been accessible to individuals of "high-to-moderate and moderate socioeconomic backgrounds,"[112] not just the very rich, such adornment indicates at least some degree of available resources. The three individuals inside the house also carried a small collection of coins and additional valuables. Even if this hoard represented the family's entire life savings, it still suggests that this was a family with at least some savings.

Rather than labeling the inhabitants of this house as either "elites" or "non-elites," we might position them somewhere in the middle of the economic spectrum. In many ways, this family lived comfortably. The truly poor likely lived not in houses like this one but in apartment blocks or small residential quarters behind workshops.[113] Some of the architectural idiosyncracies of this house may also result from its location; as part of the so-called *Altstadt*, or "old city," Regio

---

110. Quoting Fant et al. 2013: 206.

111. Ward 2016: 172.

112. Quoting Ward 2016: 173.

113. *On apartments and rental units*, see Bergmann 2012: 237–239. *On lower-class dwellings at Pompeii*, see Packer 1975, esp. 141; Clarke 1991: 79–80. *On the range of housing available in Roman urban contexts*, including non-elite apartments and shop residences, see Ellis 2000: 73–89.

VIII has a somewhat irregular layout that occasionally produced unusual-looking house plots.[114]

Nonetheless, this family did not command the resources of much larger dwellings such as the Casa di Octavius Quartio or the Casa dell'Efebo. While they could afford to furnish their (smaller) garden in relatively lavish style, they did not follow suit elsewhere. Rather than expending funds to create a luxurious atmosphere throughout the dwelling, they concentrated their display efforts in one area: the garden on the eastern side of the house. The garden thus assumes special importance as a site for self-presentation, social competition, and social performance. Within this space, Egyptian imagery played a prominent (but by no means exclusive) role.

"Familiar Alterity": Egyptian and Bacchic Imagery

The basic features of garden (b) closely resemble many of the Nilotic garden assemblages we have considered. As at the Casa di Octavius Quartio, Egyptian and "Egyptianizing" sculptures appear among other objects in a garden setting that boasts dining facilities (the outdoor triclinium and the two marble tables) and water features (the table fountain and the water channel along the pluteus). No painted or mosaic Nilescapes are attested at the Casa di Acceptus e Euhodia. However, many features of such landscapes did appear in three dimensions. The Egyptian faiences and "Egyptianizing" pseudo-faience and marble sculptures prompted viewers to imagine themselves inside an Egyptian landscape, within which the water features became their own private "Nile." In proximity to these water features, the achondroplastic faience and pseudo-faience statuettes would have recalled the *pygmaei* of Nilescapes. Visitors to the Casa di Acceptus e Euhodia had no crocodiles or hippopotami to fight, but river creatures did appear in more harmless form: the marble frog and tortoise. Finally, the garden triclinium invites guests to engage in the outdoor drinking and banqueting that so often characterize Nilotic scenes, completing the imagined landscape through their own social performance.

As in the other houses we have examined, the Egyptian themes at the Casa di Acceptus e Euhodia shared space with more traditionally "Greco-Roman" imagery. Particularly notable in this garden are Bacchic motifs: one herm representing Silenus, another wearing an ivy crown, a third wearing headgear paralleled in images of Bacchus, and an *oscillum* depicting a satyr. On one level, this pairing of Egyptian and Bacchic imagery recalls the imagined relationship

---

114. On the intense debates over the development of the so-called *Altstadt*, see now Van Roggen 2017.

of Egyptian deities to the Bacchic *thiasos*. Many Greeks and Romans identified Osiris with Dionysus/Bacchus, and a statue of Bacchus appears in the Iseum at Pompeii.[115] On another level, however, the juxtaposition of Egyptian and Bacchic images suggests the perceived appropriateness of both themes to garden contexts.

As discussed in chapter 3,[116] both Egyptian and Bacchic imagery evoked liminal spaces at the edge of (idealized Roman conceptions of) civilized Roman society. Wyler has shown that for Roman audiences, Bacchic imagery could produce sensations of what she calls "familiar alterity"[117]; people might imagine the Bacchic realm as uncanny, uncontrollable, and potentially dangerous, but representations of this realm were still familiar features of domestic life. I have argued in the preceding chapters that in Flavian Pompeii, images of Egypt could carry a similar cultural and emotional charge. Whether depicted on decorated surfaces or in statuary, Nilotic landscapes turned domestic gardens into spaces both faraway and familiar, uncanny and homely.

Style, Function, and Value: "Pharaonizing" in the Garden

The garden assemblage from the Casa di Acceptus e Euhodia thus has much in common with many other assemblages considered in this book. However, this garden also presents viewers with at least two figures unattested in painted or mosaic Nilescapes: Bes and a bearded figure wearing a *nemes* crown.[118] Similar representations appear in other domestic statuary asssemblages. As we have seen, faiences of Bes and a *nemes*-crowned pharaoh were found in the garden of the Casa di Octavius Quartio. A marble bearded herm with a *nemes* crown comes from the garden of House II 1.12, a faience statuette of Bes may come from the Casa di Marcus Lucretius, and two beardless *nemes*-crowned herms come from another private house.[119] However, neither Bes nor *nemes*-wearing figures are

115. On the identification of Osiris and Dionysus, see Hdt. 2.42, 2.144; Barrett 2011: 382–383. Carrella et al. (2008: 149) and Trentin (2014: 245) relate the "Isiac" and Bacchic imagery at the Casa di Acceptus e Euhodia to the syncretism of Dionysus and Osiris. See also note 114, chapter 3, this volume.

116. See chapter 3, this volume, "Safety and Danger in the Domestic *Locus Amoenus*," and "Landscapes, Identities and Viewer Responses."

117. On Dionysus as "familiar alterity," see Wyler 2006: 226.

118. The image of Sarapis is also rare in Nilotic scenes, but not unattested: see, e.g., the fresco from the Casa delle Amazzoni (Versluys 2002: no. 43).

119. On the herms (MANN 10108, 10109, and 10473), see de Vos 1983: 60 (proveniencing MANN 10108 and 10109 to II 11.13) and De Caro 2006b: nos. III.119–121 (proveniencing

present in two-dimensional Nilotic scenes from Pompeii. Bes does not appear in Pompeian paintings outside of the Iseum, and while *nemes*-crowned figures do appear in some Pompeian frescoes, they are entirely unattested in painted Nilotic landscapes. Indeed, in the rare cases when Pompeian frescoes do depict *nemes*-crowned, fully anthropomorphic figures, they appear not as "real" people but as *objects*: statues or paintings on *pinakes*.[120]

It would be tempting to conclude from these iconographic discrepancies that two-dimensional and three-dimensional Aegyptiaca have different functions. Indeed, many scholars draw category distinctions between (two-dimensional) "Nilotica" and other "Aegyptiaca."[121] These *formal* and *iconographic* categories are often further treated (implicitly or explicitly) as *functional* categories, with Nilotic frescoes and mosaics imagined to function as "decoration" and statuettes of gods and pharaohs seen as more likely to function in "religion."[122]

However, many of the differences between garden statuary and Nilotic paintings pertain less to *religious content* than to *style*. Nilotic frescoes frequently contain as many, if not more, representations of divine images than do "Egyptianizing" garden statuary assemblages. For example, the Casa dell'Efebo frescoes depict statues of several Egyptian deities, including Isis (figure 1.3, A), Apis (figure 1.4, K), and another theriomorphic deity of uncertain identity

MANN 10108 and 10109 to II 1.13); both sources provenience MANN 10473 to II 1.12. Carrella et al. (2008: 149–150) note another, unpublished example in the Naples Museum. On the association of the faience statuette with the Casa di Marcus Lucretius, see Kekulé 1880: 61; Dwyer 1982: 31; Di Gioia 2006: no. 5.6 (followed by Mol 2015: 479). The function of II 1.12 (the so-called Complesso dei Riti Magici) in its last phase remains disputed. Some see it as a private sanctuary because some garden artifacts evoke the cult of Sabazius (e.g., Pace 1997; Charles-Laforge 2006), but a garden shrine is also compatible with a private house.

120. On representations of statues, see Pearson 2015: 65. See, e.g., the *nemes*-crowned anthropomorphic statues and figures on *pinakes* in the garden paintings at the Casa del Frutteto and Casa del Bracciale d'Oro (De Caro 2006b: 164–165, 188–191; Mol 2015: figs. 4.21–4.22, 4.24, 4.26; Pearson 2015: 60–68); the Casa del Bracciale d'Oro paintings also depict *nemes*-crowned sphinx statues. The small *nemes*-crowned figures in the Villa dei Misteri tablinum (Versluys 2002: no. 67, verified through personal observation; see further appendix B, this volume) may also represent images on *pinakes*, since they appear inside small squares. At other sites, *nemes*-crowned figures also appear on representations of textiles, as at the fourth century CE Basilica of Junius Bassus (Pearson 2015: 46).

121. See, e.g., Malaise's construction of three separate categories: Aegyptiaca, Pharaonica, and Nilotica (2005: 201–214; and see further note 98, chapter 1, this volume). On the influence of Malaise's approach, see note 99, chapter 1, this volume. Versluys's study of Nilotica acknowledges three-dimensional parallels but focuses in practice largely on two-dimensional material (see note 1, this chapter).

122. On late twentieth-century interpretive tendencies to divorce "Nilotica" from "religious" Aegyptiaca, see chapter 1, this volume, "Material Culture, Identity, and 'Authenticity'"; chapter 2, "Navigating 'Nilotic Scenes.'"

(figure 1.4, J), as well as numerous temples, shrines, altars, and aediculae.[123] Even the Casa del Medico frescoes, which otherwise downplay pious themes, include an image of a sanctuary (figure 2.4, A). Additionally, Egyptian and "Egyptianizing" garden statuary does not by any means exclusively depict deities. Other popular subjects include river animals, *nemes*-crowned figures identifiable in Egypt as pharaohs (who received worship in Egyptian tradition, but not in Roman Italy), and human dwarfs or pygmies (who, as we have seen, are the probable intended referents of so-called Pataikos statuettes at Pompeii).[124]

I argue that in garden contexts, the biggest difference between two-dimensional Nilotic scenes and "Egyptianizing" sculptural assemblages is not their relative amount of "religious" content, but their *use or avoidance of "Pharaonizing" stylistic conventions*. These stylistic choices then affected the iconographic choices available. Some motifs (e.g., representations of Bes or of pharaoh-like figures) had more limited precedents outside Egyptian visual traditions and hence became largely limited to "Pharaonizing" art, while other motifs (e.g., representations of pygmies, or of Isis) had well-established antecedents in earlier Hellenistic, as well as Egyptian, visual culture and could thus also become common in non-"Pharaonizing" imagery.

Approaching "Pharaonizing": An Example from the Casa del Bracciale d'Oro

To understand why Bes and *nemes*-crowned figures appear in garden statuary but not in garden paintings, we need to ask why Pompeian consumers might consider "Pharaonizing" style appropriate for statuary but not for Nilotic surface decoration. The results may shed light not only on Roman receptions of Egyptian objects and images but also on the construction of *decor* in Roman visual culture. Hölscher's influential study of Roman art as a "semantic system" investigates Roman associations of meaning with specific temporal styles (styles associated with earlier time periods), but he focuses on styles derived from various periods of *Greek* art.[125] An analysis of Roman "Pharaonizing" provides a broader view of Roman retrospective art, revealing allusions to distinct *Egyptian* as well as Greek temporal styles in Roman visual culture.

---

123. See chapter 3, this volume, "The Nilescapes of the Casa dell'Efebo Garden Triclinium."

124. See appendix C, this volume. On Pompeian receptions of *nemes*-crowned figures, see later this chapter, "Imagining Pharaoh? *Nemes* Crowns and Familiar Foreignness."

125. Hölscher 2004; see also note 24, chapter 1, this volume.

As Pearson has recently demonstrated, most "Pharaonizing" Roman art takes its models not from Egyptian wall painting but from imported luxury goods.[126] I therefore take Pearson's conclusion as a point of departure, investigating the implications of Italian–Egyptian trade for the development of multiple stylistic and formal conventions. Egyptian statuettes in traditional styles were available to Pompeian consumers for purchase—and to Italian craftsmen for inspiration—partly because of the fact that their small size afforded easy transport. Because Italian craftsmen had ready access to imported small-scale Egyptian statuary, they were then able to produce similar objects emulating Egyptian styles and forms. In the absence of similar access to traditional Egyptian examples of landscape painting, however, the producers of Nilotic frescoes employed Roman rather than Egyptian styles (see further, this chapter, "Style, Medium, and *Decor*"). Two-dimensional and three-dimensional Roman "Nilotica" thus drew on very different stylistic and formal models.

By "Pharaonizing," I refer to the purposeful and emically recognizable employment of stylistic attributes that emulate conventions, proportions, postures, and motifs originating in Pharaonic-period Egyptian art.[127] In archaeological literature on technology, "style" can represent the totality of choices, both conscious and unconscious, that producers make throughout the manufacturing process.[128] Sackett refers to the results of producers' conscious choices as "active style" and the results of their unconscious choices as "passive style."[129] Active style is often a poor indicator of cultural or ethnic origin, as people can purposefully emulate it.[130] Such appears to be the case for artifacts produced in "Egyptian" or "Pharaonic" style in Italy, as there is seldom any evidence that their craftsmen had Egyptian cultural or ethnic backgrounds.[131] "Pharaonizing" style in Italy thus employs active, purposeful emulation.

In painting, an example of such emulation appears at the Casa del Bracciale d'Oro[132] (figure 6.15). In the midst of a lush painted garden sits a seated male

126. Pearson 2015, esp. 46–47. As she points out, "numerous subjects that are ubiquitous on Egyptian walls *never* appear on Roman walls" (2015: 46).

127. Recent discussions include Mol 2015: 224–254; 2017: 181–183; Pearson 2015; Swetnam-Burland 2015: 53–63.

128. Dietler and Herbich 1998: 235–244.

129. Sackett 1990.

130. Dietler and Herbich 1989: 156–157.

131. Davies 2011: 359; Swetnam-Burland 2007: 113–116; 2015: 53–63.

132. On the Egyptian imagery in this house, see most recently Pearson 2015: 60–68; Poole 2016: nos. 120–124. On the house, see Sampaolo 1996a.

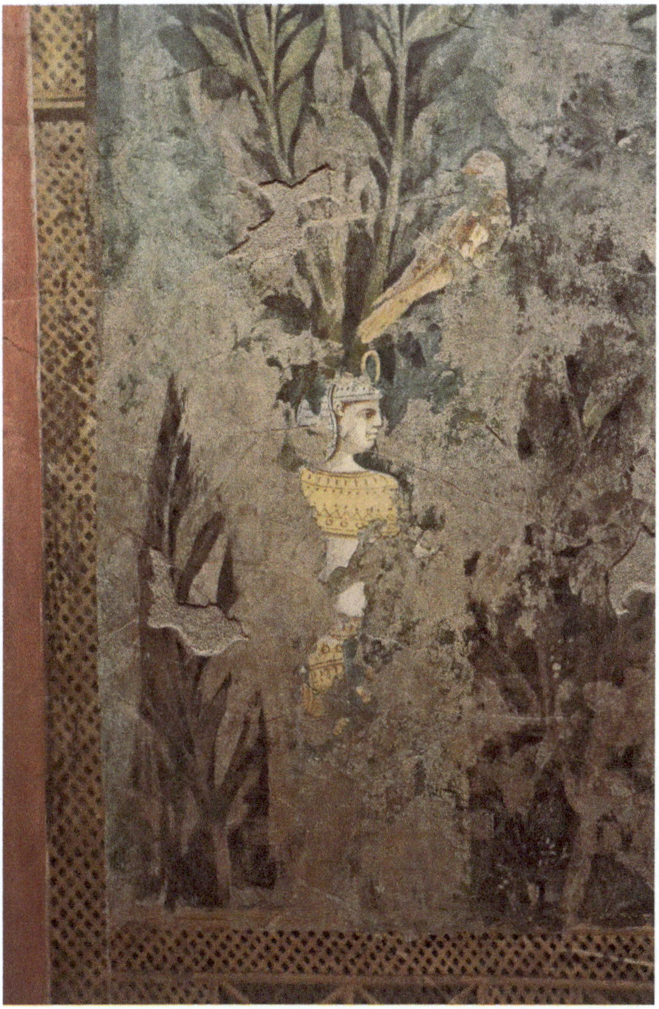

FIGURE 6.15 Detail of a painting from Triclinium 31 at the Casa del Bracciale d'Oro (VI 17 [Ins.Occ.] 42) depicting a garden scene (MANN 59467 d). A statue in "Pharaonizing" style wears an Egyptian bag wig, a kilt, and a capelet-like garment that appears to be a reinterpretation of an Egyptian *wesekh* collar. The triclinium opened onto the garden (see figure 2.39) and also contained a fountain with a simple Nilotic mosaic (Versluys 2002: no. 48).

Photo: C.E. Barrett. Reproduced by permission of the Ministero dei Beni e delle Attività Culturali e del Turismo—Museo Archeologico Nazionale di Napoli.

figure wearing Egyptian costume and depicted according to Egyptian artistic conventions. This figure is depicted not as a living person, but as a statue on a plinth.[133] As in Pharaonic art, his face and legs are in profile, but his torso faces the viewer. On his head is a coiffure whose rounded form resembles the "bag wig" sometimes worn by kings and elites in Egyptian statuary.[134] Here, however, the artist appears to have reinterpreted the Egyptian wig as a type of rigid crown or helmet. The artist must have seen statues or statuettes (or depictions of statues or statuettes) wearing the bag wig, but he appears unaware of the material of this headgear. A yellow band, decorated with geometric shapes, envelops the collarbone and shoulders. The Egyptian prototype here is the broad necklace known as the *wesekh* collar, but the artist appears to have reinterpreted this object as well, rendering it as a shawl or capelet. Further geometric shapes adorn the kilt.

The stylistic and iconographic models for this painting of an Egyptian-style statue likely come from both statuary and two-dimensional images on small imported goods. The form, pose, and costume have close parallels in statues and statuettes from Ptolemaic and Roman Egypt, while the mixture of frontal and profile portrayal imitates Egyptian conventions for two-dimensional representation. Italians could have encountered such two-dimensional portrayals on portable consumer goods from Egypt, such as textiles, inlaid vessels, and plaques of glass, ivory, and stone.[135] The painter at the Casa del Bracciale d'Oro probably did not have personal experience with traditional Egyptian costumes; his reinterpretation of the bag wig and *wesekh* collar suggests that he had never seen anyone actually wearing either item. However, he was clearly familiar with both the general appearance of Egyptian seated statues and some basic conventions of Egyptian two-dimensional art.

A close parallel to this painting comes from the Casa del Frutteto, where wall paintings similarly depicted Egyptian-style statues in a lush garden setting.[136] These "Pharaonizing" pictorial representations of statuary point to the high value

133. See note 120, this chapter.

134. The rounded form resembles the bag wig more closely than the "blue crown," with which the headgear has previously been identified (e.g., in Pearson 2015: 61). The blue color and linear striations parallel some depictions of bag wigs from Egypt (cf. Corcoran 1995: 146–147, 149, figs. 12, 14).

135. On Roman art that probably imitates figural imagery from Egyptian textiles, see Pearson 2015: 43–46, esp. 46. Inlaid obsidian vessels with "Pharaonic"-style figural decoration from Stabiae are either Egyptian imports or emulations of Egyptian products (De Caro 2006b: no. III.136; Pearson 2015: 41–42). On Egyptian glass beads, glass tiles, and ivory and stone plaques in Italy, see Pearson 2015: 40–43. On Egyptian imports in Italy, see note 55, chapter 1, this volume.

136. See note 120, this chapter.

that Pompeians attributed to actual statuary in Pharaonic style. The depiction of such statues in garden settings further testifies to the perceived appropriateness of Egyptian or "Egyptianizing" statuary within gardens.[137]

Statuary in Pharaonic Egyptian style appears in multiple contexts in Roman Italy, from sanctuaries (including the Pompeii Iseum) to private houses.[138] Some such objects were imports from Egypt, others were locally made, and others were imported but heavily reworked in Italy.[139] The most monumental imports, such as obelisks, required huge resources and imperial patronage to move.[140] However, small statuettes and figurines were easily portable. Small-scale Egyptian statuary was exported widely in the Hellenistic and early imperial periods,[141] as the imported faiences at Pompeii attest. Egyptian statuettes, and subsequent Greek and Roman emulations thereof, were a familiar part of the visual landscape of first century CE Pompeii. The employment of Egyptian-looking statuary in the garden of the Casa di Acceptus e Euhodia thus draws on a broader cultural context in which people valued such objects highly as consumer goods and considered them appropriate for display in gardens.

"Pharaonizing" and Beyond: Eclecticism at the Casa di Acceptus e Euhodia

As with the painted statue at the Casa del Bracciale d'Oro, some garden statuettes from the Casa di Acceptus e Euhodia make visible allusions to Pharaonic styles. These features are most visible on the *nemes*-crowned herm (figure 6.11b) and the statuettes of Bes (figures 6.1, 6.2). Both the *nemes* crown and the depictions of Bes have clear stylistic, formal, and iconographic prototypes in Egyptian traditions long predating the Ptolemaic period. However, other elements of the statuary assemblage cite additional temporal and cultural styles, including Greek styles of the Archaic and Hellenistic periods. Some objects, such as the *nemes*-crowned herm and the faience pygmy, are innovative combinations of multiple stylistic traditions. Rather than an exclusively "Pharaonizing" or "Egyptianizing"

137. Cf. Pearson's statement that the Casa del Bracciale d'Oro "shows that Egyptian subjects of *all* sorts, not only Nilotic scenes, were thought to be well suited to gardens" (2015: 60; emphasis in original).

138. See numerous examples in Arslan 1997; De Caro 2006b; Poole 2016.

139. See note 9, chapter 1.

140. On Roman appropriations of obelisks, see note 141, chapter 1.

141. See chapter 2, this volume, "Encounter, Communication, and Consumption."

assemblage, the garden presented visitors with an eclectic synthesis of eastern Mediterranean visual cultures.

The *nemes* crown is one of the oldest features of Egyptian royal iconography, traceable to the 1st Dynasty,[142] and this *nemes*-crowned herm (Fig. 92b) also emulates certain stylistic features of Pharaonic art. The schematized, nearly linear locks of the hair and beard have a formal, "hieratic" quality very different from the loose curls of the *modius*-crowned herm (figure 6.11a). The almond-shaped eyes, outlined in relief, have close parallels in Late Period and Ptolemaic Egyptian sculpture.[143] Similar parallels exist for the brow ridge: nearly flush with the forehead and nasal bridge, but sharply angled beneath the eyebrows.[144] Also "Pharaonizing" is the positioning of the ears, sticking straight out from the head.

However, the formal, material, and stylistic references of this sculpture are not solely Egyptian. The herm form alludes to Greek sculptural traditions.[145] The *nemes* lappets are also unusually thin and linear, recalling the locks of hair or ribbons that often drape over the shoulders of Greek and Roman herms. The use of marble further evokes Greco-Roman precedents; marble is rare in Egypt, and Egyptian craftsmen traditionally worked other stones.[146] Additionally, the flowing beard deviates from Egyptian conventions. In Egyptian art, *nemes*-wearing pharaohs often wear a small false beard, but not a natural beard. The long, thick beard on this herm references Archaic Greek and archaizing bearded herms. In view of such references, the stiff and "hieratic" appearance of this herm takes on a potential double meaning. Besides recalling Egyptian style, these features could also remind viewers of Archaic Greek statuary.[147]

The mixture of Pharaonic and other stylistic allusions on the *nemes*-crowned herm contrasts with the *modius*-crowned herm, which also wears headgear alluding to "Egypt" but employs a very different stylistic vocabulary. This herm's hair and curling beard are comparatively sinuous and flowing, and the facial features are more rounded, soft, and naturalistic. The choice of crown also has different cultural and temporal associations. The *modius* crown recalls the god

142. Patch 2011: no. 117.

143. Cf. Carrella et al. 2008: 149. On late Egyptian sculpture, see Bothmer 1960; Barrett 2015b, s.v. "Stone Statuary." For parallels, see, e.g., Bothmer 1960: nos. 99, 109, 114.

144. Compare multiple examples in Bothmer 1960; see also Barrett 2015b, s.v. "Stone Statuary."

145. Cf. Carrella et al. 2008: 149. On herms, see chapter 3, this volume, "Interactive Statuary: Social Performance and Service," with notes 125–126.

146. On stones and techniques employed in Egyptian statuary, see Peck 1999; Aston et al. 2000.

147. For the combination of archaizing and "Pharaonizing" features in a single work, cf. the statue of Isis from the Pompeii Iseum (Swetnam-Burland 2015: 58–60).

Sarapis, whose iconography derives from the early Ptolemaic period rather than Pharaonic antiquity. Sarapis also had strong associations with Alexandria, that most "Greek" of Ptolemaic cities.[148]

In both iconography and style, the *nemes*-crowned herm would have recalled deep antiquity for Roman viewers: both the ancient past of Pharaonic Egypt, whose culture Herodotus (2.142) claimed to be over 11,000 years old, and the ancient Greek past. To Roman audiences, the styles that we call "late Archaic" and "Severe Style" evoked venerability, sacredness, and timelessness.[149] In contrast, the *modius*-crowned herm alludes to a much more recent past. Both the iconographic reference to Sarapis and the soft, rounded features recall Hellenistic prototypes. Juxtaposed, the two herms evoke temporally and geographically diverse traditions and images. The garden becomes not only a microcosm of Rome's eastern empire but also an embodiment of the vast time depth of the (Greek and Egyptian) East, whose antiquity suggests *mores maiorum* not only Roman but also pan-Mediterranean.

Like the *nemes*-crowned herm, the two Bes statuettes also visibly refer to Egyptian antecedents. Their green-glazed surface recalls Egyptian manufacturing techniques. Both statuettes squat with hands on hips, a characteristic pose of Bes and related dwarf-gods since the Middle Kingdom.[150] The leopard skin on MANN 117178 (figure 6.2) and the feathered crown on MANN 116665 (figure 6.1) are also traditional attributes.[151] Additionally, MANN 116665 has a broad face, dramatically arched eyelids, snub nose, protruding tongue, lined skin, and a mane of hair or fur hanging from his cheekbones. These features derive from ancient representations of Bes and Bes-like figures as semi-leonine. In the 18th Dynasty, Bes lost some animalistic features and became an achondroplastic humanoid,[152] but his face retained certain leonine characteristics.

Since one of these statuettes is a probable Egyptian import, it might seem redundant to describe it as looking "Egyptian." By 79 CE, however, Egypt had been under Greco-Roman rule for over four hundred years. Ptolemaic Egypt was deeply integrated into eastern Mediterranean trade networks, and Egyptian workshops often produced objects whose form and style drew on "Hellenistic"

148. On Sarapis's origins, see Huss 1994: 58–68; Alvar 2008: 52–62.

149. Hölscher 2004: 65; Hallett 2012.

150. On the iconography of Bes and other Egyptian dwarf-gods, see note 85, chapter 2, this volume.

151. The Bes image acquired a plumed headdress by the 18th Dynasty and a leopard skin by the late Third Intermediate Period (Romano 1979: 1.215–216).

152. Romano 1989: 1.216.

conventions.[153] However, the Bes statuettes from the Casa di Acceptus e Euhodia draw chiefly on indigenous Egyptian iconographic traditions. Furthermore, the locally produced glazed terracotta statuette employs the same "Pharaonizing" iconography as the actual import.

Yet despite the "Egyptian" associations of faience and green-blue glaze, the third glazed object from this house—the faience pygmy (MANN 116666, figure 6.3), another probable import from Roman Egypt—displays equally close connections to Greek and Roman iconographies and styles. Precedents for this object come from both Egyptian and Greco-Roman images of pygmies and dwarves.[154] Although the faience pygmy does not share the bifrontality of the Bes statuettes, he resembles them in height, material, hands-on-hips posture, and achondroplastic proportions. The animal skin around his shoulders rhymes visually with the leopard skin on the shoulders of MANN 117178. These visual echoes would have encouraged viewers to perceive relationships between the three figures. However, the naturalistic rendering of the pygmy's musculature draws on Greek mimetic art. Additionally, the animal-skin cape recalls not only Egyptian Bes images but also Greek and Roman representations of Herakles/Hercules. The appearance of the great hero's attribute on a pygmy may have struck Roman viewers as comical, perhaps reminding them of the myth in which Hercules fought off a group of attacking pygmies by wrapping them in his lion skin.[155]

Furthermore, the pygmy's phallus is erect and hollow, and if a pipe were inserted into the figurine's open base, it could function as a fountain. If this figure were originally attached to one of the fountains in the garden, the pygmy would appear to be urinating or ejaculating. Similar activities are well attested for pygmies in Nilotic frescoes.[156]

Although scholars have previously identified this figurine as "Ptah-Pataikos," I see no evidence that either the producers or the consumers of the object intended this identification.[157] The existence of some Egyptian representations of

153. On Hellenistic visual and material *koine*, see chapter 5, this volume, "From Hellenistic *Koine* to Roman *Imperium*." On material culture in Hellenistic and Roman Egypt, see Barrett 2015, s.v. "Archaeology and Material Culture."

154. See chapter 2, this volume, "Adaptation, Emulation, and Eclecticism in Nilotic Imagery."

155. Philostr. *Imag.* 2.22; Amm. Marc. 22.12.4.

156. On defecating and urinating pygmies, see Clarke 2007b: 91–96; 2007c; Jansen 2011: 166. I suspect some Nilotic figures may be ejaculating rather than urinating (see, e.g., the emissions from the dancing figures and the penis-boat at the Casa dello Scultore: Versluys 2002: no. 60, fig. 83; Sampaolo 1998b), but these activities are difficult to differentiate iconographically.

157. For previous identifications, see note 69, this chapter. On the figure's hollow penis, note that ejaculation was in fact an appropriate activity for Egyptian creator gods (Barrett 2015a; see

Ptah as a dwarf does not make every Egyptian dwarf into Ptah.[158] In the absence of clear divine attributes, I identify this figurine more generically as a dwarf or pygmy. In both Egypt and Italy, such images possessed multiple affordances.[159] Some viewers might indeed have thought of Egyptian dwarf-deities, especially given the proximity of two images of Bes.[160] Within the context of an Italian garden, however, the most relevant iconographic parallel for many viewers would have been the *pygmaei* of Nilotic scenes.

Multiple other objects from the garden—the other herms, animal statuettes, satyr mask, and monopodium tables—derive from Greco-Roman formal and stylistic traditions. The eclecticism of this assemblage and its contents demonstrates the difficulty of drawing clear divisions between categories of Roman retrospective art: "Egyptianizing," "Classicizing," "Archaizing," and so on. Objects like the *nemes*-crowned herm and faience pygmy work simultaneously within multiple visual traditions. Even as this assemblage embodies Roman appropriation of earlier forms and styles, it also demonstrates the creative potential of emulative art. At the level of production, this creativity generated new forms like the *nemes*-crowned herm, which draw on antecedents from multiple cultures and time periods. Additionally, the Casa di Acceptus e Euhodia shows that consumption could also produce innovation: consumers created new meanings by compiling existing objects into new assemblages.

Style, Medium, and *Decor*

Contrary to older assessments of Nilotic scenes as devoid of Egyptian precedents, much of the iconographic *content* of Nilotic mosaics and paintings has antecedents in earlier Egyptian visual culture.[161] However, two-dimensional Nilotic scenes seldom employ "Pharaonizing" *styles*. In contrast, portable goods such as statuettes frequently do employ such styles. Furthermore, as the Casa del Bracciale d'Oro demonstrates, painted depictions of three-dimensional

---

also note 2, chapter 2, this volume). However, Memphite traditions typically presented Ptah as creating the universe through speech, not masturbation (as opposed to, e.g., the Heliopolitan Atum). On Egyptian creation accounts, see Allen 1988.

158. See chapter 2, this volume, "Egyptian Antecedents and Affordances," and this chapter, "Model Landscapes: Water Features, Statuary, and 'Egyptian' Installations."

159. See chapter 2, this volume, "Adaptation, Emulation, and Eclecticism in Nilotic Imagery."

160. On Bes's own complex relationship to divinity at Pompeii, see this chapter, "Dwarf Deities, Pygmies, and Sileni."

161. See chapter 2, this volume, "Egyptian Antecedents and Affordances."

Aegyptiaca (e.g., statues or textiles) often portray those objects according to "Pharaonizing" conventions.[162] According to Pompeian constructions of *decor*, then, "Pharaonizing" style was appropriate for Egyptian statuary or depictions of statuary, but not for two-dimensional Nilescapes. Accordingly, motifs closely associated with "Pharaonizing" styles—e.g., the *nemes* crown or the Bes image—appear in garden statuary assemblages but not in Nilotic paintings or mosaics.[163]

This difference in stylistic conventions likely stems partly from considerations of portability. Because statuettes were relatively small and portable, they were better fitted than wall paintings to long-distance trade. Private consumers in Italy could and did acquire "real" Egyptian sculpture. Imported Egyptian statuary was also displayed in public locations, including sanctuaries and imperial dedications.[164] As a result, many first-century Italians likely had some basic familiarity with the appearance of Egyptian sculpture.[165] Consumers could recognize some key stylistic features of Egyptian statuary, and Italian sculptors could emulate those features. "Pharaonizing" styles thus appear in garden statuary partly because many of those statuettes were themselves Egyptian imports (e.g., figures 6.2, 6.3) or inspired by imported objects (e.g., figure 6.1).

What forms of value did Pompeian consumers attach to objects in Egyptian *style*, whether imported or locally made? Long-distance imports might be valued for their faraway origins or the labor involved in their transport. A large proportion of what Dwyer calls "collectible singletons" at Pompeii—high-quality statues displayed as showpieces—were probably produced outside the city,[166] suggesting that consumers might seek out imported luxuries. Pearson's research on depictions of Egyptian luxury goods in Roman paintings indicates that Egyptian imports commanded particular value.[167] If imported statuettes signaled wealth and status, then we might predict that locally produced "Pharaonizing" statuettes enabled consumers to project wealth beyond their means, rather like

162. Pearson 2015.

163. The painting of Bes at the Iseum (De Caro 1992: no. 1.72) does not include "Nilotic" features. In later Roman art, the *nemes* crown appears on statues of Antinous with "Pharaonizing" pose and costume but in otherwise Classicizing style: see Roullet 1972: figs. 116–120. However, pre-Hadrianic art typically associates the *nemes* crown with "Pharaonizing" style (see earlier this chapter, "'Pharaonizing' and Beyond: Eclecticism at the Casa di Acceptus e Euhodia").

164. See note 17, chapter 1, this volume.

165. On the visual distinctiveness and recognizability of Egyptian statuary, compare Mazurek's discussion (2016: 54–58) of imported Egyptian statues as a visible means of legitimization for Hellenistic Isiac sanctuaries.

166. Dwyer 2012: 314.

167. Pearson 2015.

ancient versions of knockoff designer handbags. However, this hypothesis fails to address the appeal of objects visibly combining Egyptian and other stylistic features. If someone wanted a convincing "fake" Egyptian statuette, the *nemes*-crowned herm from the Casa di Acceptus e Euhodia is an unlikely candidate. Instead of an attempt to imitate Egyptian precedents as closely as possible, this artifact suggests a purposeful attempt to combine images from multiple origins.

In fact, the value of a statuary assemblage like that of the Casa di Acceptus e Euhodia may have derived partly from the cultural prestige of eclectic, emulative, retrospective art. Roman elites valued the possession of objects in multiple temporal and regional styles, and the ability to discourse intelligently on them, as markers of sophistication.[168] This garden assemblage thus asserted not only its owners' economic resources but also their (claims to) education and cultivation.

However, if "Pharaonizing" styles commanded such cultural value, then why do they seldom appear in two-dimensional Nilescapes? Again, portability appears decisive. Egyptian art boasts a long tradition of landscape depictions,[169] but these are best attested on monumental, immovable media such as temple and tomb walls. While Italian soldiers, tourists, and pilgrims in Egypt could have viewed such images,[170] it is unlikely that many examples were ever taken to Italy to serve as models for craftsmen. In the absence of such models, the producers of Nilescapes took their formal and stylistic cues from contemporary Roman landscape art. Accordingly, Nilotic paintings typically employ the formal structure, perspectival conventions, and brushwork of Roman "sacral-idyllic" landscapes.[171]

Mosaics, however, were more portable than paintings. Mosaic *emblemata* were typically purchased pre-made in trays, not manufactured in situ.[172] Alexandria was a major center for mosaic production, and mosaics in some Pompeian houses have parallels in the Egyptian Delta.[173] Given that two of the earliest Nilotic

---

168. See chapter 3, this volume, "Performing Cosmopolitanism," with note 139.

169. For an overview, see Sist 2002. Some elements of landscape depiction certainly appear on portable objects from Egypt, such as "marsh bowls" or "pilgrim flasks" (see note 134, chapter 2, this volume), but these are not as detailed as the larger-scale images on Egyptian walls. For Egyptian representations of garden landscapes, see Hugonot 1989; Wilkinson 1998.

170. See note 124, chapter 1, and note 144, chapter 2, this volume.

171. See chapter 2, this volume, "Antecedents and Affordances in Greek and Roman Visual Culture."

172. See note 67, chapter 5, this volume.

173. See chapter 5, this volume, "From Hellenistic *Koine* to Roman *Imperium*."

mosaics come from Ptolemaic Egypt,[174] it is possible that some Campanian consumers of Nilotic mosaics could have known of (or indeed purchased) similar Egyptian representations. Even if this were the case, however, Egyptian-made mosaics would not have been in "Pharaonizing" style. Mosaic production in Ptolemaic and Roman Egypt was a practice imported from the Greek world, and mosaics in Egypt—including Nilotic mosaics—employed stylistic conventions derived from Hellenistic art.[175]

The absence of "Pharaonizing" features in Nilotic paintings and mosaics thus appears inextricable from the particular histories and material affordances of those media. Egyptian landscape paintings and reliefs in Pharaonic styles were largely nontransportable, and Egyptian mosaics with landscape themes were themselves already products of a Hellenistic visual *koine*.

The iconographic repertoire of Egyptian and "Egyptianizing" garden statuettes overlaps to some degree with two-dimensional Nilotica, especially in depicting dwarflike beings and river animals. However, the different affordances of statuary, wall paintings, and mosaics—and their correspondingly different entanglements with long-distance trade—encouraged consumers to associate these media with different characteristic styles. Two-dimensional Nilescapes typically depict "Egypt" in styles derived from Greek and Roman art. However, statuary (and paintings depicting statuary) employ a greater range of stylistic strategies, including direct emulation of Pharaonic styles.

Isis Cult and Identity: Beyond "Isiac" vs. "Non-Isiac"
Visual Culture, Cult, and Identity at the Casa di Acceptus e Euhodia

The sculptural assemblage from the Casa di Acceptus e Euhodia reveals multiple parallels with multimedia assemblages from other houses considered in this book, including the Casa dell'Efebo, Casa del Medico, Casa del Fauno, and Casa di Octavius Quartio. Unlike those other houses,[176] however, the Casa di Acceptus e Euhodia also provides fairly strong evidence for a domestic Isis cult. This evidence thus problematizes the widespread assumption that "Isiacs" and "non-Isiacs" must

174. On Ptolemaic Nilotic mosaics, see note 23, chapter 2, this volume. However, earlier assumptions about the general dependence of Roman mosaics on Alexandrian precedents are probably overstated (Dunbabin 1999: 22–23).

175. Daszewski 1985: 98–100. On Nilotic scenes from Egypt, see Versluys 2002: nos. 99–101.

176. With the possible exception of the latest phase at the Casa del Fauno; see chapter 5, this volume, "From Hellenistic *Koine* to Roman *Imperium*," with note 79.

have engaged in fundamentally different ways with Aegyptiaca. Instead, the Casa di Acceptus e Euhodia suggests that Isis worshipers could employ Aegyptiaca in essentially similar ways as other Pompeians. The assemblage further points to the limitations of binary constructions of religious identities as "Isiac" or "non-Isiac."

A lararium in the kitchen originally contained a painting whose central figure has been identified as Isis-Fortuna.[177] Now entirely faded, this painting survives only in a drawing by Discanno (figure 6.16) and detailed descriptions by Mau and Boyce.[178] On a red-bordered white ground, the painting depicted a female figure in chiton and himation, framed by hanging banners or garlands and flanked by vegetal growths. Her left hand held a cornucopia, and her right hand held a rudder resting on a globe. Her hair was long and dark, and in Discanno's drawing, it falls on her shoulders in two heavy corkscrew curls. According to the drawing, she wore a headband or diadem surmounted by a tapering vertical element and flanked by plant fronds.

The cornucopia, rudder, and globe are common attributes of Fortuna, with whom Sogliano and Mau initially identified the central figure.[179] In a subsequent publication, Mau identified the vertical element in the headdress as a lotus flower and characterized the goddess as both Fortuna and "eine Erscheinungsform der Isis."[180] Boyce's later publication labeled the figure as "Isis-Fortuna," an identification followed by most subsequent authors.[181] Images of goddesses with the combined attributes of Isis and Fortuna are common at Pompeii, and while their headgear varies, many other examples also have lotus crowns.[182] I concur with the general consensus that this crown incorporates a lotus, although I suggest that Discanno's drawing likely depicts not a lotus flower but a lotus *leaf*—another coiffure element attested in divine statuettes from Hellenistic and Roman Egypt.[183]

The lotus elements in the crowns of Roman depictions of Isis almost certainly reinterpret the so-called *basileion* crown worn by Ptolemaic queens and Isis in

177. Sogliano 1882: 280; Mau 1884: 130; 1908: 362; Boyce 1937: no. 372; Fröhlich 1991: no. I.97; Bragantini 1998c: 653; Beaurin 2008: no. 9. Descriptions of paint colors follow the reports of Boyce and Mau.

178. Mau 1884: 130; Boyce 1937: 78, no. 372.

179. Sogliano 1882: 280; Mau 1884: 130. On the iconography of Fortuna, see Arya 2002.

180. Mau 1908: 362.

181. Boyce 1937: 78; Tran tam Tinh 1964: 134, no. 27; Fröhlich 1991: 292; Kastenmeier 2007: 160; Beaurin 2008: 284, no. 9; Carrella et al. 2008: 149.

182. Cf. numerous examples in Beaurin 2008.

183. For example, statuettes of Hermes-Thoth often feature a single vertical lotus leaf above the forehead (Török 1995: 84–86).

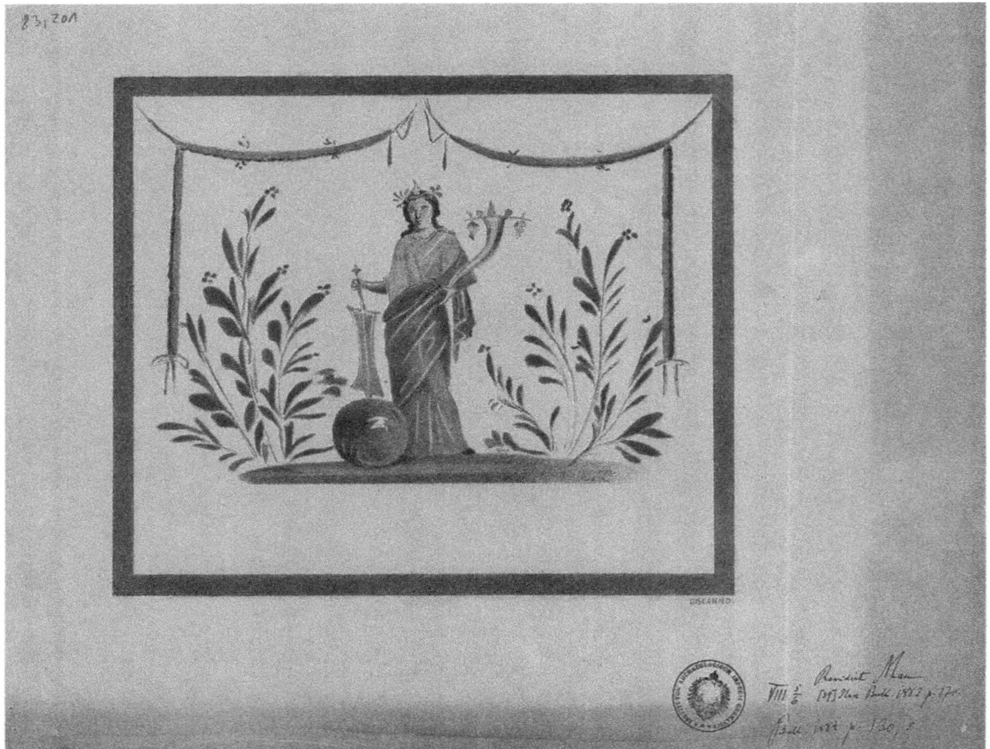

FIGURE 6.16 Drawing by G. Discanno, recording the now-lost lararium painting from the kitchen (e) of the Casa di Acceptus e Euhodia. Dimensions of original painting: 59 cm high, 70 cm wide (Boyce 1937: 78).

Source: DAI-Rom, Archivio, A-VII-32-195. Courtesy of the Deutsches Archäologisches Institut, Rome.

Hellenistic art: a sun disc, bovine horns, and two feather plumes, sometimes flanked by two ears of wheat.[184] After Egypt's conquest, the end of the Ptolemaic monarchy untethered this crown from its earlier political associations, and Roman artists often appear to have reconceptualized its central elements. Because of Isis's increasing association with the moon, the cow horns were commonly reinterpreted as a lunar crescent (e.g., Ov. *Met.* 9.688–689). The vertical plumes often became in Italy a sprouting lotus bud, leaf, or flower: a formally similar thin vertical shape, but one that Italians more readily understood as signifying "Egypt."[185] In this lararium painting, the vegetal fronds bordering the crown appear to recall the wheat sheaves of Hellenistic *basileion* crowns in their placement and bilateral symmetry.

Another, previously unremarked, "Isiac" feature of the painting is the goddess's hairstyle. While long, unbound hair can characterize multiple deities in Roman art, corkscrew curls have particular associations with Isis. Both in Egypt (both Ptolemaic and Roman-period) and elsewhere in the Hellenistic and Roman world, including at Pompeii, Isis commonly wears corkscrew curls that fall onto her shoulders.[186]

Eva Mol has recently questioned whether Pompeians actually identified images of the "Isis-Fortuna" type with Isis: Might viewers have simply identified such images as Fortuna without recognizing their Egyptian associations?[187] However, several considerations weigh against this position. Representations of Isis with lotus crown and corkscrew curls were familiar features of Pompeii's visual landscape.[188] Since lotus crowns also appear in Roman representations of other Egyptian figures, including Sarapis, Harpocrates, and Bes,[189] such headgear

184. On this crown and other iconographic elements shared between Isis and Ptolemaic queens, see Barrett 2011: 136–137, 148–157.

185. Representations of the original *basileion* crown also continued in Italy (multiple examples in De Caro 2006b). Images of Isis-Fortuna at Pompeii sometimes wear a lotus crown, sometimes a *basileion* crown (Beaurin 2008: 269–270).

186. Barrett 2011: 148–157. At Pompeii, see a statue and a painting from the Iseum (De Caro 1992: no. 1.63; 2006b: no. II.56), as well as numerous figurines and other small-scale objects from houses and shops (e.g., De Caro 2006b: nos. III.25, III.26, III.35, III.36, III.44, III.45, III.55).

187. Mol 2015: 173–176.

188. *On corkscrew curls*, see note 186, this chapter. *On lotus crowns*, see notes 185 and 189, this chapter.

189. E.g., Roullet 1972: nos. 164, 315 (Canopic jars/Osiris Canopus); De Caro 1992: no. 1.5 (Harpocrates, wearing what the catalogue identifies as a uraeus crown, but which may alternatively be a lotus), 1.71 (Osiris), 1.72 (Bes); 2006b: II.39 (Isis), II.51 (river deities in the Iseum), II.52 no. 1.77 (ibis), III.22 (Isis and Sarapis), III.57 (Harpocrates); Seiler 1992: fig. 271 (Isis and Sarapis); Versluys 2002: no. 43 (Isis, Sarapis, and Harpocrates).

appears to have commonly signified Egyptian origins. Images of Isis holding a cornucopia, a common attribute of Fortuna or Tyche (among other goddesses), often appear together with images of other Egyptian gods. For example, a garden shrine painting at the Casa delle Amazzoni depicts a cornucopia-holding Isis with Sarapis and Harpocrates.[190] A cornucopia-holding Isis accompanies Harpocrates and Anubis on three faience lamps from Pompeii, including one from the Casa degli Amorini Dorati garden, which also contained a shrine to Egyptian deities.[191] The Nilotic frescoes of the Casa dell'Efebo depict a statue of Isis-Fortuna (figure 1.3, A) holding a cornucopia and probably also a rudder.[192] While many goddesses can hold cornucopiae, the rudder more distinctively identifies Fortuna.[193] The statue in this painting also wears a lotus crown very similar to that at the Casa di Acceptus e Euhodia.

A Greek graffito, "Εἰσιτύχη σώζουσα," near the entrance to the Pompeii Iseum testifies to local awareness of the identification of Isis with Tyche/Fortuna.[194] The syncretism of these deities was common in Hellenistic and Roman religion, within which Isis was frequently identified with a whole pantheon of goddesses.[195] The representation of Isis with features of Tyche or Fortuna, including the cornucopia, was already common in the Hellenistic period. Isis's identification with Tyche is further attested in inscriptions and possibly also material culture from Hellenistic Delos, a site where many Campanian merchants may have first encountered Isis worship.[196] In the Roman Empire, representations of goddesses with features of both Isis and Fortuna became ubiquitous throughout Italy.[197]

---

190. Boyce 1937: no. 141; Bragantini 1993: 174–175; Versluys 2002: no. 43.

191. Tran tam Tinh 1964: 170–171; De Caro 2006b: nos. III.105–III.107; Beaurin 2008: 288, nos. 24, 28 (and see no. 27 for another example from Boscoreale); on the house more generally, see Seiler 1992; Petersen 2012.

192. See chapter 3, this volume, "The Nilescapes of the Casa dell'Efebo Garden Triclinium."

193. Beaurin characterizes Isis-Fortuna as carrying a rudder as well as a cornucopia (2008: 270), but the iconography of Fortuna is in fact somewhat variable (Arya 2002: 68–93), and not all of the examples in Beaurin's catalog of Isis-Fortuna images carry rudders (2008).

194. *RICIS* 504/0216. References to "Isityche" also appear in Latin (*RICIS* 503/0602) and Greek (*RICIS* 501/0139, 515/1001) inscriptions from other sites in Italy (respectively, Praeneste, Rome, and Mama d'Avio).

195. See especially the "Isis aretalogies," for which see bibliography in Barrett 2011: 129–130n410.

196. For Isis and Tyche on Delos, see *RICIS* 202/0283, 202/0284; Barrett 2011: 223–225, 235 with n857, 244–246. For Campanian merchants on Delos, see note 62, chapter 1.

197. See numerous examples in Arslan 1997; De Caro 2006b. For discussion, see Beaurin 2008.

I thus argue that most Pompeian viewers would indeed have identified the goddess in figure 6.16 with Isis, as well as Fortuna. We do not know what precise name or names the house inhabitants used for this deity in daily practice: "Isis-Fortuna," "Isityche," "Isis," "Fortuna," or all of the above. Whatever terminology they employed, however, the lararium painting recognizably presents its central figure as *both* (1) an Egyptian goddess, as indicated by the lotus headdress, and (2) a goddess with power over fate, as indicated by the cornucopia, globe, and rudder.

The lararium painting does not necessarily imply that the inhabitants of this house were initiates into Isiac "mysteries," or even that they considered themselves "Isiaci" in any formal or official capacity. However, it does suggest that in at least some capacity, Isis was part of the regular religious experience of this *familia*. She occupied the visual center of a household shrine—the only shrine recorded for this house—and presided over domestic rituals there.

Furthermore, religious activities at lararia like this one may have involved multiple members of the *familia*. Older scholarship often assumed that kitchen shrines were intended for slaves only, while the free members of the *familia* would worship at shrines elsewhere in the house.[198] However, this segregated image of Roman domestic cult has been subject to increasing critique. Giacobello has recently argued that some rituals at kitchen lararia would have been overseen by the paterfamilias and incorporated his entire family, while slaves would have been responsible for performing other, daily rites.[199] While scholars continue to debate whether slaves and free individuals would have worshiped together at household shrines, a recent re-investigation concludes that there is no solid evidence for spatial separation between free and servile domestic cults.[200] Additionally, since the kitchen lararium is the only household shrine recorded at the Casa di Acceptus e Euhodia, it is hard to envision how any form of cultic segregation could have functioned there in practice.

Yet regardless of which members of the *familia* would have directly participated in rituals at this shrine, we should still be cautious of projecting "Isiac" or "Egyptian" interpretations onto other, potentially unrelated features

---

198. On the significance and cultic use of shrines in different areas of the house, see further note 85, chapter 3, this volume.

199. Giacobello 2008: 100–101, 110–116. On the involvement of the paterfamilias and other free members of his family in cult at lararia, see most recently Flower 2017: 28–30, 48–49, 58, 75.

200. Schörner 2016.

of the Casa di Acceptus e Euhodia. Because of the lararium painting and garden Aegyptiaca, scholars have long looked for Egyptian connections elsewhere at this house, even attempting to identify the owners as Alexandrians. Such extrapolations go beyond the evidence.

The house takes its modern name from electoral dipinti on its outside wall (*CIL* IV 3595), in which individuals named Euhodia and Acceptus recommend political candidates. Much speculation has accrued to this supposed couple, who are often assumed to have lived here.[201] Taking their names as characteristic of slaves, Mau and Eschebach further speculate that this couple were freedmen from Alexandria.[202] However, such conclusions rest on precarious assumptions. Electoral dipinti were not always located outside their commissioners' homes, and personal names do not always reliably indicate free, freed, or servile status.[203] I therefore take as uncertain both the identities of Acceptus and Euhodia and their relationship to House VIII 5.39, and I retain the house name as purely conventional.

Also sometimes adduced as evidence for an Alexandrian connection is the table leg inscribed in Greek as "of Sarapion." Because this name refers to Sarapis, Mau and Eschebach take it as evidence that the house owners were Alexandrians.[204] However, theophoric names referring to Isis and Sarapis were common in the Roman Empire and do not necessarily signify ethnic or geographic origins.[205]

Mau also invoked the family's supposed Alexandrian origins to explain the irregular form of the Casa di Acceptus e Euhodia.[206] However, considerations of location (in an old section of the city, characterized by multiple irregular street and building forms) and economic resources provide sufficient explanation for these irregularities. Whatever the background of the house's last inhabitants, it is unlikely that they could have obtained a vacant plot in one of the busiest areas of Pompeii in order to construct a new house entirely to their specifications. If they had sufficient wealth to do this, then they surely could afford a more architecturally impressive dwelling.

---

201. Mau 1884: 127; 1908: 360–361; Della Corte 1965: 214; Eschebach 1984: 50.

202. Mau 1908: 362; Eschebach 1984: 50.

203. *On problems with attempts to identify house owners at Pompeii*, see Mouritsen 1988: 13–27. *On problems with identifying freedmen* in the absence of explicit assertions of status, see Petersen 2006.

204. Mau 1908: 362; Eschebach 1984: 50.

205. See, e.g., a Cyrenaican Jew named Serapion in a Neronian inscription (*SEG* 17, no. 823). On non-Egyptians with Egyptian theophorous names in the Hellenistic period, see Barrett 2011: n50, n1715.

206. Mau 1908: 362.

(a) (b)

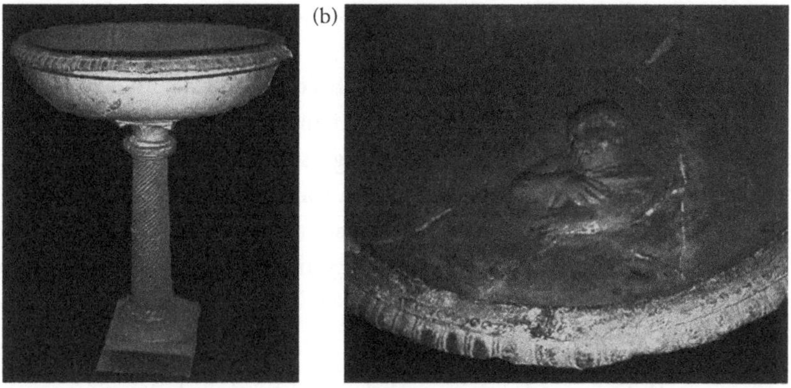

FIGURE 6.17 Footed marble basin, from the indoor triclinium (see f, figure 6.7) of the Casa di Acceptus e Euhodia. Basin diameter: 81 cm (Carrella et al. 2008: 151). MANN 120308. (a) Basin viewed from the side. (b) Image in relief inside the basin, depicting the head and torso of a young boy.

Sources: Photos after Carrella et al. 2008: figs. C 48 and C 48 decorazione. Images courtesy of "L'Erma" di Bretschneider.

Finally, earlier commentators have taken the house's proximity to the Iseum as consistent with the owners' likely worship of Isis.[207] However, there is little evidence to suggest that religious identity shaped patterns of residence in Pompeii. Most other houses in Regio VIII—including, in the same insula, the Casa del Medico—show no evidence for the worship of Egyptian gods.

Another object with a somewhat dubious "Isiac" connection is a footed marble basin from the indoor triclinium (figure 6.17).[208] The interior of this basin bore a relief image of a young boy with a long plait of hair. Since this plait resembles the "Horus-lock" of Egyptian children, Carrella et al. identify the boy as a devotee of Isis.[209] However, Carrella et al. wisely refrain from speculating on the relationship between this object and the religious identity of its owners. The past thirty years of research have abundantly demonstrated the danger of assuming that only Isis worshipers were interested in representations of Isis worshipers.[210] As we shall see, it may also be misleading to assume that

207. De Vos 1983: 62; Bragantini 2006: 167n1; 2012: 23n16.

208. Carrella et al. 2008: 151–152 (interpreting the room as an atrium – see this chapter, n. 37); for an earlier interpretation of this object, see note 104, this chapter. For the findspot, see Sogliano 1882.

209. Carrella et al. 2008: 152. On sidelocks as juvenile coiffures in Roman Egypt, see Bergmann 2016.

210. See chapter 1, this volume, "Material Culture, Identity, and 'Authenticity'".

when Isis worshipers did own such objects, they valued them *solely* because of their "Isiac" associations.

The assemblage from the Casa di Acceptus e Euhodia does not provide sufficient evidence to identify the geographic origin or legal status of its inhabitants. However, the lararium iconography does suggest that Isis played some role in the religious practice of this household. At least some inhabitants of the house were thus probably "Isis worshipers," in some capacity. Yet what did it actually mean in practice to be an "Isis worshiper," and what forms of religious identification did "Isiac" activities entail?

Religious Identities and the Problem(s) of "Isiaci"

Historically, scholarship has often grouped all "Isiaci" together as a monolithic category. However, the reality appears much more complex. Instead of a binary distinction between "Isiacs" and "non-Isiacs," Pompeian evidence suggests a spectrum of possible forms and degrees of participation in the Isis cult. The Casa di Acceptus e Euhodia assemblage provides an opportunity to develop more nuanced accounts of "Isiac" practices and identities.

Frequently lost in scholarly debate on the relationship between Aegyptiaca and "Isiac cult" is the question of what "Isiac" actually means. To what degree did the worship of Egyptian gods necessarily entail the conscious adoption of a new religious *identity*, as is often assumed, and what varieties and degrees of participation were available? Scholars often implicitly divide the users of Aegyptiaca into two groups: those who worshiped Egyptian gods and those who did not. However, much evidence suggests a broader spectrum of types and degrees of engagement with gods of Egyptian origin.

When discussing Pompeians' reception of Egyptian cults, it is important to distinguish *practices* related to the worship of originally Egyptian gods, *beliefs and attitudes* toward those gods, and *religious identities* based on those beliefs and practices. Ritual practices, initiations, beliefs, attitudes, and membership in religious associations do not constitute a single, invariant Isiac "bundle"[211] but, rather, a variety of options with which individuals might potentially engage in different ways. We must be careful not to conflate at least four different possible conditions:

1. *Initiation* into "mysteries" of Isis or related deities.
2. Other forms of *ritual practice* directed at Isis or related deities (either in temple or domestic contexts).

---

211. On "bundling," see Keane 2003: 414.

3. *Beliefs* about the existence, nature, and activities of Isis or related deities.
4. Generally positive *opinions* concerning Isis or related deities—if and when one happened to think of them, which need not necessarily be often.

Each of these conditions can be further nuanced, as people might display different degrees of intensity and frequency in practice, belief, and attitudes. One person might frequently participate in rituals at the Iseum, while another might do so only sporadically; one person might hold extremely detailed beliefs about Isis's personality and activities, while another might only conceive of her in vague terms as one of many deities inhabiting the universe.

Certainly, some relationship exists between ritual action and belief. As recent research demonstrates, older characterizations of Roman religion as a belief-free set of practices no longer appear justified.[212] Participation in rituals typically implies at least the minimal belief that such rituals might conceivably be effective,[213] and such a conviction often relies on additional beliefs concerning the existence and powers of the deities whom the rituals address. Ritual practice thus usually implies at least some associated beliefs, although that relationship does not necessarily go both ways; many people may have believed that Isis *existed* without feeling personally compelled to perform rituals at the Iseum.

However, neither initiation nor ritual practice necessarily implies the other. Initiation into ancient mysteries did not always require continued ritual activity afterward, and most deities with "mystery cults" also received other forms of worship from noninitiates.[214] Although scholars often assume that initiation and subsequent ritual performance were more closely interwoven in Isis worship[215] than in other "mystery cults," this assumption is questionable, as we shall see. Additionally, holding positive opinions about Egyptian deities does not

212. King 2003; Mackey 2009.

213. Some people may also participate in rituals out of perceived social pressure, but this is unlikely to be the sole motivation for all individuals involved, especially with an "elective" cult like that of Isis. On "elective cults," see Price 1999: 108–125; Beard et al. 1998: I.272–275.

214. Initiations associated with particular sanctuaries (Eleusis, Samothrace, etc.) typically did not require initiates to "maintain a continuing relationship with the sanctuary" (Bowden 2010: 168). On the existence of other, non-initiatory forms of worship for "mystery" deities, see Burkert 1987: 10–11; Jaccottet 2003; 2006, esp. 219.

215. Scholars often assume, based on Apuleius (*Met.* 11), that Isiac initiation required more intense devotion afterwards (e.g., Cancik 2007: 42–43; Bowden 2010: 177). However, taking Apuleius's Lucius as a typical worshiper is highly problematic, as discussed later in this chapter.

necessarily imply either ritual performance or initiation within those deities' cults. In the Flavian period, Isis and Sarapis were widely worshiped recipients of public cult and imperial largesse. While first-century and later Roman attitudes toward Egyptian gods (especially animal gods) were not invariably enthusiastic, much evidence suggests that in Flavian Italy, it was common to hold positive opinions of Isis and Sarapis as beneficent, potentially helpful deities.[216] As a result, many people—conceivably even a majority of Pompeians—likely assumed the existence of, and were generally positively disposed toward, certain deities of Egyptian origin. However, we have no reason to think that such opinions invariably required their holders to seek out initiation into Isiac mysteries, participate in ceremonies at the Isis temple, or erect domestic Isis shrines.

The scholarly reaction against "Isiac" interpretations of Aegyptiaca has produced the important realization that possession of "Egyptianizing" material culture need not imply initiation or ritual participation in Isiac cults. However, we should not forget that people could hold positive beliefs about Egyptian deities without engaging in rituals for them. Claims of evidence for such beliefs or positive opinions do not automatically entail equally strong claims of evidence for initiation or ritual performance. A textual example comes from Statius's *Silvae* 3.2, a poem whose Campanian setting and Flavian date make for useful comparison with material from Pompeii. Bidding farewell to a military officer about to visit Egypt, the speaker invokes Isis to guide the young man's intellectual inquiries during his travels.[217] The poem presumes Isis's existence and represents her as benevolent, but it provides no indication that either the speaker or the officer are (or are expected to become) initiates into Isiac mysteries or active participants in Isiac rituals.

Furthermore, we should not assume that any of the above relationships to the Isis cult invariably required adopting a self-declared, consciously-held *religious identity*.[218] People who occasionally prayed or sacrificed to Isis might not necessarily consider themselves to have acquired a new identity—to have

216. See chapter 1, this volume, "Flavian Pompeii as Case Study." On ongoing concerns about theriomorphic deities, see e.g., Juv. *Sat.* 15 (see most recently Swetnam-Burland 2015: 167–181); Lucian, *Deorum Concilium* 10–11 (though Momus's mockery of Anubis and Apis is partially countered by Zeus, who says their cult involves *ainigmata* at which the uninformed should not laugh).

217. Stat. *Silv.* 3.2.107–120. See further, chapter 3, this volume, "Performing Cosmopolitanism."

218. On recent reconsiderations of religious identities in antiquity, see Belayche and Mimouni 2009; Rebillard and Rüpke 2015. A forthcoming volume on the role of individual agents in Isis cult (Gasparini and Veymiers 2018) was not yet published at the time of this writing and could not be consulted for the present project.

become "Isiac"—through these actions. Furthermore, even if some persons did embrace such an identity, we should not assume they foregrounded this identity in every situation. In other words, engagement with "Isiac" beliefs or practices need not have produced an unvarying or unceasing state of what Brubaker calls "groupness."[219] Parallels from other ancient religious groups, from "Orphics" to Christians, suggest more flexible relationships between identities and ritual performances and indicate that people did not treat their religious identities as salient in all contexts.[220]

Some worshipers of Isis and/or Sarapis probably did understand cult practices as an important source of identity and (at least in some contexts) a generator of (at least intermittent[221]) "groupness." Evidence comes from Apuleius's *Metamorphoses*, whose protagonist's Isiac and Osirian initiations transform his life; funerary monuments depicting the deceased with objects and garments associated with Isis worship; and epigraphic evidence for associations devoted to Egyptian deities. However, each of these sources also raises interpretive problems. Additionally, none of them demonstrates that *everyone* who ever worshiped Egyptian-derived deities, in any capacity, was a self-described *Isiacus* or viewed such actions as producing community membership and "groupness."

Apuleius's goals and opinions concerning Isis worship are by no means transparent, and interpretations of the ending of the *Metamorphoses* remain controversial.[222] However, regardless of whether we interpret Book 11 as sincere, tongue-in-cheek, or somewhere in between, we should hesitate before taking the character of Lucius as paradigmatic for regular worshipers. Irrespective of the degree to which Apuleius's portrayal of initiation rituals may or may not draw on actual practices,[223] Lucius's experience of transformation into a donkey and back is—to say the least—very individually specific. The intensity of his subsequent devotion to Isis is largely a response to these traumatic events. While epigraphic evidence confirms the worship of Isis and Sarapis as saviors who rescued people

219. Brubaker 2002: 168–169; 2004: 4; and see further chapter 1, this volume, "Material Culture, Identity, and 'Authenticity.'"

220. On Christians, see Rebillard 2009, 2012. Not all people buried with "Orphic" gold leaves necessarily identified as *Orphikoi* or followed an "Orphic" lifestyle (Burkert 1982). On "Orphic" lifestyles, see Burkert 1987: 46–47; Bremmer 2014: 66–80.

221. On religiously inspired groupness as intermittent, see Rebillard 2012.

222. See most recently Keulen and Egelhaaf-Gaiser 2012; Steinhauer 2017: 63–75. For edition and commentary, see Keulen and Egelhaaf-Gaiser 2015.

223. Bremmer (2014: 110–125) and Steinhauer 2017 provide the most recent analyses of "Isiac" initiation rituals.

from misfortune,[224] Lucius's particular misfortune is more unusual than most, and his response may be correspondingly more dramatic.

Isis worship may also appear as a source of social identity on some funerary monuments. In Greece and Italy, certain funerary stelae depict individuals, especially women, carrying sistra and wearing fringed, knotted robes associated with Isis's priests and priestesses.[225] These funerary monuments are generally, and undoubtedly correctly, seen as expressions of religious identity.[226] However, they may also broadcast other messages: for example, suggesting the deceased served as a priest or priestess.[227] The intended statement may thus be not only "This person was a worshiper of Isis" (an assertion of religious affiliation) but also "This person held a priestly position" (a statement also implying claims to certain forms of status or prestige). This implication may help explain the frequency of "Isiac" motifs on female tombstones, as women had more opportunities for priestly roles in this cult than in many other forms of Roman ritual.[228]

Voluntary associations provide another occasion for Isiac group identity, and at Pompeii, epigraphic references to *Isiaci* certainly demonstrate some form of collective identity associated with Isis. However, it remains unclear who these *Isiaci* were and what they did. At Pompeii, *Isiaci* support candidates in two electoral graffiti.[229] At other sites and in literary references, *Isiaci* usually appear to be members of a religious association,[230] but such associations might take different forms in different places. As Malaise points out, Latin has fewer designations than Greek for cult associations devoted to Egyptian-derived deities; epigraphic

224. Sfameni Gasparro 2007; Fassa 2013: 130–133.

225. Bricault and Versluys 2014a: 23n94 (with earlier bibliography); Poole 2016: no. 57. Male examples are also attested, though rarer: a painted funerary stele from Hellenistic Demetrias depicts a male Isis priest (Stamatopoulou 2008), and a funerary relief from fourth century CE Ostia depicts a boy wearing a "Horus-lock" (Arslan 1997: no. V.38).

226. Bricault and Versluys 2014a: 23.

227. Martzavou suggests these individuals may have been ritual practitioners (2011: 71, 73). The Demetrias funerary stele (see note 225, this chapter) identifies its central figure as a *hiereus Isidos* (Stamatopoulou 2008: 249).

228. Heyob 1975: 81–110 (reiterating, however, that most "Isiac" cult officiants were male); Alvar 2017: 39.

229. *CIL* IV 787, 1011 = *RICIS* 504/0209, 504/0210.

230. Malaise 2005: 25–31; cf. Burkert 1987: 47–48. It is unclear when the term applies to priests as opposed to worshipers. The *Oxford Latin Dictionary* (s.v. "Isiacus") defines the masculine substantive of the Latin adjective as "a worshipper of Isis," and Rüpke suggests that *Isiacae* are probably "einfache Kultmitglieder" (2006: 14). Vidman (1970: 91) and Heyob (1975: 108–109) interpret *Isiacus* and *Isiaca* as potentially referring to either priests or worshipers.

evidence testifies to, for example, Greek organizations of *Sarapiastai*, but not Roman organizations of *Sarapiaci*.[231] However, lack of terminological variety need not imply lack of variety in groups' activities or forms of organization, and the term *Isiacus* may conceal various context-specific or locally specific meanings. Such local variation appears generally characteristic of Roman voluntary associations.[232] In Rome, the term *Isiacus* might even have a geographic meaning, describing someone who lived in a neighborhood near the temple of Isis and Sarapis.[233]

The equivalent Greek term, *Isiakos*, also appears somewhat fluid in meaning. In various contexts, it may refer to Isis worshipers, initiates, or priests.[234] One of the most famous attestations, Plutarch's discussion of the true *Isiakos*, is particularly problematic because of the overtly polemical nature of the passage. In the course of a Platonizing explanation of Egyptian myths, Plutarch asserts that "just as letting one's beard grow and wearing a worn garment do not make philosophers, neither do linen-wearing and shavings make *Isiakoi*; but an *Isiakos* is one who, when he has truly received what is shown and done concerning these gods according to custom, inquires and studies through reason about the truth in them."[235] Plutarch's "real" *Isiakos* is one who does not just make an outward show of identity but also philosophizes about the deeper meaning of his actions. The outward displays that Plutarch describes as insufficient—wearing a special garment and shaving (here, surely implying head-shaving)—were markers of priestly status in Roman Egypt and characterize mystery initiands and initiates in

---

231. Malaise 2005: 26. On Greek religious associations for Egyptian gods, see Malaise 2005: 27–29. For Latin terms for associations of Isis priests and/or worshipers, see Bricault 2012. Cf. Vidman 1970: 66–94; Burkert 1987: 38–39.

232. Rüpke 2006: 23.

233. Malaise 2005: 26. A geographic meaning for the Pompeian *Isiaci* is unlikely, however (*pace* de Vos 1994: 131–132), as the Pompeii Iseum did not give its name to the associated *vicus* (Gasparini 2014: 293).

234. See, e.g., *PGM* IV.3092; the practitioner of this spell is to dress in the clothing of an *Isiakos*—perhaps a priestly costume (Betz 1992: 98n414), although a special costume also characterizes "Isiac" initiates in Apuleius' text (*Met.* 11.10). The *LSJ* defines *Isiakos* as "priest of Isis" (*LSJ Online*, s.v. Ἰσιακός), but it is sometimes unclear whether *Isiakos* describes priests, initiates, or other worshipers: e.g., Dioscorides Pedanius, *De materia medica* 3.23.5. Josephus's reference to the deed τῶν Ἰσιακῶν (*AJ* 18.65) is also ambiguous, as his story involves both priests and non-priests (*AJ* 18.65–80).

235. ὄυτε γὰρ φιλοσόφους πωγωνοτροφίαι, ὦ Κλέα, καὶ τριβωνοφορίαι ποιοῦσιν ὄυτ' Ἰσιακοὺς αἱ λι νοστολίαι καὶ ξυρήσεις: ἀλλ' Ἰσιακός ἐστιν ὡς ἀληθῶς ὁ τὰ δεικνύμενα καὶ δρώμενα περὶ τοὺς θεοὺς τούτους, ὅταν νόμῳ παραλάβῃ, λόγῳ ζητῶν καὶ φιλοσοφῶν περὶ τῆς ἐν αὐτοῖς ἀληθείας (Plut. *De Is. et Os.* 3 = *Mor.* 352C).

Apuleius's *Metamorphoses.*[236] However, the polemical construction of Plutarch's statement itself demonstrates that other people might have quite different definitions of *Isiakoi.*[237] Indeed, Plutarch's perspective is likely atypical in multiple ways: besides criticizing people who engage in rituals without thinking about their meaning, he is also claiming Isis and Osiris as legitimate subjects for Greek philosophy and interpreting their rites in terms of Platonic thought.[238] Rather than providing evidence of universal consensus on the one true way to be an *Isiakos*, his account flags the term as contested even in antiquity.

At Pompeii, scholars typically treat the *Isiaci* of the electoral inscriptions as a collective term for Isis worshipers.[239] However, this identification raises multiple problems. First, the category of "Isis worshipers" usually goes undefined, its meaning left up to readers' assumptions; and second, the treatment of *Isiaci* as a synonym for *all* Pompeian "Isis worshipers" presumes a broad form of collective identity that may not be applicable. Bricault and Gasparini are likely correct to associate the *Isiaci* with a voluntary cult association.[240] However, we cannot assume that everyone who engaged in some form of Isis worship at Pompeii necessarily belonged to that particular association. There is no evidence that the *Isiaci* of the electoral inscriptions included everyone at Pompeii who was initiated into Isis mysteries, frequented the Isis temple, owned a domestic shrine to Egyptian-derived deities, or even belonged to a voluntary cult association for such deities. While the *Isiaci* inscriptions may indeed indicate that some Pompeians viewed Isis worship as a source of collective identity, they cannot prove this was the case for everyone involved in some form of Isis worship at Pompeii.

In fact, parallels at other sites suggest that worshipers at the Iseum may *not* necessarily have seen themselves as belonging to a single group. Elsewhere in the Greco-Roman world, inscriptions testify to multiple different cult associations active in and around sanctuaries of Isis and Sarapis.[241] Some associations who dedicated at such sanctuaries were even devoted to gods

236. Apul. *Met.* 11.10; see further note 262, this chapter.

237. See Martzavou 2011.

238. On his famous claim that "Ἑλληνικὸν γὰρ ἡ Ἶσις" (Plut. *De Is. et Os.* 2 = *Mor.* 351F), see Richter 2001; Finkelpearl 2012.

239. E.g., van Andringa 2009: 326–328; Swetnam-Burland 2015: 112.

240. Bricault 2012: 100; Gasparini 2014: 294.

241. *Thessaloniki:* Steimle 2006. *Delos:* Roussel 1915–16; Vidman 1970: 67–74; Baslez 1977: 197–203; Martzavou 2011: 65–67.

without any apparent Egyptian background.[242] While the smaller number of Latin terms for such associations may suggest fewer groups, it is dangerous to assume *ex silentio* that only one homogeneous association of Isis worshipers existed at Pompeii.

Another potential source of common identity for Pompeian Isis worshipers—namely, their supposed shared status as mystery initiates—also proves problematic on closer inspection. It is generally accepted that by the first century CE, if not earlier, some Greco-Roman worshipers restructured Isis cult to include initiations modeled after the Eleusinian Mysteries.[243] Scholars often treat such mystery rites as synonymous with Roman Isis cult, with all Isis worshipers presumed to be initiates.[244] However, as Bremmer puts it: "Nothing could be further from the truth."[245] Although Apuleius presents initiation as central to Lucius's experience of Isis worship, evidence for the actual practice of Isis mysteries is surprisingly rare and geographically localized.[246] Epigraphic evidence testifies to such mysteries at some sites, including Rome (where Apuleius's hero was initiated),[247] but it is by no means clear that Isis cults were uniformly initiatory throughout the Roman world. The worship of Isis and Sarapis was organized locally and took different forms, from site to site; no transregional authority imposed homogeneous practices across the Mediterranean.[248] Bricault and Veymiers wisely stress local

242. E.g., a dedication by the *mystai* of Zeus Dionysos Gongylos at the temple of the Egyptian gods in Thessaloniki (*IG* X 2.1, 16; Steimle 2006: 32–35).

243. Burkert 1987: 40–41; Sfameni Gasparro 2006: 192; Alvar 2008: 10; Belayche and Rebillard 2008: 142–143; Bricault 2013: 430–431; Versluys 2013b: 254; Bremmer 2014: 110–114. On the question of Pharaonic Egyptian "mysteries," see Hornung 1994; Assmann and Bommas 2002; Barrett 2011: 388n1600. Egyptian religion of the Pharaonic and Greco-Roman periods did involve a concept of mysterious knowledge reserved for people with advanced religious training, but does not seem to have provided any means for the public to pay a fee to undergo a ritual that would impart such knowledge. On initiatory ceremonies for priests in Egypt, see note 263, this chapter.

244. Bommas 2012: 181; Petersen 2006: 48; Hornung 1994: 287; Powers 2006: 116; Versluys 2013b: 253–254.

245. Bremmer 2014: 114.

246. Vidman 1970: 125–138; Burkert 1987: 40; Sfameni Gasparro 2006: 192; Bowden 2010: 168, 178; Bricault 2013: 428–445; Bremmer 2014: 114; Gordon 2014: 666; Steinhauer 2017.

247. Bremmer 2014: 114.

248. Sfameni Gasparro 2006: 193; Nagel 2013: 174. Isis worship was organized differently from site to site (Van Haeperen 2006: 44). At Delos, different Sarapis sanctuaries appointed priests in different ways (Martzavou 2011: 65–67). The essays in Bricault and Versluys 2014b demonstrate much regional variation in Isis cults' interactions with local sociopolitical structures; see esp. Gasparini 2014.

variability in Isis's worship outside Egypt, warning against overgeneralizing from a few well-known sites or texts.[249] Such local variation is generally typical of cult associations, leading Rüpke to warn that "Conceptualizing all instances of the veneration of Isis as a coherent religion or 'cult' is as problematical as postulating the coherence of an actor 'Christianity.'"[250]

There is, in fact, no direct evidence that initiation was ever part of Isis worship at Pompeii. Campania has produced no epigraphic references to *initia, mysteria, teletai,* or *orgia* of Isis.[251] The architecture of the Pompeii Iseum similarly provides no clear evidence for initiatory practices. Scholars have proposed that certain rooms could be used for initiations or restricted to initiates,[252] but such suggestions remain speculative. While it is certainly possible that activities at the Iseum included initiations, we should be wary of a priori assumptions.

Whether or not initiations occurred at Pompeii, non-initiates could have worshiped Isis, Sarapis, and company in other ways.[253] There was no mechanism for preventing people of any initiatory status from, for example, directing prayers or private offerings toward these deities, wearing amulets depicting Egyptian gods, or setting up statuettes of those gods in household shrines. Within temples and beyond, texts and images suggest that many Isiac rites were relatively visible and accessible. Even Apuleius's *Metamorphoses* does not suggest that initiation is necessary for all forms of cultic participation. Only some participants in Apuleius's Isis procession are described as initiates, and a crowd of non-initiates gathers to enjoy the festival.[254] The fact that literary texts can record certain cult acts—for example, various festivals and their preparatory rites, or libations of Nile water

---

249. Bricault and Veymiers 2012: 2; cf. Sfameni Gasparro 2006: 192–193; Bowden 2010: 164–165; Gasparini 2014.

250. Rüpke 2013a: 4–5. On such local variation, see Rüpke 2006: 23. On many religious groups' small-scale, fluid organization, cf. Cancik 2007; Rüpke 2007b.

251. For Campanian inscriptions pertaining to Isis and her circle, see *RICIS* 504/0101–504/0802. For ancient terminology for initiation and mysteries, see Pirenne-Delforge and Scarpi 2006.

252. Golvin 1994: 244; Coarelli 2006: 63; cf. Gasparini (2008: 86–87) on enclosed spaces in Isea more generally.

253. As also observed by Burkert 1987: 41; Bowden 2010: 164; Bremmer 2014: 113–114; Alvar 2017: 30–31.

254. Apul. *Met.* 11.7–17. Explicit references to initiates include "turbae sacris divinis initiatae" (Apul. *Met.* 11.10) and "qui venerandis penetralibus pridem fuerant initiati" (Apul. *Met.* 11.17).

within temples[255]—demonstrates that not all rituals were concealed by initiatory secrecy. Two famous paintings of "Egyptianizing" rituals from Herculaneum depict large crowds in temple courtyards, locations of apparently unrestricted access.[256] As Alvar notes, the inclusion of "occasional participants or casual spectators" differentiates Isiac rituals from some other "mystery cults."[257] Many private "Egyptianizing" rites also seemingly lacked initiatory prerequisites; for example, there is no suggestion that the woman consulting Egyptian horoscopes in Juvenal's sixth *Satire* (6.528–529) is some sort of initiate.

Indeed, people may sometimes have used Isis temples for activities unrelated to Egyptian gods. The sanctuary at Pompeii appears to have been open to the public, and people may even have conducted business transactions there.[258] In Rome, the Iseum Campense staff may have actively tried to attract non-"Isiac" visitors. Bommas suggests that the temple's impressive collection of Aegyptiaca was publicly accessible and could have helped advertise the cult to potential worshipers.[259] Similarly implying the public accessibility and popularity of such sanctuaries is Ovid's suggestion of an Isis temple as a good place to find a date.[260]

It is possible that, at least in some places, Isiac initiation was only necessary for aspiring priests.[261] In Apuleius's novel, Lucius prepares for initiation by shaving his head and wearing special garments—that is, dressing like an Egyptian priest—and obeying behavioral restrictions resembling those of priests in Roman

---

255. *Festivals and processions:* Merkelbach 1995: 153–158; Quack 2003. "Isiac" festivals entered the official Roman calendar in the first century CE—an addition often attributed to Caligula, though Scheid argues that we cannot rule out a Vespasianic date (2004: 310). *Sexual abstinence before a festival*: Ov. *Am.* 1.8.73–76, 3.9.33–34; Prop. 2.33A, 2.28A.61–62; Tib. 1.3.25–26. *Sprinkling Nile water inside a temple:* Juv. *Sat.* 6.528–529. *Daily rituals within temples:* Gasparini 2008: 85–86; 2013: 201–205.

256. The paintings may come from a private house, although precise contextual information is lost (Tran tam Tinh 1971: 28–49; Swetnam-Burland 2015: 12–14). The crowds wear specialized dress associated with Isis cult (Alvar 2008: 311). However, temple courtyards were generally accessible spaces, both in Egypt (see note 74, chapter 2, this volume) and seemingly also in Italian sanctuaries of Isis and Sarapis (see further discussion below, this chapter).

257. Alvar 2008: 345. Cf. also Gasparini 2008: 86.

258. Swetnam-Burland 2002: 112–123, esp. 113; Petersen 2006: 54. Vitruvius recommends locating Isis and Sarapis temples in market districts (Vitr. 1.7.1; see discussions in Swetnam-Burland 2002: 113; Bommas 2012: 189).

259. Bommas 2012.

260. Ov. *Ars am.* 1.77–78, 3.393.

261. Cf. Malaise 1986: 355; Burkert 1987: 40; Bowden 2010: 177; Nielsen 2014: 56, 218; and (focusing specifically on Apuleius's Lucius) Steinhauer 2017: 67–75.

Egypt.[262] Within Egypt, most people did not undergo formal religious initiation, but there were initiatory ceremonies for priests from at least the Third Intermediate Period on.[263] Alvar argues against a priestly status for Greco-Roman Isis initiates, pointing out that Apuleius's Isiac procession distinguishes between initiates and priests.[264] However, these passages might simply distinguish between different degrees of what Martzavou calls "sacerdotisation."[265] In Roman Egypt, there were many grades of priestly status, including not only full-time priests but also temporary practitioners who served the temple on a rotating basis while also maintaining other employment.[266] The latter arrangement actually resembles the ultimate situation of Apuleius's Lucius, who serves the temple as a *pastophorus* but also works as an advocate.[267] Rather than a strict division between "priests" and "non-priests," the Greco-Roman Isis cult likely involved a spectrum of religious practitioners with various levels of commitment and specialization.[268]

Participation in (at least some aspects of) Isis worship did not necessarily require initiation, nor did all Pompeian Isis worshipers necessarily embrace one

262. Apul. *Met.* 11.19, 11.21, 11.23, 11.28, 11.30. On evidence for behavioral restrictions on priests in Roman Egypt, see the *Gnomon* of the Idios Logos (see esp. §71; *BGU* V 1210, ed. Schubart 1919; *POxy.* 42.3014); the hieroglyphic cult regulations at Esna Temple (Sauneron 1962: 340–349); and the (likely idealizing) accounts of Chaeremon (Porph. *Abst.* 4.6–8; van der Horst 1984: 16–23, 56–61) and Plutarch (*De Is. et Os.* 4–8 = *Mor.* 352C–354B). See Bowden 2010: 168; Bremmer 2014: 117.

263. Kruchten 1989; Moyer 2011: 258–260; Jasnow and Zauzich 2005: 54–61; Quack 2007b: 250–262. On "mysteries" in Egypt, see note 243, this chapter.

264. Alvar 2008: 343–344; Apul. *Met.* 11.10, 11.17. However, Alvar himself notes that it is "somewhat baffling to find that the *pastophori*, who certainly included some who had been initiated (if indeed they were not all initiates), are only summoned afterwards" to enter the inner sanctuary, rather than entering with the initiates (2008: 344n560, referring to Apul. *Met.* 11.17).

265. Martzavou 2011: 81.

266. On Egyptian priests, see Otto 1905–1908; Dunand and Zivie-Coche 2004: 102–104, 121–122, 210–213, 296–297; Clarysse 2010; Teeter 2011: 16–38. The Roman-period Egyptian temple manual known as the *Book of the Temple* discusses the many priests associated with an ideal temple (Quack 2004: 19–24). These ritual specialists range from "les plus élevés de la hiérarchie jusqu'aux couches les plus basses de la population," most of whom work "un système de quatre phylés tournantes pour le service mensuel" (Quack 2004: 20). Although the *Gnomon* of the Idios Logos forbids "priests" from holding other occupations (§71; see also notes 262 and 280, this chapter), many lower-ranking positions would not have merited the Greek title of *hiereus*, so this restriction would have been irrelevant to them. For example, the *Gnomon* (§82) explicitly excludes *pastophoroi* from the category of *hiereus*.

267. Apul. *Met.* 11.30. On *pastophori*/*pastophoroi*, see Bricault 2012: 92–93; Hoffmann and Quack 2014.

268. Martzavou 2011.

collective identity. Additionally, to the extent that people did conceive of *Isiacus* (or something like it) as an available religious identity, it did not exclude other religious identities.[269] Roman religion was profoundly eclectic, expansive, and pluralistic, and membership in multiple religious communities was completely feasible.[270] As Rüpke points out of elective cults in general, "the intense internal interaction of the communities thus created, the degree and range of exclusivity have been grossly overvalued."[271] Dedications to Isis together with "non-Egyptian" gods are common, and in domestic shrines, Isis and her circle often rub shoulders with deities of non-Egyptian origins.[272] Priests of Isis or Sarapis could also hold priesthoods for other deities, and many people who held other cultic positions, such as *seviri augustales* or *flamines*, also made dedications to Isis and/or Sarapis.[273]

Accordingly, it is misleading to speak of a distinct Isis "religion" in the first century CE.[274] The first-century empire presented individuals with numerous religious options, but most of these options were not yet separate "religions" in modern terms.[275] At most they were separate cults, though Rüpke suggests that

269. Beard et al. (1998: 286) suggest that some Isis worshipers may have been more exclusive, and Price reads Apuleius's novel as suggesting that "some people were exclusively attached to the cult of Isis" (2000: 300). However, there is no indication that Lucius (or anyone else) was expected to deny the existence, or eschew the cults, of other deities. Both Egyptian religion and Greco-Roman adaptations of Isis cult employed a henotheistic theology in which other deities were actually *aspects* of Isis (Apul. *Met.* 11.5; Nagel 2017). Such henotheistic approaches to Isis are attested in Campania: e.g., Arrius Balbinus's dedication to Isis as "una quae es omnia" at Capua (*RICIS* 504/0601).

270. On religious pluralism, see Ando 2003; Belayche and Rebillard 2008; Rebillard and Sotinel 2010a; Rebillard 2012, esp. 93–94.

271. Rüpke 2013a: 17.

272. *Images from domestic shrines*: see De Caro 2006b: nos. III.19–21; Kaufmann-Heinimann 2007: 155; van Andringa 2009: 259; Petersen 2012. *Domestic pantheons' eclecticism*: see van Andringa 2009: 256–269; Petersen 2012. *Joint dedications*: for an example from Campania (Puteoli), see a "hearing ear" plaque dedicated to Juno Sospes, the Mother of the Gods, and Isis (De Caro 2006b: no. II.9).

273. *On priests of Egyptian gods holding other priesthoods*, see Burkert 1987: 49; Mikalson 1998: 237–238; Bowden 2010: 178. *On holders of other priestly positions dedicating to Egyptian gods*, see Alvar and Muñiz 2004: 89–90; Bricault and Veymiers 2014a: 21–22; Gasparini 2014: 267–272.

274. For a critique of such terminology, see Burkert 1987: 3–4; Bricault 2000a: 92; Malaise 2005: 29; Dunand 2010; Versluys 2010: 28–29.

275. *Pace* Beard et al. 1998; Price 2000; and agreeing instead with North 1992: 187–192; Alvar 2008: 2–4; 2017: 36–37 (on "eastern" cults as, at most, "proto-religions"); Rüpke 2013b: 7–8, 20. *On gradual processes of "religionization" in antiquity*, see Rüpke 2013b: 25–34.

even "cult" implies too much transregional coherence.[276] To the extent that participation in shared rituals provided a source of identity, that identity was nonexclusive and potentially compatible with many other identities: social, political, cultural, ethnic, religious, and more.

Furthermore, most of Isis worshipers' supposedly distinctive behaviors—unusual modes of dress, grooming, and prohibitions on activities such as sexual intercourse[277]—were occasional and temporary, not year-round. Scholars sometimes imply that Isis worshipers followed a different lifestyle from other Romans.[278] However, most of the behavioral restrictions were confined to brief purificatory periods before major festivals.[279] Even priests may have only maintained their characteristic dress and shaved heads in certain settings and at certain times.[280] Swetnam-Burland points out that funerary portraits of priests in Roman Egypt typically depict them with the same hairstyles and clothes as other people.[281] Similarly, Apuleius's Lucius has to reshave his head for his second and third initiations, implying that his hair has grown out in between.[282] For most of the year, the lifestyle of even the most devout Isis worshipers would have been hard to distinguish from that of anybody else in Italy.

---

276. Rüpke 2007c: 7; 2013a: 4–6; cf. Gordon 2014: 659.

277. For sexual prohibitions, see Alvar 2008: 177–192.

278. E.g., Beard et al. 1998: 1.289. Heyob emphasizes Isis's supposed associations with chastity, "asceticism," and "morality" (1975: 111–127; cf. Mol 2013: 127, 2015: 387n 925). Alvar acknowledges that the chastity requirements may have been temporary, but asserts that "there can be no doubt that sexual abstinence was understood both as a mark of piety and as a technique to attain moral purity over a life-time" (2008: 178–179).

279. Bremmer 2014: 120. Chaeremon characterizes the sexual and dietary abstinence of Egyptian priests as a *temporary* purification before important rituals (Porph. *Abst.* 4.7; van der Horst 1984: 20–21). According to the second century CE hieroglyphic cult regulations of Esna, temple personnel abstained from sex and impure foods for 1–9 days before various rites (Sauneron 1962: 340–349). Temporary abstinence from sex characterized not only "Isiac" festivals but also some other Roman festivals, such as that of Ceres (e.g., Ov. *Am.* 3.10).

280. Swetnam-Burland (2011: 347) suggests that this choice deviates from the behavioral requirements in the *Gnomon* of the Idios Logos. This is possible, though one might also note that many priests did not serve year-round but, rather, on a rotating, several-month basis (see note 266, this chapter), and they may not have needed to wear linen or shave their heads when off duty. The *Gnomon* states that these restrictions apply even when the priests are not participating in divine processions ("ὀυδὲ ἐὰν [ξε]νωθῶσιν τοῦ θ[ε]ί[ο]υ ψαου," §71), but this does not necessarily mean when they are off duty altogether; see Schubart (1919: 79) on ψαου. The *Gnomon*'s restrictions also may not have applied to lower-ranking officiants (see note 266, this chapter).

281. Swetnam-Burland 2011: 347. Corcoran challenges the identification of some mummy portraits as priests (1995: 71–72), but Roman-period Egyptian statues of priests—identified as such by hieroglyphic inscriptions—sometimes also display full heads of hair (e.g., Riggs 2005: fig. 37).

282. Apul. *Met.* 11.28, 11.30; discussed by Swetnam-Burland 2011: 347.

In place of binary statements that Italians either *did* or *did not* worship Isis and her circle, evidence suggests an enormous spectrum of overlapping possible attitudes, beliefs, and practices related to deities of Egyptian origin. Not all worshipers of Isis (and perhaps none at Pompeii) were initiates; not all shared identical practices or beliefs; not all necessarily self-identified as "Isiacs"; and to the extent that "Isiac" religious identities and modes of "groupness" existed, they were nonexclusive and frequently co-occurred with other religious identities.

These conclusions have substantial ramifications for the interpretation of material culture depicting, or alluding to, Egyptian deities. When we encounter such images in Pompeian homes, it is not enough to simply ask whether their owners were "Isis worshipers," as though "Isis worshipers" were a monolithic collective entity that functioned as a unitary social actor. Instead of one true way to be an "Isiac," there were numerous different ways for people (and material culture) to interact with deities of Egyptian origin.

Positioning the Inhabitants of the Casa di Acceptus e Euhodia on the "Isiac" Spectrum

Characterizing "Isiac cult" as a spectrum of behaviors, rather than a binary condition, enables us to reconsider religious identity at the Casa di Acceptus e Euhodia. The inhabitants of this house likely did fall somewhere on the "Isiac" spectrum. The central image from the lararium provides evidence for the incorporation of an Egyptian-derived deity into regular ritual practices. Such ritual activity in turn suggests that some members of this *familia* likely possessed at least some minimal beliefs about Isis: for example, that she existed, that she had power to influence the world (and perhaps more specifically power over fate, as indicated by her syncretism with Fortuna), and that it was possible to appeal to her through domestic cult practice.

However, we *cannot* automatically conclude that any inhabitants of this house underwent initiation into "Isiac mysteries." We cannot even be certain that the inhabitants of the Casa di Acceptus e Euhodia took part, regularly or at all, in rituals at the Iseum. It is certainly possible that some members of this *familia* visited the Iseum, but we have no direct evidence for their performance of any rituals besides those at their own lararium.

Viewership and Religious Identities

At least some inhabitants of the Casa di Acceptus e Euhodia thus appear to have participated, at least to some degree, in the domestic worship of a (partially) Egyptian deity. However, we should not be too quick to pin our entire understanding of the garden installation on this conclusion. Most scholarship on Roman Aegyptiaca presumes that "Isiacs" would have responded to Egyptian or "Egyptianizing" imagery in a fundamentally different way from non-"Isiacs." As we have seen, however, the category of "Isiacs" was likely less coherent than typically assumed. Multiple forms of "Isiac" belief and practice were integrated into life at Pompeii and did not necessarily require formal membership in a distinct group.

Additionally, it would be misleading to assume that an "Isiac"—however one chooses to define that term—need only have one possible response to Egyptian imagery. Textual evidence for differing practices among Isis worshipers, as well as the general absence of any transregional authority enforcing "Isiac" conformity, should remind us that Isis worshipers were not a monolithic bloc. While some people on the "Isiac" spectrum might conceivably see Nilotic imagery as a reference to sacred festival rites, others might disapprove of the pygmies' frequent excesses.[283] Others might not give salience to their "Isiac" identity at all (if indeed they ever so self-identified) when viewing Nilotic or Egyptian images.

When house owners had a personal interest in Egyptian cults, that interest might well inform their conversation about domestic Aegyptiaca. For example, if guests remarked on the Nilotic resonances of statuary at the Casa di Acceptus e Euhodia, their host might conceivably have responded with personal views about the marvels of the Nile and its significance in Egyptian religion. Such topics were frequently discussed in Greco-Roman literature on Egypt, and there is little reason to think worshipers considered their discussion inappropriate.[284] Precisely because such topics

283. Some of the more extreme *tḫy*-festival celebrations may also have been controversial within Egypt; Jasnow and Smith (2010–11) interpret a Demotic papyrus from the first or second century CE as a defense of orgiastic festival behavior against its critics. Bryan (2014: 114, 118) notes some evidence for Egyptian disapproval of excessive *tḫy*-festivities in the 18th Dynasty, as well as the Greco-Roman period.

284. On the place of such topics in Greco-Roman intellectual discourse, see chapter 3, this volume, "Performing Cosmopolitanism." Some Isis worshipers, particularly those who had undergone initiations, may have considered some specific religious topics inappropriate for conversation. For example, Lucian's satirical *Assembly of the Gods* (11) suggests that the reasons for animal worship are known only to initiates. However, this text's purpose is overtly humorous, and as discussed earlier, there is no definitive evidence for initiation at the Pompeii Iseum. Supporters of Isis worship did not always feel constrained on the subject of animal cults (e.g., Plut.. *De Is. et Os.* 70–76 = *Mor.* 379D–382C), and Diodorus portrays most Egyptians (aside from some priests) as quite happy to expound on the practice (Diod. Sic. 1.86–88; for parallels to some of his sources' theories, cf. Plutarch, cited earlier, and Cic. *Nat. D.* 1.101). Although Egyptian sources characterize some subjects (e.g., the death of Osiris) as inappropriate for casual discussion, they do not generally attempt to forbid discussions of religious practice altogether (Barrett 2011n418, on Osiris; cf. Hdt. 2.170–171).

arose frequently in Greco-Roman literature, however, people need not have personally identified as "Isiac" to discuss them. People of any religious persuasion could be expected to engage in informed discourse about the visual culture around them.

Furthermore, whatever the religious activities of the house inhabitants, they would not have been the only viewers of the garden assemblage. In addition to family members and slaves, the sculptures' audience would have included visiting guests, a group whose religious composition might have varied from one day to the next. There is no evidence that Isis worshipers insisted on socializing only with other Isis worshipers, and much evidence in fact suggests that this was not the case.[285] Regardless of religious identity or position on the "Isiac" spectrum, the users of the Casa di Acceptus e Euhodia garden space might have responded to its imagery in a potentially wide variety of ways.

The Valuation of Domestic Material Culture: Beyond "Religious" vs. "Decorative"

Valuation, Function, and Iconography

Existing scholarship frequently interprets Egyptian themes in garden statuary either as expressions of the house owners' "Isiac" religious identity or as examples of nonreligious "exoticizing" decoration. This binary choice presumes a mutually exclusive relationship between "religious" and "decorative" functions. Exemplifying "religious" interpretations at the Casa di Acceptus e Euhodia, Mau took both the lararium painting and the faience statuettes as evidence that the house owners were Isis-worshiping Alexandrians, and Bragantini has more recently argued that the image of Isis-Fortuna on the lararium may help explain the "statuette di divinità egizie" in the garden.[286] Exemplifying "decorative" interpretations, Alfano writes of garden statuary of "aquatic animals and extravagant-looking gods, such as Bes and Ptah-Pateco, always used for decorative and non-religious purposes."[287] Mol similarly asserts that "A serious adherent of the Isis cult would not 'fool about' with Egypt as exotic garden display element."[288] Besides leaving the category of

285. For example, Propertius and Tibullus characterize the girlfriends of their poems' speakers—but not the speakers themselves—as Isis worshipers. Propertius humorously calls for driving Isis out of Rome so her rituals cannot enjoin Cynthia to chastity (Prop. 2.33A). Though Tibullus's speaker calls on Isis to fulfill Delia's prayers to cure him, he seems somewhat dubious, and he characterizes Isis specifically as *Delia's* goddess: "quid tua nunc Isis mihi, Delia?" (Tib. 1.3.23).

286. Mau 1908: 362; Bragantini 1998c: 648.

287. Alfano 2001b: 340.

288. Mol 2013: 127.

"serious adherent of the Isis cult" undefined, this statement presumes that "serious adherence" is incompatible with the simultaneous use of Egyptian motifs to convey playfulness, exoticism, or luxury. The result is the further entrenching of a priori binaries: not only are people either "Isiacs" or "non-Isiacs," but material culture also serves either "religious" functions or "decorative" ones.

As we have seen, however, there are many reasons to question the existence of a meaningful emic distinction between "religious" and "decorative" material culture.[289] Scholarly debates about "religious" versus "decorative" functions for domestic statuettes, "Egyptianizing" or otherwise, draw more on modern, etic concepts of "religion" and "decoration" than on ancient, emic concepts.[290] A high degree of technical quality could make a figurine effective not only for displaying its owner's taste but also as a pleasing gift for the gods; a figurine's ritual functions need not stop viewers from admiring its workmanship; and even if a divine image did not receive regular offerings, it might still evoke thoughts of piety in viewers or reassure them of divine protection. Furthermore, seeking a single totalizing "meaning" for any figurine is misleading, as different individuals might respond differently to the same object depending on background or context. Beyond binary oppositions of "cultic" and "decorative" uses, material and textual evidence indicates that figurines could play numerous active roles within Hellenistic and Roman houses: facilitating human–divine encounters, protecting the home, impressing or amusing viewers, prompting guests to contemplate their own social performance, or even activating magical spells.[291] By participating in a wide range of domestic interactions, such objects inextricably entangled the human, divine, and material worlds.

To this end, we must be careful to distinguish between objects' *function(s)* and their perceived *value(s)*. I here employ David Graeber's definition of economic or object value as "the degree to which objects are desired," based on culturally mediated "ideas about what [people] *ought* to want."[292] The reasons people wish to own an object—in other words, the reasons they attribute value to it—may draw on many factors above and beyond the object's function(s). For example, a statuette in a lararium might *function* as a recipient of domestic cult, but its owners might *value* it for numerous additional reasons. People might ascribe value to the statuette because of its aesthetic appeal; its visual allusions to earlier

289. See chapter 1, this volume, "Material Culture, Identity, and 'Authenticity.'"

290. See further discussion in Barrett, in press (a).

291. Barrett 2015d, and in press (a).

292. Graeber 2001: 1–3. Archaeological approaches to value: Bailey and Mills 1998; Barrett 2009; Papadopoulos and Urton 2012.

or more monumental artworks; its valuable raw materials; its place of manufacture; its connection to family history; or the socially advantageous impression of piety that it conveyed to visiting guests—among many other possibilities. Yet none of these sources of value should be assumed, a priori, to diminish the importance of the object for domestic cult or make it somehow "less religious." Valuation is not zero-sum. Additionally, different viewers might attribute very different *values* to an object without necessarily changing its *functions*. The various members of a household might have different reasons for valuing (or failing to value) the objects furnishing the house, even if they used those objects in similar ways.

To understand what forms of value ancient Pompeians would have attached to the garden statuettes at the Casa di Acceptus e Euhodia, we first need to examine the perceived affordances of those statuettes. One important question concerns ancient perceptions of their iconography. Certainly, Pompeians did not perceive divine images as inappropriate for domestic gardens, as numerous images of Greco-Roman gods come from garden contexts.[293] However, scholars disagree on whether ancient Pompeians would have recognized some of the Casa di Acceptus e Euhodia statuettes as deities. For Mau and Bragantini, the statuettes of dwarflike beings represent deities connected with Isis cult. On the other hand, Mol argues that Roman viewers had little knowledge of Egyptian dwarf deities and might have reinterpreted both the Bes and so-called Pataikos statuettes as human dwarfs or "grotesques."[294] An analysis of these statuettes' valuation thus needs to begin at the basic level of iconography.

Much of the statuary from the Casa di Acceptus e Euhodia garden depicts figures widely recognizable in Roman visual culture: for example, animals and members of the Bacchic *thiasos*. Also widely recognizable would have been the *modius* crown associating one herm with Sarapis. Images of Sarapis were widespread in multiple media in Flavian Italy, and he received cult in private as well as public contexts at Pompeii.[295] Most viewers in Italy would thus have been familiar with images of Sarapis, whatever their opinions on his cult.

However, not all Pompeian Aegyptiaca represent figures so well-known in Italy. The Casa di Acceptus e Euhodia garden also contained several depictions

293. E.g., Jashemski 1979–93: 1:115–140; see also chapter 3, this volume.

294. Mol 2013: 127; 2015: 182–219, esp. 212–213.

295. On the most prominent members of the "*gens* isiaque" outside Egypt, see Malaise 2005: 33–78, 222; 2007: 22–27. On the presence of four especially prominent Egyptian deities—Isis, Sarapis, Anubis, and Harpocrates—in the Pompeii Iseum, as well as in private contexts such as the Casa degli Amorini Dorati, see van Andringa 2012: 106–107.

of Egyptian figures whose Italian interpretations are more ambiguous. The three glazed statuettes of dwarflike figures fall into this category, as does the *nemes*-wearing herm.

Dwarf Deities, Pygmies, and Sileni

Mol's study of Pompeian Aegyptiaca provides the most recent discussion of Roman receptions of Egyptian dwarf deities. In her interpretation, Egyptians would have understood both the Bes statuettes and the so-called Pataikos statuette as divine images. However, she argues that Bes and Pataikos lost their divine status in Italy, where consumers reinterpreted their images as human dwarfs or pygmies and valued them for apotropaic reasons.[296] Mol further argues that while Bes sometimes retained associations with Egypt, these associations need not always have been present: "in many . . . examples a mental connection to Egypt was absent."[297]

Mol's study makes the important point that images need not retain the same meanings in Italy that they originally possessed in Egypt. However, as I have argued, there is actually no evidence that Egyptians would themselves have interpreted the being in figure 6.3 as "Ptah-Pataikos" or any other god. This figurine's closest parallels are not Egyptian depictions of Ptah as a dwarf but, rather, Egyptian depictions of probable human dwarfs. Pompeian uses of such figurines to represent human dwarfs or pygmies are actually thus largely *consistent* with their "original" Egyptian meanings.

I further propose that many Pompeians likely *did* associate images of Bes (as opposed to the misnamed "Pataikos") with the divine sphere—but that they may *not* have identified the Bes figure as a deity or recipient of cult, instead interpreting him as an attendant on other deities. Pompeian audiences did not necessarily know Bes's name (which actually varied in Egypt as well[298]), and they may not have known that Egyptians worshiped him as a god. However, images of Bes at Pompeii do frequently appear in "Isiac" contexts or in close association with images of Isis.[299] This contextual evidence suggests a local perception of Bes

296. Mol 2015: 182–219. However, recent research suggests shaky foundations for apotropaic explanations of Roman dwarf imagery: see note 186, chapter 2, this volume.

297. Mol 2013: 127; 2015: 212–219 (quoting 219).

298. The so-called Bes image was actually identified with multiple Egyptian dwarf-gods (see note 85, chapter 2, this volume).

299. *Pace* Mol (2013: 127; 2015: 182–219), who emphasizes separation between these spheres. On the relationship between Bes and the "Isiac" circle, see also Malaise 2004; 2005: 79–80 (concluding that Bes was "plutôt un compagnon qu'un member de la souche isiaque": 2005: 80).

as a member of Isis's entourage—an association recalling Bes's longstanding role in Egypt as an attendant on female deities.[300]

Iconographic and contextual evidence also suggests that Greeks and Romans often identified Bes with Silenus and satyrs, the attendants of Dionysus/Bacchus. As we will see, images of Bes at Pompeii sometimes appear in close association with Bacchic imagery, as at the Casa di Acceptus e Euhodia. Pompeians may thus have perceived Bes as a potential member of the entourages of both Isis and Bacchus. Reimagined as a sort of Egyptian satyr or Silenus, Bes forges another link between two divine spheres already frequently associated at Pompeii: Egyptian and Bacchic. His presence in the garden of the Casa di Acceptus e Euhodia thus evokes not only the perceived appropriateness of dwarflike figures within Nilotic landscapes but also the perceived appropriateness of both Egyptian and Bacchic themes within garden settings.

Images of Bes appear in multiple media at Pompeii: figurines, painting, jewelry, embellishments on lampstands and sistra, and even coins. Seven or possibly eight faience or pseudo-faience figurines of Bes are recorded. Four come from domestic contexts (three identifiable as gardens), one possible example comes from the Iseum, and the others come from unknown findspots.[301] A terracotta lampstand from a house or shop (I 18.4) depicts a standing Bes wearing an Egyptian *shendyt* kilt.[302] The only attested painting of Bes at Pompeii comes from the so-called sacrarium of the Iseum (a room seemingly used for storing cult objects).[303] One bronze statuette of Bes from Pompeii has an unknown context.[304] The handles of three or possibly four bronze sistra, one from a shop and the others from uncertain contexts, depict Bes holding up a Hathor head.[305] Bes also appears on two necklaces composed of multiple amulets. One, from a shop (I 10.17), included (among other motifs) two Bes amulets, one Isis-Fortuna, one Harpocrates, and a

300. See chapter 2, this volume, "Egyptian Antecedents and Affordances."

301. See notes 14 and 15, this chapter.

302. Tran tam Tinh 1964: 165, no. 114; D'Ambrosio and Borriello 1990: no. 78; Malaise 2004: 282; Mol 2015: 184.

303. De Caro 1992: no. 1.72. On the "sacrarium," see Egelhaaf-Gaiser 2000: 188; Moormann 2007: 148–149; Swetnam-Burland 2015: 109–112.

304. De Caro 2006b: III.49.

305. Tran tam Tinh 1964: 181–183, nos. 173, 181; Malaise 2004: 288 (suggesting that the sistra are imports rather than local products); De Caro 2006b: nos. III.1 (from VII 4.13, a shop), III.8 (from the "campagna pompeiana"), III.13 (context unknown), and probably also the poorly preserved III.9 (context unknown); Mol 2015: 184. Tran tam Tinh also describes a sistrum from the Casa di C. Vibius Italus as depicting Bes (1964: no. 172; followed by Mol 2015: 184), but Bes does not in fact appear on this object (see De Caro 2006b: no. III.5).

lotus flower.[306] Another, from a house (V 3.11), included amulets in various semi-precious materials with a wide array of themes: some Egyptian (Bes, Harpocrates, Bastet, multiple scarabs, and a naked female figure with parallels in Egypt), some pertaining to Roman apotropaic traditions (phalli and hands making apotropaic gestures), among other motifs.[307] This necklace came from a chest that also contained three bronze statuettes, two depicting Harpocrates and one depicting Venus, and a bronze sistrum, as well as other domestic valuables.[308] Eight other amulets of Bes, mostly made of blue frit or glass paste, have also been found at Pompeii. Most come from unknown contexts, except for one from a shop (I 1.4) and one from a house (V 1.7).[309]

The contexts of several of these objects associate Bes with Isis. The Iseum contained both a painting of Bes and a now-lost figurine that may have depicted Bes.[310] Bes also appears on three sistra, instruments employed in Isis cult.[311] These sistra further associate Bes with Egyptian deities by depicting him holding up the head of Hathor, who was often identified with Isis in Ptolemaic and Roman Egypt, and whose "Pharaonizing" depiction would have clearly identified her as an Egyptian figure.[312] Two necklaces associated amulets of Bes with amulets of other Egyptian deities, including Isis and Harpocrates. One of these necklaces was also stored with two statuettes of Harpocrates.

While most Pompeian sculpted or painted images of Bes are formally and stylistically "Pharaonizing," two—the Iseum painting and a bronze statuette—are rather different.[313] The bronze statuette depicts a naked, *modius*-crowned figure squatting with hands on hips. The pose is characteristic of Bes, and the smallness of the limbs relative to the trunk suggests dwarfism. However, the face is fully anthropomorphic rather than leonine and depicts a broad-cheeked, bearded, middle-aged male.

306. Tran tam Tinh 1964: 167, no. 117; Malaise 2004: 282; Mol 2015: 184.

307. De Vos 1983: 66–67; De Caro 2006b: no. III.137. The naked female figure is squatting with her hands on her hips. On the Egyptian associations of such figures, see Barrett 2011: 208–220.

308. Paribeni 1902: 370–371; Boyce 1937: 108, no. 2.

309. De Caro 2006b: nos. III.88–III.95; cf. Pannuti 1983: 72–74, 182–184. The editors classify four of these (nos. III.92–III.95) as possibly the so-called Pataikos rather than Bes (following Pannuti 1983: 73–74, 183–184), but traces of Bes's characteristic cheek mane are visible on all four. However, since the modeling of these four figurines is quite indistinct, Pompeian consumers might certainly have interpreted them as more generic "pygmy" types.

310. On the lost figurine, see note 15, this chapter.

311. On sistra in the Greco-Roman world, see Barrett 2011: 129n408.

312. On Isis's association with Hathor, see Barrett 2011: 128–131; Nagel 2017.

313. *Iseum painting:* De Caro 1992: no. 1.72. *Bronze statuette:* De Caro 2006b: no. III.49.

Similar features also appear on the Iseum painting, which depicts a naked, lotus-crowned Bes seated with hands on hips. This Bes also has a fully anthropomorphic face, and the traditional mane on his cheeks has become a beard.

At Pompeii, the earliest precedents for these representations of Bes as anthropomorphic and bearded come from imported coinage of the second century BCE. Widely circulated and imitated in Samnite Pompeii, a series of coins from Ebusus (modern Ibiza) depicts a Bes figure with a human head and thick beard.[314] Viewers in Ebusus would have seen a Punic adaptation of the Egyptian god; but in Campania, such images may have already evoked other associations.

In the Greco-Roman world, these anthropomorphic, bearded beings have numerous parallels in Hellenistic and Roman images of an otherwise Bes-like figure whose broad, bearded, anthropomorphic face resembles that of Silenus.[315] Examples include, among others, plastic vases from Hellenistic Delos and Crete; early Hellenistic potstands from Locri; terracotta figurines from Ptolemaic Athribis; two statues of Bes-like satyrs or Sileni at Hadrian's villa at Tivoli (figure 6.18); and terracotta and bronze plastic vases from Roman Dalmatia.[316] Scholars often identify such figures as "Bes-Silenus."[317] This term is a purely scholarly convention, as we do not know what names ancient viewers actually applied to these figures. However, viewers familiar with Greco-Roman visual culture would likely have recognized these figures as both "Egyptian" and "Bacchic," simultaneously resembling both Egyptian dwarfs and Sileni.

Textual evidence also suggests that Greeks and Romans interpreted Egyptian dwarf deities as variants on Sileni and satyrs. Diodorus Siculus's reference to "satyr people" from Ethiopia who danced and sang for Osiris (1.18.4) may recall the roles of Egyptian dwarfs and dwarf-deities as musicians and dancers in religious

---

314. See Frey-Kupper 2014: 100–103; Hobbs 2017. In the second to first centuries BCE, the bronze coinage circulating at Pompeii consisted of imported coins from other cities and local imitations of those imports. Most of the Ebusan Bes coins (and local imitations thereof) at Pompeii date to 150–101 BCE, but some stayed in circulation into the first century CE (Hobbs 2017: 353, 354). The coins from Ebusus (Hobbs 2017) reflect a Carthaginian adaptation of an originally Pharaonic-style Bes image.

315. While Volokhine (2010: 255) notes the "Silenus"-like features of the Iseum painting, these features have gone previously unobserved for the statuette (De Caro 2006b: no. III.49).

316. *Delian vase:* Hatzidakis 2004: 386, no. 83.N; Barrett 2011: 278–279, 582–583, fig. D77. *Cretan vases (Ierapetra and Mochlos):* Vogeikoff-Brogan 2016. *Locri potstands:* Bagnasco 1992. *Athribis figurines:* Mysliwiec 1998: 12–13. *Tivoli statues:* On the resemblance to both satyrs and Bes, see Kuttner 2003: 141, and note 17, this chapter. *Dalmatian figurines:* Bugarski-Mesdjian 2007: 314–316. A bearded, squatting figure at the upper left of the *Ariccia relief* also resembles this type (see most recently Veymiers 2018, though without reference to the Silenus-like features).

317. On such depictions, see Bagnasco 1992; Volokhine 2010: 248–255; Barrett 2011: 278–279.

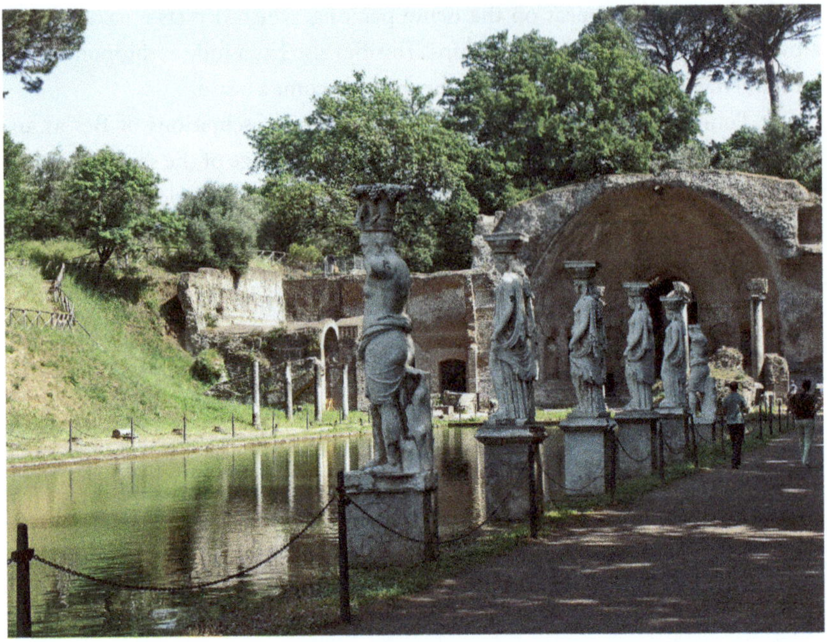

FIGURE 6.18 Statuary on the banks of the *euripus* in the so-called Canopus area of Hadrian's villa, facing south (modern casts; marble originals stored elsewhere). Sculpture visible in this photo includes replicas of the Erechtheion caryatids, flanked by two Sileni or satyrs with some features reminiscent of Bes (see Kuttner 2003: 141). At the end of the *euripus* is a large, vaulted banquet hall.

Photo: Mary Ann Sullivan, reproduced with permission.

festivals.[318] Plutarch (*De Is. et Os.* 14) and Philostratus (*V A* 6.27) also speak of "satyrs" living, respectively, in Egypt and Ethiopia. Ever since Herodotus, many Greeks conceptualized Osiris as the Egyptian Dionysus, and Classical and later Greek iconography often depicted dwarfs with satyr-like features and assimilated them to the Dionysiac sphere.[319] Just as satyrs and Sileni were attendants on Dionysus, so too were Egyptian dwarfs and dwarf-gods often attendants on deities (especially Hathor and other goddesses).[320] In another parallel to the

318. Darnell 1995: 57n58; Volokhine 2010: 248–249; Barrett 2011: 279.

319. *Osiris and Dionysus:* see note 115, this chapter. *Greco-Roman dwarfs and pygmies as satyr-like or Dionysiac:* Dasen 1993: 236–241, 243–244, 246; 2013: 263–264; Garmaise 1996: 76, 105–106, 144–145; Sparkes 2000: 88.

320. See chapter 2, this volume, "Egyptian Antecedents and Affordances."

Dionysiac *thiasos*, dwarfs and dwarf-gods appear in Egyptian art as "grotesque"-featured but cheerful dancers, musicians, and merrymakers.[321]

At Pompeii, then, Bes figures may possess both "Isiac" and Bacchic associations. Such associations do not mean that every image of Bes necessarily played a role in religious *rituals*. A ritual connection is plausible for the Iseum images and probably also for the sistra decorated with Bes figures, since sistra were used in cult performances. However, the functions and perceived valuations of many of the other artifacts likely varied widely. Nonetheless, both iconographic and contextual evidence suggests that Pompeians could perceive the Bes image as a potential participant in the entourages of both Isis and Bacchus, and that such images were potentially appropriate within both "Isiac" and Bacchic milieus.

I say "the Bes image" rather than "Bes" because it is by no means clear that Pompeians attached the name "Bes" (or any other Egyptian name) to these figures. As Malaise points out, none of approximately 1,700 "Isiac" inscriptions recorded throughout the Greco-Roman world refers directly to the name "Bes."[322] Volokhine is thus probably correct to suggest that Roman audiences felt no need to learn a new name for a being they thought they already knew: "il est fort probable qu'ils y voyaient simplement un Silène égyptien."[323] Many Romans likely perceived Bes images as simply "Egyptian"-flavored representations of Sileni or satyrs.[324]

Reenvisioned within the Bacchic *thiasos*, Bes (or, perhaps, "Egyptian Silenus") becomes a visible embodiment of the connection between Egyptian and Bacchic imagery at Pompeii. The identification of Osiris (and, to some degree, Sarapis) with Dionysus/Bacchus provides one explanation for this connection.[325] More broadly, however, Egyptian and Bacchic themes shared additional cultural associations in Roman Italy. Egyptian and Bacchic motifs often appear together both in sanctuaries and in domestic contexts, and both were considered particularly appropriate for gardens.[326] Both Egyptian and

---

321. Volokhine 2010: 248–255.

322. Malaise 2004: 281 (though with some discussion of an obscure title attested at Thessaloniki, *besartes*: see *RICIS* 113/0537, 113/0561); followed by Volokhine 2010: 255; Mol 2015: 191.

323. Volokhine 2010: 255. Mol similarly considers it "unlikely that the word Bes was ever used within the context of Pompeii" (2015: 191).

324. Such a reinterpretation would have numerous parallels in the *interpretatio graeca* (and *romana*) of many other Egyptian deities as alternative names and forms for members of the Greco-Roman pantheon.

325. On Sarapis and Dionysus, see Plutarch (*De Is. et Os.* 28). On Osiris and Dionysus, see note 115, this chapter.

326. On Bacchic statuary in Roman gardens, see note 51, chapter 3, this volume.

Bacchic imagery could evoke spheres simultaneously within and beyond the safety of normative practices and ordered society.[327] Such images might produce a pleasurable *frisson* for viewers who associated them with the outré, faraway, or mysterious. Paradoxically, however, their ability to produce this result also depended precisely on their very familiarity and ubiquity.

At the Casa di Acceptus e Euhodia, the two figurines of Bes (a term I will continue to use for convenience, but which here reflects strictly etic usage) were thus doubly appropriate to their garden setting. First, they enhanced the Nilotic qualities of the assemblage. Their "Pharaonizing" style and blue-green glaze signaled "Egypt," and their dwarflike proportions resembled Nilotic *pygmaei*. Second, their Bacchic associations complemented the Bacchic themes of the satyr *oscillum* and three of the garden's five herms. Simultaneously Bacchic and Egyptian, these figurines represent the ultimate incarnation of "familiar alterity."[328]

Imagining Pharaoh? *Nemes* Crowns and Familiar Foreignness

Like the Bes images and the pygmy statuette, the *nemes*-crowned herm also prompted viewers to think of "Egypt." Besides employing "Pharaonizing" style, this object wears Egyptian costume: a *nemes* crown with *uraeus*. However, the specific implications of this headgear require further analysis. While Egyptian viewers recognized *nemes*-wearing figures as kings, Italian viewers may have interpreted this crown more generally as a visual reference to Egypt.

Because the *nemes* crown is strongly associated with Egyptian themes in Roman art, Roman viewers would almost certainly have understood this headdress as "Egyptian." *Nemes*-crowned figures are fairly common in Italy, both in sanctuaries of Egyptian deities and in domestic contexts. The majority are either sphinxes or standing male figures, almost always stylistically "Pharaonizing." They appear in statuary, wall paintings, and other media.[329] The *nemes* crown also

---

327. See chapter 3, this volume, "Safety and Danger in the Domestic *Locus Amoenus*," and "Landscapes, Identities, and Viewer Responses."

328. Here adapting a term that Wyler (2006: 226) uses for Bacchic imagery.

329. *From Pompeii Iseum:* A terracotta sphinx (De Caro 1992: no. 5.2; 2006b: no. II.72) and wall paintings of sphinxes (De Caro 2006b: no. II.42, 1.64). *From possible Iseum at Beneventum:* Statues and statuette of a pharaoh, perhaps Domitian (De Caro 2006b: nos. II.92, II.93, II.96). *Artifacts from Pompeian houses:* De Caro 2006b: nos. III.25 (statuette of Isis), III.119–121 (herm heads), III.133 (sphinx table leg), III.134 (sphinx), and the statuette from the Casa

appears on at least five other Pompeian herms: three from houses, one from the Iseum, and one unprovenienced.[330]

In Egypt, the *nemes* was a kingly crown, and statues of Roman emperors in Egypt often wear this crown.[331] In Italy, too, a *nemes*-crowned statue from Beneventum may represent Domitian as pharaoh.[332] However, Italian viewers may not have interpreted the *nemes* as exclusive to pharaohs. Possibly suggestive of broader applications are Roman images of *nemes*-crowned sphinxes, at least one *nemes*-wearing Isis statuette from Pompeii, and second-century statues of a *nemes*-crowned Antinous;[333] even after his deification, Antinous was not a pharaoh.

If any visitors to the Casa di Acceptus e Euhodia did understand the *nemes* as the crown of Pharaoh, then the *nemes*-crowned herm would have represented a figure both profoundly Egyptian and profoundly Italian. Egypt's actual pharaoh in the first century CE was, of course, the Roman emperor. The *nemes* crown might thus simultaneously evoke a faraway land and the central figure of Roman *imperium*, playing on the "exotic" qualities of Egyptian imagery while also naturalizing that imagery through connections to more familiar aspects of Roman life.

However, for viewers who did not interpret the *nemes* as a specifically royal headdress, this herm would have evoked somewhat different associations. The *nemes, uraeus*, and "Pharaonizing" style would still have recalled Egypt, and many viewers would be familiar with Roman images of Egyptian sphinxes and striding figures wearing similar headgear. In this case, however, the *nemes* crown

---

di Octavius Quartio (see this chapter, "Garden Luxury: The Casa di Octavius Quartio"). *Wall paintings at Pompeii:* Standing figure at the Casa del Frutteto (Mol 2015: fig. 4.21); sphinxes at the Casa del Bracciale d'Oro (Pearson 2015: fig. 118); vignettes of Egyptian deities at the Casa dei Misteri (Elsner 2006: fig. 8.1). *Statues of standing pharaonic figures from other sites:* Roullet 1972: nos. 152, 153, 155, 156, 159, 162–167a, 168, 169, 171, 172, 177. *Sphinxes from other sites:* Roullet 1972: nos. 279, 280, 284–288, 296–298, 300–301b, 310, 311–313; Versluys 2017a: 287–288. *Hadrianic or later statues of Antinous:* see note 163, this chapter. *Wall paintings at other sites:* De Caro 2006b: no. III.61 (vignettes of Egyptian deities). *Other artifacts from other sites:* Roullet 1972: nos. 53 (figure on fragmentary wall relief), 324 (assorted deities on "Mensa Isiaca"), 327 (pharaonic figure on a water clock); Pearson 2015: fig. 103 (bronze table leg of winged sphinx).

330. *From houses:* A bearded herm from II 1.12 and two beardless herms from another house (see note 119, this chapter). *From Iseum (bearded):* De Caro 1992: 70, no. 3.9. Carrella et al. also mention another, unpublished example, bearded, of unknown provenance, in the Naples Museum (2008: 150).

331. On imperial portraits in Egypt, see Kiss 1984; Brophy 2014.

332. See note 329, this chapter.

333. See notes 163 and 329, this chapter. Sphinxes could be identified with the deified king in Egypt, but there is little evidence that Italian viewers were aware of this theological point.

appears not on a sphinx or striding statue but on a herm: a form that, for Greeks and Romans, specifically suggested domestic protection and well-being.[334] Once again, the foreign turns into the familiar. The garden not only domesticates Egyptian imagery but even transforms it into the embodiment of domestic security.

The Garden as Ensemble: Egyptian, Bacchic, and Pompeian

The garden of the Casa di Acceptus e Euhodia presented its viewers with a range of "Egyptian" elements: the *nemes*-crowned herm, *modius*-crowned herm, faience and pseudo-faience Bes statuettes (potentially understandable as Egyptian "Sileni"), and faience pygmy. Together with the water features and fountain statuettes of river animals, these visual references to Egypt evoke a Nilotic landscape within the garden.

As in other Pompeian houses, however, these Egyptian themes do not appear in isolation. Bacchic imagery provides another especially prominent theme. If the Egyptian objects transport visitors to the banks of the Nile, then the Bacchic herms and satyr *oscillum* transform the garden into a wild landscape fit for Bacchus and his entourage. Some objects, such as the Bes figurines, are at home in both Egyptian and Bacchic settings. The juxtaposition of these Egyptian "Sileni" with a Silenus herm stages a confrontation between figures that Pompeians might have understood as "Egyptian" and "Greek" representations of the same being. Since literary evidence indicates that people sometimes identified Sarapis with Dionysus (e.g., Plut. *De Is. et Os.* 28), the *modius*-wearing herm may also represent a figure simultaneously Egyptian and Bacchic. In combining elements of both Egyptian and Bacchic landscapes, the garden assemblage interweaves the two.

The assemblage thus invites its users to consider—and construct—relationships between Egyptian and Greco-Roman traditions. The materials employed include marble (characteristic of Greek and Italian production) and faience (both genuine and imitation, and evocative of Egyptian production), and the regional and temporal styles of the objects allude to antecedents both "Greek" (Archaic through Hellenistic) and "Egyptian" (Pharaonic and Ptolemaic[335]). Moreover, the

---

334. On the associations of herms, see chapter 3, this volume, "Interactive Statuary: Social Performance and Service."

335. On allusions to both "Pharaonic" and "Ptolemaic" styles in Roman art, cf. Elsner (2004: xviii); Faraone 2019: 148–176. I use this terminology with some caution, since the actual visual culture of Ptolemaic Egypt frequently employed traditional, "Pharaonic" styles as well as more "Hellenistic" styles (see, e.g., Robins 1997: 231–251). However, there is little evidence

juxtaposition of materials, styles, and images creates new cultural combinations. Some faience objects employ "Hellenizing" stylistic features, as with the naturalistically rendered faience pygmy, and some marble objects employ "Pharaonizing" stylistic features, as with the *nemes*-wearing herm. The *nemes*-crowned herm and faience pygmy further allude to visual antecedents both Egyptian and Greco-Roman, while the *modius*-wearing herm recalls a deity, Sarapis, who himself embodies attempts to reconcile Egyptian traditions with Greek iconography. Rather than drawing a neat line between "Egyptian" and "Greco-Roman" material culture, the garden assemblage deconstructs and scrambles these categories to such an extent that they become mutually constitutive.

Aegyptiaca, Identity, and Material Culture in the "Isiac" Household

The material, stylistic, and iconographic eclecticism of the Casa di Acceptus e Euhodia garden is in many ways typical of first-century Pompeian domestic contexts. The evocation of a "Nilotic" landscape in the garden, and the juxtaposition of Egyptian imagery with a range of other visual and cultural references, is very much in keeping with the garden assemblages of the Casa dell'Efebo, Casa del Medico, Casa del Fauno, and Casa di Octavius Quartio.

However, the Casa di Acceptus e Euhodia also presents some evidence for the incorporation of Isis into household cult. According to dominant interpretations of Roman Aegyptiaca, this evidence of "Isiac" ritual practice should predict significant qualitative differences in the ways the inhabitants of this house used Egyptian imagery. However, such qualitative differences are difficult if not impossible to detect here. Although at least some members of this *familia* probably did fall somewhere on the "Isiac" spectrum, their engagement with images of "Egypt" appears entirely compatible with other Pompeians' engagement with similar images.

The Casa di Acceptus e Euhodia thus suggests continuity, not discontinuity, between the domestic material culture of "Isis worshipers" and other Romans. This finding complements other evidence suggesting a broad spectrum, rather than a binary divide, between "Isiacs" and others. Attempts to isolate a distinctive

that Roman audiences would have been able to discriminate visually between "Pharaonic"-style Egyptian artifacts from different periods. The visual difference between two Ptolemaic-period artifacts, one in indigenous "Pharaonic" style and one in "Greek" style, would probably have struck most Pompeians as more readily apparent than the visual difference between a "Pharaonic"-style artifact from Ptolemaic Egypt and one from the New Kingdom.

material culture signature for "Isiac" households are thus misguided,[336] as are attempts to identify some forms of Aegyptiaca as incompatible with "Isiac" practice. Those Pompeians who worshiped Isis (in whatever capacity and to whatever degree) appear to have employed domestic Aegyptiaca, and engaged with images of "Egypt," in ways essentially similar to their neighbors.

336. Cf. Rebillard's recent call to abandon attempts "to identify the signature of a religion in material culture," a project he sees as fatally tainted by the logical fallacy of "groupism" (2015, quoting 431). On "groupism," see chapter 1, this volume, "Material Culture, Identity, and 'Authenticity.'" Similarly, on Hellenistic Delos, the material culture of houses that contained figurines of Egyptian deities was otherwise indistinguishable from the material culture of houses without such figurines (Barrett 2011).

7

Conclusions

EGYPTIAN LANDSCAPES IN CONTEXT

Epilogue: From Pompeii to Tivoli

About half a century after the destruction of Pompeii, the emperor Hadrian's villa at Tivoli constructed an image of the Roman Empire in which Egyptian material culture played a foundational legitimizing role.[1] Numerous sculptures in "Pharaonic" styles, some imported from Egypt and some made in Italy, inserted Egyptian visual culture into imperial space.[2] Multiple statues depicted Hadrian's companion Antinous, deified after drowning in the Nile.[3] Some scholars even propose the villa as the original site of an obelisk, commissioned by Hadrian, whose hieroglyphic texts praise Hadrian as pharaoh and identify Antinous with Osiris.[4]

Hadrian's villa also included an outdoor water feature that reproduced on a grand scale the little *euripi* and *nili* of earlier Pompeian houses. In one area of the complex, a dining platform overlooked a truly monumental artificial river (figure 6.18). Sculpture on the banks celebrated the empire's broad expanse,

---

1. For a recent discussion, see Versluys 2012.

2. On the villa and its sculpture, see MacDonald and Pinto 1995. For Egyptian allusions, see recently Adembri and Mari 2006 (though note some controversy regarding the proposed "Antinoeion"; see note 4, this chapter).

3. MacDonald and Pinto 1995: 147, 149–151. On sculptures of Antinous attributed to the villa, see Roullet 1972: nos. 96–102; and see Raeder 1983 on the statuary generally.

4. On Hadrian's obelisk, see Derchain 1987; Meyer 1994; Grenier 2008: 1–35. Much controversy continues to attend the original location of the obelisk. Proposed sites include various locations at Hadrian's villa (Mari and Sgalambro 2007), in Rome, or even at Antinoopolis in Egypt. See the recent overview of Renberg (2010: 181–191), who argues against identifying part of Hadrian's villa as an "Antinoeion."

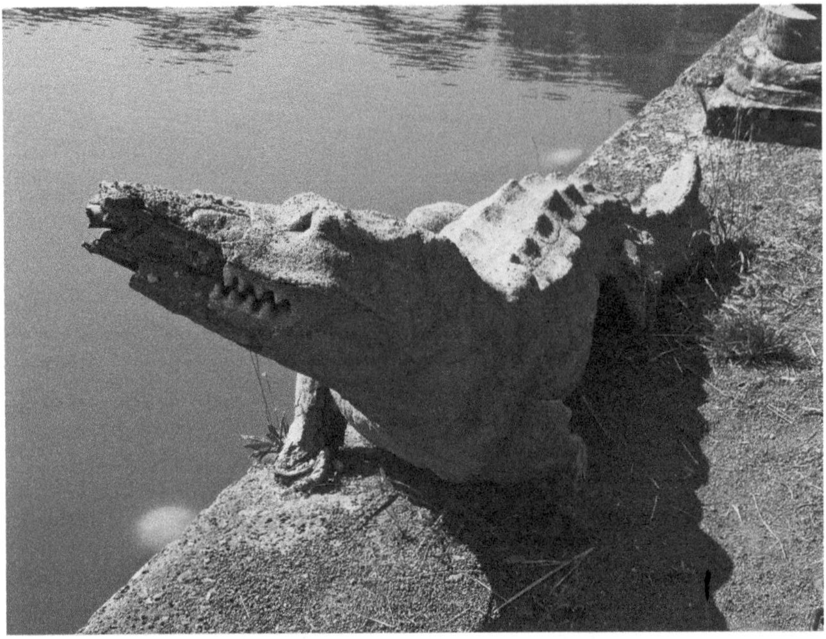

FIGURE 7.1 Crocodile statue on the bank of the *euripus* in Hadrian's villa (modern cast; marble original stored elsewhere).
Photo: Mary Ann Sullivan, reproduced with permission.

evoking Egypt (e.g., the crocodile in figure 7.1), as well as Italy and the Greek-speaking provinces of Achaea and Asia.[5] The modern designation of this area as Hadrian's "Canopus" depends on a late and notoriously unreliable text, the *Historia Augusta* (*Hist. Aug.*, Life of Hadrian 26.5), and I treat this label as purely conventional. Whatever Hadrian himself called this ensemble, however, its basic components are familiar from Pompeii: a prominent water feature, located within a planted garden,[6] associated with an outdoor dining installation, and decorated with an eclectic assemblage of material culture alluding to Egyptian, as well as Greek and Italian, styles, themes, and landscapes.

Nilotic and other Egyptian imagery shared space in Hadrian's villa with numerous visual allusions to other parts of the eastern Roman Empire. If the *Historia Augusta* does preserve older traditions about the villa, then the entire site was effectively a model of the Greek East. The text claims that Hadrian named

5. On the "Scenic Triclinium" and "Scenic Canal," see MacDonald and Pinto 1995: 108–116. For an analysis of the multiple geographic and cultural allusions of the statuary assemblage, see Kuttner 2003: 135–146.

6. On plantings alongside the canal, see Jashemski et al. 1992; MacDonald and Pinto 1995: 178.

parts of his villa after "the names of provinces and places of the greatest renown, calling them, for instance, Lyceum, Academia, Prytaneum, Canopus, Poecile and Tempe."[7] Whether or not we can actually reconcile these names with specific parts of the site, the sculptural assemblage similarly presents the villa as an empire in miniature: a microcosm of Hadrian's eastern realms that, as von Stackelberg puts it, "redefined the Mediterranean as an imperial panopticon."[8]

The use of Greek (and Greek-looking) material culture to legitimize Roman rule was by Hadrian's time an ancient practice, already well established in Republican Italy.[9] However, imperial uses of Aegyptiaca also had an impressive pedigree, dating back to the start of Augustus's principate. Egyptian and "Egyptianizing" material culture helped legitimize Augustus's rule through, for example, coins marked with a crocodile and the legend *Aegypto Capta*, or the incorporation of an Egyptian obelisk into an architectural complex that also included Augustus's tomb and the Ara Pacis.[10] These visual and material allusions to Egypt presented Augustus as conqueror of Cleopatra and restorer of peace and stability to Rome. Several generations later, the Flavian emperors also used Egypt and Aegyptiaca to legitimize their rule, but they presented Egypt less as conquered enemy than as productive bulwark of Roman prosperity. Vespasian was first acclaimed emperor in Egypt, and his reign may have seen the construction of the Iseum Campense and the entry of Isis and Sarapis into the Roman *sacra publica*.[11] Vespasianic coins depict the Iseum Campense, and Vespasian and Titus held vigil there after Titus's victory in Judaea.[12] Domitian even one-upped Augustus by not merely importing an existing Egyptian obelisk but also commissioning an entirely new obelisk with personalized hieroglyphic texts.[13]

Hadrian's use of Egyptian imagery thus enabled him to position himself as the leader of a new Golden Age and the worthy successor to both Augustus and the

7. *provinciarum et locorum celeberrima nomina inscriberet, velut Lyceum, Academian, Prytaneum, Canopum, Poicilen, Tempe vocaret* (*Hist. Aug.*, Life of Hadrian 26.5; trans. Magie 1921: 79).

8. Von Stackelberg 2009: 85; cf. Spencer (2010: 174) on Tivoli as "synecdoche for empire."

9. On Roman emulations of Greek art, see chapter 1, this volume, "Roman Emulation of Earlier Visual Cultures," with note 156.

10. Barrett 2017a. On the obelisk, see chapter 1, this volume, "Object Biographies." On Egyptian appropriations within the Ara Pacis itself, see Trimble 2018.

11. See notes 64 and 78, chapter 1, this volume.

12. Joseph. *BJ* 7.123; Gasparini 2009: 349.

13. See note 153, chapter 1, this volume.

Flavian emperors.[14] Rather than the Augustan image of Egypt as defeated threat, however, the Tivoli villa builds on the Flavian image of Egypt as valued contributor to shared prosperity. Two statues of river gods watched over Hadrian's model river: the Tiber and the Nile.[15] Similarly paired statues of Nile and Tiber, also of possibly Hadrianic date, appear at the Iseum Campense in Rome.[16] At both sites, the pairing of the two river gods—of equal size and in similar postures—suggests not a one-sided domination but, rather, a productive partnership of equals.

By representing Egypt together with Greek culture as foundational to the Roman state, Hadrian was thus building on earlier imperial precedents. However, the Nilotic imagery of his so-called Canopus takes inspiration not only from public monuments or imperial commissions but also from domestic material culture. In scale, Hadrian's vast installation dwarfs the Pompeian gardens considered in this book. Instead of miniature statuettes, his water garden boasts monumental statues. Nonetheless, their iconography has older parallels in domestic assemblages like those of the Casa di Octavius Quartio and the Casa di Acceptus e Euhodia. Amid statues of Greco-Roman gods, mythological figures, and emulations of famous artworks such as the Erechtheion caryatids, visitors to Hadrian's *euripus* also encountered a range of objects with Nilotic themes: a rearing crocodile, two statues of Bes-Silenus whose heads carry fruit baskets that resemble *modius* crowns,[17] and a statue of the Nile god (figures 6.18, 7.1).

These Nilotic themes find their closest antecedents not in the water-triclinia of earlier imperial villas, such as those at Sperlonga or Baia, but in the comparatively modest homes of private citizens at sites like Pompeii.[18] This trajectory upends

14. Versluys 2012: 34.

15. Raeder 1983: nos. 86, 87; Kuttner 2003: 137, 142.

16. Swetnam-Burland 2009; 2015: 155–167.

17. Kuttner identifies the basket-carrying statues as "satyrs resembling the Egyptian god Bes" (2003: 140). As evidence for an intended allusion to Bes, she cites the figures' "dwarf physiognomy characterized by long torso and short legs," their relative shortness compared to the adjacent copies of the Erechtheion porch maidens, and their "unusual wrap skirts [that] suggest Egyptian linen kilts" (2003: 141n141). To her astute observations, I would add that the statues' bearded faces resemble images of Bes-Silenus (on whom, see chapter 6, this volume, "Dwarf Deities, Pygmies, and Sileni"), and the baskets on their heads recall Roman depictions of Egyptian deities, including Bes, wearing *modius* crowns (e.g., De Caro 2006b: no. III.49; and on the *modius* crown more generally, see note 83, chapter 6, this volume).

18. Cf. Hales's remark that the mixture of geographic and cultural allusions in Pompeian garden statuary "predates and anticipates the empire as imagined by Hadrian at Tivoli" (2003: 156–157). *On the ownership of the Sperlonga villa*, see most recently Squire 2009: 202. *On the Baia site*, see Tocco Sciarelli 1983. *Nilotic frescoes do appear in (non-imperial) villas contemporary with Pompeii*: e.g., the Villa San Marco, Stabiae (Versluys 2002: no. 72), and the Villa dei Misteri and Villa di Diomede at Pompeii (Versluys 2002: nos. 67, 68).

top-down narratives of change in Roman domestic assemblages. Following Zanker's influential characterization of Pompeian houses as imitations of villa architecture,[19] scholars have traditionally assumed that emulation moved down the socioeconomic ladder rather than upward. Recent scholarship has problematized many older assumptions about the intersections of class, wealth, and status in Roman households, deconstructing received conceptions of "freedman style" and critiquing the imposition of anachronistic standards of taste onto ancient material culture.[20] The trajectory of Nilotic imagery raises another challenge to standard models. In place of unidirectional, top-down models of culture change, Nilotic garden installations suggest a multidirectional emulative process in which innovation might occur at the popular level, as well as the imperial. Domestic material culture emerges not just as a response to larger political or social developments but also as a potential agent in broader societal change.

Plural Audiences, Plural Responses

Within Pompeii, multiple features of Nilotic landscapes (both two-dimensional and three-dimensional) suggest that first-century Italians could still associate "Egypt" with the exotic, strange, and marvelous. On one level, both the human and animal populations of Nilescapes present spectacles of *difference*. The *pygmaei* are portrayed as both somatically and (at least to some degree) culturally different from the average Pompeian viewer, and the crocodiles and hippopotami appear as threatening *monstra*. The river itself appears as a curiosity and marvel, whose unknown sources and seemingly miraculous fertility were among the greatest mysteries of ancient science. The exoticizing aspects of Pompeian Nilotica are by no means always negative or anti-Egyptian; the unusual fecundity of the Nile riverbanks was crucial to Roman prosperity, and the pygmies appear not only as orgiastic banqueters but also as pious celebrants of rural cults. However, these images of *mirabilia* and *monstra* demonstrate that cultural constructions of Egypt as "Other" had not entirely disappeared by the late first century. People in Flavian Italy could still imagine Egypt as a sometimes tempting, sometimes frightening land of fantasy.

However, the contexts of Pompeian Nilescapes often complicate such constructions of "Self" and "Other," confronting viewers not only with difference but also with similarity. Nilotic imagery is frequently associated with garden settings, water features, and dining facilities. In such settings, Egyptian

19. Zanker 1979, 1998.

20. See note 32, chapter 6, this volume.

landscapes become mirrors for the hosts' and guests' own activities. The carousing *pygmaei* become not only objects of spectacle but also potential commentaries on (or prompts for) their viewers' own activities. The social performance of the viewers becomes a participatory response to the visual culture around them. Iconographically, Nilotic landscapes engage with ideas of "Egypt," but socially, they engage with practices and performances in Pompeii.

An unresolved tension thus animates many of the images of Egypt examined in this book. Some aspects of these assemblages emphasize the foreignness, unknowability, and strangeness of Nilotic landscapes, while others highlight parallels with their Pompeian settings and prompt viewers to consider their own relationships to the figures portrayed. Pompeian houses thus construct an image of "Egypt" that is not exclusively oppositional to "Rome" but also potentially compatible with (or even included within) "Romanness." This flexible approach to the relationship of "Roman" and "provincial" identities tracks broader cultural changes in the early empire and parallels the increasingly multifaceted and nuanced representations of Egypt in Flavian-period literature.[21]

However, rather than making a single statement about Egypt, Rome, or anything else, these domestic assemblages create much room for individual viewers' responses and interpretations. Different viewers might have reacted in quite different ways, depending on their own backgrounds and the degree of their knowledge about Egyptian culture.[22] For individuals who were (or considered themselves) knowledgeable about Egypt and its customs (through participation in Egyptian cults, reading, trade, travel, or other sources), representations of the flooding Nile might represent an opportunity to discourse on Egyptian culture and natural phenomena. Simultaneously, individuals involved in commerce might appreciate allusions to the wealth enabled by the Egyptian grain trade.[23] Both these perspectives were likely well represented in Campania, where Egyptian grain and other imports passed through Puteoli, and temples and household shrines to Egyptian gods were familiar sights.[24] Other visitors might have reflected, either disapprovingly or aspirationally, on the luxurious lifestyle that many Romans associated with Egypt—a lifestyle to which that of the Bay of Naples was sometimes compared (e.g., Sen. *Ep.* 51.1). For those viewers, a Nilotic

21. See also chapter 1, this volume, "Material Culture, Identity, and 'Authenticity'"; chapter 4, "Landscapes, Identities, and Viewer Responses."

22. Cf. Versluys and Meyboom 2000: 126; Malaise 2005: 211–212; Meyboom and Versluys 2007: 208; Versluys 2007: 14.

23. Cf. Versluys and Meyboom 2000: 126; Malaise 2005: 211.

24. See chapter 1, this volume, "Flavian Pompeii as Case Study."

water-triclinium might represent a more geographically specific, Ptolemaically flavored variant of the allusions to Hellenistic luxury that appear elsewhere in Roman domestic material culture.[25] Others might see Nilotic frescoes as a more geographically localized version of "sacral-idyllic" imagery, a humorous riff on the mores and etiquette of their own dining practices, or an opportunity to debate contemporary theories about the origins and marvels of the Nile.

A similar multiplicity of functions, values, and significations also characterizes Egyptian or Egyptian-style statuary from garden contexts. Statuettes of Egyptian deities, dwarflike beings, and river animals might be valuable for their ability to evoke "Egypt," but they might also be valued for their costly materials, their stylistic references to Pharaonic antiquity, their associations with Egyptian cults, or their thematic rapports with Bacchic imagery. These sources of value should be understood as potentially complementary and overlapping, not exclusive. As the Casa di Acceptus e Euhodia demonstrates, individuals who worshiped Isis in some capacity did not necessarily use or value Aegyptiaca in ways different from other Pompeians. Binary constructions of "decorative" versus "religious" value fail to address the full complexity of viewers' possible responses to material culture.

Domesticating Empire: The Domus *as* Orbis

While military and political factors were crucial to establishing Roman domination in much of Europe, the Near East, and northern Africa, it was in large part the movement, circulation, and emulation of objects that made the empire into *oikoumene*. The Roman Empire was not just something that a reified group called "the Romans" did to other people; it was a network of people, places, objects, and power relations that materially transformed all that it entangled. Within this vast net of entanglements, everyday dwelling practices[26] provided individuals with a way to internalize, perform, and construct their own experience of empire. Through the domestic use and display of internationalizing objects, individuals helped create a material *koine* connecting them with the rest of the Roman world. Images of Egypt in Pompeian houses are thus also images of—and assertions about—Pompeii and its relationships with the wider Roman world.

As early as Augustus's principate, Ovid could assert that "to other peoples is given land with a fixed limit; the extent of the city (*urbis*) of Rome is identical

25. Cf. Dunbabin 2003: 61; Bragantini 2006: 163.

26. On dwelling, see Ingold 2000: 172–188.

with the world (*orbis*)."[27] In many ways, the imperial *urbs* did become the *orbis*. The population of Flavian Rome included individuals (both free and enslaved) from all over the empire; its temples contained not only Greco-Roman deities but also gods from Egypt, the Near East, and (in at least one case) the Celtic West; and its streets, parks, and temples displayed numerous objects appropriated from the provinces.[28] Early imperial literature celebrates the display of foreign *mirabilia* at Rome.[29] Artworks and other objects appropriated from conquered peoples were displayed in fora, porticoes, and temples such as the Flavian Templum Pacis.[30] Trade brought a steady supply of international products to Rome, which Aelius Aristides (*Or.* 26.11–13) would later characterize as taking in so many international goods that the city contained everything that humans ever made or cultivated.[31] Exotic animals performed in the amphitheaters of the capital, and foreign plants—"the literal fruits of empire"—grew in public parks such as Pompey's Porticus and probably also the botanical gardens of the Templum Pacis.[32] Even humans from distant lands were collectible, sold as luxuries in Rome's slave markets.[33] Encyclopedic texts and maps provided an intellectual counterpart to these processes of physical collection by compiling and systematizing knowledge about conquered territories.[34] Even unavoidable daily practices, such as eating, reminded Romans of all the ways in which the *orbis* came to the *urbs*. In addition to the highly valued imported delicacies that graced elite feasts, the basic staple of imported grain embodied Roman dependence on the provinces, including Egypt.[35]

---

27. *gentibus est aliis tellus data limite certo / Romanae spatium est urbis et orbis idem* (Ov. *Fast.* 2.683–684).

28. *On immigration*, see Morley 2003; Moatti 2013. *On the incorporation of foreign gods as imperial strategy*, see, inter alia, Ando 2003; 2007: 444; Orlin 2010. In addition to numerous eastern gods, at least one western deity is attested in Rome: the horse goddess Epona (Apul. *Met.* 3.27; Juv. *Sat.* 8.156–157; Min. Fel. *Oct.* 28.7).

29. Naas 2011.

30. On the display in Rome of appropriated material culture from conquered lands, as well as personifications and other visual representations of those lands, see Edwards 2003; Miles 2008; Rutledge 2012; Loar et al. 2018.

31. See discussion in Ando 2003: 326.

32. Quoting Pollard 2009: 329. On exotic plants in Rome and Italy, see Plin. *HN* 12.11–12, 12.20, 12.28–29, 12.94, 12.111–112, 15.102, 27.2–3; Pollard 2009; Rutledge 2012: 213–215; Marzano 2014; Zarmakoupi 2014: 111–112.

33. International slaves as collectible luxury goods: Dalby 2000: 268.

34. The most famous such encyclopedic text is Pliny's *Natural History*: see note 152, chapter 3, this volume. *On mapping in the early empire*, see Evans 2003: 255–256; Rutledge 2012: 199–206.

35. On imports in elite feasting, see Plin. *HN* 15.105; Gowers 1993: 20; Dalby 2000: 243–257.

Even as the *urbs* became the *orbis* in Rome, individuals in Pompeii were also turning the *domus* into a miniature *orbis*. In each of the houses examined in this book, Aegyptiaca shared space with objects and images recalling other regions of the Roman world. Pompeian house owners may have wanted to turn their gardens into the banks of the Nile, but they also wanted to turn them into *gymnasia* evocative of Greek learning, *paradeisoi* reminiscent of the estates of Eastern dynasts, and (reassuringly tame-looking versions of) wildernesses suitable for roving satyrs, Pans, and other uncanny beings. The gardens in this book embody a vision of the *locus amoenus* that draws on a long tradition in Greek and Roman literature and culture.[36] However, they also update that tradition for the needs of an increasingly globalizing world. The "pleasant place" of Pompeian fantasy is as likely located in Egyptian Canopus as it is in the Thessalian Vale of Tempe or the Campanian countryside—and many Pompeian gardens may indeed recall all three of these at once.

Ex Oriente Lux(uria)? Imagining the (Eastern) Empire in Italy

Not as visible in Pompeian gardens, however, are images of the empire west of Italy. Many raw materials, foods, drinks, and slaves from western provinces were highly valued luxury goods.[37] However, this enthusiasm for western commodities does not seem to have translated into an equal enthusiasm for stylistically distinctive, finished craft products.[38] *Materials* from the western empire surely contributed to domestic display, but *stylistic or iconographic* allusions to those provinces are less common in domestic settings. House owners may have wanted to demonstrate their sophistication through displays of artwork in Greek and Egyptian styles, but they were not collecting, for example, La Tène metalwork.[39] Similarly, Nilotic scenes demonstrate interest in images of faraway lands, but this interest does not seem to have extended to representations of, for example,

36. See especially chapter 3, this volume.

37. See Dalby 2000: 82–117. On the value and desirability of imported stones from western (as well as eastern) provinces, see most recently Müskens 2017: 39–41.

38. Though exceptions did exist: for example, literary references attest to people in Italy wearing Gaulish-made cloaks (Mart. 6.11; Dalby 2000: 98–99). However, Martial (6.11) presents Gaulish cloaks as relatively low-status clothing and contrasts them with more overtly luxurious imported textiles from Tyre. Terra sigillata pottery produced in the western provinces also circulated widely (see most recently Van Oyen 2016, including pp. 86–90 on Pompeii).

39. On the material culture of the "Celtic" West, see now Farley and Hunter 2015.

Roman Britain. If Pompeian gardens present the *domus* as *orbis*, that *orbis* is in certain ways heavily weighted toward the east.

This eastern bias was not limited to the domestic sphere. Certainly, imperial dedications and other "official" material culture might include imagery related to western provinces, such as personifications of their peoples as bound captives.[40] However, appropriated foreign objects selected for permanent public display in Rome more often came from the Greek-speaking, eastern empire. It was these objects that Romans appear to have valued in ways most analogous to modern conceptions of "art."[41] The emulation of Italian craftsmen, the purchasing power of Italian consumers, and the praise of Italian literati were largely directed toward finished craft products from the eastern, rather than the western, provinces.

"Aegyptiaca" and "Hellenism" in Roman Retrospective Art

While these provinces corresponded to the "Greek" portion of the empire, the material culture in question did not come from exclusively "Greek" backgrounds. Egyptian material culture was seemingly the most popular alternative option, but Italians also engaged with other eastern visual traditions. In the Republican period, Italian artisans' emulation of Greek material culture sometimes also involved (at least indirect) replication of Classical producers' own engagement with Persian art and artifacts.[42] Jewish and Near Eastern, as well as Egyptian, motifs appear on "magical gems" popular throughout the Roman Empire.[43] Similarly, in the cults of Cybele and Mithras, Italians frequently employed visual culture asserting these deities' respective claims to Anatolian and Persian origins.[44]

Despite these examples, however, much Roman artistic emulation and adaptation—and most of the geographic and cultural references in the gardens

40. On Roman representations of "barbarians," see Ferris 2000, 2011.

41. On ancient versus modern constructions of "art," see Platt and Squire 2010.

42. E.g., Papini 2015: 98–99.

43. Barrett 2015c, and in press (b).

44. On these cults, see Alvar 2008. On iconography, see, e.g., the funerary stele of a Roman *gallus* displaying "exotic" gear (Beard et al. 1998: 2:211, no. 8.7c), and note the use of "Phrygian caps" for the iconography of Mithras as well as Attis and *galli* (Alvar 2008: 38, 81, 165). Cf. the use of Egyptian and Egyptian-looking material culture in Isis cult (note 116, chapter 1, this volume). On Roman strategies of emphasizing the "exotic" features of foreign gods in order to better appropriate them, see Orlin 2010.

here examined—dealt with material culture that Roman viewers would have recognized as (in some capacity) "Greek," "Egyptian," or "Italic."[45] While objects and images in various "Greek" temporal styles were particularly ubiquitous and highly valued in Pompeian homes, this study has argued for the frequent inclusion of "Egyptian" objects and images alongside them.

I do not propose that all forms of Roman "Classicism"[46] were identical, or that the perceived value of Egyptian visual culture was interchangeable with that of Greek visual culture. The histories of Roman–Egyptian and Roman–Greek interactions were very different, and Roman engagements with Aegyptiaca had historically responded to cultural needs and consumer interests that only partly overlapped with those shaping Roman engagements with objects perceived as "Greek." However, the popularity of Egyptian landscapes, styles, and objects in first-century Italian households—and the frequent association of such Aegyptiaca with visual allusions to various forms of "Greek" culture—suggest a more prominent role for Egypt than is often acknowledged in scholarship on Roman emulative art.

So domestic material culture enabled house owners to recast the *domus* as *orbis*—but with a heavy emphasis on the eastern half of the *orbis*, and within that, a particular focus on "Greek" and "Egyptian" geographic and cultural references. To some degree, the eastern empire may stand as synecdoche for empire more generally. However, the preference for eastern allusions also evokes more specific cultural associations. Rather than objects from *all over* the empire, these domestic assemblages collect objects from, or visual allusions to, parts of the empire that Romans perceived as either cultural ancestors or co-participants in a shared *koine* of international "high culture."[47] The "internationalizing" elements of these Pompeian gardens emphasize images, styles, and objects that were widely recognizable, evocative of *luxuria* and *otium*, and associated with cultures considered to be ancient, wealthy, and sophisticated.

45. On Roman emulation and incorporation of pre-"Roman" Italic arts, see Elsner 2006: 271; Papini 2015.

46. I here use "Classicism" in Elsner's broadened sense, referring to "emulation of any earlier set of visual styles, forms, or iconographies, which in the very fact of their being borrowed are established as in some sense canonical (or classic)" (Elsner 2006: 271).

47. Compare Allison's comment that in the Casa della Caccia Antica, "the entire combination of Greek, Egyptian and Roman motifs would seem to allude to sophistication and civilization" (2002: 77). Pearson similarly argues that Egyptian motifs in Roman wall painting play a parallel role to Greek motifs, as both allude to the consumption of luxury goods (2015: 12).

Otium, Luxuria, Paideia, and Cosmopolitan Material Culture

Considerations of access and recognizability may explain part of this preference. In addition to the longstanding connections between Italy and Greece, Egyptian imports had circulated in Italy ever since the early Iron Age.[48] Luxury goods and crafts from the eastern Mediterranean were widely popular in Italy. In contrast, exports from the western provinces included fewer finished craft products designed for domestic furnishing. As discussed earlier, while foods, drinks, slaves, and objects made of raw materials from western provinces would have been very much on display in wealthy Italian households,[49] distinctively "Western"-looking *styles* of visual and material culture would have been less accessible or recognizable. Additionally, by the late first century, well-heeled Italian travelers regularly visited the eastern Mediterranean—including Egypt—for educational and touristic purposes.[50] Both the material culture and the landscapes of the eastern Mediterranean would thus have been familiar to many Italians of means.

Other contributing factors were cultural. Like many aspects of Greek culture, Egyptian culture might attract criticism in some Roman contexts (especially those associated with *negotium* and the public sphere) and admiration in others (especially contexts associated with *otium* and intellectual endeavors). Latin texts certainly offer many negative stereotypes about Egyptians: for example, as debauched, untrustworthy, or unduly attached to strange-looking animal gods. However, Roman literature also inherited earlier Greek perceptions of Egyptians as wise, pious, skilled in medicine and the arts, and adept in all forms of esoteric knowledge.[51] The resulting image of "Egyptians" as intelligent, knowledgeable, and sophisticated, but also potentially tricksy and loose of morals, displays some overlap with Roman stereotypes of "Greeks."

Some of this overlap may speak to more general conceptions of "eastern" versus "western" foreigners. Roman literature frequently associates eastern and southern peoples with, on the one hand, cultural sophistication and intelligence,

48. The earliest Egyptian or "Egyptianizing" imports in Italy date back to the ninth or eighth centuries BCE (De Salvia 2006; Sannibale 2014).

49. See note 37, this chapter.

50. See note 124, chapter 1, and note 144, chapter 2, this volume.

51. See chapter 3, this volume, "Performing Cosmopolitanism." Egyptians' reputation for esoteric knowledge could have both a positive valence (Egyptians as pious, wise philosophers) and a negative one (Egyptians as potentially dangerous magicians). See, e.g., the fictional representation of an Egyptian priest in Heliodorus's *Aethiopica* (3.16); in this Greek novel, the priest acknowledges both kinds of Egyptian expertise and seeks to identify himself with the former. On Egypt as land of magic, see Frankfurter 1998b: 217–221; Hornung 2001: 55–63.

but, on the other hand, dissipation, deceitfulness, effeteness, and excessive luxury. In contrast, peoples from western and northern regions were frequently imagined as something closer to "noble savages": strong, manly, and in some ways virtuous, but also unsophisticated, crude, and barbarous.[52]

Additionally, we should not forget that the "Egypt" which Rome conquered in 30 BCE was also one of the great "Greek" polities of the eastern Mediterranean. Three centuries of Ptolemaic rule had fostered extensive interaction between Greek and Egyptian individuals and practices,[53] and many Romans perceived Ptolemaic Alexandria as the epicenter of "Greek" intellectual discourse and cultural sophistication. The conquest of Egypt represented not only the acquisition of *Aegyptus* itself but also the culmination of Rome's incorporation of the Hellenistic Greek world. The historical and cultural circumstances of Ptolemaic Egypt may thus have encouraged associations between "Egyptian" and "Greek" culture for some Romans.

Visual allusions to "eastern" sophistication thus provided opportunities not only to display material wealth but also to perform membership (or at least aspirational membership) in an empire-wide cultural and intellectual elite. Roman performances of cosmopolitanism had long emphasized "Greek" high culture and *paideia*. The evidence of Pompeian households suggests that by the late first century, images of Egypt and Egyptian culture could also play a role in such performances.

Socioeconomic Status and Strategies of Domestic Display

The houses investigated in this book vary widely in size, contents, and resources. The vast Casa del Fauno clearly belonged to one of the wealthiest families in Pompeii, and the large size and relatively luxurious furnishings of the Casa dell'Efebo also suggest substantial means.[54] In contrast, the Casa del Medico is small, lacks features associated with patronage, and contained relatively

52. On Roman ethnographic accounts of westerners, northerners, and the various paradigms employed to explain ethnic difference, see Krebs 2011 (coining "Borealism" as a parallel to "Orientalism"); Woolf 2011. Almagor and Skinner (2013) survey scholarly approaches to ancient ethnography. On ancient ideas about foreign cultures and the effect of geography on peoples, see Romm 2010; Kennedy et al. 2013; Campbell 2016; Kennedy and Jones-Lewis 2016.

53. *Contra* the emphasis on cultural separation in much Ptolemaic scholarship of the later twentieth century, much recent research explores "frame switching," hybridization, and ethnic flexibility in Ptolemaic Egypt; see Barrett 2015b, s.v. "Ethnicity and Ethnic Identity."

54. Regardless of whether or not the final owner of the Casa dell'Efebo was a freedman, as is often supposed (see most recently Merrills 2017: 108, 142). For concerns regarding the common

minimal domestic decoration. The Casa di Acceptus e Euhodia presents a complex case: roughly similar in size and architectural irregularity to the Casa del Medico, it nonetheless boasted a surprisingly extensive statuary assemblage. It is unlikely that any of these houses belonged to truly poor individuals, for whom any sort of *domus* was probably out of reach. However, a wide gulf separates the palatial Casa del Fauno from the modest Casa del Medico.

Despite their varying socioeconomic conditions, however, the inhabitants of all four houses appear to have sought—to the best of their abilities and resources—to use images of "Egypt" in analogous ways. As the comparison of the Casa dell'Efebo and Casa del Medico demonstrates (chapters 3–4), the iconography of Nilescapes may differ from house to house, potentially activating different cultural associations for "Egypt." However, the *contextualization* of those images displays substantial similarities. In both wealthy and modest houses, Egyptian landscapes frequently appear in close association with garden settings, dining activities, and water features. The visual and artifactual assemblages in these spaces construct connections between an imagined "Egypt," the real-life activities associated with Roman gardens, and a range of additional geographical and cultural referents.

Art historians sometimes postulate a division between "elite" and "non-elite" forms of Roman visual culture, with "non-elite" art displaying very different goals, attitudes, and cultural preoccupations.[55] However, the use of Egyptian imagery in Pompeian gardens suggests more continuity than rupture in domestic strategies of self-presentation. The owners of the houses in this study could not all boast the same claims to wealth, but they appear to have pursued similar strategies for seeking to *claim* and *perform* (at least aspirational) elite identities.

We cannot know today whether, or to what degree, other Pompeians judged them successful in such performances. Petronius's account of Trimalchio's dinner (Petron. *Sat.* 26–79) may not testify to objective differences between the social performances of old-money aristocrats and striving arrivistes, as was once thought[56]—but his satire certainly does attest to the potentially brutal social policing that could attend any attempt at self-presentation. Within these four Pompeian houses, however, trans-Mediterranean material culture—including

identification of the owner as the freedman Publius Cornelius Tages, see note 3, chapter 3, this volume.

55. E.g., Bianchi Bandinelli 1967; 1970: 51–105; Clarke 2003. For recent reconsiderations of the supposed category of "plebeian art," see Clarke 2003: 2–9; De Angelis et al. 2012; Petersen 2015.

56. Petersen 2006; see also note 32, chapter 6, this volume.

images of Egypt—provided tools for individuals to do their best to construct the social identities they wanted.

Inside, Looking Out: Between Domus *and Landscape*

Gardens were useful sites to display international-looking material culture partly because they were places for *otium*, or cultured leisure. Such leisure ideally involved not only relaxation and pleasure but also intellectual contemplation (e.g., Sen., *De Otio*). Eastern imports, emulations of eastern goods, and images of eastern landscapes were thus appropriate (*decorus*) to gardens on multiple levels. They evoked the *luxuria* of the imagined "East" while also enabling house owners to present themselves as cultured, sophisticated cosmopolitans.

Gardens were also appropriate locations for Nilotic imagery because of their conceptualization as artificial "landscapes," simultaneously inside and outside the *domus*. The visual culture of Roman gardens frequently assimilates them to outdoor landscapes of many kinds: pastoral, mythological, royal, agricultural, wild, Italian, Greek, and Egyptian. Common themes for paintings in gardens include "sacral-idyllic" landscapes, outdoor scenes from Greco-Roman mythology (often set in landscapes conceptualized as "Greek"), gods in outdoor settings, *paradeisoi* recalling royal and lordly estates, animals fighting each other in rocky landscapes, and, indeed, Nilotic scenes depicting imagined Egyptian landscapes.[57] Garden statuary also typically emphasizes "outdoor" themes: for example, Bacchic imagery, Nilotic and other Egyptian themes, wild animals, fishermen and other individuals engaged in outdoor activity (such as children pulling thorns out of their feet), figures from mythological narratives that unfold outdoors (as with Hermaphroditus), and mythological beings associated with outdoor landscapes (like river deities, nymphs, and Muses, who roamed Mt. Parnassos or Mt. Helicon).[58]

In evoking landscapes beyond the household, such images turn gardens into a form of domestic space that is simultaneously a model of *non*-domestic space. Gardens thus become appropriate sites for imagery that in other domestic contexts is rare or marginalized. Certainly, Egyptian and "Egyptianizing" motifs are not limited to gardens; they can appear in almost every type of room

57. *On geographical allusions in Roman landscape painting*, see Bergmann 2001; O'Sullivan 2007: 517–518. *On the themes of paintings in gardens*, see Jashemski 1979–93: 1:55–87; Leach 2004: 124–132; Bergmann 2008.

58. On common subjects for garden statuary, see Hill 1981.

in Pompeian houses.[59] Outside of gardens, however, Egyptian images are typically small subsidiary motifs or marginalia within larger compositions.[60] When Egyptian themes form the central focus of a painted wall or mosaic, or occupy the majority of space on a wall, they do so almost exclusively within gardens.[61] Additionally, when multiple sculptures within a single room depict Egyptian themes, that room is usually a form of garden space.[62]

Beyond "Isis Cult" vs. "Egyptomania": Context, Framing, and Assemblage

The major exception concerns domestic shrines. Lararium paintings of Isis, and lararium statuettes depicting Egyptian-derived gods, appear in multiple room types—including gardens, but without any particular or exclusive association with gardens.[63] Some recent studies have posited a strict separation between ownership of garden Aegyptiaca and worship of Egyptian deities, claiming not only that garden Aegyptiaca are not *evidence* for Isis worship but also that they indicate the *absence* of Isis worship.[64] However, such assertions go beyond the evidence. A number of houses feature both garden Aegyptiaca and evidence for some form of Isis worship.[65] The Casa di Acceptus e Euhodia contained both a lararium with a painting of Isis-Fortuna and an assemblage of Egyptian and "Egyptianizing" garden statuary (see chapter 6). Shrines to Egyptian deities are attested in multiple gardens at Pompeii.[66] Outside of shrines, statues and paintings of Isis also appear in several garden contexts.[67] Nilotic garden imagery

---

59. For examples, see most recently Mol 2015; Pearson 2015.

60. See chapter 5, this volume, "Introduction," with note 1.

61. See chapter 5, this volume, "The Garden as Margin."

62. See chapter 6, this volume.

63. Beaurin 2008.

64. Mol 2013: 127 ("A serious adherent of the Isis cult would not 'fool about' with Egypt as exotic garden display element"); 2015: 310 ("the green-glazed statuettes of Bes, Ptah-Pataikos, and a variety of animals were never attested in houses of those assured to be devotees of the Isis cult. On the other hand objects directly linked to the Isis cult (statues of Isis, Harpocrates, Anubis, and Serapis) were never seen in the spaces of the house destined for leisure activities (e.g., the garden) . . . ").

65. See note 208, chapter 2, this volume.

66. See appendix D, this volume.

67. See appendices B and C, this volume. The painted Isis-Fortuna statuette in the Nilotic frieze at the Casa dell'Efebo also demonstrates, *pace* Mol (2015: 285), that depictions of Isis

does not constitute evidence for "Isiac" religious practices, but neither does it constitute evidence for their absence.

Rather than either returning to "Isiac" explanations for all Aegyptiaca or asserting a mutually exclusive relationship between "decorative" and "religious" Aegyptiaca, this book has sought to problematize the binary categories underlying older interpretations. For ancient consumers, the "decorative" and "religious" values of images were often complementary, not incompatible. Additionally, the available range of possible interactions with Egyptian deities at Pompeii was much wider and more nuanced than that implied by binary constructions of "Isiacs" versus "non-Isiacs." Rather than a stark difference between the material culture of "Isiacs" and that of others, garden Nilotica point to a fundamental *continuity* in domestic display strategies, regardless of where people fell along the "Isiac" spectrum.

In place of totalizing explanations of Nilotic and other Egyptian imagery as evidence for either ("religious," "authentic") Isis worship or ("decorative," "exoticizing") Egyptomania, this book has centered the concepts of context, framing, and assemblage. The total assemblage within a domestic space acts as a perceptual and conceptual frame for its individual components. Guests at, for example, the Casa dell'Efebo would not have approached the Nilotic frescoes solely as two-dimensional, abstracted images. Instead, they would have seen the frescoes as components of a dining installation within a specific type of garden space, associated with certain water features, and intervisible with a collection of additional artifacts and paintings. To understand how viewers made meaning with domestic images of Egypt, we need to contextualize them within larger assemblages.

Order and Chaos in the Garden

Conceptualized as existing both within and outside the *domus*, gardens were appropriate settings for images of the faraway, the outré, and the dangerous. As we have seen, the visual culture of gardens emphasizes images of the outdoors. However, these images of the *outdoors* are also frequently images of a world *beyond human control*. While "sacral-idyllic" landscapes may seem to present reassuring scenes of order and piety, other forms of garden imagery have more unsettling

cult can appear in Nilotic scenes. Another such example may appear in a garden shrine painting at the Casa delle Amazzoni, if the river behind the painted statues of Isis, Sarapis, and Harpocrates is to be taken as the Nile (Boyce 1937: no. 141; Bragantini 1993: 174–175; Versluys 2002: no. 43).

implications. Images of wild animals and landscapes, violent mythological scenes, and dangerous supernatural beings evoke the limits of human order and control. Even seemingly charming images, like garden statuettes of children pulling thorns from their feet,[68] can also convey messages about the danger and pain that natural landscapes may pose to humans.

In other words, Roman gardens are often—among other things—depictions of barely restrained chaos. However, other aspects of the assemblages reassure visitors that these chaotic forces are indeed controlled. Miniaturization makes dangerous imagery less threatening—an observation that applies not only to garden statuettes but indeed to the garden itself as a "landscape" in miniature. Even if gardens depict the world beyond the *domus*, they are in reality still part of that *domus*. Their landscapes are not "real" but artificial, created in order to please humans, and—unlike the actual world outside the *domus*—totally subject to the control of the paterfamilias. By including potentially dangerous images within a carefully framed domestic space and representing themselves as the possessors and controllers of those images, Roman house owners asserted their power over the wild forces that the images represented.

On one level, the association of Nilotic imagery with liminal spaces and margins reinforces perceptions of Egypt as exotic "Otherworld," land of fantasy and danger. The iconography of Nilotic scenes often bolsters such perceptions by emphasizing images that Roman viewers would have considered unusual, monstrous, or marvelous. Yet even as domestic visual culture represents Egyptian landscapes as distant and fantastic, that visual culture itself inserts those landscapes into the everyday life of the *domus*. Once-exotic imagery becomes commonplace, familiar, and literally domesticized.

The very "marginality" of gardens thus also appears as a source of power, in which images at the (conceptual) periphery of domestic space become capable of reframing individuals' perception of that larger space. Because gardens were appropriate places for fantastic or outré imagery, they were appropriate for depictions of distant lands; but because they were also forms of household space, they also became sites for the domesticization and familiarization of foreign imagery. By the second century, Hadrian's villa at Tivoli demonstrates that viewers could understand such images as the opposite of chaotic forces. In the imperial space of Hadrian's villa, Nilotic imagery becomes one embodiment of Roman order, peace, and prosperity.

68. On the "Spinario" and related types in gardens, see, e.g., Hill 1981: 91–92; Higgs 2016; Zanker 2016: 94. The "Spinario" theme took many forms, with the central figure sometimes idealized, sometimes represented as a hunchback or "grotesque" (e.g., Trentin 2015: 24–25).

Domestic Worlds: From Household to Oikoumene

Just as garden space framed and recast people's experience of the *domus*, the edges of the empire framed and recast people's experience of the center. The garden assemblages in this book present the Flavian *domus* as, among other things, a theater for the performance of cosmopolitanism. The *domus* thus becomes a model *orbis*, as well as a workshop where individuals could work out through practice what it meant to be part of a global empire.

Studies of Italian Aegyptiaca sometimes frame their subject in terms of interactions between "Romans" and "Egyptians," eliding distinctions between Italy and Rome. However, Pompeii entered the Roman Empire through conquest no less than Egypt. At the time of the eruption, Pompeii had been a Roman *colonia* for over 150 years, and Egypt had been a province for over 100 years. Juridically, the people of Pompeii were Roman citizens, but we should not assume that this change necessarily caused local Campanian identities to vanish. Members of old Samnite families still retained prominence after the Social War, and locally specific history continued to shape later developments in Pompeii.[69] The inhabitants of the city were *both* Pompeians *and* Romans, inhabiting both identities simultaneously.

This local context may have some relevance to the ways that Pompeian houses construct the experience of empire. Rather than presenting a narrative of straightforward domination of provincial groups, much Pompeian (and other Flavian) material culture instead emphasizes their incorporation and collaboration in a shared endeavor. Certainly, we might look askance in the twenty-first century at such attempts to gloss over the unequal power relations inherent to imperial rule. However, the people whose relationship to "the Roman Empire" required negotiation were not only Egyptians (real or imagined) but also Italians—including the inhabitants of Pompeii, and even the inhabitants of Rome itself. The acquisition of empire entailed vast social, political, and economic changes in Rome and throughout Italy, and the long-term effects of those changes were still playing out in the late first century. Politically, the provinces were subject to Rome, but Rome was also dependent on the provinces for its basic sustenance and maintenance. Representations of those provinces in Italian houses thus represent not only self-congratulatory paeans to the glories of imperialism (although many of them surely are, among other things, that); they also represent attempts to work through what it actually meant to live in a changing, profoundly interconnected, and increasingly interdependent world.

---

69. See chapter 1, this volume, "Flavian Pompeii as Case Study."

Similar issues may underlie the popularity of Aegyptiaca outside Italy. Many of the "Romans" who collected Egyptian imagery came not from Rome or even Italy but from other provinces throughout the empire, including the western empire. The material culture of, for example, Suebi or Britons may not have inspired as much emulation in Rome as did the material culture of Greeks and Egyptians, but they too ultimately became consumers of Roman Aegyptiaca.[70]

Material culture evoking trans-Mediterranean cultural *koine* provided a way for individuals to engage with the experience of empire through daily practice. Through the domestic performance of cosmopolitanism, people could assert their participation in an empire-wide, transregional elite sphere. Domestic images of Egyptian landscapes thus testify to the individual appropriation and transformation of imperial ideology. While imperial and "official" material culture sought to construct a politically motivated image of empire as a harmonious unity of originally disparate components, Pompeian houses emphasize individuals' social performance as cosmopolitan, sophisticated citizens of that empire. Such performance concerned not only ethnic or legal identities (e.g., "Italian" or "Roman") but also claims to socioeconomic and cultural status (e.g., "elite," "wealthy," "educated," "worldly").

Representations of Egyptian landscapes in Pompeian houses thus ultimately testify to the power and agency of domestic material culture. The objects and images with which Pompeians shared their homes were not just illustrations to discourses on identity that "really" took place in some other sphere, whether political or literary. Material culture afforded individuals another, equally important sphere in which to negotiate their relationships to the wider world. On one level, domestic material culture enabled and facilitated the action of individual human agency: it constituted a tool that people could use in social performance and self-presentation. On another level, however, domestic material culture exerted agency over its users. The Egyptian landscapes examined in this book were part of the everyday experience of the people who shared their houses. These domestic landscapes confronted those individuals on a daily basis, demanding their gaze as a matter of routine and inviting them to engage with the tensions and cultural problems that these images raised. The Egyptian landscapes of the Casa dell'Efebo, Casa del Medico, Casa del Fauno, and Casa di Acceptus e Euhodia did not tell their viewers what to think about them. However, they did insist upon their viewers' attention and interaction. By demanding engagement with the imagined "Egypts" that they constructed, these garden assemblages called their human inhabitants to reflect on the relationships between reality and

70. See, e.g., Bricault 2004; Barrett, in press (b).

representation, foreign and familiar, and individual and empire. In so doing, they asserted a vital role for Egypt in the Roman world and for Aegyptiaca within a shared imperial *koine* of material culture. Through participatory multimedia assemblages evoking landscapes simultaneously local and international, the houses examined in this book made the breadth of *oikoumene* compatible with the familiarity of home.

Appendices

APPENDIX A

Contexts of Nilotic Paintings and Mosaics from Domestic Contexts at Pompeii

Pompeii has produced 39 Nilotic paintings and mosaics from household contexts, of which 35 come from well-documented proveniences within those houses.[1] The table that follows presents contextual information from the 35 Nilotic scenes from Pompeii for which *all of the following* forms of data exist: (1) house of origin, (2) room of origin, and (3) overall house plan.[2] The individual units listed in this table are not houses, but rooms. That is, if a house contains two separate rooms with Nilotic decoration, each of those rooms receives a separate entry in the table.

A 36th Nilotic installation comes from a known house (the Casa dello Scultore) and a known room (the peristyle garden), but derives from a stratum lower than that to which the rest of the house has been excavated.[3] Accordingly, it is not possible to assess the intervisibility or interaccessibility of these paintings with other rooms in the house.

Three additional houses contained Nilotic paintings or mosaics, but their room of origin is unknown. These are the Casa dei Fiori, Casa del Cinghiale I, and Villa di Diomede.[4] The painting fragments from the Casa dei Fiori came

1. On possible commercial reuse of some of these houses, see note 205, chapter 2, this volume.

2. Major sources for the data in this table include, inter alia, Versluys's catalog of Nilotica (2002), Jashemski's catalog of Pompeian gardens (1979–93), Penelope Allison's database of finds from selected houses (2004b), relevant entries in the series *Pompei: pitture e mosaici* and *Häuser in Pompeji*, and personal examination of in situ remains.

3. Maiuri 1956; Sampaolo 1998b.

4. Versluys 2002: nos. 44, 58, 68.

from sub-floor fill below a room adjacent to the atrium.[5] The mosaic from the Casa del Cinghiale I may have come from an exedra adjacent to the peristyle, but this provenience is uncertain.[6] The paintings from the Villa di Diomede are now preserved only in historical drawings and their findspots are poorly documented.[7]

Another eight Nilotic scenes from Pompeii come from nondomestic contexts. Four come from public baths (Terme del Sarno, Terme Suburbanae, and two rooms at the Terme Stabiane); three come from temples (Temple of Isis, both portico and *ekklesiasterion*; Temple of Apollo, peristyle); and one comes from a tomb (tomb of C. Vestorius Priscus).[8]

Check marks and gray background indicate contextual associations (e.g., intervisibility or interaccessibility with gardens, water features, or dining features). In cases where the nature of the association might not be immediately obvious, some additional explanatory text has been provided.

---

5. See Bonghi Jovini 1984: 325, and 43, 66, 391 on the stratum.

6. Bragantini 1998a: 363.

7. See discussion in Versluys 2002: 158.

8. Versluys 2002: nos. 49, 53, 55, 61, 66, 69.

Building	Number in Versluys' catalog of Nilotica (2002)	Medium	Room type	Date (CE unless otherwise specified)*	Association with gardens?		
					In garden?	Intervisible? (If in garden: intervisible with another one?)	Accessible? (If in garden: accessible to another one?)
House I 2.24	31	painting	tablinum	0-10			
House I 6.2 (Casa del Criptoportico)	32	painting and mosaic	bath (caldarium)	c. 30 BC			
House I 6.15 (Casa dei Ceii)	33	painting	garden	c. 70	(✓)		
House I 7.1 (Casa di Paquius Proculus)	34	mosaic	triclinium	c. 30 BC		(✓)	(✓)
House I 7.10-12 (Casa dell'Efebo)	35	painting	garden	c. 70	(✓)	(✓) (garden of attached house I 7.19)	(✓) (garden of attached house I 7.19)
House I 10.4 (Casa del Menandro)	36	painting	atrium	c. 50-62		(✓) (peri-style visible through tablinum)	
House I 10.4 (Casa del Menandro)	37	mosaic	oecus	c. 30 BC		(✓)	(✓)
House II 4.2 (Praedia di Julia Felix)	38	painting	biclinium	c. 70		(✓)	(✓)

Association with water features?				Association with dining features?			
Attached to water feature?	In same room?	Intervisible? (If in room with one: intervisible to another room with one?)	Accessible? (If in room with one: accessible to another room with one?)	Attached to dining feature?	In same room?	Intervisible? (If in room with one: intervisible to another room with one?)	Accessible? (If in room with one: intervisible to another room with one?)
(✓) (bath)							
	(✓) (water channel for rainwater - fictively coming from painted fountain in Nilotic scene)						
		(✓) (peristyle: water-triclinium, pool, fountain, puteal)	(✓)		(✓)	(✓)	(✓)
(✓) (water-triclinium)		(✓) (in garden of house I 7.19)	(✓) (in garden of house I 7.19)	(✓) (on banks of triclinium)			
	(✓) (impluvium)	(✓)	(✓)			(✓) (peristyle was likely used for dining and may have contained wooden triclinium: see von Stackelberg 2009: 109)	
		(✓) (water fountain, pool, channel, and cistern in peristyle)			(✓)	(✓) (wooden triclinium in peristyle)	
		(✓)	(✓)		(✓)		

Building	Number in Versluys' catalog of Nilotica (2002)	Medium	Room type	Date (CE unless otherwise specified)*	Association with gardens?		
					In garden?	Intervisible? (If in garden: intervisible with another one?)	Accessible? (If in garden: accessible to another one?)
House II 4.2 (Praedia di Julia Felix)	38	painting	triclinium	c. 70		(✓)	(✓)
House II 9.2 (Casa del Gemmario)	39	painting	garden	c. 70	(✓)		
House II 9.4 (Casa del Larario Fiorito)	40	painting	garden	c. 70	(✓)		
House V 2.i (Casa delle Nozze d'Argento)	41	painting (and statuary)	garden (peristyle)	c. 62-79	(✓)	(✓)	(✓)
House VI 2.4 (Casa di Sallustio)	42	painting	garden	c. 70	(✓)		
House VI 2.4 (Casa di Sallustio)	42	painting	garden (partial peristyle)	c. 70	(✓)		
House VI 2.14 (Casa delle Amazzoni)	43	painting	garden (domestic shrine in garden)	c. 70	(✓)		
House VI 7.23 (Casa di Apollo)	45	painting	garden	c. 70	(✓)		
House VI 9.6-7 (Casa dei Dioscuri)	46	painting	tablinum	c. 70		(✓)	(✓)
House VI 12.2 (Casa del Fauno)	47	mosaic	exedra	early 1st c. BC (see Chapter 5)		(✓)	(✓)

Association with water features?				Association with dining features?			
Attached to water feature?	In same room?	Intervisible? (If in room with one: intervisible to another room with one?)	Accessible? (If in room with one: accessible to another room with one?)	Attached to dining feature?	In same room?	Intervisible? (If in room with one: intervisible to another room with one?)	Accessible? (If in room with one: intervisible to another room with one?)
	(✓) (water-triclinium)	(✓)	(✓)		(✓)		
				(✓) (on banks of triclinium)			
	(✓) (cisterns and water channel to collect runoff)			(✓) (on banks of triclinium)			
	(✓) (gutter and water pipes)	(✓)	(✓)			(✓)	(✓)
	(✓) (pool between benches, water channel, fountain, basins, cistern)				(✓)		
	(✓) (pool and fountain, cistern, gutter)					(✓)	(✓)
	(✓) (pool)					(✓)	(✓)
	(✓) (fountain and waterfall stairs in garden)				(✓) (probable summer triclinium)		
		(✓) (impluvium of atrium; cistern in garden 45)	(✓) (impluvium of atrium; cistern in garden 45)				
		(✓) (fountain, pool, basins in peristyle)	(✓) (fountain, pool, basins in peristyle)		(✓)		

Building	Number in Versluys' catalog of Nilotica (2002)	Medium	Room type	Date (CE unless otherwise specified)*	Association with gardens?		
					In garden?	Intervisible? (If in garden: intervisible with another one?)	Accessible? (If in garden: accessible to another one?)
House VI 17.42 (Casa del Bracciale d'Oro)	48	mosaic	triclinium	35-45		(✓)	(✓)
House VII 2.25 (Casa delle Quadrighe)	50	painting	garden ("viridarium")	c. 70	(✓)	(✓) (peristyle)	
House VII 2.25 (Casa delle Quadrighe)	50	painting	garden (partial peristyle)	c. 70	(✓)	(✓) ("viri-darium")	
House VII 4.48 (Casa della Caccia Antica)	51	painting	tablinum	71-79		(✓)	(✓)
House VII 4.48 (Casa della Caccia Antica)	51	painting (and crocodile statue)	garden (partial peristyle)	71-79	(✓)		
House VII 4.56 (Casa del Granduca/ della Fontana)	52	mosaic	garden (partial peristyle)	0-40	(✓)		
VII 16 (Insula Occidentalis), 17 (Casa di Maius Castricius)	54	painting	garden, planted as a vineyard (Jashemski 1993: 204)	c. 80-20 BCE (following Bragantini 1997 in assigning a Second Style date; pace Versluys 2002: no. 54)	(✓)		

Association with water features?				Association with dining features?			
Attached to water feature?	In same room?	Intervisible? (If in room with one: intervisible to another room with one?)	Accessible? (If in room with one: accessible to another room with one?)	Attached to dining feature?	In same room?	Intervisible? (If in room with one: intervisible to another room with one?)	Accessible? (If in room with one: intervisible to another room with one?)
(✓) (fountain of a water-triclinium)					(✓)		
	(✓) (water channel for rainwater)					(✓) (triclinium visible through window)	
	(✓) (cistern)					(✓)	(✓)
		(✓) (water features in peristyle; impluvium in atrium)	(✓) (water features in peristyle; impluvium in atrium)				
	(✓) (pool in center of garden, water channel around edge of garden beneath inter-columniation wall of portico)						
(✓) (fountain with water stairs; pool divided into basins)	(✓) (cistern)				(✓) (table)		

Building	Number in Versluys' catalog of Nilotica (2002)	Medium	Room type	Date (CE unless otherwise specified)*	Association with gardens?		
					In garden?	Intervisible? (If in garden: intervisible with another one?)	Accessible? (If in garden: accessible to another one?)
House VIII 2.28 (Casa con Ninfeo)	56	painting	triclinium	c. 70			
House VIII 2.34-35 (Casa delle colombe a mosaico)	57	painting	garden (planted terrace)	c. 70	(✓)		
VIII 5.24 (Casa del Medico)	59	painting	garden (partial peristyle)	c. 70	(✓)		
House IX 5.9 (Casa dei Pigmei)	62	painting	small room of uncertain purpose	c. 70		(✓)	(✓)
House IX 5.14-16	63	painting	atrium	c. 70			
House IX 6.f-g	64	painting	triclinium (probable)	c. 70		(✓)	x
House IX 8.3-7 (Casa del Centenario)	65	painting	bath (frigidarium)	c. 70			
House IX 8.3-7 (Casa del Centenario)	65	painting	garden ("viridarium")	c. 70	(✓)	(✓)	
Villa dei Misteri	67	painting	atrium	80-70 BC		(✓) (peristyle)	(✓) (peristyle)
Villa dei Misteri	67	painting	tablinum	c. 10 BC		x (peristyle)	

*Dates follow Versluys 2002, except where otherwise specified.

Association with water features?				Association with dining features?			
Attached to water feature?	In same room?	Intervisible? (If in room with one: intervisible to another room with one?)	Accessible? (If in room with one: accessible to another room with one?)	Attached to dining feature?	In same room?	Intervisible? (If in room with one: intervisible to another room with one?)	Accessible? (If in room with one: intervisible to another room with one?)
(✓) (nymphaeum)					(✓)		
(✓) (fountain)						(✓)	(✓)
	(✓) (broad water channel)					(✓)	(✓)
		(✓) (cistern, puteal, gutter)	(✓)				
(✓) (impluvium)							
					(✓)		
(✓) (bath)							
	(✓) (nymphaeum with water-stairs, pool, fountain)	(✓) (in peristyle: pool, cisterns, water channel)					
	(✓) (impluvium)						
		x (impluvium)	x (impluvium)				

APPENDIX B

Iconographic Motifs in Nilotic Paintings and Mosaics from Pompeii

Pompeii has produced 49 Nilotic paintings and mosaics, of which 47 come from known contexts, 39 of those domestic (see appendix A). The following table lists some of the most common iconographic motifs in these Nilotic scenes and records the relative frequency with which they appear.[1] However, many images are poorly preserved, fragmentary, or known only from archival drawings or excavators' reports, so absence of evidence is not evidence of absence. Many scenes may originally have contained motifs that have not survived. Another potential biasing factor comes from the reluctance of many early scholars to discuss sexual imagery in detail.[2] This table can thus provide only a partial sample of the iconography of Pompeian Nilotica.

1. Major sources for the data in this table include Versluys's catalog of Nilotica (2002), relevant entries in the series *Pompei: pitture e mosaici* and *Häuser in Pompeji*, other sources as cited, and personal examination of the majority of the relevant frescoes and mosaics both in situ and in the Naples Museum. Note that *pace* Versluys (2002: 159), the pygmies from the tomb of Vestorius Priscus are not ithyphallic or macrophallic; see further note 76, chapter 4, this volume. I also do not follow Meyboom and Versluys (2007: 183) and Jacobelli (1995: 88n28) in identifying sexual imagery in the Nilotic fresco from the garden of II 9.4; see further note 241, chapter 2, this volume.

2. See, e.g., one 19th-century publication (Roux and Barré 1870: 47): "Peut-être cette caricature renferme-t-elle quelque basse obscénité; mais nous n'avons ni l'intention ni le loisir de la deviner." Other publications refer to the existence of erotic imagery without expanding on its content: see note 74, chapter 4, this volume.

Catalog Number	Context Type	Building	Number in Versluys' Catalog of Nilotica (2002)	Medium	Pygmies	Hostile Encounters Between Humans and River Beasts	Graphic Pygmy-Animal Violence	Explicit Sexual Encounters	Ithyphallic Macrophall Figures Not Engaged in Sex
1	Domestic	House I 2.24	31	painting					
2	Domestic	House I 6.2 (Casa del Criptoportico)	32	mosaic and painting					(✓)
3	Domestic	House I 6.15 (Casa dei Ceii)	33	painting	(✓)	(✓)			
4	Domestic	House I 7.1 (Casa di Paquius Proculus)	34	mosaic	(✓)				
5	Domestic	House I 7.10-12 (Casa dell'Efebo)	35	painting	(✓)	(✓)		(✓)	
6	Domestic	House I 10.4 (Casa del Menandro), atrium	36	painting	(✓)				
7	Domestic	House I 10.4 (Casa del Menandro), oecus	37	mosaic	(✓)				
8	Domestic	House II 4.2 (Praedia di Julia Felix), biclinium	38	painting					

Defecation, Urination, or Ejaculation	Reed Huts	Alcohol	Dance or Music	Crossed Sticks	Nile Flora	Fauna	Temples, Shrines, Deities, or Statues of Deities	Boats
					(✓) (water plants)	(✓) (ibis)		
		(✓) (figures carry amphorae)			(✓) (water plants)	(✓) (duck, crocodile)		
		(✓) (boat filled with amphorae; one figure carries a jug)			(✓) (lotuses)	(✓) (ducks, crocodile, hippo, ibis)	(✓)	(✓)
Not explicitly depicted, but compare the pose of the pygmy leaning backward off the boat with that of the pygmy defecating off a boat at the tomb of C. Vestorius Priscus.			(✓) (pygmy dancing in front of boat)		(✓) (palm tree, lotus)	(✓) (crocodile, hippo, duck, fishes)	(✓)	(✓)
		(✓)	(✓) (aulos)		(✓) (palm trees, lotus)	(✓) (crocodile, ducks, hippo, ibises)	(✓)	(✓)
	(✓)					(✓) (crocodile)	(✓)	
			(✓) (two pygmies seem to be dancing in the boats)		(✓) (lotuses)	(✓) (ducks)		(✓)
					(✓) (lotuses)	(✓) (ducks)		

Catalog Number	Context Type	Building	Number in Versluys' Catalog of Nilotica (2002)	Medium	Pygmies	Hostile Encounters Between Humans and River Beasts	Graphic Pygmy-Animal Violence	Explicit Sexual Encounters	Ithyphallic c Macrophall Figures Not Engaged in Sex
9	Domestic	House II 4.2 (Praedia di Julia Felix), triclinium	38	painting	(✓)	(✓)			
10	Domestic	House II 9.2 (Casa del Gemmario)	39	painting	(✓)	(✓)		(✓) ("scenette erotiche": Sodo 1988: 200)	(✓)
11	Domestic	House II 9.4 (Casa del Larario Fiorito)	40	painting	(✓)				
12	Domestic	House V 2.i (Casa delle Nozze d'Argento)	41	painting					
13	Domestic	House VI 2.4 (Casa di Sallustio), garden (peristyle)	42	painting				(✓) (seemingly censored in the 19th-century drawing, but reconstructable from the figures' poses)	
14	Domestic	House VI 2.4 (Casa di Sallustio), garden	42	painting					
15	Domestic	House VI 2.14 (Casa delle Amazzoni)	43	painting					

Defecation, Urination, or Ejaculation	Reed Huts	Alcohol	Dance or Music	Crossed Sticks	Nile Flora	Fauna	Temples, Shrines, Deities, or Statues of Deities	Boats
	(✓)		(✓) (pygmies dance on boat)	(✓) (recorded in an old drawing; see Whitehouse 1977: 55)	(✓) (lotuses, reeds)	(✓) (crocodiles, hippos, ducks)		(✓)
					(✓) (rushes)	(✓) (crane)		(✓)
					(✓) (water plants)			
					(✓) (lotuses)	(✓) (ducks; also an ibis or swan)		
	(✓)	(✓) (amphorae)	Possibly; in the 19th-century drawing, one figure appears to be either running or dancing.			(✓) (ibis, dog)		(✓)
					(✓) (lotuses)	(✓) (ducks)		
					(✓) (lotuses, palms, other plants)	(✓) (various birds)	(✓)	

Catalog Number	Context Type	Building	Number in Versluys' Catalog of Nilotica (2002)	Medium	Pygmies	Hostile Encounters Between Humans and River Beasts	Graphic Pygmy-Animal Violence	Explicit Sexual Encounters	Ithyphallic Macrophal Figures No Engaged in Sex
16	Domestic	House VI 5.9–10, 19 (Casa dei Fiori)	44	painting				(✓) (the scene is only partially preserved, but clearly depicted a Nilotic *symplegma*; see Versluys 2002: 117)	
17	Domestic	House VI 7.23 (Casa di Apollo)	45	painting	(✓)				
18	Domestic	House VI 9.6-7 (Casa dei Dioscuri)	46	painting					
19	Domestic	House VI 12.2 (Casa del Fauno)	47	mosaic					
20	Domestic	House VI 17.42 (Casa del Bracciale d'Oro)	48	mosaic					
21	Domestic	House VII 2.25 (Casa delle Quadrighe), garden ("viridarium")	50	painting	(✓)	(✓)			
22	Domestic	House VII 2.25 (Casa delle Quadrighe), garden (partial peristyle)	50	painting	(✓) (as well as non-pygmy figures)			(✓)	

Defecation, Urination, or Ejaculation	Reed Huts	Alcohol	Dance or Music	Crossed Sticks	Nile Flora	Fauna	Temples, Shrines, Deities, or Statues of Deities	Boats
			(✓) (dancing, aulos)					
								(✓)
		(✓)	(✓) (aulos)		(✓) (palm trees)	(✓) (ibis, dog)		
					(✓) (lotuses)	(✓) (ducks and other birds, cobra, ichneumon, ibises, crocodile, frog, hippo)		
					(✓) (lotuses)	(✓) (ducks)		
						(✓) (crocodile, ibis, water birds)		x
	(✓)				(✓) (lotuses)	(✓) (crocodile, hippo)		(✓)

Catalog Number	Context Type	Building	Number in Versluys' Catalog of Nilotica (2002)	Medium	Pygmies	Hostile Encounters Between Humans and River Beasts	Graphic Pygmy-Animal Violence	Explicit Sexual Encounters	Ithyphallic o Macrophall Figures Not Engaged in Sex
23	Domestic	House VII 4.48 (Casa della Caccia Antica), tablinum	51	painting	(✓)	(✓)			
24	Domestic	House VII 4.48 (Casa della Caccia Antica), garden ("viridarium")	51	painting	(✓)	(✓)			
25	Domestic	House VII 4.56 (Casa del Granduca/ della Fontana)	52	mosaic	(✓)	(✓)			
26	Domestic	VII 16 (Insula Occidentalis), 17 (Casa di Maius Castricius)	54	painting	(✓) (as well as non-pygmy figures)	(✓)	(✓)		
27	Domestic	House VIII 2.28 (Casa con Ninfeo)	56	painting	(✓)			(✓) ("Pigmei in atteggiamenti osceni": Sogliano 1888: 512)	
28	Domestic	House VIII 2.34-35 (Casa delle Colombe a Mosaico)	57	painting	(✓)	(✓)			
29	Domestic	House VIII 3.8-9 (Casa del Cinghiale I)	58	mosaic					

Defecation, Urination, or Ejaculation	Reed Huts	Alcohol	Dance or Music	Crossed Sticks	Nile Flora	Fauna	Temples, Shrines, Deities, or Statues of Deities	Boats
	(✓)					(✓) (crocodiles, hippos, donkey, dog)		
						(✓) (crocodile)	(✓)	(✓)
	(✓)				(✓) (palm)	(✓) (hippo)	No, but note that the pediment of the aedicular fountain may depict a stylized *basileion* crown.	(✓)
						(✓) (crocodile, ducks, snake, lion, unidentified quadrupeds, hippo, unidentified bipedal creature)	(✓) (probable image of Anubis)	
						(✓) (uraeus serpent, duck)		
	(✓)					(✓) (crocodile, hippo)		
					(✓) (lotuses)	(✓) (ducks)		

Catalog Number	Context Type	Building	Number in Versluys' Catalog of Nilotica (2002)	Medium	Pygmies	Hostile Encounters Between Humans and River Beasts	Graphic Pygmy-Animal Violence	Explicit Sexual Encounters	Ithyphallic Macrophal Figures No Engaged in Sex
30	Domestic	VIII 5.24 (Casa del Medico)	59	painting	(✓)	(✓)	(✓)	(✓)	
31	Domestic	VIII 7.24 (Casa dello Scultore)	60	painting	(✓) (as well as non-pygmy figures)	(✓)	(✓)		(✓)
32	Domestic	House IX 5.9 (Casa dei Pigmei)	62	painting	(✓)	(✓)			
33	Domestic	House IX 5.14-16	63	painting	(✓)	(✓)		(✓)	
34	Domestic	House IX 6.f-g	64	painting	(✓)				
35	Domestic	House IX 8.3-7 (Casa del Centenario)	65	painting	(✓)				
36	Domestic	House IX 8.3-7 (Casa del Centenario)	65	painting					
37	Domestic	Villa dei Misteri, atrium	67	painting					
38	Domestic	Villa dei Misteri, tablinum	67	painting					
39	Domestic	Villa di Diomede	68	painting	(✓)				

Defecation, Urination, or Ejaculation	Reed Huts	Alcohol	Dance or Music	Crossed Sticks	Nile Flora	Fauna	Temples, Shrines, Deities, or Statues of Deities	Boats
	(✓)	(✓)	(✓) (aulos, dancing figure with crossed sticks)	(✓)		(✓) (crocodiles, hippos, ibises)	(✓)	(✓)
(✓)		(✓) (probable drinking party, though not well-preserved; table with drinks and food)	(✓) (aulos, crossed sticks, dancing figures)	(✓)	(✓) (lotus, palms, papyrus, water plants)	(✓) (ducks, crocodile, possible stork)		(✓)
	(✓)	(✓) (amphorae on a boat)	(✓) (figure with crossed sticks appears to be dancing)	(✓)	(✓) (palm, water plants, papyrus)	(✓) (hippo, ducks, crocodile)	(✓)	(✓)
						(✓) (water birds)		(✓)
					(✓) (palms)		(✓) (statue identified as Fortuna)	
						(✓) (ibis, snake, hippo, crocodiles, ducks)		
					(✓) (lotuses)	(✓) (ducks, fish, and probable crocodile)		
	(✓)				(✓) (palm trees)	(✓) (goats)	(✓)	(✓)
					(✓) (lotuses, other water plants)	(✓) (ducks, ibis, other birds)	(✓) (small Nilotic scenes alternate with images of Egyptian deities)	
							(✓) (pyramid)	

Catalog Number	Context Type	Building	Number in Versluys' Catalog of Nilotica (2002)	Medium	Pygmies	Hostile Encounters Between Humans and River Beasts	Graphic Pygmy-Animal Violence	Explicit Sexual Encounters	Ithyphallic or Macrophallic Figures Not Engaged in Sex
40	Public baths	Building VII 1.8 (Terme Stabiane), room F	49	painting	(✓)	(✓)			
41	Public baths	Building VII 1.8 (Terme Stabiane), room G	49	painting and mosaic		(✓)		(✓)	(✓)
42	Public baths	Building VIII 2.17 (Terme del Sarno)	55	painting	(✓)	(✓)			
43	Public baths	Terme Suburbanae	66	painting	(✓)	(✓)			(✓)
44	Temple	Building VII 7.32 (Temple of Apollo)	53	painting	(✓)	(✓)	(✓)		
45	Temple	Building VIII 7. 28 (Temple of Isis), "*ekklesi-asterion*"	61	painting					
46	Temple	Building VIII 7.28 (Temple of Isis), portico	61	painting	(✓)				

Defecation, Urination, or Ejaculation	Reed Huts	Alcohol	Dance or Music	Crossed Sticks	Nile Flora	Fauna	Temples, Shrines, Deities, or Statues of Deities	Boats
(✓) (see euphemistic description in Minervini 1856: 34, "curvando la persona par che cerchi in altro modo offendere il coccodrillo")					(✓) (water plants)	(✓) (crocodile, duck, goat, small birds)		
			(✓) (aulos, and a woman with "un timpano"; see Eschebach 1979: 88.)			(✓) (crocodiles, water birds)	(✓)	
(✓)	(✓)				(✓) (palms)	(✓) (ibis, possible crocodile or hippo)		(✓)
	(✓)				(✓) (lotuses, water plants)	(✓) (hippo, crocodile, ducks)	(✓)	
	(✓)		(✓) (pygmy dancing with crossed sticks)	(✓)	(✓) (palms)	(✓) (crocodiles, birds, hippo)	(✓)	
						(✓) (cows, ibis, falcon)	(✓)	
	(✓)				(✓) (lotuses, palms)	(✓) (crocodile, ducks, ibises)		

Catalog Number	Context Type	Building	Number in Versluys' Catalog of Nilotica (2002)	Medium	Pygmies	Hostile Encounters Between Humans and River Beasts	Graphic Pygmy-Animal Violence	Explicit Sexual Encounters	Ithyphallic Macrophall Figures No Engaged in Sex
47	Tomb	Tomb of C. Vestorius Priscus	69	painting	(✓)				
48	Unknown	Unknown	n/a; see Roux and Barré (1870: 5.46-47, pl. 22) and Reinach (1922: 161 fig. 1).	painting	(✓)			(✓) (censored in early drawings, but reconstructable from the figures' poses; see Trentin 2015: 99, and note 2 in this Appendix).	
49	Unknown (MANN 8573)	Possibly from House II 4.2 (Praedia di Julia Felix), but location is disputed; see Versluys 2002: 110.	No separate entry; discussed in Versluys 2002: 110.	painting	(✓)				

* One pygmy stands with a bent leg and holds an object which Roux and Barré render as a garland, but which was likely originally the crossed stic

Defecation, Urination, or Ejaculation	Reed Huts	Alcohol	Dance or Music	Crossed Sticks	Nile Flora	Fauna	Temples, Shrines, Deities, or Statues of Deities	Boats
(✓)			(✓) (pygmies dancing on boat with crossed sticks, aulos)	(✓)		(✓) (fish, dolphin)		(✓)
			(✓) (probably)*	(✓) (probably)*	(✓) (lotuses, water plants)	(✓) (fish)		(✓)
							(✓)	

often held by dancing pygmies.

APPENDIX C

Egyptian and "Egyptianizing" Sculpture and "Small Finds" from Pompeian Gardens

The following table provides a list of three-dimensional Aegyptiaca from Pompeian gardens—that is, statues, statuettes, and "small finds" either imported from Egypt or emulating Egyptian styles or motifs.[1] For three-dimensional Aegyptiaca from other contexts (as well as unknown contexts) at Pompeii, see most recently Mol 2015. For Pompeian wall paintings that may depict Egyptian or Egyptianizing garden sculpture, see most recently Pearson 2015.

House	Description	Sources
I 13.16	• Baboon, faience	de Vos 1983: 63.
I 2.20–21	• Bust of Isis wearing a *modius* crown, terracotta	Kekulé 1880: 41 (noting the findspot in a "viridario"); De Caro 2006b: no. III.34.
I 2.24	• Bust of Isis wearing a *modius* crown, terracotta	Kekulé 1880: 41 (noting the findspot "im Garten").

1. Two other statuettes from Pompeian houses come from rooms other than gardens, but their iconography, size, and materials have suggested to scholars that they might originally have been displayed in gardens and then moved elsewhere for safekeeping (e.g., during renovations or during pre-eruption tremors). These include a marble pygmy statuette from an indoor room at the Casa di Oppius Gratus (IX 6.5–7; see Jashemski 1979–93: 2:238; Carrella et al. 2008: 178, no. D 22) and an "Isis-Io" herm from the atrium of the Casa della Fortuna (IX 7.23; see Dwyer 1982: 78; Jashemski 1979–93: 2:240–242; De Caro 2006b: no. III.44). Note also that I do not follow Allison or Mol in attributing an Isis statue to the garden of the Casa dell'Efebo: see note 7, chapter 3, this volume.

House	Description	Sources
II 1.12 (Complesso dei Riti Magici)	• *Nemes*-wearing, bearded herm, marble	de Vos 1983: 60, 61 fig. 19; De Caro 2006b: no. III.121.
II 2.2 (Casa di Octavius Quartio)	• *Nemes*-wearing male sphinx, marble, from upper terrace. • At least 7 faience or pseudo-faience statuettes, including one striding figure wearing a *nemes* crown and *shendyt* kilt; one figure of Bes; and possibly a seated baboon, all from small peristyle garden.	Spinazzola 1953; de Vos 1983: 63; De Caro 2006b: no. III.134; Di Gioia 2006: 117; Tronchin 2006, 2011b; von Stackelberg 2009: 101–125; Mol 2015: 392–446; 2016.
II 4.2 (*Praedia* di Julia Felix)	• Statue of ibis and lizard, marble, adjacent to *euripus*. • Statuette of Harpocrates, silver, from this house's garden shrine to Isis. • Crescent moon (probably alluding to the headdress of Isis), silver, from this house's garden shrine to Isis.	Appleton 1987: no. 139; Parslow 1989: 451 (identifying the bird as a heron), 456; 2013.
III 4.2–3 (Casa di C. Arrius Crescens/Casa del Moralista)	• Statuette of Isis, marble, found together with equipment for making offerings: an incense burner, a small column with a receptacle for incense, and a brazier.	Jashemski 1979–93: 2:102–103; Arslan 1997: 444; De Caro 2006b: no. III.45; Petersen 2006: 46.
V 2.1 (Casa delle Nozze d'Argento)	• Four fountain statuettes (two crocodiles and two frogs), faience[2]	Ehrhardt 2004: 224; De Caro 2006b: no. III.125; Di Gioia 2006: nos. 8.1, 8.2, 9.1; Barrett 2017a.
VI 9.6–7 (Casa dei Dioscuri)	• Statuette of Isis, marble	Boyce 1937: 50, no. 176; Tran tam Tinh 1964: 155, no. 76; Petersen 2006: 46; Mol 2015: 170.

2. Chemical testing suggests at least one is an Egyptian import (see note 132, chapter 1, this volume).

House	Description	Sources
VI 16.7 (Casa degli Amorini Dorati)	• Statuette of Horus, calcite (Egyptian alabaster), probably associated with this house's garden shrine to Egyptian deities • Lamp depicting Isis, Harpocrates, and Anubis, probably associated with this house's garden shrine to Egyptian deities	Seiler 1992: 37–49; De Caro 2006b: nos. III.105, III.115; Di Gioia 2006: 104, no. 54; Beaurin 2008: no. 7*bis*; Petersen 2012; Mol 2013; 2015: 332–391.
VII 2.18 (Casa di C. Vibius Italus)	• Statuette of Isis-Fortuna, bronze	Boyce 1937: no. 252; Tran tam Tinh 1964: no. 78; Beaurin 2008: no. 8; Mol 2015: 476.
VII 4.48 (Casa della Caccia antica)	• Crocodile statue, black stone	See discussion in note 203, chapter 2, this volume.
VII 4.56 (Casa del Granduca/della Fontana)	• Two statuettes of rams, faience or pseudo-faience	Kekulé 1880: 61; de Vos 1983: 63; Appleton 1987: 129–130, 151 (cf. Di Gioia 2006: 129–130, no. 12).
VII 4.59	• Two oscilla, now lost, depicting masks described as "Egyptianizing"	Dwyer 1981: 268, nos. 27, 28; Trentin 2014: 245.
VIII 5.37 (Casa delle Pareti Rosse)	• Herm of Sarapis (according to Sogliano's report)	Sogliano 1882: 439; Jashemski 1979–93: 2:no. 444.
VIII 5.39 (Casa di Acceptus e Euhodia)	• Two bifrontal figurines of Bes, one faience and one pseudo-faience • Figurine of a pygmy, faience • *Nemes*-crowned herm, marble • *Modius*-crowned herm, marble	de Vos 1983: 62; Appleton 1987: 160; Bragantini 2006: 167n1; 2012: 23n13; De Caro 2006b: nos. III.126, III.127; Di Gioia 2006: 112, 116, 137–138; Carrella et al. 2008: 149–150, nos. C 43, C 44; Mol 2015: 119, 184; Poole 2016: 151.
IX 3.2	• Statuette of Isis-Fortuna, bronze	Boyce 1937: no. 406; Tran tam Tinh 1964: no. 89; Beaurin 2008: 285, no. 10.

House	Description	Sources
IX 3.5 (Casa di Marcus Lucretius)	• Two ibis statues, marble • Head of a Bes figurine, faience[3]	Dwyer 1982: 17–52; de Vos 1983: 62; Appleton 1987: nos. 135, 136; Hales 2003: 155–156, 160; Di Gioia 2006: no. 5.6; Mol 2015: 479.
IX 7.12	• Statuette of a ram, faience or pseudo-faience • Three containers or cylindrical bases, faience or pseudo-faience	De Vos 1983: 61–62.
Villa delle Colonne del Mosaico	• Statuette of a frog or toad, faience or pseudo-faience • Statuette of a pygmy (characterized by excavators as "Ptah"), faience or pseudo-faience	Schulz 1838: 191–193; Kekulé 1880: 61; Appleton 1987: no. 79; Jashemski 1979–93: 2:278.

3. Kekulé also reports a second faience or pseudo-faience statuette from this house (1880: 61), though without reference to specific findspot. The excavation records describe this object as a "cercopiteco," a term that early excavators often applied to Bes images as well as actual simians (cf. other examples in Kekulé 1880: 61).

APPENDIX D

Pompeian Garden Shrines with Images of Isis and Related Deities

The following table is a list of shrines that (1) were located within Pompeian gardens and (2) included depictions of Egyptian-derived deities. Such shrines typically take the form of aediculae or wall niches, with one exception: the free-standing statuette of Isis from the Casa del Moralista (III 4.2–3). While statuettes of this type certainly need not always serve "religious" functions, I have in this case opted to include this object because associated finds testify to the ongoing performance of rituals (incense burning and sacrifice) in front of the statuette.

House	**Description**	**Sources**
I 2.20–21	Wall niche containing an assemblage of bronze and terracotta figurines; one of the figurines was a clay bust of Isis wearing a *modius*.	Kekulé 1880: 41 (discussing the context); De Caro 2006b: no. III.34.
II 4.2 (*Praedia* di Julia Felix)	Shrine with paintings of Isis, Anubis, Sarapis (?), Fortuna, and Harpocrates. The shrine contained a large assemblage that included multiple statuettes, amulets, lamps, vessels, and a bronze tripod. Several of these objects also depict Egyptian imagery; see appendix C.	Parslow 2013; and see also the entry in appendix C.

House	Description	Sources
III 4.2–3 (Casa di C. Arrius Crescens/Casa del Moralista)	Statuette of Isis, marble, found together with equipment for making offerings: an incense burner, a small column with a receptacle for incense, and a brazier.	Jashemski 1979–93: 2:102–103; Arslan 1997: 444; De Caro 2006b: no. III.45; Petersen 2006: 46.
VI 2.14 (Casa delle Amazzoni)	Domestic shrine in a now-lost wall niche (probably made of stucco), painted with images of Isis-Fortuna, Sarapis, and Harpocrates. The wall behind the shrine bore a large painting of a river landscape with palm trees and birds; in context, the shrine appeared to be standing on the riverbank. The presence of the Egyptian deities and the palm trees gave the riverscape a "Nilotic" character. A brick altar stood in front of the shrine.	Boyce 1937: no. 141; Bragantini 1993: 174–175; Versluys 2002: no. 43; Beaurin 2008: no. 6.
VI 9.6–7 (Casa dei Dioscuri)	Aedicular shrine painted with Dionysiac symbols (a panther and grapes). A statuette, identified as Isis, was also reported; Boyce and Petersen suggest that this statuette may have come from the shrine.	Boyce 1937: 50, no. 176; Tran tam Tinh 1964: 155, no. 76; Petersen 2006: 46; Mol 2015: 170.
VI 16.7 (Casa degli Amorini Dorati)	Aedicular shrine with painted images of Isis, Sarapis, Harpocrates, Anubis, and ritual objects used in Isis worship. Other Aegyptiaca from the garden may have been associated with the shrine; see appendix C.	Seiler 1992: 37–49; Beaurin 2008: no. 7; Petersen 2012; Mol 2015: 332–392; and see also the entry in appendix C.
VII 2.18 (Casa di C. Vibius Italus)	Aedicular wall niche with a bronze statuette of Isis-Fortuna.	Boyce 1937: no. 252; Tran tam Tinh 1964: no. 78; Beaurin 2008: no. 8; Mol 2015: 476.
IX 3.2	Wall niche located beneath paintings of Lares, serpents, and altar. The niche contained bronze statuettes of multiple deities: Isis-Fortuna, a Genius, a Lar, Hercules, and Jupiter.	Boyce 1937: no. 406; Tran tam Tinh 1964: no. 89; Beaurin 2008: no. 10.

References

Note: Abbreviations used are those of the *American Journal of Archaeology*, with two additions: *PPM* for the series *Pompei: Pitture e mosaici*, ed. I. Baldassarre; and *SAK* for the journal *Studien zur Altägyptischen Kultur*.

Adembri, B., and Z. Mari, eds. 2006. *Suggestioni egizie a Villa Adriana*. Verona.

Adriani, A. 1940–50. "Trouvaille à Ras el Soda." *Annuaire du Musée Gréco-Romain* 3: 28–46.

Alfano, C. 2001a. "Egyptian Influences in Italy." In *Cleopatra of Egypt*, ed. S. Walker and P. Higgs, 276–291. Princeton.

Alfano, C. 2001b. "Glazed Terracotta Statuette of a Crocodile." In *Cleopatra of Egypt*, ed. Walker and Higgs, 340. Princeton.

Alföldi, A. 1965–66. "Die alexandrinischen Götter und die Vota publica am Jahresbeginn." *JAC* 8–9: 53–87.

Allen, J.P. 1988. *Genesis in Egypt*. New Haven.

Allen, Marti Lu. 1985. *The Terracotta Figurines from Karanis*. 2 vols. Ph.D. dissertation, University of Michigan.

Allison, P.M. 1997. "VII 4, 48. Casa della Caccia antica." In *PPM* VII:6–43. Rome.

Allison, P.M. 2002. "Analyses of the Wall-Paintings." In *Casa della Caccia antica (VII 4, 48)*, by P.M. Allison and F.B. Sear, 66–86. Munich.

Allison, P.M. 2004a. *Pompeian Households*. Los Angeles.

Allison, P.M. 2004b. *Pompeian Households: An On-Line Companion*. In *The Stoa*, ed. R. Scaife. www.stoa.org/projects/ph/index.html.

Allison, P.M. 2006. *The Insula of the Menander at Pompeii*. Vol. 3. Oxford.

Allison, P.M. 2007. "Domestic Spaces and Activities." In *The World of Pompeii*, ed. J.J. Dobbins and P.W. Foss, 269–278. New York.

Allison, P.M., and F.B. Sear. 2002. *Casa della Caccia Antica (VII 4, 48)*. Munich.

Almagor, E., and J. Skinner. 2013. "Introduction." In *Ancient Ethnography*, ed. E. Almagor and J. Skinner, 1–22. London/New York.

Alston, R. 1995. *Soldier and Society in Roman Egypt*. New York.

Altenmüller, H. 1975. "Bes." *LÄ* 1, 720–724. Wiesbaden.

Alvar, J. 2008. *Romanizing Oriental Gods*. Trans. R. Gordon. Leiden.

Alvar, J. 2017. "The 'Romanization' of 'Oriental Cults.'" In *Entangled Worlds*, ed. S. Nagel et al., 23–46. Tübingen.

Alvar, J., and E. Muñiz. 2004. "Les cultes égyptiens dans les provinces romaines d'Hispanie." In *Isis en Occident*, ed. L. Bricault, 69–94. Leiden.

Amedick, R. 1993. "Stibadia in Herculaneum und Pompeji." In *Ercolano 1738–1988*, ed. L. Franchi dell'Orto, 179–192. Rome.

Amedick, R. 1994. "Herakles im Speisesaale." *RM* 101: 103–119.

Amenta, A. 2002. "Progetto per un 'Corpus' dei patechi delle collezioni egiziana in Italia e relativo studio della figura del 'Grande Nano' nei testi magici egiziani." *Aegyptus* 82: 157–168.

Amenta, A. 2011. "Iconografia del 'Pateco su coccodrilli' su una gemma magica." In *Aegyptiaca et Coptica*, ed. P. Buzi et al., 1–14. Oxford.

Andersson, E.B. 1990. "Fountains and the Roman Dwelling: Casa del Torello in Pompeii." *JdI* 105: 207–236.

Ando, C. 2003. "A Religion for the Empire." In *Flavian Rome*, ed. A.J. Boyle and W.J. Dominik, 323–344. Leiden.

Ando, C. 2007. "Exporting Roman Religion." In *A Companion to Roman Religion*, ed. J. Rüpke, 429–445. Malden/Oxford.

Andreae, B. 1977. *Das Alexandermosaik aus Pompeji*. Recklinghausen.

Andreae, B. 2003. *Antike Bildmosaiken*. Mainz am Rhein.

Andreau, J. 1974. *Les affaires de Monsieur Jucundus*. Rome.

Appadurai, A., ed. 1986. *The Social Life of Things: Commodities in Cultural Perspective*. Cambridge.

Appleton, G. 1987. *Animal Sculpture from Roman Gardens Buried by Vesuvius*. Ph.D. dissertation, University of Newcastle upon Tyne.

Arnold, D. 1999. *Temples of the Last Pharaohs*. Oxford.

Arslan, E.A., ed. 1997. *Iside*. Milan.

Arya, D.A. 2002. *The Goddess Fortuna in Imperial Rome*. Ph.D. dissertation, University of Texas at Austin.

Ashton, S.-A. 2004. *Roman Egyptomania*. London.

Assmann, J. 2000. "Ägypten als Gegenwelt." In *Gegenwelten zu den Kulturen Griechenlands und Roms in der Antike*, ed. T. Hölscher, 67–83. Munich/Leipzig.

Assmann, J., and M. Bommas, eds. 2002. *Ägyptische Mysterien?* Munich.

Aston, B.G., et al. 2000. "Stone." In *Ancient Egyptian Materials and Technology*, ed. P.T. Nicholson and I. Shaw, 5–75. Cambridge.

Ault, B., and L. Nevett. 1999. "Digging Houses: Archaeologies of Classical and Hellenistic Greek Domestic Assemblages." In *The Archaeology of Household Activities*, ed. P. Allison, 43–56. London.

Bachelard, G. 1964. *The Poetics of Space*. Trans. M. Jolas. New York.

Bagnasco, M.B. 1992. "Bes-Sileno. Un'iconografia tra mondo egizio e greco: nuovi documenti." In *Sesto Congresso Internazionale di Egittologia*, 41–49. Turin.

Bailey, D.M. 2008. *Catalogue of Terracottas in the British Museum*. Vol. 4. London.

Bailey, D.W. 2005. *Prehistoric Figurines*. New York.

Bailey, D., with S. Mills, eds. 1998. *The Archaeology of Value*. Oxford.

Baines, J. 1970. "*Bnbn*: Mythological and Linguistic Notes." *Orientalia* 39: 389–404.

Baines, J. 1985. *Fecundity Figures*. Chicago.

Baines, J. 1990. "Restricted Knowledge, Hierarchy, and Decorum: Modern Perceptions and Ancient Institutions." *JARCE* 27: 1–23.

Bal, M. 2002. *Travelling Concepts in the Humanities*. Toronto.

Balch, D.L. 2008. *Roman Domestic Art and Early House Churches*. Tübingen.

Ballabriga, A. 1981. "Le malheur des nains: Quelques aspects du combat des grues contre les Pygmées dans la littérature grecque." *RÉA* 83: 57–74.

Ballet, P. 2000. "Terres cuites isiaques de l'Égypte hellénistique et romaine. État de la recherche et des publications." In *De Memphis à Rome*, ed. L. Bricault, 91–110.

Barbet, A. 2009. *La peinture murale romaine*. 2nd ed. Paris.

Barrenechea, F. 2010. "Didactic Aggressions in the Nile Excursus of Lucan's *Bellum Civile*." *AJP* 131: 259–284.

Barrett, C.E. 2009. "The Perceived Value of Minoan and Minoanizing Pottery in Egypt." *JMA* 22: 211–234.

Barrett, C.E. 2011. *Egyptianizing Figurines from Delos*. Leiden.

Barrett, C.E. 2013. "'Campana' Relief with Roman Nilotic Landscape." *Echoes of Egypt*. http://echoesofegypt.peabody.yale.edu/overview/campana-relief-roman-nilotic-landscape.

Barrett, C.E. 2015a. "Harpocrates on Rheneia: Two Egyptian Figurines from the Necropolis of Delos." In *Figurines de terre cuite en Méditerranée orientale grecque et romaine*, ed. E. Laflı and A. Muller, 187–200. Villeneuve d'Ascq.

Barrett, C.E. 2015b. "Hellenistic and Roman Egypt." In *Oxford Bibliographies in Classics*, ed. D. Clayman. www.oxfordbibliographies.com/view/document/obo-9780195389661/obo-9780195389661-0189.xml.

Barrett, C.E. 2015c. "Plaster Perspectives on 'Magical Gems': Rethinking the Meaning of 'Magic' in Cornell's Dactyliotheca." In *Cornell Collection of Antiquities*, ed. A. Alexandridis and V. Platt. https://antiquities.library.cornell.edu/gems/plaster-perspectives-on-magical-gems#_ftn63.

Barrett, C.E. 2015d. "Terracotta Figurines and the Archaeology of Ritual: Domestic Cult in Greco-Roman Egypt." In *Figurines grecques en contexte*, ed. S. Huysecom-Haxhi et al., 401–420. Villeneuve d'Ascq.

Barrett, C.E. 2017a. "Egypt in Roman Visual and Material Culture." In *Oxford Handbooks Online in Classical Studies*, ed. G. Williams et al. http://www.oxfordhandbooks.com/view/10.1093/oxfordhb/9780199935390.001.0001/oxfordhb-9780199935390-e-18. Oxford.

Barrett, C.E. 2017b. "Recontextualizing Nilotic Scenes: Interactive Landscapes in the Garden of the Casa dell'Efebo, Pompeii." *AJA* 121: 293–332.

Barrett, C.E. 2018a. "Acrobat with Crocodile." In *Beyond the Nile: Egypt and the Classical World*, ed. J. Spier et al., 276–277. Los Angeles.

Barrett, C.E. 2018b. "Battle Between Pygmies, Crocodiles, and Hippopotami." In *Beyond the Nile: Egypt and the Classical World*, ed. J. Spier et al., 252. Los Angeles.

Barrett, C.E. 2018c. "Campana Relief with Nilotic Scene." In *Beyond the Nile: Egypt and the Classical World*, ed. J. Spier et al., 272. Los Angeles.

Barrett, C.E. 2018d. "Nilotic Scenes in Roman Art." In *Beyond the Nile: Egypt and the Classical World*, ed. J. Spier et al., 250. Los Angeles.

Barrett, C.E. In press (a). "The Affordances of Terracotta Figurines in Domestic Contexts: Reconsidering the Gap between Material and Ritual." In *Stuff of the Gods*, ed. M. Haysom et al. Stockholm.

Barrett, C.E. In press (b). "Magical Gems as Material Texts: Contextual Analysis of Greco-Roman Amulets." In *Textual Archaeology of the Ancient Near East*, ed. Y. Heffron. Cambridge.

Barry, F. 2011. *Painting in Stone*. Ph.D. dissertation, Columbia University.

Bartman, E. 1991. "Sculptural Collecting and Display in the Private Realm." In *Roman Art in the Private Sphere*, ed. E.K. Gazda, 71–88. Ann Arbor.

Bartman, E. 2002. "Eros's Flame: Images of Sexy Boys in Roman Ideal Sculpture." In *The Ancient Art of Emulation*, ed. E.K. Gazda, 249–272. Ann Arbor.

Bartsch, S., and J. Elsner. 2007. "Introduction: Eight Ways of Looking at an Ekphrasis." *CP* 102: i–vi.

Baslez, M.-F. 1977. *Recherches sur les conditions de pénétration et de diffusion des religions orinetales a Délos*. Paris.

Bassani, M. 2008. *Sacraria*. Rome.

Beagon, M. 1992. *Roman Nature: The Thought of Pliny the Elder*. Oxford.

Beard, M. 2014. *Laughter in Ancient Rome*. Berkeley.

Beard, M., J. North, and S. Price. 1998. *Religions of Rome*. 2 vols. Cambridge.

Beaurin, L. 2008. "Isis-Fortuna à Pompéi. Le succès d'une déesse intégrée." *Oebalus* 3: 267–293.

Bechi, G. 1832. "Relazione degli scavi di Pompei. Da maggio 1831 a maggio 1832." *Real Museo Borbonico* 8: 1–15.

Bechi, G. 1833. "Relazione degli scavi di Pompei. Da maggio 1832 ad aprile 1833." *Real Museo Borbonico* 9: 1–7.

Beckmann, M. 2010. "How the Alexander Mosaic Was Used." Paper read at the 2010 Annual Meeting of the Archaeological Institute of America, January 6–9, Anaheim.

Bek, L. 1980. *Towards Paradise on Earth*. Odense.

Bek, L. 1983. "Questiones Convivales: The Idea of the Triclinium and the Staging of Convivial Ceremony from Rome to Byzantium." *AnalRom* 12: 81–107.

Belayche, N., and S.C. Mimouni, eds. 2009. *Entre lignes de partage et territoires de passage*. Paris/Leuven.

Belayche, N., and E. Rebillard. 2008. "'Cultes orientaux' et pluralisme religieux. Introduction thématique." In *Religioni in contatto nel Mediterraneo anitico*, ed. C. Bonnet et al., 137–150. Pisa/Rome.

Bell, L. 1985. "Luxor Temple and the Cult of the Royal Ka." *JNES* 44: 251–294.

Bell, L. 1997. "The New Kingdom 'Divine' Temple: The Example of Luxor." In *Temples of Ancient Egypt*, ed. B.E. Shafer, 127–184. Ithaca.

Bendinelli, G. 1941. *Le pitture del Colombario di Villa Pamphili*. Rome.

Bergmann, B. 1991. "Painted Perspectives of a Villa Visit: Landscape as Status and Metaphor." In *Roman Art in the Private Sphere*, ed. E.K. Gazda, 49–70. Ann Arbor.

Bergmann, B. 1992. "Exploring the Grove: Pastoral Space on Roman Walls." In *The Pastoral Landscape*, ed. J.D. Hunt, 20–46. Washington, DC.

Bergmann, B. 1995. "Greek Masterpieces and Roman Recreative Fictions." *HSCP* 97: 79–120.

Bergmann, B. 1999. "Rhythms of Recognition: Mythological Encounters in Roman Landscape Painting." In *Im Spiegel des Mythos*, ed. F. de Angelis and S. Muth, 81–108. Wiesbaden.

Bergmann, B. 2001. "Meanwhile, Back in Italy. . . Creating Landscapes of Allusion." In *Pausanias: Travel and Memory in Roman Greece*, ed. S.E. Alcock et al., 154–166. Oxford.

Bergmann, B. 2002a. "Playing with Boundaries: Painted Architecture in Roman Interiors." *The Built Surface* 1: 15–46.

Bergmann, B. 2002b. "Art and Nature in the Villa at Oplontis." In *Pompeian Brothels, Pompeii's Ancient History, Mirrors and Mysteries, Art and Nature at Oplontis, The Herculaneum "Basilica,"* ed. T. McGinn, 87–120. Portsmouth.

Bergmann, B. 2007. "A Painted Garland: Weaving Words and Images in the House of the Epigrams in Pompeii." In *Art and Inscriptions in the Ancient World*, ed. Z. Newby and R. Leader-Newby, 60–101. Cambridge.

Bergmann, B. 2008. "Staging the Supernatural: Interior Gardens of Pompeian Houses." In *Pompeii and the Roman Villa*, ed. C. Mattusch, 53–70. New York.

Bergmann, B. 2012. "Housing and Households (Roman World)." In *Classical Archaeology*, ed. S.E. Alcock and R. Osborne, 228–248. 2nd ed. Malden.

Bergmann, B. 2014. "The Concept of Boundary in the Roman Garden." In *Le jardin dans l'Antiquité*, ed. K. Coleman and P. Derron, 245–289. Geneva.

Bergmann, B. 2016. "The Gardens and Garden Paintings of Villa A." In *Leisure and Luxury in the Age of Nero*, ed. E.K. Gazda and J.R. Clarke, 96–110. Ann Arbor.

Bergmann, B. 2018. "Frescoes in Roman Gardens." In *Gardens of the Roman Empire*, ed. W.F. Jashemski et al., 278–316. Cambridge.

Bergmann, M. 2016. "*Mallokouria*: Portraits of Local Elite Boys in Roman Egypt." In *Beyond Boundaries*, ed. S.E. Alcock et al., 156–173. Los Angeles.

Berlandini, J. 1995. "Ptah-démiurge et l'exaltation du ciel." *RÉg* 46: 9–41.

Bernand, E. 1988. "Pèlerins dans l'Égypte gréco-romaine." In *Mélanges Pierre Lévêque*, ed. M.M. Mactoux and E. Genys, 1:49–63. Paris.

Berry, J. 1997. "Household Artefacts: Re-Interpreting Roman Domestic Space." In *Domestic Space in the Roman World*, ed. R. Laurence and A. Wallace-Hadrill, 183–195. Portsmouth.

Berry, J. 2007. "*Instrumentum Domesticum*—A Case Study." In *The World of Pompeii*, ed. J.J. Dobbins and P.W. Foss, 292–301. New York.

Betz, H.D, ed. 1992. *The Greek Magical Papyri in Translation*. 2nd ed. Chicago.

Bhabha, H.K. 1994. *The Location of Culture*. London/New York.

Bianchi, R.S. 1988a. "Symplegma." In *Cleopatra's Egypt*, dir. R.A. Fazzini, 241. Mainz am Rhein.

Bianchi, R.S. 1988b. "Vessel with Appliqués." In *Cleopatra's Egypt*, dir. R.A. Fazzini, 226. Mainz am Rhein.

Bianchi Bandinelli, R. 1967. "Arte plebea." *DialArch* 1: 7–19.

Bianchi Bandinelli, R. 1970. *Rome: The Center of Power*. Trans. P. Green. New York.

Bielfeldt, R. 2014. "Lichtblicke—Sehstrahlen. Zur Präsenz römischer Figuren- und Bildlampen." In *Ding und Mensch in der Antike*, ed. R. Bielfeldt, 195–238. Heidelberg.

Bishop, M.C., and J.C.N. Coulston. 1993. *Roman Military Equipment from the Punic Wars to the Fall of Rome*. London.

Blanc, N., H. Eristov, and M. Fincker. 2000. "*A fundamento restituit?* Réfections dans le temple d'Isis à Pompéi." *RA*: 227–309.

Bliquez, L.J. 1994. *Roman Surgical Instruments and Other Minor Objects in the National Archaeological Museum of Naples*. Mainz am Rhein.

Boissel, I. 2007. *Égypte dans les mosaïques de l'occident romain*. Ph.D. dissertation, Université de Reims Champagne-Ardenne.

Bommas, M. 2012. "The Iseum Campense as a Memory Site." In *Memory and Urban Religion in the Ancient World*, ed. M. Bommas et al., 177–212. London.

Bommas, M. 2013a. "Pyramids in Ancient Rome: Images without Cult?" In *Kult und Bild*, ed. M.M. Luisell et al., 91–110. Würzburg.

Bommas, M. 2013b. "Temples for Egyptian Gods within an Urban Landscape: The Roman *Iseum Campense* and the Red Hall of Pergamon as Case Studies." In *Cities and Gods*, ed. T. Kaizer et al., 59–68. Leuven.

Bonghi Jovini, M., ed. 1984. *Ricerche a Pompei.* Rome.

Bonneau, D. 1964. *La crue du Nil, divinité égyptienne*. Paris.

Bonneau, D. 1991. "Continuité et discontinuité notionale dans la terminologie religieuse du Nil d'après la documentation grecque." In *Mélanges Étienne Bernand*, ed. N. Fick and J.-C. Carrière, 23–35. Paris.

Bonnet, C., and L. Bricault. 2013. "Introduction." In *Panthée,* ed. L. Bricault and C. Bonnet, 1–16. Leiden.

Bonnet, C., S. Ribichini, and D. Steuernagel, eds. 2008. *Religioni in contatto nel Mediterraneo antico*. Pisa/Rome.

Bonnet, C., J. Rüpke, and P. Scarpi, eds. 2006. *Religions orientales – culti misterici.* Stuttgart.

Boschetti, C. 2003. "I mosaici di fattura alessandrina in area campana: Un'analisi tecnica." *Journal of Intercultural and Interdisciplinary Archaeology* 1: C/01. http://v01.jiia.it/JIIA.it/Sezione_II/JIIA_01/Boschetti_C01/Articolo_C01/Boschetti_articolo_pag1.html.

Boschetti, C. 2011. "Vitreous Materials in Early Mosaics in Italy: Faience, Egyptian Blue, and Glass." *Journal of Glass Studies* 53: 59–91.

Bothmer, B.C. 1960. *Egyptian Sculpture of the Late Period.* New York.

Bounia, A. 2004. *The Nature of Classical Collecting.* Burlington.

Boutantin, C. 2014. *Terres cuites et culte domestique.* Leiden.

Bowden, H. 2010. *Mystery Cults in the Ancient World.* London.

Bowes, K. 2015. "At Home." In *A Companion to the Archaeology of Religion in the Ancient World,* ed. R. Raja and J. Rüpke, 209–219. Malden.

Boyce, G.K. 1937. "Corpus of the Lararia of Pompeii." *MAAR* 14: 5–112.

Bradley, K. 1998. "The Roman Family at Dinner." In *Meals in a Social Context,* ed. I. Nielsen and H.S. Nielsen, 36–55. Aarhus.

Bragantini, I. 1991. "III 4, 2. Casa del Moralista o di C. Arrius Crescens." In *PPM* III:406–434. Rome.

Bragantini, I. 1993. "VI 2, 14. Casa delle Amazzoni." In *PPM* IV:168–197. Rome.

Bragantini, I. 1997. "VII 16 (*Ins. Occ.*), 17. Casa di Ma. Castricius." In *PPM* VII:887–946. Rome.

Bragantini, I. 1998a. "VIII 3, 8–9. Casa del Cinghiale I." In *PPM* VIII:362–384. Rome.

Bragantini, I. 1998b. "VIII 5, 24. Casa del Medico." In *PPM* VIII:604–610. Rome.

Bragantini, I. 1998c. "VIII 5, 39. Casa di Acceptus e Euhodia." In *PPM* VIII:648–655. Rome.

Bragantini, I. 1999. "IX 3, 5.24. Casa di Marcus Lucretius." In *PPM* IX: 141–313. Rome.

Bragantini, I. 2006. "Il culto di Iside e l'egittomania antica in Campania." In *Egittomania,* ed. S. De Caro, 159–167. Milan.

Bragantini, I. 2012. "The Cult of Isis and Ancient Egyptomania in Campania." In *Contested Spaces,* ed. D.L. Balch and A. Weissenrieder, 21–33. Tübingen.

Bragantini, I., and M. de Vos. 1982. *Le decorazioni della villa romana della Farnesina.* Rome.

Bragantini, I., and R. Pirelli. 2006–2007. "Osservazioni sul fregio della villa romana della Farnesina." *Annali di archeologia e storia antica* 13–14: 221–231.

Bremmer, J.N. 1998. "'Religion,' 'Ritual' and the Opposition 'Sacred vs. Profane': Notes towards a Terminological 'Genealogy.'" In *Ansichten griechischer Rituale,* ed. F. Graf, 9–32. Stuttgart.

Bremmer, J.N. 2014. *Initiation into the Mysteries of the Ancient World.* Berlin.

Bresciani, E. 1992. *Il mito dell'Occhio del Sole.* Brescia.

Breton, E. 1870. *Pompeia: décrite et dessinée.* 3rd ed. Paris.

Bricault, L. 2000a. "Bilan et perspectives dans les études isiaques." In *La Grande Dea tra passato e presente,* ed. E. Leospo and D. Taverna, 91–96. Turin.

Bricault, L., ed. 2000b. *De Memphis à Rome.* Leiden.

Bricault, L., ed. 2004. *Isis en Occident*. Leiden.

Bricault, L. 2005. *Recueil des inscriptions concernant les cultes isiaques (RICIS)*. 3 vols. Paris.

Bricault, L. 2008. *Bibliotheca Isiaca I*. Bordeaux.

Bricault, L. 2012. "Associations isiaques d'Occident." In *Demeter, Isis, Vesta, and Cybele*, ed. A. Mastrocinque and C.G. Scibona, 91–104. Stuttgart.

Bricault, L. 2013. *Les cultes isiaques dans le monde gréco-romain*. Paris.

Bricault, L., and C. Bonnet, eds. 2013. *Panthée: Religious Transformations in the Graeco-Roman Empire*. Leiden.

Bricault, L., and M.J. Versluys, eds. 2010. *Isis on the Nile*. Leiden.

Bricault, L., and M.J. Versluys, eds. 2012. *Egyptian Gods in the Hellenistic and Roman Mediterranean*. Palermo.

Bricault, L., and M.J. Versluys. 2014a. "Isis and Empires." In *Politics, Power, and the Cults of Isis*, ed. L. Bricault and M.J. Versluys, 3–35. Leiden.

Bricault, L., and M.J. Versluys, eds. 2014b. *Power, Politics and the Cults of Isis*. Leiden.

Bricault, L., M.J. Versluys, and P.G.P. Meyboom, eds. 2007. *Nile into Tiber*. Leiden.

Bricault, L., and R. Veymiers. 2011. *Bibliotheca Isiaca II*. Bordeaux.

Bricault, L., and R. Veymiers. 2012. "Quinze ans après. Les études isiaques (1997–2012): un premier bilan." In *Egyptian Gods in the Hellenistic ad Roman Mediterranean*, ed. Bricault and Versluys, 1–24. Palermo.

Brilliant, R. 1963. *Gesture and Rank in Roman Art*. New Haven.

Brophy, E. 2014. "Two-for-One: Looking at Imperial Egyptian Statues with Roman Features in Egypt." In *Current Research in Egyptology 2013*, ed. K. Accetta et al., 22–33. Oxford/Philadelphia.

Brose, M. 2009. "Ostrakon BM EA 50727: 'An unknown literary text'?" *SAK* 38: 57–81.

Brubaker, R. 2002. "Ethnicity without Groups." *European Journal of Sociology* 43: 163–189.

Brubaker, R. 2004. *Ethnicity without Groups*. Cambridge.

Bruneau, P. 1970. "La vaisselle." In *L'Îlot de la Maison des Comédiens*, ed. U.T. Bezerra de Meneses et al., 263–265. Paris.

Brunner, H. 1954. "Die theologische Bedeutung der Trunkenheit." *ZÄS* 79: 81–83.

Brunner, H. 1964. *Die Geburt des Gottkönigs*. Wiesbaden.

Bruun, C. 2016. "L'acqua come elemento di lusso nella cultura romana: da Varrone alla *Historia Augusta*." *MÉFRA* 128: 1–26. https://mefra.revues.org/3250.

Bryan, B.M. 2014. "Hatshepsut and Cultic Revelries in the New Kingdom." In *Creativity and Innovation in the Reign of Hatshepsut*, ed. J.M. Galán et al., 93–124. Chicago.

Budin, S. 2011. *Images of Woman and Child from the Bronze Age*. Cambridge.

Budischovsky, M.-C. 2004. "Témoignages de dévotion isiaque et traces culturelles le long du *limes* danubien." In *Isis en Occident*, ed. L. Bricault, 171–192. Leiden.

Bugarski-Mesdjian, A. 2007. "Traces d'Égypte en Dalmatie romaine: Culte, mode et pouvoir." In *Nile into Tiber*, ed. L. Bricault et al., 289–328. Leiden.

Bülow Clausen, K. 2012. "Domitian between Isis and Minerva: The Dialogue between the 'Egyptian' and 'Graeco-Roman' Aspects of the Sanctuary of Isis at Beneventum." In *Egyptian Gods in the Hellenistic and Roman Mediterranean*, ed. L. Bricault and M.J. Versluys, 93–122. Palermo.

Bülow Clausen, K. 2015. *The Flavian* Isea *in Beneventum and Rome*. Ph.D. thesis, University of Copenhagen.

Bulté, J. 1991. *Talismans égyptiens d'heureuse maternité*. Paris.

Bulté, J. 1998. "À propos d'une tête de patèque de Tanis: Sân 93–351, OAE 3423." In *Tanis*, eds. P. Brissaud and C. Zivie-Coche, 379–389. Paris.

Burke, P.J. 2003. "Relationships Among Multiple Identities." In *Advances in Identity Theory and Research*, ed. P.J. Burke et al., 195–214. New York.

Burke, P.J., and J.E. Stets. 2009. *Identity Theory*. Oxford.

Burkert, W. 1982. "Craft versus Sect: The Problem of Orphics and Pythagoreans." In *Jewish and Christian Self-Definition*, ed. B.F. Meyer and E.P. Sanders, 3: 1–22. Philadelphia.

Burkert, W. 1987. *Ancient Mystery Cults*. Cambridge.

Bussels, S. 2012. *The Animated Image*. Berlin.

Butler, H.E., ed. and trans. 1967. *Propertius*. Cambridge.

Calabrò, A. 2012. "Pompeian *Cauponae* in Their Spatial Context: Interaction between Bars, Inns, and Houses." In *Privata Luxuria*, ed. A. Anguissola, 73–94. Munich.

Camille, M. 1992. *Image on the Edge: The Margins of Medieval Art*. London.

Campbell, G. 2016. "Anthropology: Knowledge of Man." In *A Companion to Science, Technology, and Medicine in Ancient Greece and Rome*, ed. G.L. Irby, 1: 500–515. Malden/Oxford/Chichester.

Cancik, H. 2007. "Haus, Schule, Gemeinde: Zur Organisation von 'fremder Religion' in Rom (1.–3. Jh. n. Chr.)." In *Gruppenreligionen im römischen Reich*, ed. J. Rüpke, 31–48. Tübingen.

Capel, A.K., and G.E. Markoe, eds. 1996. *Mistress of the House, Mistress of Heaven*. New York.

Cappel, A. 1994. *Untersuchungen zu Pygmäendarstellungen in der römischen Dekorationskunst*. Inauguraldissertation, Julius-Maximilians-Universität zu Würzburg.

Capriotti Vittozzi, G. 2002a. "L'Egitto fuori dall'Egitto." In *Tra le palme del Piceno*, ed. A. Roccati and G. Capriotti Vittozzi, 26–28. Poggibonsi.

Capriotti Vittozzi, G. 2002b. "Il fiume e gli acquitrini." In *Tra le palme del Piceno*, ed. A. Roccati and G. Capriotti Vittozzi, 110–116. Poggibonsi.

Capriotti Vittozzi, G. 2014. "The Flavians: Pharaonic Kingship between Egypt and Rome." In *Power, Politics and the Cults of Isis*, ed. L. Bricault and M.J. Versluys, 237–259. Leiden.

Carbado, D.W., et al., eds. 2013. "Intersectionality" (Special issue of *Du Bois Review*). *Du Bois Review* 10.2.

Carettoni, G. 1983a. "La decorazione pittorica della Casa di Augusto sul Palatino." *RM* 90: 373–416.
Carettoni, G. 1983b. *Das Haus des Augustus auf dem Palatin*. Mainz am Rhein.
Carey, S. 2002. "A Tradition of Adventures in the Imperial Grotto." *GaR* 49: 44–61.
Carey, S. 2003. *Pliny's Catalogue of Culture*. Oxford.
Carrella, M., L.A. D'Acunto, N. Inserra, and C. Serpe. 2008. *Marmora Pompeiana nel Museo Archeologico Nazionale di Napoli*. Rome.
Carroll, M. 2010. "Exploring the Sanctuary of Venus and Its Sacred Grove: Politics, Cult and Identity in Roman Pompeii." *PBSR* 78: 63–106, 347–51.
Casson, L. 1980. "The Role of the State in Rome's Grain Trade." In *The Seaborne Commerce of Ancient Rome*, ed. J.H. D'Arms and E.C. Koppf, 21–29. Rome.
Castrén, P. 1975. *Ordo Populusque Pompeianus*. Rome.
Castriota, D. 1995. *The Ara Pacis Augustae and the Imagery of Abundance in Later Greek and Early Roman Imperial Art*. Princeton.
Charles-Laforge, M.-O. 2006. "Le 'complexe des rites magiques' et le culte de Sabazios à Pompéi." In *Contributi di archeologia vesuviana* II, 161–184. Rome.
Chevallier, R. 1991. *L'artiste, le collectionneur et le faussaire*. Paris.
Christensen, A.M. 2006. *From Palaces to Pompeii*. Ph.D. dissertation, Florida State University.
Cibu, S., and B. Rémy. 2004. "Isis et les dieux égyptiens dans les provinces alpines au Haut-Empire (Alpes maritimes, cottiennes, graies, poenines, Rétie, Norique)." In *Isis en Occident*, ed. L. Bricault, 137–170. Leiden.
Clarke, J.R. 1991. *The Houses of Roman Italy 100 B.C.–A.D. 250*. Berkeley.
Clarke, J.R. 1998. *Looking at Lovemaking*. Berkeley.
Clarke, J.R. 2003. *Art in the Lives of Ordinary Romans*. Berkeley.
Clarke, J.R. 2006. "High and Low: Mocking Philosophers in the Tavern of the Seven Sages, Ostia." In *The Art of Citizens, Soldiers and Freedmen in the Roman World*, ed. E. D'Ambra and G.P.R. Métraux, 47–57. Oxford.
Clarke, J.R. 2007a. "A Compendium of Pygmy Imagery in the Casa del Medico at Pompeii: Content, Context, and Viewers." In *Circulación de temas y sistemas decorativos en la pintura mural antigua*, ed. C. Guiral Pelegrín, 219–222. Calatayud.
Clarke, J.R. 2007b. *Looking at Laughter*. Berkeley.
Clarke, J.R. 2007c. "Three Uses of the Pygmy and the Aethiops at Pompeii: Decorating, 'Othering,' and Warding Off Demons." In *Nile into Tiber*, ed. L. Bricault et al., 155–169. Leiden.
Clarke, J.R. 2014. "Roman Visual Humor." In *Encyclopedia of Humor Studies*, ed. S. Attardo, 652–656. Thousand Oaks.
Clarysse, W. 2010. "Egyptian Temples and Priests: Graeco-Roman." In *A Companion to Ancient Egypt*, ed. A.B. Lloyd, 1:274–290. Malden.
Coarelli, F. 2006. "Iside." In *Egittomania*, ed. S. De Caro, 59–67. Milan.
Coarelli, F., D. Musti, and H. Solin, eds. 1982. *Delo e l'Italia*. Rome.

Cody, J.M. 2003. "Conquerors and Conquered on Flavian Coins." In *Flavian Rome*, ed. A.J. Boyle and W.J. Dominik, 103–124. Leiden.
Cohen, A. 1997. *The Alexander Mosaic: Stories of Victory and Defeat*. Cambridge.
Cohen, A. 2010. *Art in the Era of Alexander the Great*. Cambridge.
Colburn, H.P. 2016. "Roman Collecting and the Biographies of Egyptian Late Period Statues." *WorldArch.* 48: 226–238.
Collins, D. 2008. *Magic in the Ancient Greek World*. Malden.
Collombert, P., and Y. Volokhine. 2005. "*De aegyptiacis rebus doctorum verecundia* ou 'Let's Talk About Sex!'" *Egypte: Afrique & Orient* 40: 45–56.
Cooley, A.E., and M.G.L. Cooley. 2014. *Pompeii and Herculaneum: A Sourcebook*. 2nd edn. Abingdon/New York.
Corcoran, L.H. 1995. *Portrait Mummies from Roman Egypt*. Chicago.
Corlàita Scagliarini, D. 1974–76. "Spazio e decorazione nella pittura pompeiana." *Palladio* 23–25: 3–44.
Crenshaw, K. 1989. "Demarginalizing the Intersection of Race and Sex: A Black Feminist Critique of Antidiscrimination Doctrine, Feminist Theory and Antiracist Politics." *University of Chicago Legal Forum* 1989: 139–168. https://legal-forum.uchicago.edu/page/issues.
Culler, J. 1988. *Framing the Sign*. Norman.
Curran, B. 1996. "Review of *Egyptomania: Egypt in Western Art, 1730–1930*." *ArtB* 78: 739–745.
Dalby, A. 2000. *Empire of Pleasures*. New York.
D'Ambrosio, A., and M. Borriello. 1990. *Le terrecotte figurate di Pompei*. Rome.
D'Arms, J.H. 1974. "Puteoli in the Second Century of the Roman Empire: A Social and Economic Study." *JRS* 64: 104–124.
Darnell, J.C. 1994. "Plate 26. Priestesses and Priests Chanting a Song: Detail of Plate 18" (transliteration and translation). In *Reliefs and Inscriptions at Luxor Temple*, ed. Epigraphic Survey, 1:12–14. Chicago.
Darnell, J.C. 1995. "Hathor Returns to Medamûd." *SAK* 22: 47–94.
Darnell, J.C. 2010a. "A Midsummer Night's Succubus—The Herdsman's Encounters in P. Berlin 3024, 'The Pleasures of Fishing and Fowling, The Songs of the Drinking Place, and the Ancient Egyptian Love Poetry.'" In *Opening the Tablet Box*, ed. S.C. Melville and A.L. Slotsky, 99–140. Leiden.
Darnell, J.C. 2010b. "Opet Festival." In *UCLA Encyclopedia of Egyptology*, ed. J. Dieleman and W. Wendrich. http://escholarship.org/uc/item/4739r3fr.
Darnell, J.C. 2010c. "A Pharaonic *De Profundis* from the Western Desert Hinterland of Naqada." In *Honi soit qui mal y pense*, ed. Hermann Knuf et al., 39–48. Leuven.
Darnell, J.C. 2016. "The Rituals of Love in Ancient Egypt: Festival Songs of the Eighteenth Dynasty and the Ramesside Love Poetry." *Die Welt des Orients* 46: 22–61.
Dasen, V. 1988. "Dwarfism in Egypt and Classical Antiquity: Iconography and Medical History." *Medical History* 32: 253–276.

Dasen, V. 1993. *Dwarfs in Ancient Egypt and Greece*. Oxford.

Dasen, V. 1994. "Pygmaioi." In *LIMC* VII:594–601.

Dasen, V. 2000. "Squatting Comasts and Scarab-Beetles." In *Periplous*, ed. G.R. Tsetskhladze et al., 89–97. London.

Dasen, V. 2009. "D'un monde à l'autre. La chasse des pygmées dans l'iconographie impériale." In *Chasses antiques*, ed. J. Trinquier and C. Vendries, 215–233. Rennes.

Dasen, V. 2013. "Des artistes différents? Nains danseurs et musiciens dans le monde hellénistique et romain." In *Le statut du musicien dans la Méditerranée ancienne*, ed. S. Emerit, 259–277. Cairo.

Daszewski, W.A. 1985. *Corpus of Mosaics from Egypt*. Vol. 1. Mainz am Rhein.

Daszewski, W.A. 2002. "Il paesaggio nell'Egitto greco-romano." In *Tra le palme del Piceno*, ed. A. Roccati and G. Capriotti Vittozzi, 22–25. Poggibonsi.

Daumas, F. 1958. *Les mammisis des temples égyptiens*. Paris.

Daumas, F. 1968. "Les propylées du temple d'Hathor à Philae et le culte de la déesse." *ZÄS* 95: 1–17.

Davies, N. de G. 1933. *The Tombs of Menkheperrasoneb, Amenmose, and Another (Nos. 86, 112, 42, 226)*. London.

Davies, P.J.E. 2011. "*Aegyptiaca* in Rome: *Adventus* and *Romanitas*." In *Cultural Identity in the Ancient Mediterranean*, ed. E.S. Gruen, 354–370. Los Angeles.

de Angelis, F., J.-A. Dickmann, F. Pirson, and R. von den Hoff, eds. 2012. *Kunst von unten?* Wiesbaden.

De Caro, S. 1983. "La Campania e l'Egitto in età ellenistica e romana." In *Civiltà dell'antico Egitto in Campania*, ed. E. Pozzi, 53–58. Naples.

De Caro, S, ed. 1992. *Alla ricerca di Iside*. Rome.

De Caro, S. 2006a. "L'Egitto in Campania: Le ragioni di una mostra." In *Egittomania*, ed. S. De Caro, 13–17. Milan.

De Caro, S, ed. 2006b. *Egittomania: Iside e il mistero*. Milan.

De Cenival, Françoise. 1988. *Le mythe de l'oeil du soleil*. Sommerhausen.

Dee, J.H. 1978. "Elegy 4.8: A Propertian Comedy." *TAPA* 108: 41–53.

DeFelice, J. 2007. "Inns and Taverns." In *The World of Pompeii*, ed. J.J. Dobbins and P.W. Foss, 474–486. New York.

Degardin, J.-C. 2010. "Le fonctionnement du toit du temple de Khonsou à Karnak." *Cahiers de Karnak* 13: 227–241.

de Haan, N. 2001. "*Si aquae copia patiatur*: Pompeian Private Baths and the Use of Water." In *Water Use and Hydraulics in the Roman City*, ed. A.O. Koloski-Ostrow, 41–50. Dubuque.

Delia, D. 1991. *Alexandrian Citizenship during the Roman Principate*. Atlanta.

Della Corte, M. 1930. *Pompei, i nuovi scavi e l'anfiteatro*. Pompeii.

Della Corte, M. 1931–32. "I MM. Lorei Tiburtini di Pompei." *Atti e memorie della Società tiburtina di storia e d'arte* 11–12: 182–216.

Della Corte, M. 1965. *Case ed abitanti di Pompei*. 3rd. ed. Naples.

De Maria, S. 1999. "Un bronzetto da Bakchias (Fayyum) e la serie dei nani danzanti ellenistici." *Ocnus* 7: 45–68.

DePauw, M., and M.J. Smith. 2004. "Visions of Ecstasy. Cultic Revelry before the Goddess Ai/Nehemanit. Ostraca Faculteit Letteren (K.U. Leuven) dem. 1–2." In *Res Severa Verum Gaudium*, ed. F. Hoffmann and H.J. Thissen, 67–94. Leuven.

Derchain, P. 1975. "La perruque et le cristal." *SAK* 2: 55–74.

Derchain, P. 1981. "Appendix K. Observations sur les erotica." In *The Sacred Animal Necropolis at North Saqqâra: The Southern Dependencies of the Main Temple Complex*, ed. G.T. Martin, 166–170. London.

Derchain, P. 1987. *Le dernier obélisque*. Brussels.

de Rossi, G.B. 1883. "Adunanze dell'Instituto, Febbraio 23." *BdI* 3: 37–38.

De Salvia, F. 2006. "Egitto faraonico e Campania pre-romana: gli *Aegyptiaca* (secoli IX-IV a.C.)." In *Egittomania*, ed. S. De Caro, 21–58. Milan.

Desroches Noblecourt, C. 1995. *Amours et fureurs de La Lointaine*. Paris.

Dessales, H. 2006. "*Castella privata*: Water Towers and Tanks in Roman Dwellings." In *Cura Aquarum in Ephesus*, ed. G. Wiplinger, 363–370. Leuven.

Dessales, H. 2013. *Le partage de l'eau*. Rome.

de Vos, A. 1990. "I 7, 11. Casa dell'Efebo o di P. Cornelius Tages." In *PPM* I: 619–727. Rome.

de Vos, M. 1980. *L'egittomania in pitture e mosaici romano-campani della prima età imperiale*. Leiden.

de Vos, M. 1983. "'Egittomania' nelle case di Pompei ed Ercolano." In *Civiltà dell'antico Egitto in Campania*, ed. E. Pozzi, 59–71. Naples.

de Vos, M. 1990. "I 2, 24. Caupona." In *PPM* I:53–57. Rome.

de Vos, M. 1991. "II 2, 2. Casa di D. Octavius Quartio." In *PPM* III:42–111. Rome.

de Vos, M. 1994. "Aegyptiaca Romana." *La parola del passato* 49: 130–159.

Dickmann, J.-A. 1997. "The Peristyle and the Transformation of Domestic Space in Hellenistic Pompeii." In *Domestic Space in the Roman World*, ed. R. Laurence and A. Wallace-Hadrill, 121–136. Portsmouth.

Dickmann, J.-A. 1999. *Domus frequentata*. Munich.

Dickmann, J.-A. 2015. "Crucial Contexts: A Closer Reading of the Household of the Casa del Menandro at Pompeii." In *Household Studies in Complex Societies*, ed. M. Müller, 211–228. Chicago.

Dieleman, J. 2005. *Priests, Tongues, and Rites*. Leiden.

Dieleman, J., and I.S. Moyer. 2010. "Egyptian Literature." In *A Companion to Hellenistic Literature*, ed. J.J. Clauss and M. Cuypers, 429–447. Chichester.

Dietler, M., and I. Herbich. 1989. "*Tich Matek*: The Technology of Luo Pottery Production and Definition of Ceramic Style." *WorldArch* 21: 148–163.

Dietler, M., and I. Herbich. 1998. "*Habitus*, Techniques, Style: An Integrated Approach to the Social Understanding of Material Culture and Boundaries." In *The Archaeology of Social Boundaries*, ed. M. Stark, 232–263. Washington, DC.

Dietrich, N., and M. Squire, eds. 2018. *Ornament and Figure in Graeco-Roman Art*. Berlin/Boston.

Di Gioia, E. 2006. *La ceramica invetriata in area vesuviana*. Rome.

Dillon, S. 2006. *Ancient Greek Portrait Sculpture*. Cambridge.

Donderer, M. 1990. "Das pompejanische Alexandermosaik – Ein östliches Importstück?" In *Das antike Rom und den Osten*, ed. C. Börker and M. Donderer, 19–31. Erlangen.

Draycott, Jane. 2012. "The Symbol of Cleopatra Selene: Reading Crocodiles on Coins in the Late Republic and Early Principate." *Acta Classica* 55: 43–56.

Dunand, F. 1990. *Catalogue des terres cuites gréco-romaines d'Égypte*. Paris.

Dunand, F. 1997. "La consultation oraculaire en Égypte tardive: l'oracle de Bès à Abydos." In *Oracles et prophéties dans l'antiquité*, ed. J.-G. Heintz, 65–84. Paris.

Dunand, F. 2010. "Culte d'Isis ou religion isiaque?" In *Isis on the Nile*, ed. L. Bricault and M.J. Versluys, 39–54. Leiden.

Dunand, F. 2013. "Images de dieux en dialogue." In *Panthée*, ed. L. Bricault and C. Bonnet, 191–234. Leiden.

Dunand, F., and C. Zivie-Coche. 2004. *Gods and Men in Egypt*. Trans. D. Lorton. Ithaca.

Dunbabin, K. 1991. "Triclinium and Stibadium." In *Dining in a Classical Context*, ed. W.J. Slater, 121–148. Ann Arbor.

Dunbabin, K. 1996. "Convivial Spaces: Dining and Entertainment in the Roman Villa." *JRA* 9: 66–80.

Dunbabin, K. 1999. *Mosaics of the Greek and Roman World*. Cambridge.

Dunbabin, K. 2003. *The Roman Banquet*. Cambridge.

Dunbabin, K. 2016. *Theater and Spectacle in the Art of the Roman Empire*. Ithaca.

Dunbabin, K.M.D., and M.W. Dickie. 1983. "Invida Rumpantur Pectora: The Iconography of Phthonos/Invidia in Graeco-Roman Art." *JAC* 26: 7–37.

Dupras, T.L., S.M. Wheeler, L. Williams, and P. Sheldrick. 2015. "Birth in Ancient Egypt: Timing, Trauma, and Triumph? Evidence from the Dakhleh Oasis." In *Egyptian Bioarchaeology*, ed. S. Ikram et al., 53–66. Leiden.

Dupras, T.L., L.J. Williams, S.M. Wheeler, and P.G. Sheldrick. 2016. "Life and Death in the Desert: A Bioarchaeological Study of Human Remains from the Dakhleh Oasis, Egypt." In *Mummies, Magic and Medicine in Ancient Egypt*, ed. C. Price et al., 286–304. Manchester.

Dwyer, E.J. 1981. "Pompeian Oscilla Collections." *RM* 88: 247–306.

Dwyer, E.J. 1982. *Pompeian Domestic Sculpture*. Rome.

Dwyer, E.J. 2010. *Pompeii's Living Statues*. Ann Arbor.

Dwyer, E.J. 2012. "Collectible Singletons." *Arethusa* 45: 305–318.

Dyson, S.L. 1997. "Some Random Thoughts on a Collection of Papers on Roman Archaeology." In *Sequence and Space in Pompeii*, ed. S.E. Bon and R. Jones, 150–157. Oxford.

Eaton, K. 2006. "The Festivals of Osiris and Sokar in the Month of Khoiak: The Evidence from Nineteenth Dynasty Royal Monuments at Abydos." *SAK* 35: 75–101.

Edwards, C. 2003. "Incorporating the Alien: The Art of Conquest." In *Rome the Cosmopolis*, ed. C. Edwards and G. Woolf, 44–70. Cambridge.

Egelhaaf-Gaiser, U. 2000. *Kulträume im römischen Alltag*. Stuttgart.
Ehrhardt, W. 2004. *Casa delle Nozze d'argento (V 2,i)*. Munich.
Ehrhardt, W. 2012. *Dekorations- und Wohnkontext*. Wiesbaden.
Ellis, S.P. 1997. "Late-Antique Dining: Architecture, Furnishings and Behaviour." In *Domestic Space in the Roman World*, ed. R. Laurence and A. Wallace-Hadrill, 41–51. Portsmouth.
Ellis, S.P. 2000. *Roman Housing*. London.
Elsner, J. 1995. *Art and the Roman Viewer*. Cambridge.
Elsner, J. 1999. "The Viewer in the Roman Landscape." *Apollo* 150: 13–17.
Elsner, J. 2003. "Style." In *Critical Terms for Art History*, ed. R.S. Nelson and R. Shiff, 98–109. Chicago.
Elsner, J. 2004. "Foreword." In *The Language of Imagess in Roman Art*, ed. T. Hölscher, xv–xxxi. Cambridge.
Elsner, J. 2006. "Classicism in Roman Art." In *Classical Pasts*, ed. J.I. Porter, 270–297. Princeton.
Elsner, J. 2007. *Roman Eyes*. Princeton.
Engelmann, R. 1904. "Ein neues 'Urtheil Salomonis' und die Friesbilder der Casa Tiberina." *Hermes* 39: 146–154.
Epigraphic Survey, The. 1994. *Reliefs and Inscriptions at Luxor Temple*. Vol. 1. Chicago.
Erdkamp, P. 2005. *The Grain Market in the Roman Empire*. Cambridge.
Eristov, H. 1979. "Corpus des faux-marbres peints à Pompéi." *MÉFRA* 91: 693–771.
Erler, M., and M. Stadler, eds. 2017. *Platonismus und spätägyptische Religion*. Berlin.
Eschebach, H. 1984. "Die Arzthäuser in Pompeji." *AntW* 15 (Sondernummer): 3–68.
Esposito, D. 2007. "Pompeii, Silla e la Villa dei Misteri." In *Villas, maisons, sanctuaires et tombeaux tardo-républicains*, ed. B. Perrier, 441–465. Rome.
Evans, R. 2003. "Containment and Corruption: The Discourse of Flavian Empire." In *Flavian Rome*, ed. A.J. Boyle and W.J. Dominik, 255–276. Leiden.
Ewigleben, C., and J. von Grumbkow, eds. 1991. *Götter, Gräber und Grotesken*. Hamburg.
Faber, A. 2009. "Die stratigraphischen Befunde und Funde der Ausgrabungen in den Jahren 1961 bis 1963." In *Die Casa del Fauno in Pompeji*, ed. A. Faber and A. Hoffmann, 1:55–103. Wiesbaden.
Faber, A., and A. Hoffmann. 2009. *Die Casa del Fauno in Pompeji (VI 12)*. Vol. 1. Wiesbaden.
Fant, J.C., and S.J. Barker. 2016. "The Cost of Luxury: Procurement and Labor for the Marble Décor of Villa A." In *Leisure and Luxury in the Age of Nero*, ed. E.K. Gazda and J.R. Clarke, 126–132. Ann Arbor.
Fant, J.C., B. Russell, and S.J. Barber. 2013. "Marble Use and Reuse at Pompeii and Herculaneum: The Evidence from the Bars." *PBSR* 81: 181–209.
Faraone, C. 2018. *The Transformation of Greek Amulets in Roman Imperial Times*. Philadelphia.
Farley, J., and F. Hunter. 2015. *Celts: Art and Identity*. London.
Farrar, L. 1996. *Gardens of Italy and the Western Provinces of the Roman Empire*. Oxford.

Farrar, L. 1998. *Ancient Roman Gardens.* Gloucestershire.
Fassa, E. 2013. "Shifting Conceptions of the Divine: Sarapis as Part of Ptolemaic Egypt's Social Imaginary." In *Shifting Social Imaginaries in the Hellenistic Period*, ed. E. Stavrianopoulos, 115–142. Leiden.
Fear, A. 2011. "The Roman's Burden." In *Pliny the Elder*, ed. R. Gibson and R. Morello, 21–34. Leiden.
Feder, T. 2008. "Solomon, Socrates and Aristotle: In Earliest Biblical Painting, Greek Philosophers Admire King's Wisdom." *Biblical Archaeology Review* 34: 32–36.
Feil, E. 1992. "From the Classical *Religio* to the Modern *Religion*." In *Religion in History*, ed. M. Despland and G. Vallée, 31–44. Waterloo.
Ferris, I.M. 2000. *Enemies of Rome*. Stroud.
Ferris, I.M. 2011. "The Pity of War: Representations of Gauls and Germans in Roman Art." In *Cultural Identity in the Ancient Mediterranean*, ed. E.S. Gruen, 185–201. Los Angeles.
Finkelpearl, E. 2012. "Egyptian Religion in *Met.* 11 and's *DIO*: Culture, Philosophy, and the Ineffable." In *Aspects of Apuleius' Golden Ass*, ed. Keulen and U. Egelhaaf-Gaiser, 3:183–201. Leiden.
Finnestad, R.B. 1985. *Image of the World and Symbol of the Creator*. Wiesbaden.
Finnestad, R.B. 1997. "Temples of the Ptolemaic and Roman Periods." In *Temples of Ancient Egypt*, ed. B.E. Shafer, 185–237. Ithaca.
Fiorelli, G. 1860–64. *Pompeianarum Antiquitatum Historia*. 3 vols. Naples.
Fiorelli, G. 1882. "Pompei." *NSc* 420–423.
Fiorelli, G. 1875. *Descrizione di Pompei*. Naples.
Fischer, J. 1994. *Griechisch-römische Terrakotten aus Ägypten*. Tübingen.
Fischer, J. 1998. "Der Zwerg, der Phallos und der Buckel: Groteskfiguren aus dem ptolemäischen Ägypten." *ChrÉg* 78: 327–361.
Fischer, J. 1999. "Der Zwerg, der Phallos und Harpokrates: Groteskfiguren aus dem ptolemäischen Ägypten." In *Ägypter – Griechen – Römer*, ed. H. Felber and S. Pfisterer-Haas, 27–46. Leipzig.
Fischer-Bovet, C. 2014. *Army and Society in Ptolemaic Egypt*. Cambridge.
Fisher, K.D. 2009. "Placing Social Interaction: An Integrative Approach to Analyzing Past Built Environments." *JAnthArch* 28: 439–457.
Fjeldhagen, M. 1995. *Graeco-Roman Terracottas from Egypt*. Copenhagen.
Flohr, M. 2010. Review of A. Faber and A. Hoffmann, *Die Casa del Fauno in Pompeji (VI 12). Göttinger Forum für Altertumswissenschaft* 13: 1165–1171.
Flower, H.I. 2017. *The Dancing Lares and the Serpent in the Garden*. Princeton.
Foertmeyer, V.A. 1989. *Tourism in Graeco-Roman Egypt*. Ph.D. dissertation, Princeton University.
Foss, P. 1997. "Watchful *Lares*: Roman Household Organization and the Rituals of Cooking and Dining." In *Domestic Space in the Roman World*, ed. R. Laurence and A. Wallace-Hadrill, 196–218. Portsmouth.
Fowden, G. 1993. *The Egyptian Hermes*. Princeton.

Frankfurter, D., ed. 1998a. *Pilgrimage and Holy Space in Late Antique Egypt*. Leiden.
Frankfurter, D. 1998b. *Religion in Roman Egypt*. Princeton.
Fredrick, D. 1995. "Beyond the Atrium to Ariadne: Erotic Painting and Visual Pleasure in the Roman House." *ClAnt* 14: 266–287.
Fredrick, D., ed. 2002. *The Roman Gaze*. Baltimore.
Fredrick, D. 2003. "Grasping the Pangolin: Sensuous Ambiguity in Roman Dining." *Arethusa* 36: 309–343.
Frey-Kupper, S. 2014. "Coins and Their Use in the Punic Mediterranean: Case Studies from Carthage to Italy from the Fourth to the First Century BCE." In *The Punic Mediterranean*, ed. J.C. Quinn and N.C. Vella, 76–110. Cambridge.
Friedman, F., ed. 1998. *Gifts of the Nile: Ancient Egyptian Faience*. Providence.
Fröhlich, T. 1991. *Lararien- und Fassadenbilder in den Vesuvstädten*. Mainz am Rhein.
Fuentes, N. 1987. "The Roman Military Tunic." In *Roman Military Equipment*, ed. M. Dawson, 41–76. Oxford.
Fuhrmann, H. 1931. *Philoxenos von Eretria*. Göttingen.
Fullerton, M.D. 1997. "Imitation and Intertextuality in Roman Art." *JRA* 10: 427–440.
Fullerton, M.D. 2003. "'*Der Stil der Nachahmer*': A Brief Historiography of Stylistic Retrospection." In *Ancient Art and Its Historiography*, ed. A.A. Donohue and M.D. Fullerton, 92–117. Cambridge.
Gabolde, M. 1995. "L'inondation sous les pieds d'Amon." *BIFAO* 95: 235–258.
Gabra, S., et al. 1941. *Rapport sur les fouilles d'Hermoupolis Ouest (Touna el-Gebel)*. Cairo.
Gahlin, L. 2007. "Private Religion." In *The Egyptian World*, ed. T. Wilkinson, 325–339. New York.
Gahtan, M.W., and D. Pegazzano. 2015. *Museum Archetypes and Collecting in the Ancient World*. Leiden.
Gaifman, M. 2012. *Aniconism in Greek Antiquity*. Oxford.
Garland, R. 1995. *The Eye of the Beholder*. Ithaca.
Garmaise, M. 1996. *Studies in the Representation of Dwarfs in Hellenistic and Roman Art*. Ph.D. dissertation, McMaster University.
Gasparini, V. 2008. "Santuari isiaci in Italia: Criteri e contesti di diffusione." In *Religioni in contatto nei Mediterraneo antico*, ed. C. Bonnet et al., 65–87. Pisa/Rome.
Gasparini, V. 2009. "I culti egizi." In *Divus Vespasianus*, ed. F. Coarelli, 348–353. Milan.
Gasparini, V. 2013. "Staging Religion: Cultic Performances in (and around) the Temple of Isis in Pompeii." In *Memory and Religious Experience in the Greco-Roman World*, ed. N. Cusumano et al., 185–212. Stuttgart.
Gasparini, V. 2014. "Les cultes isiaques et les pouvoirs locaux en Italie." In *Power, Politics and the Cults of Isis*, ed. L. Bricault and M.J. Versluys, 260–299. Leiden.
Gasparini, V., and R. Veymiers, eds. 2018. *Individuals and Materials in the Greco-Roman Cults of Isis*. Leiden/Boston.
Gasparri, C. 1986a. "Dionysos." In *LIMC* III:414–514.
Gasparri, C. 1986b. "Dionysos/Bacchus." In *LIMC* III:540–566.
Gazda, E.K., ed. 1991. *Roman Art in the Private Sphere*. Ann Arbor.

Gazda, E.K, ed. 2002. *The Ancient Art of Emulation*. Ann Arbor.

Gazda, E.K., and J.R. Clarke, eds. 2016. *Leisure and Luxury in the Age of Nero*. Ann Arbor.

Geissler, B. 2011. "Arzthäuser in Pompeji." *Kölner und Bonner Archaeologica* 1: 7–36.

Gesztelyi, T. 1989. "Zur Frage der Darstellungen des sog. Salomourteils." *Acta Classica Universitatis Scientiarium Debreceniensis* 25: 73–84.

Giacobello, F. 2008. *Larari pompeiani*. Milan.

Gibson, J. 2015 (1986). *The Ecological Approach to Visual Perception*. New York.

Giordano, C., and I. Kahn. 2001. *The Jews in Pompeii, Herculaneum, Stabiae and in the Cities of Campania Felix*. Trans. W.F. Jashemski. 3rd ed. Rome.

Golvin, J.-C. 1994. "L'architecture de l'Iseum de Pompéi et les caractéristiques des édifices isiaques romains." In *Hommages à Jean Leclant*, ed. C. Berger et al., 3:235–246. Cairo.

Gordon, R.L. 2014. "Coming to Terms with the 'Oriental Religions of the Roman Empire.'" *Numen* 61: 657–672.

Gowers, E. 1993. *The Loaded Table*. Oxford.

Gowers, E. 2000. "Vegetable Love: Virgil, Columella, and Garden Poetry." *Ramus* 29: 127–148.

Graeber, D. 2001. *Toward an Anthropological Theory of Value*. New York.

Grant, M. 1975. *Eros in Pompeii*. New York.

Graves-Brown, C. 2010. *Dancing for Hathor*. London/New York.

Grenier, J.-C. 2008. *L'Osiris Antinoos*. Montpellier.

Griesbach, J. 2013. *Griechisch-ägyptisch*. Regensburg.

Grimal, P. 1969. *Les jardins romains*. 2nd ed. Paris.

Gruen, E.S. 2011. *Rethinking the Other in Antiquity*. Princeton.

Gunderson, J. 2013. *Inscribing Pompeii: A Reevaluation of the Jewish Epigraphic Data*. M.A. thesis, University of Kansas.

Gutbub, A. 1961. "Un emprunt aux textes des pyramides dans l'hymne à Hathor, dame de l'ivresse." In *Mélanges Maspero*. Fasc. 4, 1:31–72. Cairo.

Györy, H. 2003. "A Pataikos with Hawks on the Shoulders." *BMusHongr* 98: 11–29, 87–103.

Habachi, L. 1977. *The Obelisks of Egypt: Skyscrapers of the Past*. New York.

Hales, S. 2003. *The Roman House and Social Identity*. Cambridge.

Hallett, C.H. 2012. "The Archaic Style in Sculpture in the Eyes of Ancient and Modern Viewers." In *Making Sense of Greek Art*, ed. V. Coltman, 70–100. Exeter.

Hamilakis, Y. 2002. "The Past as Oral History: Towards an Archaeology of the Senses." In *Thinking through the Body*, ed. Y. Hamilakis et al., 121–136. New York.

Hamilakis, Y. 2013. *Archaeology and the Senses*. Cambridge.

Hannerz, U. 1990. "Cosmopolitans and Locals in World Culture." *Theory, Culture & Society* 7: 237–251.

Hansen, W. 2002. *Ariadne's Thread*. Ithaca.

Harari, M. 2004. "A Short History of Pygmies in Greece and Italy." In *Greek Identity in the Western Mediterranean*, ed. K. Lomas, 163–190. Leiden.

Hardiman, C. 2013. "'Alexandrianism' Again: Regionalism, Alexandria, and Aesthetics." In *Belonging and Isolation in the Hellenistic World*, ed. S.L. Ager and R.A. Faber, 199–222. Toronto.

Harrisson, J. 2012. "Isis in the Greco-Roman World: Cultural Memory and Imagination." In *Memory and Urban Religion in the Ancient World*, ed. M. Bommas et al., 213–236. London.

Haselberger, L. 2011. "A Debate on the Horologium of Augustus: Controversy and Clarifications." *JRA* 24: 47–73.

Haselberger, L., ed. 2014. *The Horologium of Augustus: Debate and Context*. Portsmouth.

Hatzidakis, P.J. 2004. "Ειδωλιόμορφα σκεύη από τη Δήλο." In *Στ' Επιστημονική Συνάντηση για την Ελληνιστική Καραμική*, ed. S. Drougou et al., 367–392, pls. 159–178. Athens.

Hawass, Z. 2000. *Valley of the Golden Mummies*. New York.

Heyob, S.K. 1975. *The Cult of Isis among Women in the Graeco-Roman World*. Leiden.

Heyworth, S.J. 2007a. *Cynthia*. Oxford.

Heyworth, S.J., ed. 2007b. *Sexti Properti Elegi*. Oxford.

Hicks, D. 2010. "The Material-Cultural Turn: Event and Effect." In *The Oxford Handbook of Material Culture Studies*, ed. D. Hicks and M.C. Beaudry, 25–98. Oxford.

Higbie, C. 2017. *Collectors, Scholars, and Forgers in the Ancient World*. Oxford.

Higgs, P. 2016. "Statue of a Boy Removing a Thorn (The Spinario)." In *Pergamon and the Hellenistic Kingdoms of the Ancient World*, ed. C.A. Picón and S. Hemingway, 277–278. New York.

Hill, D.K. 1981. "Some Sculpture from Roman Domestic Gardens." In *Ancient Roman Gardens*, ed. E.B. MacDougall and W.F. Jashemski, 81–94. Washington, DC.

Hillier, B. 1996. *Space Is the Machine*. Cambridge.

Hillier, B. 2014. "Spatial Analysis and Cultural Information: The Need for Theory as Well as Method in Space Syntax Analysis." In *Spatial Analysis and Social Spaces*, ed. E. Paliou et al., 19–48. Berlin.

Hillier, B., and J. Hanson. 1984. *The Social Logic of Space*. Cambridge.

Hinterholler, M. 2007. "Typologie und stilistische Entwicklung der sakral-idyllischen Landschaftsmalerei in Rom und Kampanien während des Zweiten und Dritten pompejanischen Stils." *Römische historische Mitteilungen* 49: 17–69.

Hobbs, R. 2017. "Bes, Butting Bulls, and Bars: The Life of Coinage at Pompeii." In *The Economy of Pompeii*, ed. M. Flohr and A. Wilson, 339–362. Oxford.

Hodder, I. 2012. *Entangled*. Malden, MA.

Hodske, J. 2007. *Mythologische Bildthemen in den Häusern Pompejis*. Ruhpolding.

Hoffmann, A. 1986. "VI 12 Casa del Fauno." *AA*: 493–495.

Hoffmann, A. 1996. "Die Casa del Fauno in Pompeji. Ein Haus wie ein Palast. (Kurzfassung)." In *Basileia: Die Paläste der hellenistischen Könige*, ed. W. Hoepfner and G. Brands, 258–259. Mainz am Rhein.

Hoffmann, A. 2009. "Bauhistorische Analyse." In *Die Casa del Fauno in Pompeji*, ed. A. Faber and A. Hoffmann, 19–54. Wiesbaden.

Hoffmann, A., and M. De Vos. 1994. "VI 12, 2. Casa del Fauno." In *PPM* V:80–141. Rome.
Hoffmann, F., and J.F. Quack. 2007. *Anthologie der demotischen Literatur*. Berlin.
Hoffmann, F., and J.F. Quack. 2014. "Pastophoros." In *A Good Scribe and an Exceedingly Wise Man*, ed. A.M. Dodson et al., 127–156. London.
Hoffmann, P. 1993. *Der Isis-Tempel in Pompeji*. Münster/Hamburg.
Hölscher, T. 2004. *The Language of Images in Roman Art*. Trans. A. Snodgrass and A. Künzl-Snodgrass. Cambridge.
Hölscher, T. 2009. "Architectural Sculpture: Messages? Programs? Towards Rehabilitating the Notion of 'Decoration.'" In *Structure, Image, Ornament*, ed. P. Schutz and R. von den Hoff, 54–67. Oxford.
Hornung, E. 1994. "Altägyptische Wurzeln der Isismysterien." In *Hommages à Jean Leclant*, ed. C. Berger et al., 3:287–294. Cairo.
Hornung, E. 2001. *The Secret Lore of Egypt*. Trans. D. Lorton. Ithaca.
Hubbard, T. 2010. "Locus Amoenus." In *The Classical Tradition*, ed. A. Grafton et al., 538. Cambridge.
Hugonot, J.-C. 1989. *Le jardin dans l'Égypte ancienne*. Frankfurt am Main.
Huss, W. 1990. "Die Quellen des Nils." *CE* 65: 334–343.
Huss, W. 1994. *Der makedonische König und die ägyptischen Priester*. Stuttgart.
Ikram, S. 1995. *Choice Cuts: Meat Production in Ancient Egypt*. Leuven.
Inconnu-Bocquillon, D. 2001. *Le mythe de la Déesse Lointaine à Philae*. Cairo.
Ingold, T. 2000. *The Perception of the Environment*. London.
Ippel, A. 1930. "Mosaikstudien." *RM* 45: 80–111.
Isager, J. 1991. *Pliny on Art and Society*. Trans. H. Rosenmeier. Odense.
Jaccottet, A.-F. 2003. *Choisir Dionysos: Les associations dionysiaques ou la face cachée du dionysisme*. 2 vols. Zurich.
Jaccottet, A.-F. 2006. "Un dieu, plusieurs mystères? Les différents visages des mystères dionysiaques." In *Religions orientales—culti misterici*, ed. C. Bonnet et al., 219–230. Stuttgart.
Jacobelli, L. 1995. *Le pitture erotiche delle Terme Suburbane di Pompei*. Rome.
Jacono, L. 1927. "Excursus: della cochlea di Archimede nella descrizione Vitruviana e in un dipinto pompeiano." *NSc* 3: 84–89.
Jansen, G.C.M. 2001. "Water Pipe Systems in the Houses of Pompeii: Distribution and Use." In *Water Use and Hydraulics in the Roman City*, ed. A.O. Koloski-Ostrow, 27–40. Dubuque.
Jansen, G.C.M. 2005. "Water and Water Technology in the Pompeian Garden." In *Omni pede stare*, ed. S.T.A.M. Mols and E.M. Moormann, 277–290. Naples.
Jansen, G.C.M. 2007. "The Water System: Supply and Drainage." In *The World of Pompeii*, ed. J.J. Dobbins and P.W. Foss, 257–268. New York.
Jansen, G.C.M. 2011. "Interpreting Images and Epigraphic Testimony." In *Roman Toilets*, ed. G.C.M. Jansen et al., 165–175. Leuven.
Jansen, G.C.M. 2018. "Water and Water Technology in Roman Gardens." In *Gardens of the Roman Empire*, ed. W.F. Jashemski et al., 402–431. Cambridge.

Jashemski, W.F. 1964. "A Pompeian Copa." *CJ* 59: 337–349.

Jashemski, W.F. 1973. "The Discovery of a Large Vineyard at Pompeii: University of Maryland Excavations, 1970." *AJA* 77: 27–41.

Jashemski, W.F. 1974. "The Discovery of a Market-Garden Orchard at Pompeii: The Garden of the 'House of the Ship Europa.'" *AJA* 78: 391–404.

Jashemski, W.F. 1977. "The Excavation of a Shop-House Garden at Pompeii (I. XX. 5)." *AJA* 81: 217–227.

Jashemski, W.F. 1979–93. *The Gardens of Pompeii, Herculaneum and the Villas Destroyed by Vesuvius*. 2 vols. New Rochelle.

Jashemski, W.F. 1981. "The Campanian Peristyle Garden." In *Ancient Roman Gardens*, ed. E.B. MacDougall and W.F. Jashemski, 29–48. Washington, DC.

Jashemski, W.F. 1994. "The Use of Water in Pompeian Gardens." In *Cura Aquarum in Campania,* ed. N. de Haan and G.C.M. Jansen, 51–58. Leiden.

Jashemski, W.F. 2018. "Produce Gardens." In *Gardens of the Roman Empire*, ed. W.F. Jashemski et al., 121–151. Cambridge.

Jashemski, W.F., and F.G. Meyer, eds. 2002. *The Natural History of Pompeii*. Cambridge.

Jashemski, W.F., F.G. Meyer, and M. Ricciardi. 2002. "Plants: Evidence from Wall Paintings, Mosaics, Sculpture, Plant Remains, Graffiti, Inscriptions, and Ancient Authors." In *Natural History of Pompeii,* ed. W.F. Jashemski and F.G. Meyer, 80–180. Cambridge.

Jashemski, W.F., E. Salza Prina Ricotti, and J. Foss. 1992. "Preliminary Excavations in the Gardens of Hadrian's Villa: The Canopus Area and the Piazza d'Oro." *AJA* 96: 579–597.

Jashemski, W.F., K.L. Gleason, K.J. Hartwick, and A.-A. Malek, eds. 2018. *Gardens of the Roman Empire*. Cambridge.

Jasnow, R., and M. Smith. 2010–2011. "'As for Those Who Have Called Me Evil, Mut Will Call Them Evil': Orgiastic Cultic Behaviour and Its Critics in Ancient Egypt." *Enchoria* 32: 9–53.

Jasnow, R., and K.-T. Zauzich. 2005. *The Ancient Egyptian Book of Thoth*. 2 vols. Wiesbaden.

Jewell, E. 2013. *Beyond the Apotropaion and Towards the Somatic Context: The Pygmy Somatotype in Roman Wall Painting*. MA thesis, Columbia University.

Jewell, E. Forthcoming. "The Pygmy Motif as Somatic Spectacle: Somatic Dialogues in the House of the Menander, Pompeii."

Johnston, P. 1985. *Ship and Boat Models in Ancient Greece*. Annapolis.

Jones, C.P. 1991. "Dinner Theater." In *Dining in a Classical Context*, ed. W.J. Slater, 185–198. Ann Arbor.

Jones, D. 2006. *The Bankers of Puteoli*. Gloucestershire.

Jones, H.L., ed. and trans. 1960. *The Geography of Strabo*. Vol. 1. Cambridge.

Jördens, A. 2012. "Status and Citizenship." In *The Oxford Handbook of Roman Egypt*, ed. C. Riggs, 247–259. Oxford.

Junker, H. 1911. *Der Auszug der Hathor-Tefnut aus Nubien*. Berlin.

Junker, H. 1917. *Die Onurislegende*. Vienna.

Kapossy, B. 1969. *Brunnenfiguren der hellenistischen und römischen Zeit*. Zürich.

Kastenmeier, P. 2007. *I luoghi del lavoro domestico nella casa pompeiana*. Rome.

Kaufmann-Heinimann, A. 2007. "Statuettes de laraire et religion domestique à Pompéi." In *Contributi di Archeologia Vesuviana III*, ed. L. Barnabei, 151–157. Rome.

Keane, W. 2003. "Semiotics and the Social Analysis of Material Things." *Language and Communication* 23: 409–425.

Keenan-Jones, D. 2015. "Somma-Vesuvian Ground Movements and the Water Supply of Pompeii and the Bay of Naples." *AJA* 119: 191–215.

Kekulé, R. 1880. *Die Terracotten von Pompeji*. Stuttgart: W. Spemann.

Kemp, B.J. 1979. "Wall Paintings from the Workmen's Village at el-'Amarna." *JEA* 65: 47–53.

Kemp, B.J. 2006. *Ancient Egypt: Anatomy of a Civilization*. 2nd ed. New York.

Kennedy, R.F., and M. Jones-Lewis, eds. 2016. *Identity and the Environment in the Classical and Medieval Worlds*. London/New York.

Kennedy, R.F., C.S. Roy, and M.L. Goldman, eds. and trans. 2013. *The Routledge Handbook of Identity and the Environment in the Classical and Medieval Worlds*. Indianapolis.

Kessler, D., and P. Brose, eds. 2008. *Ägyptens letzte Pyramide: Das Grab des Seuta(s) in Tuna el-Gebel*. Haar.

Keulen, W., and U. Egelhaaf-Gaiser. 2012. *Aspects of Apuleius' Golden Ass. Vol. 3: The Isis Book*. Leiden.

Keulen, W., and U. Egelhaaf-Gaiser, eds. 2015. *Apuleius Metamorphoses XI*. Leiden: Brill.

Kimel, D.V. Forthcoming. *The Evolution of the Roman Orgy*. Ph.D. dissertation, Yale University.

King, A. 2002. "Mammals: Evidence from Wall Paintings, Scupture, Mosaics, Faunal Remains, and Ancient Literary Sources." In *The Natural History of Pompeii,* ed. W.F. Jashemski and F.G. Meyer, 401–450. Cambridge.

King, C. 2003. "The Organization of Roman Religious Beliefs." *ClAnt* 22: 275–312.

Kiss, Z. 1984. *Études sur le portrait imperial romain en Égypte*. Warsaw.

Kleibl, K. 2009. *Iseion*. Worms.

Klotz, D. 2012a. *Caesar in the City of Amun*. Turnhout.

Klotz, D. 2012b. "The Lecherous Pseudo-Anubis of Josephus and the 'Tomb of 1897' at Akhmim." In *Et in Aegypto et ad Aegyptum*, ed. A. Gasse et al., 383–396. Montpellier.

Knappett, C. 2012. "Materiality." In *Archaeological Theory Today*, ed. I. Hodder, 188–207. 2nd ed. Cambridge.

Kockel, V. 1986. "Archäologische Funde und Forschungen in den Vesuvstädten II." *AA* 443–569.

Kockelmann, H. 2008. *Praising the Goddess*. Berlin.

Koloski-Ostrow, A.O., and E.M. Moormann. 2011. "Case Study: The Paintings of Philosophers in the Baths of the Seven Sages in Ostia." In *Roman Toilets*, ed. G.C.M. Jansen et al., 178–181. Leuven.

Koltsida, A. 2007. *Social Aspects of Ancient Egyptian Domestic Architecture*. Oxford.

König, J. and G. Woolf. 2013. "Encyclopaedism in the Roman Empire." In *Encyclopaedism from Antiquity to the Renaissance*, ed. J. König and G. Woolf, 23–63. Cambridge.

Kopytoff, I. 1986. "The Cultural Biography of Things: Commoditization as Process." In *The Social Life of Things*, ed. A. Appadurai, 64–94. Cambridge.

Kousser, R.M. 2008. *Hellenistic and Roman Ideal Sculpture*. Cambridge.

Krebs, C.B. 2011. "Borealism: Caesar, Seneca, Tacitus, and the Roman Discourse about the Germanic North." In *Cultural Identity in the Ancient Mediterranean*, ed. E. Gruen, 202–221. Los Angeles.

Kruchten, J.-M. 1989. *Les annales des prêtres de Karnak*. Leuven.

Kuttner, A. 1995. *Dynasty and Empire in the Age of Augustus*. Berkeley.

Kuttner, A. 1999. "Looking Outside Inside: Ancient Roman Garden Rooms." *SHG* 19: 7–35.

Kuttner, A. 2003. "Delight and Danger: Motion in the Roman Water Garden at Sperlonga and Tivoli." In *Landscape Design and the Experience of Motion*, ed. M. Conan, 103–156. Washington, DC.

Labrique, F. 2011. "La salle aux Bès géants à Mouftella: une lecture de pieds." In *La XXVIe dynastie*, ed. D. Devauchelle, 185–196. Paris.

Laehn, T.R. 2013. *Pliny's Defense of Empire*. New York.

Lahire, B. 2011. *The Plural Actor*. Trans. D. Fernbach. Malden.

Laidlaw, A. 1973. "A Reconstruction of the First Style Decoration in the Alexander Exedra of the House of the Faun." In *Neue Forschungen in Pompeji und den anderen vom Vesuvausbrauch 79 n. Chr. verschütteten Städten*, ed. B. Andreae and H. Kyrieleis, 39–52. Recklinghausen.

Laidlaw, A. 1985. *The First Style in Pompeii*. Rome.

Langin-Hooper, S. 2015. "Fascination with the Tiny: Social Negotiation through Miniatures in Hellenistic Babylonia." *WorldArch* 47: 60–79.

La Rocca, E. 1986. "Il lusso come espressione di potere." In *Le tranquille dimore degli dei. La residenza imperiale degli* Horti Lamiani, ed. M. Cima and E. La Rocca, 3–36. Venice.

Latour, B. 2005. *Reassembling the Social: An Introduction to Actor-Network-Theory*. Oxford.

Laurence, R. 2007. *Roman Pompeii: Space and Society*. 2nd ed. New York.

Lavagne, H. 1988. *Operosa antra*. Paris.

Lazer, E. 2009. *Resurrecting Pompeii*. New York.

Leach, E. 1997. "*Oecus* on *Ibycus*: Investigating the Vocabulary of the Roman House." In *Sequence and Space in Ancient Pompeii*, ed. R. Jones and S. Bon, 50–71. Oxford.

Leach, E. 2004. *The Social Life of Painting in Ancient Rome and on the Bay of Naples*. Cambridge.

Leach, E. 2010. "*Litora picta*… *nativis lapillis*: Campanian Mosaic Fountains and Their Contexts." In *Atti del X congresso internazionale dell'AIPMA*, ed. I. Bragantini. 1:65–76. Naples.

Leclant, J. 1984. "Un aspect des influences alexandrines en Gaule: les scènes nilotiques exhumées en France." In *Alessandria e il mondo ellenistico-romano*, ed. N. Bonacasa and A. di Vita. 3:440–444. Rome.

Leclant, J. 2010 [1976]. "Egypt, Land of Africa, in the Greco-Roman World." In *The Image of the Black in Western Art*, ed. L. Bugner. 2nd ed. 1:275–287. Cambridge, MA.

Leemreize, M. 2014. "The Egyptian Past in the Roman Present." In *Valuing the Past in the Greco-Roman World*, ed. J. Ker and C. Pieper, 56–82. Leiden.

Leemreize, M. 2016. *Framing Egypt: Roman Literary Perceptions of Egypt from Cicero to Juvenal*. Ph.D. dissertation, Leiden University.

Leja, M. 1993. *Reframing Abstract Expressionism*. New Haven.

Lembke, K. 1994. *Das Iseum Campense in Rom*. Heidelberg.

L'Enfant, J. 1999. "Roman Landscape Painting and the Rise of the Picturesque." In *Life of the Average Roman*, ed. M.R. DeMaine and R.M. Taylor, 85–104. White Bear Lake.

Lenski, N. 2013. "Working Models: Functional Art and Roman Conceptions of Slavery." In *Roman Slavery and Roman Material Culture*, ed. M. George, 129–157. Toronto.

Levi, D. 1941. "The Evil Eye and the Lucky Hunchback." In *Antioch-on-the-Orontes*, ed. R. Stillwell, 3:220–232. Princeton.

Ling, R. 1977. "Studius and the Beginnings of Roman Landscape Painting." *JRS* 67: 1–16.

Ling, R. 1991. *Roman Painting*. Cambridge.

Ling, R. 1993. "The Paintings of the Columbarium of Villa Doria Pamphili." In *Functional and Spatial Analysis of Wall Painting*, ed. E.M. Moormann, 127–135. Leiden.

Lissarrague, F. 1995. "Identity and Otherness: The Case of Attic Head Vases and Plastic Vases." *Source: Notes in the History of Art* 15: 4–9.

Loar, M., C. MacDonald, and D. Padilla Peralta, eds. 2018. *Rome, Empire of Plunder*. Cambridge.

Lorenz, K. 2008. *Bilder machen Räume: Mythenbilder in pompeianischen Häusern*. Berlin.

Lucas, G. 2012. *Understanding the Archaeological Record*. Cambridge.

MacDonald, W.L., and J.A. Pinto. 1995. *Hadrian's Villa and Its Legacy*. New Haven.

Mackey, J.L. 2009. *Rethinking Roman Religion*. Ph.D. dissertation, Princeton University.

Maehler, H. 2003. "Roman Poets on Egypt." In *Ancient Perspectives on Egypt*, ed. R. Matthews and C. Römer, 203–215. London.

Magie, D., trans. 1921. *Historia Augusta*. Vol. 1. Cambridge, MA.

Maiuri, A. 1925. "La raffigurazione del 'placentarius' in quattro bronzetti pompeiani." *BdA* series II, fasc. 6: 268–275.

Maiuri, A. 1926. "L'efebo di Via dell'Abbondanza a Pompei." *BdA* 5: 337–353.

Maiuri, A. 1927. "Pompei—Relazione sui lavori di scavo dal marzo 1924 al marzo 1926." *NSc* 3: 3–83.

Maiuri, A. 1929. "Pompei—Relazione sui lavori di scavo dall'aprile 1926 al dicembre 1927." *NSc* 5: 354–438.

Maiuri, A. 1938. *Le pitture delle case di "M. Fabius Amandio" del "Sacerdos Amandus" e di "P. Cornelius Teges."* Rome.

Maiuri, A. 1956. "Una nuova pittura nilotica a Pompei." *MemLinc* 7: 5–80.

Malaise, M. 1972. *Les conditions de pénétration et de diffusion des cultes égyptiens en Italie.* Leiden.

Malaise, M. 1986. "Les charactéristiques et la question des antécédents de l'initiation isiaque." In *L'expression du sacré dans les grandes religions*, ed. J. Ries, 3:355–362. Louvain-la-Neuve.

Malaise, M. 2003. "La signification des scènes nilotiques dans la culture romaine." *ChrÉg* 78: 308–325.

Malaise, M. 2004. "Bes et la famille isiaque." *ChrÉg* 79: 266–292.

Malaise, M. 2005. *Pour une terminologie et une analyse des cultes isiaques*. Brussels.

Malaise, M. 2007. "La diffusion des cultes isiaques: Un problème de terminologie et de critique." In *Nile into Tiber*, ed. L. Bricault et al., 19–39. Leiden.

Mangone, A., et al. 2011. "A Multianalytical Study of Archaeological Faience from the Vesuvian Area as a Valid Tool to Investigate Provenance and Technological Features." *New Journal of Chemistry* 35: 2723–2922.

Manniche, L. 1977. "Some Aspects of Ancient Egyptian Sexual Life." *Acta Orientalia* 38: 11–23.

Manniche, L. 2003. "The So-Called Scenes of Daily Life in the Private Tombs of the Eighteenth Dynasty: An Overview." In *The Theban Necropolis*, ed. N. Strudwick and J.H. Taylor, 42–45. London.

Manolaraki, E. 2011. "*Noscendi Nilum Cupido:* The Nile Digression in Book 10." In *Brill's Companion to Lucan,* ed. P. Asso, 153–182. Leiden.

Manolaraki, E. 2013. *Noscendi Nilum Cupido*. Berlin.

Mari, Z., and S. Sgalambro. 2007. "The Antinoeion of Hadrian's Villa: Interpretation and Architectural Reconstruction." *AJA* 111: 83–104.

Martha, J. 1885. *Manuel d'archéologie étrusque et romaine*. Paris.

Martzavou, P. 2010. "Les cultes isiaques et les Italiens entre Délos, Thessalonique et l'Eubée." *Pallas* 84: 181–205.

Martzavou, P. 2011. "Priests and Priestly Roles in the Isiac Cults: Women as Agents in Religious Change in Late Hellenistic and Roman Athens." In *Ritual Dynamics in the Ancient Mediterranean,* ed. A. Chaniotis, 61–84. Stuttgart.

Marvin, M. 1989. "Copying in Roman Sculpture: The Replica Series." In *Retaining the Original: Multiple Originals, Copies, and Reproductions,* ed. K. Preciado, 29–45. Washington.

Marvin, M. 2008. *The Language of the Muses*. Los Angeles.

Marzano, A. 2014. "Roman Gardens, Military Conquests, and Elite Self-Representation." In *Le jardin dans l'antiquité*, ed. K. Coleman and P. Derron, 195–244. Geneva.

Masséglia, J. 2015. *Body Language in Hellenistic Art and Society*. Oxford.

Mathieu, B. 2001. "Travaux de l'Institut français d'archéologie orientale en 2000–2001." *BIFAO* 101: 449–610.

Mau, A. 1883a. "Gli scavi di Pompei 1881–1883." *BdI* 8–9: 170–177.

Mau, A. 1883b. "Scavi di Pompei." *BdI* 12: 225–234.

Mau, A. 1884. "Scavi di Pompei." *BdI* 6: 126–139.

Mau, A. 1908. *Pompeji in Leben und Kunst*. 2nd ed. Leipzig.

Mayer, E. 2012. "From Silver Cups to Garden Gnomes: Toward a Contextual Reception of Standardized Images." *Arethusa* 45: 283–304.

Mazurek, L.A. 2016. "Material and Textual Narratives of Authenticity? Creating Cabotage and Memory in the Hellenistic Eastern Mediterranean." In *Across the Corrupting Sea*, ed. C. Concannon and L.A. Mazurek, 39–64. London/New York.

McAlpine, L.J. 2016. "Luxury in Fantasy and Reality: Exotic Marble in Villa A." In *Leisure and Luxury in the Age of Nero*, ed. E.K. Gazda and J.R. Clarke, 111–118. Ann Arbor.

McDaniel, W.G. 1932. "A Fresco Picturing Pygmies." *AJA* 36: 260–271.

McKay, M.J. 1985. *Pygmy Landscapes in Roman Art*. Ph.D. dissertation, University of Toronto.

Meeks, D. 1971. "Génies, anges démons en Égypte." In *Génies, anges, et démons*, 17–84. Paris.

Melillo, L. 2013. "The Ephebe from the Via dell'Abbondanza: History of a Restoration." In *The Restoration of Ancient Bronzes*, ed. E. Risser and D. Saunders, 57–63. Los Angeles.

Merkelbach, R. 1963. *Isisfeste in griechisch-römischer Zeit*. Meisenheim am Glan.

Merkelbach, R. 1995. *Isis Regina—Zeus Sarapis*. Stuttgart.

Merrills, A. 2017. *Roman Geographies of the Nile*. Cambridge.

Meskell, L. 1998. "An Archaeology of Social Relations in an Egyptian Village." *Journal of Archaeological Method and Theory* 5: 209–243.

Meyboom, P.G.P. 1977. "I mosaici pompeiani con figure di pesci." *Meded* 41: 49–93.

Meyboom, P.G.P. 1995. *The Nile Mosaic of Palestrina*. Leiden.

Meyboom, P.G.P., and H.H.J. Brouwer. 2005. "Il gatto ladro a Leida." In *Omni pede stare*, ed. S.T.A.M. Mols and E.M. Moormann, 229–242. Naples.

Meyboom, P.G.P., and M.J. Versluys. 2007. "The Meaning of Dwarfs in Nilotic Scenes." In *Nile into Tiber*, ed. L. Bricault et al., 170–208. Leiden.

Meyer, H. 1994. *Der Obelisk des Antinoos*. Munich.

Michel, D. 1980. "Pompejanische Gartenmalerei." In *Tainia*, ed. H. Cahn and E. Simon, 373–404. Mainz am Rhein.

Mikalson, J.D. 1998. *Religion in Hellenistic Athens*. Berkeley.

Miles, M.M. 2008. *Art as Plunder*. Cambridge.

Mills, B.J., and W.H. Walker. 2008. "Introduction: Memory, Materiality, and Depositional Practice." In *Memory Work*, ed. B.J. Mills and W.H. Walker, 3–24. Santa Fe.

Minas-Nerpel, M. 2013. "Ptah-Pataikos, Harpokrates, and Khepri." In *Decorum and Experience,* ed. E. Frood and A. McDonald, 147–150. Oxford.

Minervini, G. 1856. "Notizia de' più recenti scavi di Pompei. Terme alla strada Stabiana." *Bullettino Archeologico Napolitano* 5: 33–37.

Mitchell, A.G. 2008. Review of J.R. Clarke, *Looking at Laughter. Bryn Mawr Classical Review* 2008.09.55. http://bmcr.brynmawr.edu/2008/2008-09-55.html.

Mitchell, A.G. 2013. "Disparate Bodies in Ancient Artefacts: The Function of Caricature and Pathological Grotesques among Roman Terracotta Figurines." In *Disabilities in Roman Antiquity,* ed. C. Laes et al., 275–298. Leiden.

Mitchell, A.G. 2017. "The Hellenistic Turn in Bodily Representations: Venting Anxiety in Terracotta Figurines." In *Disability in Antiquity,* ed. C. Laes, 182–196. Oxon/New York.

Moatti, C. 2013. "Immigration and Cosmopolitanization." In *The Cambridge Companion to Ancient Rome,* ed. P. Erdkamp, 77–92. Cambridge.

Moatti, C. 2015. *The Birth of Critical Thinking in Republican Rome.* Trans. J. Lloyd. Cambridge.

Mol, E.M. 2013. "The Perception of Egypt in Networks of Being and Becoming: A Thing Theory Approach to Egyptianising Objects in Roman Domestic Contexts." In *TRAC 2012,* ed. A. Bokern et al., 117–132. Oxford/Oakville.

Mol, E.M. 2015. *Egypt in Material and Mind.* Ph.D. dissertation, Leiden University.

Mol, E.M. 2016a. "La casa di Octavius Quartio a Pompei." In *Il Nilo a Pompei,* ed. F. Poole, 135–162. Modena.

Mol, E.M. 2016b. "Statuetta di babbuino." In *Il Nilo a Pompei,* ed. F. Poole, 144. Modena.

Mol, E.M. 2017. "Object Ontology and Cultural Taxonomies: Examining the Agency of Style, Material and Objects in Classification through Egyptian Material Culture in Pompeii and Rome." In *Materialising Roman Histories,* ed. A. Van Oyen and M. Pitts, 169–190. Oxford/Philadelphia.

Mols, S.T.A.M. 1999. *Wooden Furniture in Herculaneum.* Amsterdam.

Montserrat, D. 1996. *Sex and Society in Graeco-Roman Egypt.* London/New York.

Moormann, E.M. 1996. "Giardini al piede del Vesuvio." *JRA* 8: 391–398.

Moormann, E.M. 2002. "Pompeii's Proprietors and Tenants under One Roof." *JRA* 15: 431–436.

Moormann, E.M. 2007. "The Temple of Isis at Pompeii." In *Nile into Tiber,* ed. L. Bricault et al., 137–154. Leiden.

Moreno, P. 2001. *Apelles: The Alexander Mosaic.* Trans. D. Stanton. Milan.

Moreno, P. 2009. "Iconografia di Alessandro nell'arte antica." In *Alessandro Magno in età moderna,* ed. F. Biasutti and A. Coppola, 373–474. Padua.

Moret, J.-M. 2012. "I nani di Antonio." In *Nani in festa,* ed. C. Bocherens, 137–162. Bari.

Morley, N. 2003. "Migration and the Metropolis." In *Rome the Cosmopolis,* ed. C. Edwards and G. Woolf, 147–157. Cambridge.

Morvillez, E. 2018. "The Garden in the *Domus.*" In *Gardens of the Roman Empire,* ed. W.F. Jashemski et al., 17–71. Cambridge.

Mouritsen, H. 1988. *Elections, Magistrates and Municipal Élite*. Rome.
Moyer, I.S. 2011. *Egypt and the Limits of Hellenism*. Cambridge.
Murphy, T. 2004. *Pliny the Elder's Natural History*. Oxford.
Müskens, S. 2014. "A New Fragment of an Architectonic Hathor-Support from Rome: Aegyptiaca Romana Reconsidered." In *Bibliotheca Isiaca III*, ed. L. Bricault and R. Veymiers, 95–116. Bordeaux.
Müskens, S. 2017. *Egypt Beyond Representation: Materials and Materiality of Aegyptiaca Romana*. Leiden.
Müskens, S., D. Braekmans, M.J. Versluys, and P. Degryse. 2017. "Egyptian Sculptures from Imperial Rome: Non-Destructive Characterization of Granitoid Stateus through Macroscopic Methodologies and In Situ XRF Analysis." *Archaeological and Anthropological Sciences*: 1–16.
Myers, K.S. 2005. "*Docta Otia*: Garden Ownership and Configurations of Leisure in Statius and Pliny the Younger." *Arethusa* 38: 103–129.
Myers, K.S. 2018. "Representations of Gardens in Roman Literature." In *Gardens of the Roman Empire*, ed. W.F. Jashemski et al., 258–277. Cambridge.
Myśliwiec, K. 1998. *Eros on the Nile*. Trans. G.L. Packer. London.
Naas, V. 2002. *Le projet encyclopédique de Pline l'ancien*. Rome.
Naas, V. 2011. "Imperialism, *Mirabilia* and Knowledge: Some Paradoxes in the *Naturalis Historia*." In *Pliny the Elder*, ed. R. Gibson and R. Morello, 57–70. Leiden.
Nachtergael, G. 1985. "Les terres cuites 'du Fayoum' dans les maisons de l'Égypte romaine." *ChrÉg* 60: 223–239.
Nachtergael, G. 1991. "Die Terrakotten des griechisch-römischen Ägypten—Eine Einführung." In *Götter, Gräber und Grotesken*, ed. C. Ewigleben and J. von Grumbkow, 19–24. Hamburg.
Nachtergael, G. 1995. "Chronique: Terres cuites de l'Égypte gréco-romaine. À propos de quatre catalogues récents." *ChrÉg* 70: 254–294.
Nagel, S. 2013. "Kult und Ritual der Isis zwischen Ägypten und Rom." In *Rituale als Ausdruck von Kulturkontakt*, ed. A.H. Pries et al., 151–176. Wiesbaden.
Nagel, S. 2017. "One for All and All for One? Isis as *una quae es(t) omnia* in the Egyptian Temples of the Graeco-Roman Period." In *Entangled Worlds*, ed. S. Nagel et al., 207–231. Tübingen.
Nagel, S., et al., eds. 2017. *Entangled Worlds: Religious Confluences between East and West in the Roman Empire*. Tübingen.
Naiden, F.S. 2003. "Supplication on Roman Coins." *AJN* 15: 41–52.
Nelson, H.H., and W.J. Murnane. 1981. *The Great Hypostyle Hall at Karnak*. Vol. 1, Part 1. Chicago.
Nelson, R.S. 2000. "Descartes's Cow and Other Domestications of the Visual." In *Visuality Before and Beyond the Renaissance*, ed. R.S. Nelson, 1–21. Cambridge.
Nenna, M.-D., and M. Seif el-Din. 1994. "La petite plastique en faïence du musée gréco-romain d'Alexandrie." *BCH* 118: 291–320.

Nenna, M.-D., and M. Seif el-Din. 2000. *La vaisselle en faïence d'époque gréco-romaine*. Cairo.

Neudecker, R. 2015. "Collecting Culture: Statues and Fragments in Roman Gardens." In *Museum Archetypes and Collecting in the Ancient World*, ed. M.W. Gahtan and D. Pegazzano, 129–136. Leiden.

Neuerburg, N. 1965. *L'architettura delle fontane e dei ninfei nell'Italia antica*. Naples.

Nevett, L.C. 2010. *Domestic Space in Classical Antiquity*. Cambridge.

Newby, Z. 2012. "The Aesthetics of Violence: Myth and Danger in Roman Domestic Landscapes." *ClAnt* 31: 349–389.

Newlands, C.E. 1984. *The Transformation of the* Locus Amoenus *in Roman Poetry*. Ph.D. dissertation, University of California, Berkeley.

Newlands, C.E. 2002. *Statius'* Silvae *and the Poetics of Empire*. Cambridge.

Nielsen, I. 2014. *Housing the Chosen*. Turnhout.

North, J.A. 1992. "The Development of Religious Pluralism." In *The Jews among Pagans and Christians in the Roman Empire*, ed. J. Lieu et al., 174–193. New York.

Noy, D. 2004. "Being an Egyptian in Rome: Strategies of Identity Formation." In *Christians as a Religious Minority in a Multicultural City*, ed. J. Zangenberg and M. Labahn, 47–55. London.

Nuber, H.U. 1972. "Kanne und Griffschale. Ihr Gebrauch im täglichen Leben und die Beigabe in Gräbern der römischen Kaiserzeit." *Bericht der römisch-germanischen Kommission* 53: 1–232.

O'Connor, D. 1996. "Sexuality, Statuary and the Afterlife; Scenes in the Tomb-Chapel of Pepyankh (Heny the Black). An Interpretive Essay." In *Studies in Honor of William Kelly Simpson*, ed. Peter der Manuelian, 2:621–633. Boston.

Oleson, J.P. 1984. *Greek and Roman Mechanical Water-Lifting Devices*. Toronto.

Omlin, Joseph. 1973. *Der Papyrus 55001 und seine satirisch-erotischen Zeichnungen und Inschriften*. Turin.

Orlin, E. 2008. "Octavian and Egyptian Cults: Redrawing the Boundaries of Romanness." *AJP* 129: 231–253.

Orlin, E. 2010. *Foreign Cults in Rome*. Oxford.

Osborne, R. 2007. "Projecting Identities in the Greek Symposion." In *Material Identities*, ed. J. Sofaer, 31–52. Malden.

O'Sullivan, T.M. 2007. "Walking with Odysseus: The Portico Frame of the Odyssey Landscapes." *AJP* 128: 497–532.

Otto, W. 1905–1908. *Priester und Tempel im hellenistischen Ägypten*. *2* vols. Leipzig/Berlin.

Pace, R. 1997. "Il 'Complesso dei Riti Magici' a Pompei. II, 1, 11–12." *RStPomp* 8: 73–97.

Packer, J.E. 1975. "Middle and Lower Class Housing in Pompeii and Herculaneum: A Preliminary Survey." In *Neue Forschungen in Pompeji*, ed. B. Andreae and H. Kyrieleis, 133–146. Recklinghausen.

Pagán, V.E. 2006. *Rome and the Literature of Gardens*. London.

Pagano, M., and R. Prisciandaro. 2006. *Studio sulle provenienze degli oggetti rinvenuti negli scavi borbonici del Regno di Napoli.* Castellammare di Stabia.
Pannuti, U. 1983. *Catalogo della collezione glittica.* Museo Archeologico Nazionale di Napoli. Vol. 1. Rome.
Papadopoulos, J.K., and G. Urton, eds. 2012. *The Construction of Value in the Ancient World.* Los Angeles.
Papantoniou, G. 2012. *Religion and Social Transformations in Cyprus.* Leiden.
Papini, M. 2015. "Republican Rome and Italic Art." In *A Companion to Roman Art*, ed. B.E. Borg, 95–113. Malden/Oxford.
Paribeni, R. 1902. "Pompei—Relazione degli scavi eseguiti duranti il mese di Maggio 1902." *NSc* 7: 369–381.
Parker, G. 2003. "Narrating Monumentality: The Piazza Navona Obelisk." *JMA* 16: 193–215.
Parker, G. 2007. "Obelisks Still in Exile: Monuments Made to Measure?" In *Nile into Tiber*, ed. L. Bricault et al., 209–222. Leiden.
Parker, G. 2018. "Monolithic Appropriation? The Lateran Obelisk Compared." In *Rome, Empire of Plunder*, ed. M. Loar et al., 137–159. Cambridge.
Parkinson, R.B. 2008. *The Painted Tomb-Chapel of Nebamun.* London.
Parlasca, K. 1975. "Hellenistische und römische Mosaiken aus Ägypten." In *La mosaïque gréco-romaine II*, ed. H. Stern and M. Le Glay, 363–369. Paris.
Parslow, C.C. 1989. *The Praedia Iuliae Felicis in Pompeii.* Ph.D. dissertation, Duke University.
Parslow, C.C. 2013. "The Sacrarium of Isis in the Praedia of Julia Felix in Pompeii in Its Archaeological and Historical Contexts." In *Rediscovering the Ancient World on the Bay of Naples, 1710–1890*, ed. C.C. Mattusch, 47–74. New Haven.
Patch, D.C., ed. 2011. *Dawn of Egyptian Art.* New York.
Pearce, S., and A. Bounia, eds. 2000. *The Collector's Voice.* Vol. 1. Burlington, VT.
Pearson, S. 2011. "An Un-Augustan Understanding of the Third Style." Paper read at the 2011 Annual Meeting of the Archaeological Institute of America, January 6–9, San Antonio.
Pearson, S. 2015. *Egyptian Airs: The Life of Luxury in Roman Wall Painting.* Ph.D. dissertation, University of California, Berkeley.
Peck, W.H. 1999. "Sculpture, Production Techniques." In *Encyclopedia of the Archaeology of Ancient Egypt*, ed. K.A. Bard, 874–876. London.
Perdrizet, P. 1921. *Les terres cuites grecques d'Égypte de la collection Fouquet.* Nancy.
Pernice, E. 1938. *Pavimente und figürliche Mosaiken.* Berlin.
Perpillou-Thomas, F. 1993. *Fêtes d'Égypte ptolémaïque et romaine, d'après la documentation papyrologique grecque.* Leuven.
Perry, E. 2005. *The Aesthetics of Emulation in the Visual Arts of Ancient Rome.* Cambridge.
Pesando, F. 1996. "Autocelebrazione aristocratica e propaganda politica in ambiente privato: La Casa del Fauno a Pompei." *Cahiers du Centre Gustave Glotz* 7: 189–228.
Petersen, L.H. 2006. *The Freedman in Roman Art and Art History.* Cambridge.

Petersen, L.H. 2012. "Collecting Gods in Roman Houses: The House of the Gilded Cupids (VI.16.7, 38) at Pompeii." *Arethusa* 45: 319–332.

Petersen, L.H. 2015. "'Arte Plebea' and Non-Elite Roman Art." In *A Companion to Roman Art*, ed. B.E. Borg, 214–230. Malden/Oxford.

Pfeiffer, S. 2010. "Ägypten in der Selbstdarstellung der Flavier." In *Tradition und Erneuerung*, ed. N. Kramer and C. Reitz, 273–288. Berlin.

Pfisterer-Haas, S. 1994. "Die bronzenen Zwergentänzer." In *Das Wrack: Der antike Schiffsfund von Mahdia*, ed. G. Hellenkemper Salies, 483–504. Köln.

Pfrommer, M. 1998. *Untersuchungen zur Chronologie und Komposition des Alexandermosaiks auf antiquarischer Grundlage*. Mainz am Rhein.

Philipp, H. 1972. *Terrakotten aus Ägypten im ägyptischen Museum Berlin*. Berlin.

Phillips, J. 1996. "A Note on Puntite Housing." *JEA* 82: 206–207.

Pinch, G. 1993. *Votive Offerings to Hathor*. Oxford.

Pirenne-Delforge, V., and P. Scarpi. 2006. "Les cultes à mystères: Introduzione." In *Religions orientales—culti misterici*, ed. C. Bonnet et al., 159–162. Stuttgart.

Pitts, M., and M.J. Versluys, eds. 2015. *Globalisation and the Roman World*. Cambridge.

Plantzos, D. 2012. "Wall- and Panel-Painting." In *A Companion to Greek Art*, ed. T.J. Smith and D. Plantzos, 1:171–185. Malden.

Platt, V. 2002. "Viewing, Desiring, Believing: Confronting the Divine in a Pompeian House." *Art History* 25: 87–112.

Platt, V. 2003. "Review of *L'histoire végétalisée*, by G. Sauron." *JRS* 93: 402–403.

Platt, V. 2009. "Where the Wild Things Are: Locating the Marvellous in Augustan Wall-Painting." In *Paradox and the Marvellous in Augustan Literature and Culture*, ed. P. Hardie, 41–74. Oxford.

Platt, V. 2018. "Of Sponges and Stones: Matter and Ornament in Roman Painting." In *Ornament and Figure in Graeco-Roman Art*, ed. N. Dietrich and M. Squire, 241–278. Berlin.

Platt, V., and M. Squire, eds. 2010. "The Art of Art History in Graeco-Roman Antiquity" (special edition of *Arethusa*). *Arethusa* 43.

Platt, V., and M. Squire. 2017a. "Framing the Visual in Greek and Roman Antiquity: An Introduction." In *The Frame in Classical Art*, ed. V. Platt and M. Squire, 3–99. Cambridge.

Platt, V., and M. Squire, eds. 2017b. *The Frame in Classical Art*. Cambridge.

Podvin, J.-L. 2007. "Les égyptiens en Occident." In *Étrangers dans la cité romaine*, ed. R. Compatangelo-Soussignan and C.-G. Schwentzel, 113–128. Rennes.

Pollard, E.A. 2009. "Pliny's Natural History and the Flavian Templum Pacis: Botanical Imperialism in First-Century C.E. Rome." *Journal of World History* 20: 309–338.

Pollitt, J.J. 1974. *The Ancient View of Greek Art*. London.

Poo, M.-C. 2009 [1995]. *Wine and Wine Offering in the Religion of Ancient Egypt*. New York.

Poole, F. 2004. "Il culto di Iside a Pompei." In *Pompei, Capri e la Penisola Sorrentina*, ed. F. Senatore, 209–243. Capri.

Poole, F., ed. 2016. *Il Nilo a Pompei*. Modena.

Porter, J.I. 2012. "The Value of Aesthetic Value." In *The Construction of Value in the Ancient World*, ed. J.K. Papadopoulos and G. Urton, 336–353. Los Angeles.

Postl, B. 1970. *Die Bedeutung des Nil in der römischen Literatur mit besonderer Berücksichtigung der wichtigsten griechischen Autoren*. Ph.D. dissertation, Universität Wien.

Powers, J.D. 2006. *Patrons, Houses and Viewers in Pompeii*. Ph.D. dissertation, University of Michigan.

Price, S. 1999. *Religions of the Ancient Greeks*. Cambridge.

Price, S. 2000. "Religions of Rome." In *Ancient Rome*, ed. J. Coulston and H. Dodge, 290–305. Oxford.

Purcell, N. 1987. "Town in Country and Country in Town." In *Ancient Roman Villa Gardens*, ed. E.B. MacDougall, 185–204. Washington, DC.

Purcell, N. 1996. "The Roman Garden as a Domestic Building." In *Roman Domestic Buildings*, ed. I.M. Barton, 121–152. Exeter.

Quack, J. 2003. "Zum ägyptischen Ritual im Iseum Campense in Rom." In *Rituale in der Vorgeschichte, Antike und Gegenwart*, ed. C. Metzner-Nebelsick, 57–66. Rahden-Westf.

Quack, J. 2004. "Organiser le culte idéal. Le Manuel du temple." *Bulletin de la Société Française d'Égyptologie* 160: 9–25.

Quack, J. 2005. *Einführung in die altägyptische Literaturgeschichte*. Vol. 3. Münster.

Quack, J. 2007a. "Das Buch vom Tempel: Die Geschichte einer sensationellen Rekonstruktion." *Ruperto Carola* 2/2007. www.uni-heidelberg.de/presse/ruca/ruca07–2/tempel.html.

Quack, J. 2007b. "Die Initiation zum Schreiberberuf im alten Ägypten." *SAK* 36: 249–295.

Quack, J. 2013. "Quelques apports récents des études démotiques à la comprehension du livre II d'Hérodote." In *Hérodote et l'Égypte*, ed. L. Coulon et al., 63–88. Lyon.

Quack, J. 2016. "Where Once Was Love, Love Is No More? What Happens to Expressions of Love in Late Period Egypt?" *Die Welt des Orients* 46: 62–89.

Quaegebeur, J. [completed by N. Cherpion]. 1999. *La naine et le bouquetin, ou l'énigme de la barque en albâtre de Toutankhamon*. Trans. F. Coppens et al. Leuven.

Quibell, J.E. 1907. *Excavations at Saqqara (1905–1906)*. Cairo.

Radin, M. 1917. "Roman Knowledge of Jewish Literature." *CJ* 13: 149–176.

Raeder, J. 1983. *Die statuarische Ausstattung der Villa Hadriana bei Tivoli*. Frankfurt am Main.

Rakob, F. 1964. "Ein Grottentriklinium in Pompeji." *RM* 71: 182–194.

Rauch, M. 1999. *Bacchische Themen und Nilbilder auf Campanareliefs*. Rahden/Westf.

Rebillard, E. 2009. *The Care of the Dead in Late Antiquity*. Trans. E.T. Rawlings and J. Routier-Pucci. Ithaca.

Rebillard, E. 2012. *Christians and Their Many Identities in Late Antiquity*. Ithaca.

Rebillard, E. 2015. "Material Culture and Religious Identity in Late Antiquity." In *A Companion to the Archaeology of Religion in the Ancient World*, ed. R. Raja and J. Rüpke, 427–436. Malden/Oxford/Chichester.

Rebillard, E., and J. Rüpke, eds. 2015. *Group Identity and Religious Individuality in Late Antiquity*. Washington, DC.

Rebillard, E., and C. Sotinel. 2010a. "Introduction." In *Les frontières du profane dans l'Antiquité tardive*, ed. E. Rebillard and C. Sotinel, 1–14. Rome.

Rebillard, E., and C. Sotinel, eds. 2010b. *Les frontières du profane dans l'Antiquité tardive*. Rome.

Reinach, S. 1922. *Répertoire des peintures grecques et romaines*. Paris.

Renberg, G.H. 2010. "Hadrian and the Oracles of Antinous (SHA *Hadr.* 1.47); with an Appendix on the So-Called Antinoeion at Hadrian's Villa and Rome's Monte Pinci Obelisk." *MAAR* 55: 159–198.

Revell, L. 2008. *Roman Imperialism and Local Identities*. Cambridge.

Ricci, C. 1993. "Egiziani a Roma nelle fonti epigrafiche di età imperiale." *Aegyptus* 73: 71–91.

Richardson, L. 1988a. *Pompeii: An Architectural History*. Baltimore.

Richardson, L. 1988b. "Water Triclinia and Biclinia in Pompeii." In *Studia Pompeiana & Classica in Honor of Wilhelmina F. Jashemski*, ed. R.I. Curtis. 1:305–316. New Rochelle.

Richardson, L. 2000. *A Catalog of Identifiable Figure Painters of Ancient Pompeii, Herculaneum, and Stabiae*. Baltimore.

Richardson, L. 2006. *Propertius: Elegies I-IV*. Norman.

Richter, B.A. 2010. "On the Heels of the Wandering Goddess: The Myth and the Festival at the Temples of the Wadi el-Hallel and Dendera." In *8. Ägyptologische Tempeltagung*, eds. H. Beinlich and M. Dolinska, 155–186. Wiesbaden.

Richter, D.S. 2001. "Plutarch on Isis and Osiris: Text, Cult, and Cultural Appropriation." *TAPA* 131: 191–216.

Ridgway, B.S. 1984. *Roman Copies of Greek Sculpture*. Ann Arbor.

Riggs, C. 2005. *The Beautiful Burial in Roman Egypt*. Oxford.

Ritner, R. 1989. "Horus on the Crocodiles: A Juncture of Religion and Magic in Late Dynastic Egypt." In *Religion and Philosophy in Ancient Egypt*, by J.P. Allen et al., 103–117. New Haven.

Ritner, R. 1993. *The Mechanics of Ancient Egyptian Magical Practice*. Chicago.

Ritner, R. 2008. "Household Religion in Ancient Egypt." In *Household and Family Religion in Antiquity*, ed. J. Bodel and S.M. Olyan, 171–196. Oxford.

Robins, G. 1997. *The Art of Ancient Egypt*. Cambridge, MA.

Robins, G. 2007. "Male Bodies and the Construction of Masculinity in New Kingdom Egyptian Art." In *Servant of Mut*, ed. S. D'Auria, 208–215. Leiden.

Robinson, H.R. 1975. *The Armour of Imperial Rome*. New York.

Robinson, M. 2002. "Domestic Burnt Offerings and Sacrifices at Roman and Pre-Roman Pompeii, Italy." *Vegetation History and Archaeobotany* 11: 93–99.

Robisek, C. 1089. *Das Bildprogramm des Mut-Tempels am Gebel Barkal.* Vienna.
Roccati, A., and G. Capriotti Vittozzi. 2002. *Tra le palme del Piceno*. Poggibonsi.
Rodgers, R. 2003. "Female Representation in Roman Art: Feminising the Provincial 'Other.'" In *Roman Imperialism and Provincial Art*, ed. S. Scott and J. Webster, 69–94. Cambridge.
Roller, M.B. 2006. *Dining Posture in Ancient Rome*. Princeton.
Romano, J.F. 1989. *The Bes-Image in Pharaonic Egypt*. 2 vols. Ph.D. dissertation, New York University.
Romm, J.S. 1992. *The Edges of the Earth in Ancient Thought*. Princeton.
Romm, J.S. 2010. "Continents, Climates, and Cultures: Greek Theories of Global Structure." In *Geography and Ethnography*, ed. K.A. Raaflaub and R.J.A. Talbert, 215–235. Malden/Oxford/Chichester.
Rose, C.B. 2013. "The Egyptianizing of Rome in the Wake of Actium." In *Orhan Bingöl'e 67. Yaš Armağanı*, ed. G. Kökdemir, 543–558. Ankara.
Roselaar, S.T. 2012. "The Role of Italians in Local Economies of the Late Roman Republic." *Revue belge de philologie et d'histoire* 90: 93–110.
Rossi, L. 2014. "Les *frequentissimi mercatores* de Pouzzoles et le blé égyptien à Rome á la fin de l'époque républicaine." *MÉFRA* 126:
Roullet, A. 1972. *The Egyptian and Egyptianizing Monuments of Imperial Rome*. Leiden.
Rouse, W.H.D., trans. 1924. *Lucretius: De Rerum Natura*. London/New York.
Roussel, P. 1915–16. *Les cultes égyptiens à Délos du IIIe au Ier siècle av. J.-C.* Paris/Nancy.
Roux, H., and M.L. Barré. 1870. *Herculanum et Pompéi*. Vol. 5. Paris.
Rowlandson, J., and A. Harker. 2004. "Roman Alexandria from the Perspective of the Papyri." In *Alexandria*, ed. A. Hirst and M.S. Silk, 79–112. Aldershot.
Rüpke, J. 2006. "Organisationsmuster religiöser Spezialisten im kultischen Spektrum Roms." In *Religions orientales—culti misterici*, ed. C. Bonnet et al., 13–26. Stuttgart.
Rüpke, J., ed. 2007a. *Gruppenreligionen im römischen Reich*. Tübingen.
Rüpke, J. 2007b. "Integrationsgeschichten: Gruppenreligionen in Rom." In *Gruppenreligionen im römischen Reich*, ed. J. Rüpke, 113–126. Tübingen.
Rüpke, J. 2007c. "Roman Religion—Religions of Rome." In *A Companion to Roman Religion*, ed. J. Rüpke, 1–9. Malden.
Rüpke, J. 2013a. "Individualization and Individuation as Concepts for Historical Research." In *The Individual in the Religions of the Ancient Mediterranean*, ed. J. Rüpke, 3–40. Oxford.
Rüpke, J. 2013b. *Religion: Antiquity and Its Legacy*. Oxford.
Rüpke, J. 2014. *From Jupiter to Christ: On the History of Religion in the Roman Imperial Period*. Trans. D.M.B. Richardson. Oxford.
Rutledge, S.H. 2012. *Ancient Rome as a Museum*. Oxford.
Ryholt, K. 2010. "Late Period Literature." In *A Companion to Ancient Egypt*, ed. A.B. Lloyd, 2:709–731. Malden/Oxford.
Sabnis, S. 2012. "Invisible Slaves, Visible Lamps: A Metaphor in Apuleius." *Arethusa* 45: 79–108.

Sackett, J. 1990. "Style and Ethnicity in Archaeology: The Case for Isochrestism." In *The Uses of Style in Archaeology*, ed. W. Conkey and C. Hastorf, 32–43. Cambridge.
Sakkarah Expedition, The. 1938. *The Mastaba of Mereruka. Part I.* Chicago.
Salmon, E.T. 1967. *Samnium and the Samnites*. Cambridge.
Salza Prina Ricotti, E. 1979. "Forme speciali di triclini." *Cronache pompeiane* 5: 102–149.
Salza Prina Ricotti, E. 1987. "The Importance of Water in Roman Garden Triclinia." In *Ancient Roman Villa Gardens*, ed. E.B. MacDougall, 135–184. Washington, DC.
Sampaolo, V. 1992. "La decorazione pittorica." In *Alla ricerca di Iside*, ed. S. De Caro, 23–39. Rome.
Sampaolo, V. 1993. "VI 2, 4. Casa di Sallustio." *PPM* IV, 87–147. Rome.
Sampaolo, V. 1996a. "VI 17 (*Ins. Occ.*), 42. Casa del Bracciale d'oro." *PPM* VI, 44–145. Rome.
Sampaolo, V. 1996b. "VII 2, 25. Casa delle Quadrighe." *PPM* VI, 683–717. Rome.
Sampaolo, V. 1998a. "VIII 2, 28. Casa con Ninfeo." *PPM* VIII, 226–240. Rome.
Sampaolo, V. 1998b. "VIII 7, 24.22. Casa dello scultore." *PPM* VIII, 718–731. Rome.
Sampaolo, V. 2006. "L'Iseo pompeiano." In *Egittomania,* ed. S. De Caro, 87–120. Milan.
Sandri, S. 2004. "Harpokrates & Co.: Zur Identifikation gräko-ägyptischer Kindgott-Terrakotten." In *Fremdheit – Eigenheit*, ed. P.C. Bol et al., 499–510. Munich.
Sandri, S. 2012. "Terracottas." In *The Oxford Handbook of Roman Egypt*, ed. C. Riggs, 630–647. Oxford.
Sannibale, Maurizio. 2014. "Levantine and Orientalizing Luxury Goods from Etruscan Tombs." In *Assyria to Iberia at the Dawn of the Classical Age*, ed. Joan Aruz et al., 313–329. New York.
Saragoza, F. 2017. "Exploring Walls: On Sacred Space in the Pompeian *Iseum*." In *Entangled Worlds*, ed. S. Nagel et al., 372–381. Tübingen.
Sauneron, S. 1962. *Les fêtes religieuses d'Esna aux derniers siècles du paganisme*. Cairo.
Sauron, G. 2000. *L'histoire végétalisée*. Paris.
Säve-Söderbergh, T. 1953. *On Egyptian Representations of Hippopotamus Hunting as a Religious Motive*. Uppsala.
Schefold, K. 1956. "Vorbilder römischer Landschaftsmalerei." *AM* 71: 211–225.
Schefold, K. 1957. *Die Wände Pompejis*. Berlin.
Schefold, K. 1960. "The Origins of Roman Landscape Painting." *ArtB* 42: 87–96.
Schefold, K. 1972. *La peinture pompéienne*. Brussels.
Scheid, J. 2004. "Quand fut construit l'Iseum Campense?" In *Orbis Antiquus*, ed. L. Ruscu et al., 308–311. Cluj-Napoca.
Schiffer, M. 1972. "Archaeological Context and Systemic Context." *AmerAnt* 37: 156–165.
Schiffer, M. 1987. *Formation Processes of the Archaeological Record*. Albuquerque.
Schilling, R. 1954. *La religion romaine de Vénus*. Paris.
Schörner, G. 2016. "Domestic Cults in the Roman House: Can Archaeology Detect Individuality? The Case of the Casa dei Vettii." In *Religiöse Praktiken in der Antike*, ed. L. Scheuermann and W. Spickermann, 111–120. Graz.

Schrijvers, P.H. 2007. "A Literary View on the Nile Mosaic at Praeneste." In *Nile into Tiber*, ed. L. Bricault et al., 223–244. Leiden.

Schubart, W. 1919. *Der Gnomon des Idios Logos*. Vol. 1. Berlin.

Schulz, E.G. 1838. "Rapporto intorno gli scavi pompeiani negli ultimi quattro anni." *AdI* 10: 148–201.

Seif el-Din, M. 2006. *Die reliefierten hellenistisch-römischen Pilgerflaschen*. Cairo.

Seiler, F. 1992. *Casa degli Amorini dorati (VI 16, 7.38)*. Munich.

Sena Chiesa, G., and A. Pontrandolfo, eds. 2015. *Mito e natura: dalla Grecia a Pompei*. Milan.

Sethe, K. 1964 [1912]. *Zur altägyptischen Sage vom Sonnenauge das in der Fremde war*. Hildesheim.

Sfameni Gasparro, G. 2006. "Misteri e culti orientali." In *Religions orientales—culti misterici*, ed. C. Bonnet et al., 181–210. Stuttgart.

Sfameni Gasparro, G. 2007. "The Hellenistic Face of Isis: Cosmic and Saviour Goddess." In *Nile into Tiber*, ed. L. Bricault et al., 40–72. Leiden.

Shackleton Bailey, D.R., ed. and trans. 2003. *Statius: Silvae*. Cambridge.

Shinnie, P.L., and J.R. Anderson. 2004. "The Excavations." In *The Capital of Kush 2*, ed. P.L. Shinnie and J.R. Anderson, 5–88. Wiesbaden.

Signoretti, M. 2010. "A Tale of Two Tongues? The *Myth of the Sun's Eye* and Its Greek Translation." In *Proceedings of the 25th International Congress of Papyrology*, ed. R. Gagos et al., 725–732. Ann Arbor.

Silliman, S. 2015. "A Requiem for Hybridity: The Problem with Frankensteins, Purées, and Mules." *Journal of Social Archaeology* 15: 277–298.

Silliman, S. 2016. "Disentangling the Archaeology of Colonialism and Indigeneity." In *Archaeology of Entanglement*, ed. L. Der and F. Fernandini, 31–48. London/New York.

Silverman, D. 1969. "Pygmies and Dwarfs in the Old Kingdom." *Serapis* 1: 53–62.

Sist, L. 2002. "Il paesaggio nell'arte egiziana." In *Tra le palme del Piceno*, ed. A. Roccati and G. Capriotti Vittozzi, 92–95. Poggibonsi.

Slane, K.W., and M.W. Dickie. 1993. "A Knidian Plastic Vase from Corinth." *Hesperia* 62: 483–505.

Small, A.M. 2007. "Urban, Suburban and Rural Religion in the Roman Period." In *The World of Pompeii*, ed. J.J. Dobbins and P.W. Foss, 184–211. New York.

Smelik, K.A.D., and E.A Hemelrijk. 1984. "'Who Knows Not What Monsters Demented Egypt Worships?' Opinions on Egyptian Animal Worship in Antiquity as Part of the Ancient Conception of Egypt." *ANRW* 2.17.4: 1852–2000.

Smith, A.T. 2015. *The Political Machine*. Princeton.

Smith, J.Z. 1998. "Religion, Religions, Religious." In *Critical Terms for Religious Studies*, ed. M.C. Taylor, 269–284. Chicago.

Smith, M.J. 1984. "Sonnenauge, demotischer Mythos vom." *LÄ* 5: 1082–1087.

Sodo, A.M. 1988. "Regio II, insula 9." *RStPomp* 2: 195–201.

Sogliano, A. 1882. "Pompei." *NSc* 240–241, 275–282, 317–324, 435–441.
Sogliano, A. 1888. "Pompei." *NSc* 509–530.
Söldner, M. 1999. "Ägyptische Bildmotive im augusteischen Rom: Ein Phänomen im Spannungsfeld von Politik, Religion und Kunst." In *Ägypter—Griechen—Römer*, ed. H. Felber and S. Pfisterer-Haas, 95–113. Leipzig.
Soprano, P. 1950. "I triclini all'aperto di Pompei." In *Pompeiana*, ed. A. Maiuri, 288–310. Napoli.
Soto Marín, I. 2018. *The Economic Integration of a Late Roman Province: Egypt from Diocletian to Anastasius*. Ph.D. dissertation, New York University, Institute for the Study of the Ancient World.
Spalinger, A. 1993. "A Chronological Analysis of the Feast of *thy*." *SAK* 20: 289–303.
Sparkes, B. 2000. "Small World: Pygmies and Co." In *Word and Image in Ancient Greece*, ed. N.K. Rutter and B.A. Sparkes, 79–98. Edinburgh.
Spencer, D. 2010. *Roman Landscape*. Cambridge.
Spier, J., T. Potts, and S.E. Cole, eds. 2018. *Beyond the Nile: Egypt and the Classical World*. Los Angeles.
Spiegelberg, W. 1994 [1917]. *Der ägyptische Mythus vom Sonnenauge: Nach dem Leidener demotischen Papyrus I 384*. Hildesheim.
Spinazzola, V. 1953. *Pompei alla luce degli scavi nuovi di Via dell'Abbondanza (anni 1910–1923)*. Ed. S. Aurigemma. Vol. 1. Rome.
Squire, M. 2009. *Image and Text in Graeco-Roman Antiquity*. Cambridge.
Squire, M. 2011. *The* Iliad *in a Nutshell*. Oxford.
Squire, M. 2012. "Greek Art through Roman Eyes." In *A Companion to Greek Art*, ed. T.J. Smith and D. Plantzos, 2:599–620. Malden.
Stadler, M.A. 2012. *Einführung in die ägyptische Religion ptolemäisch-römischer Zeit nach den demotischen religiösen Texten*. Berlin.
Stahl, J. 2011. "Physically Deformed and Disabled People." In *The Oxford Handbook of Social Relations in the Roman World*, ed. M. Peachin, 715–734. Oxford.
Stamatopoulou, M. 2008. "Ouaphres Horou, an Egyptian Priest of Isis from Demetrias." In *Essays in Classical Archaeology for Eleni Hatzivassiliou*, ed. D. Kurtz et al., 249–257. Oxford.
Steimle, C. 2006. "Das Heiligtum der ägyptischen Götter in Thessaloniki und die Vereine in seinem Umfeld." In *Religions orientales—culti misterici*, ed. C. Bonnet et al. , 27–38. Stuttgart.
Steinhauer, J. 2017. "Osiris *mystes* und Isis *orgia*—Gab es 'Mysterien' der ägyptischen Gottheiten?" In *Entangled Worlds*, ed. S. Nagel et al., 47–78. Tübingen.
Sternberg-El Hotabi, H. 1992. *Ein Hymnus an die Göttin Hathor und das Ritual "Hathor das Trankopfer darbringen" nach den Tempeltexten der griechisch-römisch Zeit*. Brussels.
Steuernagel, D. 2004. *Kult und Alltag in römischen Hafenstädten*. Munich.
Stevenson Smith, W. 1998. *The Art and Architecture of Ancient Egypt*. 3rd ed. Rev. W.K. Simpson. New Haven.

Stewart, A.F. 1993. *Faces of Power.* Berkeley.

Stewart, A.F. 1996. "The Alexandrian Style: A Mirage?" In *Alexandria and Alexandrianism*, ed. K. Hamma, 231–243. Malibu.

Stewart, A.F. 2003. "Alexander in Greek and Roman Art." In *Brill's Companion to Alexander the Great*, ed. J. Roisman, 31–66. Leiden/Boston.

Stewart, P. 1997. "Fine Art and Coarse Art: The Image of Roman Priapus." *Art History* 20: 575–588.

Stewart, P. 2003. *Statues in Roman Society*. Oxford.

Stewart, S. 1984. *On Longing*. Baltimore.

Stockhammer, P.W. 2012a. "Conceptualizing Cultural Hybridization in Archaeology." In *Conceptualizing Cultural Hybridization*, ed. P.W. Stockhammer, 43–58. Heidelberg.

Stockhammer, P.W., ed. 2012b. *Conceptualizing Cultural Hybridization.* Heidelberg.

Stockhammer, P.W. 2012c. "Entangled Pottery: Phenomena of Appropriation in the Late Bronze Age Eastern Mediterranean." In *Materiality and Social Practice*, ed. J. Maran and P.W. Stockhammer, 89–103. Oxford.

Stockhammer, P.W. 2013. "From Hybridity to Entanglement, from Essentialism to Practice." *Archaeological Review from Cambridge* 28: 11–28.

Sumner, G. 2002. *Roman Military Clothing (1).* Botley.

Swetnam-Burland, M. 2002. *Egypt in the Roman Imagination*. Ph.D. dissertation, University of Michigan.

Swetnam-Burland, M. 2007. "Egyptian Objects, Roman Contexts. A Taste for Aegyptiaca in Italy." In *Nile into Tiber*, ed. L. Bricault et al., 113–136. Leiden.

Swetnam-Burland, M. 2009. "Egypt Embodied: The Vatican Nile." *AJA* 113: 439–457.

Swetnam-Burland, M. 2010. "Aegyptus Redacta: The Egyptian Obelisk in the Augustan Campus Martius." *ArtB* 72: 135–153.

Swetnam-Burland, M. 2011. "'Egyptian' Priests in Roman Italy." In *Cultural Identity in the Ancient Mediterranean*, ed. E. Gruen, 336–353. Los Angeles.

Swetnam-Burland, M. 2012. "Nilotica and the Image of Egypt." In *The Oxford Handbook of Roman Egypt*, ed. C. Riggs, 684–697. Oxford.

Swetnam-Burland, M. 2015. *Egypt in Italy*. Cambridge.

Swift, E. 2009. *Style and Function in Roman Decoration: Living with Objects and Interiors*. Burlington.

Swift, E. 2017. *Roman Artefacts and Society*. Oxford.

Szymańska, H. 2005. *Terres cuites d'Athribis*. Turnhout.

Takács, S. 1995. *Isis and Sarapis in the Roman World.* Leiden.

Tammisto, A. 1997. *Birds in Mosaics*. Rome.

Tanner, J. 2006. *The Invention of Art History in Ancient Greece*. Cambridge.

Tassinari, S. 1993. *Il vasellame bronzeo di Pompei*. 2 vols. Rome.

Teeter, E. 2011. *Religion and Ritual in Ancient Egypt*. Cambridge.

te Velde, H. 1971. "Some Remarks on the Structure of Egyptian Divine Triads." *JEA* 57: 80–86.

Thomas, N. 1991. *Entangled Objects: Exchange, Material Culture and Colonialism in the Pacific*. Cambridge.

Thomas, R.I. 2015a. "Egyptian Late Period Figures in Terracotta and Limestone." In A. Villing et al., *Naukratis: Greeks in Egypt*. London. www.britishmuseum.org/pdf/Thomas_Egyptian_figures_final.pdf.

Thomas, R.I. 2015b. "Lamps in Terracotta and Bronze." In A. Villing et al., *Naukratis: Greeks in Egypt*. London. www.britishmuseum.org/pdf/Thomas_Lamps.pdf.

Thomas, R.I. 2015c. "Ptolemaic and Roman Figures, Models and Coffin-Fittings in Terracotta." In A. Villing et al., *Naukratis: Greeks in Egypt*. London. www.britishmuseum.org/pdf/Thomas_Ptolemaic_figures.pdf.

Thomas, R.I. 2015d. "Stone and Terracotta Figurines—An Introduction." In A. Villing et al., *Naukratis: Greeks in Egypt*. London. www.britishmuseum.org/pdf/Thomas_Figurines_SF_AV.pdf.

Thomas, R.I, and P. Higgs. 2015. "Greek and Roman Sculpture." In A. Villing et al., *Naukratis: Greeks in Egypt*. London. https://www.britishmuseum.org/pdf/Thomas_and_Higgs_Greek_and_Roman_sculpture.pdf.

Thomas, R.I., and P.T. Nicholson. 2013. "Figurines." In *Working in Memphis: The Production of Faience at Roman Period Kom Helul*, ed. P.T. Nicholson, 201–235. London.

Tocco Sciarelli, G. 1983. *Baia: Il ninfeo imperial sommerso di punta Epitaffio*. Naples.

Török, L. 1995. *Hellenistic and Roman Terracottas from Egypt*. Rome.

Tracy, J. 2014. *Lucan's Egyptian Civil War*. Cambridge.

Tran tam Tinh, V. 1964. *Essai sur le culte d'Isis à Pompéi*. Paris.

Tran tam Tinh, V. 1971. *Le culte des divinités orientales à Herculanum*. Leiden.

Tran tam Tinh, V. 1972. *Le culte des divinités orientales en Campanie*. Leiden.

Tran tam Tinh, V. 1988. "Bès." In *LIMC* IV:98–108.

Trentin, L. 2015. *The Hunchback in Hellenistic and Roman Art*. London.

Trentin, L. 2017. "The 'Other' Romans: Deformed Bodies in the Visual Arts of Rome." In *Disability in Antiquity*, ed. C. Laes, 233–247. Oxon/New York.

Trentin, S.R. 2014. *Pompeian Peristyles*. Ph.D. dissertation, University of Iowa.

Trimble, J.F. 2002. "Greek Myth, Gender, and Social Structure in a Roman House: Two Paintings of Achilles at Pompeii." In *The Ancient Art of Emulation*, ed. E.K. Gazda, 225–248. Ann Arbor.

Trimble, J. 2011. *Women and Visual Replication in Roman Imperial Art and Culture*. Cambridge.

Trimble, J.F. 2018. "Appropriating Egypt for the Ara Pacis Augustae." In *Rome, Empire of Plunder*, ed. Loar et al., 109–136. Cambridge.

Tronchin, F.C. 2006. *An Eclectic* Locus Artis: *The Casa di Octavius Quartio at Pompeii*. Ph.D. dissertation, Boston University.

Tronchin, F.C. 2011a. "Art, Nature, City, Country and the Problem of Villa Imitation." *RStPomp* 21: 63–75.

Tronchin, F.C. 2011b. "The Sculpture of the Casa di Octavius Quartio in Pompeii." In *Pompeii*, ed. E. Poehler et al., 24–40. Oxford.

Tronchin, F.C. 2012a. "Introduction: Collecting the Eclectic." *Arethusa* 45: 261–282.

Tronchin, F.C. 2012b. "Roman Collecting, Decorating, and Eclectic Practice in the Textual Sources." *Arethusa* 45: 333–346.

Trope, B.T., S. Quirke, and P. Lacovara. 2005. *Excavating Egypt: Great Discoveries from the Petrie Museum of Egyptian Archaeology*. Atlanta.

Tybout, R.A. 1989. *Aedificiorum figurae: Untersuchungen zu den Architekturdarstellungen des frühen zweiten Stils*. Amsterdam.

Tybout, R.A. 2003. "Dwarfs in Discourse: The Functions of Nilotic Scenes and Other Roman *Aegyptiaca*." *JRA* 16: 505–515.

Tybout, R.A. 2007. "Rooms with a View: Residences Built on Terraces along the Edge of Pompeii (Regions VI, VII and VIII)." In *The World of Pompeii*, ed. J.J. Dobbins and P.W. Foss, 407–420. New York.

van Aerde, M.E.J.J. 2015. *Egypt and the Augustan Cultural Revolution*. Ph.D. dissertation, Leiden University.

van Andringa, W. 2009. *Quotidien des dieux et des hommes*. Rome.

van Andringa, W. 2012. "Statues in the Temples of Pompeii: Combinations of Gods, Local Definition of Cults, and the Memory of the City." In *Historical and Religious Memory in the Ancient World*, ed. B. Dignas and R.R.R. Smith, 83–118. Oxford.

van der Horst, P.M. 1984. *Chaeremon*. Leiden.

van Eck, C., M.J. Versluys, and P. ter Keurs. 2015. "The Biography of Cultures: Style, Objects and Agency. Proposal for an Interdisciplinary Approach." *Les Cahiers de l'École du Louvre* 7: 2–22.

van Erp, S. 2014. Pataikos. *A Forgotten Amulet: The Database of the Pataikoi Amulets of the National Museum of Antiquities in Leiden*. MA thesis, Leiden University.

Vanesse, S. 2016. "Between Street Vendors, Singing Slaves, and Envy." *Chronika* 6: 15–26.

Van Haeperen, F. 2006. "Fonctions des autorités politiques et religieuses romaines en matière de 'cultes orientaux.'" In *Religions orientales—culti misterici*, ed. C. Bonnet et al., 39–52. Stuttgart.

Van Oyen, A. 2016. *How Things Make History*. Amsterdam.

Van Oyen, A. 2017. "Material Culture in the Romanization Debate." In *The Diversity of Classical Archaeology*, ed. A. Lichtenberger and R. Raja, 287–300. Turnhout.

van Roggen, J.N. 2017. *Early Evidence for Vitruvian Urban Planning: A Study of Pompeii's Altstadt*. MA thesis, Cornell University.

van Wijngaarden, W.D. 1958. *De grieks-egyptische Terracotta's in het Rijksmuseum van Oudheden*. Leiden.

Varner, E.R. 2006. "Reading Replications: Roman Rhetoric and Greek Quotations." *Art History* 29: 280–303.

Varone, A. 2002. *Erotica Pompeiana*. Rome.

Vassilika, E. 1989. *Ptolemaic Philae*. Leuven.

Vendries, C. 2013. "Questions d'iconographie musicale: L'apport des terres cuites à la connaissance de la musique dans l'Égypte hellénistique et romaine." *Greek and Roman Musical Studies* 1: 195–227.

Venit, M. 2010. "Referencing Isis in Tombs of Graeco-Roman Egypt. Tradition and Innovation." In *Isis on the Nile*, ed. L. Bricault and M.J. Versluys, 98–119. Leiden.

Venit, M. 2016. *Visualizing the Afterlife in the Tombs of Graeco-Roman Egypt.* Cambridge.

Verhoeven, U., and P. Derchain. 1985. *Le voyage de la déesse libyque.* Brussels.

Versluys, M.J. 2000. "' . . . Auf ein Paar Stücken von Musaico im Hause Massimi . . . ' Bemerkungen zu drei römischen Mosaikfragmenten in Madrid." *MM* 41: 236–252.

Versluys, M.J. 2002. *Aegyptiaca Romana: Nilotic Scenes and the Roman Views of Egypt.* Leiden.

Versluys, M.J. 2007. "Aegyptiaca Romana: The Widening Debate." In *Nile into Tiber*, ed. L. Bricault et al., 1–18. Leiden.

Versluys, M.J. 2010. "Understanding Egypt in Egypt and Beyond." In *Isis on the Nile*, ed. L. Bricault and M.J. Versluys, 7–38. Leiden.

Versluys, M.J. 2012. "Making Meaning with Egypt: Hadrian, Antinous and Rome's Cultural Renaissance." In *Egyptian Gods in the Hellenistic and Roman Mediterranean*, ed. L. Bricault and M.J. Versluys, 25–40. Palermo.

Versluys, M.J. 2013a. "Material Culture and Identity in the Late Roman Republic." In *A Companion to the Archaeology of the Roman Republic*, ed. J. DeRose Evans, 429–440. Malden.

Versluys, M.J. 2013b. "Orientalising Roman Gods." In *Panthée*, ed. L. Bricault and C. Bonnet, 235–260. Leiden.

Versluys, M.J. 2014. "Understanding Objects in Motion: An *Archaeological* Dialogue on Romanization." *Archaeological Dialogues* 21: 1–20.

Versluys, M.J. 2015a. "Haunting Traditions. The (Material) Presence of Egypt in the Roman World." In *Reinventing "the Invention of Tradition"?*, ed. D. Boschung et al., 127–160. Paderborn.

Versluys, M.J. 2015b. "Roman Visual Material Culture as Globalising *Koine*." In *Globalisation and the Roman World*, ed. M. Pitts and M.J. Versluys, 141–174. Cambridge.

Versluys, M.J. 2017a. "Egypt as Part of the Roman *Koine*: Mnemohistory and the Iseum Campense." In *Entangled Worlds*, ed. S. Nagel et al., 274–293. Tübingen.

Versluys, M.J. 2017b. "Exploring Aegyptiaca and Their Material Agency throughout Global History." In *The Routledge Handbook of Archaeology and Globalization*, ed. T. Hodos et al., 74–89. London.

Versluys, M.J. 2017c. *Visual Style and Constructing Identity in the Hellenistic World.* Cambridge.

Versluys, M.J., ed. 2019. *Beyond Egyptomania: Objects, Style and Agency.* Boston/Berlin.

Versluys, M.J., C. Bülow-Clausen, and G. Capriotti Vittozzi, eds. 2018. *The Iseum Campense from the Roman Empire to the Modern Age: Temple – Monument – Lieu de Mémoire.* Rome.

Versluys, M.J., and P.G.P. Meyboom. 2000. "Les scènes dites Nilotiques et les cultes isiaques. Une interprétation contextuelle." In *De Memphis à Rome*, ed. L. Bricault, 111–127. Leiden.

Vidman, L. 1970. *Isis und Sarapis bei den Griechen und Römern*. Berlin.

Vogeikoff-Brogan, N. 2016. "A Bes-Silenus Plastic Vase in the Ierapetra Archaeological Collection: The Egyptian Connection." In *Ηχάδιν: Τιμητικός τόμος για τη Στέλλα Δρούγου*, ed. M. Giannopoulou and Ch. Kallini, 808–822. Athens.

Volokhine, Y. 2010. "Quelques aspects de Bès dans les temples égyptiens de l'époque Gréco-Romaine." In *Isis on the Nile*, ed. L. Bricault and M.J. Versluys, 233–256. Leiden.

von Lieven, A. 2003. "Wein, Weib und Gesang—Rituale für die Gefährliche Göttin." In *Rituale in der Vorgeschichte, Antike und Gegenwart*, ed. C. Metzner-Nebelsick, 47–55. Rahden/Westf.

von Stackelberg, K.T. 2009. *The Roman Garden*. New York.

Veymiers, R. 2018. "Relief with Isiac Ceremony (the Ariccia Relief)." In *Beyond the Nile: Egypt and the Classical World*, eds. J. Spier et al., 269–270. Los Angeles

Vout, C. 2003. "Embracing Egypt." In *Rome the Cosmopolis*, ed. C. Edwards and G. Woolf, 177–202. Cambridge.

Vucetic, S. 2014. "*Roman* Sexuality or *Roman* Sexualities? Looking at Sexual Imagery on Roman Terracotta Mould-Made Lamps." In *TRAC 2013: Proceedings of the Twenty-Third Annual Theoretical Roman Archaeology Conference*, ed. H. Platts et al., 140–158. Oxford.

Walker, S. 2003. "Carry-On at Canopus: The Nilotic Mosaic from Palestrina and Roman Attitudes to Egypt." In *Ancient Perspectives on Egypt*, ed. R. Matthews and C. Roemer, 191–202. London.

Walker, S., and P. Higgs, eds. 2001. *Cleopatra of Egypt*. Princeton.

Wallace-Hadrill, A. 1994. *Houses and Society in Pompeii and Herculaneum*. Princeton.

Wallace-Hadrill, A. 2008. *Rome's Cultural Revolution*. Cambridge.

Wallace-Hadrill, A. 2011. "Pompeian Identities: Between Oscan, Samnite, Greek, Roman, and Punic." In *Cultural Identity in the Ancient Mediterranean*, ed. E.S. Gruen, 415–427. Los Angeles.

Ward, C.A. 2016. "Luxury, Adornment, and Identity: The Skeletons and Jewelry from Oplontis B." In *Leisure and Luxury in the Age of Nero*, ed. E.K. Gazda and J.R. Clarke, 171–179. Ann Arbor.

Watson, G.E. 2002. "Birds: Evidence from Wall Paintings, Mosaics, Sculpture, Skeletal Remains, and Ancient Authors." In *The Natural History of Pompeii*, ed. W.F. Jashemski and F.G. Meyer, 357–400. Cambridge.

Wente, E.F. 1984. "Some Graffiti from the Reign of Hatshepsut." *JNES* 43: 47–54.

West, S. 1969. "The Greek Version of the Legend of Tefnut." *JEA* 55: 161–183.

White, K. 1975. *Farm Equipment of the Roman World*. Cambridge.

Whitehouse, D. 2001. *Roman Glass in the Corning Museum of Glass*. Vol. 2. Corning.

Whitehouse, D. 2003. *Roman Glass in the Corning Museum of Glass*. Vol. 3. Corning.

Whitehouse, H. 1977. "*In Praedis Iuliae Felicis:* The Provenance of Some Fragments of Wall-Painting in the Museo Nazionale, Naples." *PBSR* 45: 52–68.

Whitehouse, H. 2001. *Ancient Mosaics and Wallpaintings*. London.

Whitley, J. 2007. "Letting the Stones in on the Act: Statues as Social Agents in Archaic and Classical Greece." *Kodai: Journal of Ancient History* 13/14: 185–198.

Whitley, J. 2012. "Agency in Greek Art." In *A Companion to Greek Art*, ed. T.J. Smith and D. Plantzos, 2:579–596. Malden.

Wild, R.A. 1981. *Water in the Cultic Worship of Isis and Sarapis*. Leiden.

Wilkinson, A. 1998. *The Garden in Ancient Egypt*. London.

Williams, D. 2004. "Sotades: Plastic and White." In *Greek Art in View*, ed. S. Keay and S. Moser, 95–120. Oxford.

Williams, L., S. Wheeler, T. Dupras, and P. Sheldrick. 2012. "Unusual Modes of (Re) Production? Seasonal Birth Cycle of Women from Kellis 2, Dakhleh Oasis." Poster presented at the 7th International Conference of the Dakhleh Oasis Project, June 20–24, Leiden.

Wilson, P. 1997. "Slaughtering the Crocodile at Edfu and Dendera." In *The Temple in Ancient Egypt,* ed. S. Quirke, 179–203. London.

Woolf, G. 2011. *Tales of the Barbarians: Ethnography and Empire in the Roman West*. Malden/Oxford/Chichester.

Wrede, H. 1986. *Die antike Herme*. Mainz am Rhein.

Wyler, S. 2004. "'Dionysos domesticus': Les motifs dionysiaques dans les maisons pompéiennes et romaines (IIe s. av.-1er s. ap. J.-C.)." *MÉFRA* 116: 933–951.

Wyler, S. 2006. "Roman Replications of Greek Art at the Villa della Farnesina." *Art History* 29: 213–232.

Wyler, S. 2011. "L'acculturation dionysiaque à Rome: L'invention d'une altérité." In *Identités romaines*, ed. M. Simon, 191–201. Paris.

Yerkes, S.R. 2000. "Vitruvius' *Monstra*." *JRA* 13: 234–251.

Žabkar, L.V. 1988. *Hymns to Isis in Her Temple at Philae*. Hanover/London.

Zaccaria Ruggiu, A.P. 1995. "Origine del triclinio nella casa romana." In *Splendida civitas nostra*, ed. G. Cavalieri Manasse and E. Roffia, 137–154. Rome.

Zanker, G. 2004. *Modes of Viewing in Hellenistic Poetry and Art*. Madison.

Zanker, P. 1979. "Die Villa als Vorbild des späten pompeijanischen Wohngeschmacks." *JdI* 94: 460–523.

Zanker, P. 1988. *The Power of Images in the Age of Augustus*. Trans. A. Shapiro. Ann Arbor.

Zanker, P. 1998. *Pompeii: Public and Private Life*. Trans. D.L. Schneider. Cambridge.

Zanker, P. 2016. "Greek Art in Rome: A New Center in the First Century B.C." In *Pergamon and the Hellenistic Kingdoms of the Ancient World*, ed. C.A. Picón and S. Hemingway, 92–99. New York.

Zarmakoupi, M. 2010. "The Architectural Design of the Peristylium-Garden in Early Roman Luxury Villas." In *Städtisches Wohnen im östlichen Mittelmeerraum 4. Jh. v.–1. Jh. n. Chr.*, ed. S. Ladstätter and V. Scheibelreiter, 621–631. Vienna.

Zarmakoupi, M. 2014. *Designing for Luxury on the Bay of Naples*. Oxford.

Zevi, F. 1996. "Pompei dalla città sannitica alla colonia sillana: per un'interpretazione dei dati archeologici." In *Les élites municipales de l'Italie péninsulaire des Gracques à Néron*, ed. M. Cébeillac-Gervasoni, 125–138. Naples/Rome.

Zevi, F. 1998a. "Die Casa del Fauno in Pompeji und das Alexandermosaik." *RM* 105: 21–65.

Zevi, F. 1998b. *I mosaici della Casa del fauno a Pompei*. Naples.

Zevi, F. 2006. "Pozzuoli come 'Delo minore' e i culti egizi nei Campi Flegrei." In *Egittomania*, ed. S. De Caro, 69–76. Milan.

Zorach, R. 2005. *Blood, Milk, Ink, Gold: Abundance and Excess in the French Renaissance*. Chicago.

Index

Note: Figures are indicated by an italic *f* following the page number

Printed in the USA/Agawam, MA
July 24, 2019

707775.017